MULTIPLE CRITERIA AND MULTIPLE CONSTRAINT LEVELS LINEAR PROGRAMMING

Concepts, Techniques and Applications

Multiple Criteria and Multiple Constraint Levels Linear Programming

Concepts, Techniques and Applications

Yong Shi

Distinguished Chair of Information Technology
University of Nebraska at Omaha

With the assistance of
Yi Peng

World Scientific
Singapore • New Jersey • London • Hong Kong

Published by

World Scientific Publishing Co. Pte. Ltd.

P O Box 128, Farrer Road, Singapore 912805

USA office: Suite 1B, 1060 Main Street, River Edge, NJ 07661

UK office: 57 Shelton Street, Covent Garden, London WC2H 9HE

British Library Cataloguing-in-Publication Data
A catalogue record for this book is available from the British Library.

MULTIPLE CRITERIA AND MULTIPLE CONSTRAINT LEVELS LINEAR PROGRAMMING
Concepts, Techniques and Applications

ISBN 981-02-3738-3

Printed in Singapore by Mainland Press

To all of the individuals I love

FOREWORD

This book offers an abundance of concepts related to theory and applications of multiple criteria and multiple constraint-level (MC^2) linear programming. The author put what he has done in this area over last decade and the related literature in this book. With careful reading and digesting, the reader could greatly expand his/her domains of thinking in research and practice.

Originally, simplex method focuses on single objective with single constraint level. It has made tremendous contributions to the area of operations research of management science. Unfortunately, many significant decision problems are involved with multiple criteria and with multiple level constraints and uncertainty. This makes it difficult to apply the single criterion and single constraint simplex method to solve nontrivial realistic problems. Realizing this difficulty, many scholars (including M. Zeleny, my friend and previous student, and myself) began to work on multiple criteria (MC) simplex methods in early 1970s.

In the middle of 1970s, L. Seiford, my friend and previous student, and I successfully extended MC-simplex method to (multi-criteria and multi-constraint level) MC^2-simplex method, including the duality theory and potential solutions.

In early 1980s, Y. R. Lee, my friend and previous student, and I further extended MC^2-simplex method and its application, including how to convert an infeasible solution caused by uncertainty into a feasible solution by investing more in resource level (concepts and techniques). Linear optional design and contingency plans were therefore born. However, many challenging research and application problems remained to be explored. In 1986, the current author of this book, Yong Shi, decided to tackle this difficult challenging problem as his Ph. D dissertation.

One day in my office I asked Yong, "I know you are very talented, motivated, and hard working. You have great potential to become a "top class" of scholar if you are willing to work for it. Would you like to do it? " Without hesitation, he said yes and would work hard for it. But at that time he just came from China with limited English and his skill of mathematical analysis skill was not as strong as needed to overcome analytical problems. In order to improve his English, I suggested him to memorize one "editorial" a day of the KU (University of Kansas) newspaper "KANSANS" and write it out after memorizing the contents. Yong did this consistently for a year. His English therefore have greatly improved. In order to increase his capability of mathematical analysis, I also asked him to take a number of classes in mathematical analysis. Yong worked hard and successfully went through the training. Because of his hard work and dedication, Yong successfully finished his dissertation with distinction. He has gone much farther in research on MC^2 linear programming and applications after his graduation from KU. Yong's persistence, dedication and hard work described as above are essential for this book to be completed.

Many Scholars or professionals might think "simplex method" is adequate for their problems and learn only simplex method. Others are aware of the limitation of simplex method and work hard to learn MC-simplex method. Still there are people who know the limitation of MC-simplex method. They work hard to learn MC^2-simplex method and their further topics. It takes special interest, talent and drive to learn all of these advanced concepts and techniques. The reader will not to disappointed by spending the effort to learn these advanced concepts and techniques. He or she can at least avoid the trap, spoken eloquently by A. H. Maslow, " If the only tool you have is a hammer, you tend to see every problem as a nail". In additions, in this fast changing world, academically or industrially, the more advanced tools we have, the more competent we will be able to cope with problems and competition.

It is gratifying for me to see a talented friend and previous student who has devoted so much time and efforts to put so many materials in this book. Almost half of the book is based on the research results done by my previous students including Zeleny, Seiford, Lee, Shi, and myself. The other half, including some new form of duality theory, MC^2 transportation problems, De Novo Programming, applications, etc. are based on research results done by Shi and his colleagues. Some of new developments by Shi and his colleagues are outside of my competence and beyond my comprehension.

This book contains a large number of new ideas and a number of remaining questions for further researches. It is not only suitable for graduate students but also for professors and professionals who work in management science and operations research. It will expand readers' vision about operations research and sharpen his/her mind in management science and technology.

P. L. Yu
Carl. A. Scupin Distinguished Professor
School of Business
University of Kansas
Lawrence, KS, USA
February 2001

PREFACE

The purpose of this book is to provide a comprehensive view on multiple criteria and multiple constraint-level (MC^2) linear programming that has become a major part of the field of Multiple Criteria Decision Making (MCDM). Most of the contents in the book are directly from the research and application activities that my colleagues and I have conducted over the last fifteen years. The book also contains materials developed for the university honor course on MCDM that I taught in 1993 at the University of Nebraska at Omaha. Starting from 1990, I have received a number of letters and e-mails from scholars and professionals in several countries who generously encouraged me to publish a summary of what has been done in MC^2 linear programming. This has been the major motivation for me to write this book. I started writing the book in 1997 and it has taken almost four years to complete. There are two main reasons why it has been delayed. The first one is that because MC^2 linear programming is a growing field, I have always hesitated including immature results or ongoing projects in the book. Instead, I prefer first publishing or applying the findings and then integrating them into the book. Almost one-fifth materials in the book are related to our publications after 1997. The second reason for delay has been the completing demands of my constant research projects and overwhelming academic and service activities. Nevertheless, I now feel more relaxed to see this book finally is printed.

This book introduces the concepts, techniques and applications of MC^2 linear programming, which is an extension of linear programming (LP) and multiple criteria (MC) linear programming. It can serve as a seminar or course text for both undergraduates and graduates who have a linear algebra or equivalent background. For practitioners, it will help in handling LP type problems.

The book is organized into six parts, and each part contains several chapters. In addition to the content description, each chapter includes some remarks and necessary exercises. To practice their understanding of the chapter, readers are encouraged to do the exercises. In the back of the book, a package of five MC^2 computer programs is provided for the interested readers. Two sets of software, called firstMC2 and secondMC2, are about how to solve the standard MC^2 problems discussed in Chapter 4. Then, the two sets of software called binaryMC2 and branch-boundMC2, are for solving the integer

MC^2 problems in Chapter 5. Finally, the software, tpMC2 can be used to solve the MC^2 transportation problems in Chapter 6. The details of each chapter of the book are given as follows:

Part one addresses basic concepts and techniques. It contains Chapters 1, 2, 3 and 4. Chapter 1 reviews the history of LP, MC and MC^2 problems and uses an illustrative example to explain the meaning of the three kinds of problems. While Chapters 2 and 3 go over the fundamentals of LP and MC problems, Chapter 4 covers the bases of MC^2 problems, mainly from Chapter 8 of Yu (1985).

Part two deals with special MC^2 problems. Chapter 5 discusses first MC^2 zero-one (or binary) problems, and then general MC^2 integer problems. Chapter 6 elaborates on MC^2 Transportation problems. One model is a steady problem and another one is dynamic problem.

Part three presents fuzzy MC^2 problems. The "fuzzy" MC^2 solution is a sort of compromise solution classified by a membership function. Chapter 7 describes the primal fuzzy MC^2 problems and Chapter 8 provides approaches to dual fuzzy MC^2 problems.

Part four discusses optimal linear designs. This part includes Chapters 9, 10, 11, and 12. Much discussion in these chapters is detailed, but meaningful in the sense of linear structure. Chapter 9 provides a general framework of designing optimal systems, which goes beyond finding an optimal point for a given linear system, and then discusses the primal contingency plans. Chapter 10 extends the system designs into a generalized good system. It further derives the primal contingency plans for generalized good systems. Chapters 11 and 12 elaborate on the dual contingency plans for system designs and generalized good systems, respectively. A computer-aided system that integrates all of the different linear designs is also shown in Chapter 12.

Part five provides several related approaches. Chapter 13, heavily depending on Chapter 4 of Yu (1985), has a significant inner connection with both MC and MC^2 problems, and serves as a simplified version for various application problems that have an MC^2 feature. Chapter 14 is another approach, known as De novo programming, to linear system designs, in which the resource availability is unknown. Chapter 15 shows an effective algorithm that can save computational time to search other dominated opportunities (variables) in solving the MC^2 problems.

Part six gives a variety of MC^2 related applications. Some of them (Chapters 16, 17, 18, and 21) are applications driven by academic interests. Others (Chapters 19, 20 and 22) come directly from the real-world applications. Chapter 16 deals with transfer pricing. Chapter 17 is about capital budgeting. Chapter 18 handles data allocation in information management. Chapter 19 concerns a rural telecommunication system. Chapter 20 manages credit card portfolios. Chapter 21 schedules production planning. Chapter 22 provides an agricultural policy. All of them can be explicitly and implicitly understood from the MC^2 framework. For the convenience of the reader, each chapter of this part is self-contained and self-explained.

If this book is used for an upper-level undergraduate course, Chapters 1, 2, 3, 4, 5, 6, 13, and some of Part six may be chosen. If this book is used for a graduate course, Chapters 1, 2, 3, 4, 5, 6, 7, 8, 9, and some of Part six can be covered. Chapters 1, 2, 3, 4 and Part four are also recommended for a special graduate seminar on Optimal System Designs. Furthermore, Chapters 1, 2, 3, 4, and Part six can be attractive contents for many practitioners.

I am indebted to many people around the world for their encouragement and kind support of my research on MC^2 linear programming. The first person is my Ph.D. Advisor and personal friend, P. L. Yu (University of Kansas and National Chiao Tung University of Taiwan). His guiding me to this new, challenging, and complex field of research is greatly appreciated. His insignificant advice in my academic and social life has been always remembered.

I would like to thank my colleagues, teachers and graduate students with whom I have worked joint MC^2 research projects. They are Wikil Kwak, Yi-Hsin Liu, Sufi Nazem, Justin Stolen and Shuanglin Lin (University of Nebraska at Omaha), Heeseok Lee (Korea Advanced Institute of Science and Technology), Cheng F. Lee (Rutgers University), Yoon R. Lee (Hofstra University), Dazhi Zhang (Iona College), Changqing Zhang (Pulsemd Co., San Jose, California), Zesheng He (Rational Software Co., Irvine, California), Xirui Hao (Vision Software Inc., Wilmington, North Carolina), I-Shien Chien (Creighton University), Morgan Wise and Ming Luo (First Data Resources, Omaha, Nebraska), Yachen Lin (First North American National Bank, Georgia), Yuji Liu, Guangyao Lee, Maoze Jiang, Kaisong Wu, Wei Xiao, Zhibin Liu, and Yihua Zhong (Southwest Petroleum Institute of China), Xiaowo Tan (Chinese University of Electronic Science and Technology), Jun Li (Daqing Petroleum Institute of China), Christian Haase (Technical University Berlin of Germany), and Ferenc Gyetán (Janus Pannonius University of Hungary).

I want to take this opportunity to give special thanks to Jiali Ge (Chinese Petroleum University at Beijing), Chunli Bai, Jifa Gu, Ruwei Dai, Shouyang Wang and Weixuan Xu (Chinese Academy of Sciences),

Siwei Cheng (National Science Foundation of China), and Jian Song (Chinese Academy of Engineering) for their generous support for my research activities in China.

I am grateful to Milan Zeleny (Fordham University) and Prakash Shenoy (University of Kansas) for their constant encouragement on all aspects of my research and teaching career.

In the past four years, there were a number of graduate students at the University of Nebraska at Omaha who helped me prepare this book and develop the computer software. Among them, I want to thank Haibo Huang, Amitabh Awasthi, Yi Maggie Guo, Wei Fan, Hong Fang, and Yi Peng. Miss Yi Peng has spent a semester on the proofreading of the book, and preparing the Bibliography, Indexes, and the camera-ready copies. To honor her hard work on this book, I invited her to put her name with my name in the front page.

I also want to acknowledge the constructive comments on the first draft kindly provided by P. L. Yu (University of Kansas and National Chiao Tung University of Taiwan) and John deGroot, Justin Stolen and Philip Craiger (University of Nebraska at Omaha).

Finally, I would like to offer my greatest thanks to my wife, Bailing Gong and my son, Chris Shi for their love and support of my research works. In addition to them, I owe a great deal of thanks to my parents, parents-in-law, brothers, sister, brothers-in-law, former teachers in elementary, middle, and high schools, and lifetime friends for their personal influence on my MC^2 research projects. This is why the book is dedicated to all of the individuals I love.

Yong Shi

Omaha, Nebraska
March 2001

Contents

Part One: Basic Concepts and Techniques

Chapter 1
INTRODUCTION

Linear programming (LP) is formulated with a single criterion (objective) and a single (fixed) resource availability level (right hand side). For the past five decades, linear programming has been widely applied in many real-world decision-making problems. However, like any great advancement, linear programming is not a perfect tool. For example, if a decision problem involves multiple conflicting criteria, such as maximizing profit while minimizing production cost, linear programming may have limitations in effectively addressing possible tradeoffs among the criteria. This shortcoming has been overcome by mathematical models known as multiple criteria (MC) linear programming. MC linear programming improves the value of linear programming by changing its single criterion to multiple criteria.

Even though the response to MC linear programming from practitioners is very positive, MC linear programming is not the best modeling tool. For instance, if an MC model is used to identify a best decision (solution) for investing ten million dollars in a stock market, its right hand side level (resource availability) should not be fixed. The level of which the future stock share value is going up should differ from that of which the future stock share value is going down. Can the linear programming framework be made more flexible and realistic to reflect the real world? Multiple criteria and multiple constraint-level (MC^2) linear programming is an answer. In addition to the extension of MC linear programming from linear programming, MC^2 linear programming explicitly expresses multiple (discrete) levels of resource availability. In other words, a single fixed feasible set of linear programming is replaced by several flexible feasible sets.

In this chapter, we first review the history of linear programming, MC linear programming, and MC^2 linear programming in Section 1.1. Then, to motivate the reader's interest, we use a simple example to illustrate the concepts of linear programming, MC linear programming, and MC^2 linear programming, respectively in Sections 1.2, 1.3 and 1.4. Finally, we give some remarks for the further reading in Section 1.5.

1.1 Brief History

1. The original invention of linear programming can be traced back to the late 1930's when Russian mathematician L. K. Kantorovich found that a class of linear programming models could be used for production planning. In the United States, the duality theorem which serves as a mathematical foundation of linear programming was initiated by J. von Neumann in 1944. In 1947, G. B. Dantzig formulated the first linear program for the U.S. airforce's planning process. This formulation had its origins in the economic input-output model developed by W. Leontief and his associates in 1941 and 1945. In order to solve linear programming problems, G. B. Dantzig created the well-known "simplex method", which was first tested by a diet problem based on the home economist George Stigler's data. In 1948, T.C. Koopmans and G. B. Dantzig named this new linear structure "linear programming." Later, T.C. Koopmans applied linear programming as a framework to analyze a class of economic phenomena. In the meantime, S. A. Charnes and W. W. Cooper made their fundamental contributions to model a number of business problems through the techniques of Linear Programming. Since the 1950's, linear programming has been often applied to formulate and solve various resource allocation, scheduling, distribution, oil refinement, and transportation problems. This also had a great influence on the development of computers. In 1975, the Royal Sweden Academy of Sciences awarded a Nobel Prize in economics to both L. K. Kantorovich and T.C. Koopmans.

After the initiation of the simplex method, researchers have been proposing numerous algorithms to reduce the computation complexity in solving large-scale linear programming problems. Among them, two algorithms are particularly significant. The first is the "ellipsoid method" proved by L. G. Khachian in 1979. This method theoretically shows a polynomial-time bound to find an optimal solution of a linear program, which outperforms the simplex method that uses an exponential time to find an optimal solution. However, the ellipsoid method has not been applied to the real-world applications. The second is the "interior point method" developed by N. Karmarkar in 1984. Compared to the simplex method that searches the optimal solution from the corner points along the edges of the feasible region, the interior point method finds the optimal solution through the interior points of the feasible region. Some reports have recently shown that this approach seems to be faster than the simplex method in locating an optimal solution for a number of large-scale practical linear programming problems. However, the augments of relative superiority of these methods both practically and theoretically are still going on.

As early as 1951, D. Gale, H.W. Kuhn and A.W. Tucker noticed that the basic structure of linear programming can be mathematically extended into "a linear matrix programming problem", which since 1970's has been well known as "MC linear programming." In linear matrix programming, the single objective coefficients (a vector) of linear programming are replaced by several objective coefficients (a matrix). This extension opened a new technique to study the tradeoffs of conflicting criteria in real-world problems. In 1974, M. Zeleny published the first MC linear programming book based on his Ph.D. thesis under guidance and advice of P. L. Yu. T. Gal, in his 1979's book, extended simplex methods to a series of parametric programming. The field of MC linear programming in both theory and applications has been booming for the last three decades. During the period of time, the International Society on Multiple Criteria Decision Making (MCDM) was established to exchange ideas, scholastic works, and application experiences among researchers and practitioners throughout the world. By 2000, this organization has held 15 international conferences, published a newsletter twice each year, and has about 1,300 members from 82 counties.

In 1979, P. L. Yu and L. Seiford observed that because the objective coefficients of a primal linear program is the constraint level of its dual program (by the duality theory), the multiple constraint levels can be built in the structure of linear system like multiple criteria. This led to the concept, formulation, and development of MC^2 linear programming. The author of this book began to work on this extended technique in 1985 when he joined the Ph.D. program at the University of Kansas as a graduate student and assistant of P. L. Yu. Part of motivation of the author's dissertation was from Y. R. Lee (1986). Much of the theories and many methods of this book are based on the work with P. L. Yu. In the last decade, the author, as the principle investigator and a group of professors and graduate students have been systematically working on theory and applications of MC^2 linear programming problems. Members of this group are from twelve universities in USA, China, Korea, Germany, and Hungary. The main research results of the group range from theoretical bases to broad areas of real-world applications. The theoretical bases include: relationships between linear programming, multiple criteria linear programming, and MC^2 linear programming; integer MC^2 linear programming and MC^2

transportation model; fuzzy MC^2 linear programming problems and fuzzy duality theory in MC^2 linear programming; optimal system designs and contingency plans; decision support system of solving complex system design problems; and MC^2 computer software development. The application areas include accounting, agricultural policy, finance, management information systems, production planning, and telecommunications management.

1.2 The Linear Programming Problem

The Trend Computing Company produces the variety of PC's and Laptops. Recently, the R & D department proposes two types of the new generation PC. Each type consists of hardware (monitor, hard drive, keyboard, etc.) and software (Windows, Internet Browser, Lotus Notes, etc.). Suppose that 3 units of hardware and 8 units of software are needed to assemble a type 1 PC, while 2 units of hardware and 3 units of software are used to make a type 2 PC. Each type 1 will produce $400 as profit and each type 2's profit is $200. If the *production manager* believes that the company can produce 40,000 units of hardware and license 90,000 units of software, then how many PC's of each type should the company assemble to achieve the maximal total profit?

Table 1.1 The PC Problem for LP

Resource	Type 1	Type 2	Availability Level (units)
Hardware	3	2	40,000
Software	8	3	90,000
Profit ($)	400	200	

Table 1.1 summarizes the information given in the above PC problem. In order to find the best production mix for this problem, we first define the following **decision variables**:

Let x_1 = the number of type 1 PC made, and
x_2 = the number of type 2 PC made.

Then, the **objective (criterion) function** that shows the maximal total profit can be expressed as

$$Max \quad 400\,x_1 + 200\,x_2$$

The **constraints** of hardware and software for the production mix can be represented by:

$3\,x_1 + 2x_2 \le 40{,}000$ (Hardware)
$8\,x_1 + 3x_2 \le 90{,}000$ (Software).

Since both variables x_1 and x_2 are the quantities and cannot be negative, the **nonegativity conditions** x_1 and $x_2 \ge 0$ should be considered.

Finally, a **linear program** of the PC problem (M1.1) is formulated as

$$\begin{array}{lll} Max & 400\,x_1 + 200\,x_2 & \text{(Profit)} \\ Subject\ to & 3\,x_1 + 2x_2 \le 40{,}000 & \text{(Hardware)} \\ & 8\,x_1 + 3x_2 \le 90{,}000 & \text{(Software)} \\ & x_1,\ x_2 \ge 0 & \end{array}$$

Using the **matrix representation**, the formulation is

$$Max \begin{pmatrix} 400 & 200 \end{pmatrix} \begin{pmatrix} x_1 \\ x_2 \end{pmatrix}$$

$$Subject\ to \begin{pmatrix} 3 & 2 \\ 8 & 3 \end{pmatrix} \begin{pmatrix} x_1 \\ x_2 \end{pmatrix} \le \begin{pmatrix} 40{,}000 \\ 90{,}000 \end{pmatrix}$$

$$\begin{pmatrix} x_1 \\ x_2 \end{pmatrix} \ge 0$$

In this representation,

$$\begin{pmatrix} x_1 \\ x_2 \end{pmatrix}$$

is called **decision vector**;

$$\begin{pmatrix} 400 & 200 \end{pmatrix}$$

is the **criterion coefficients**;

$$\begin{pmatrix} 3 & 2 \\ 8 & 3 \end{pmatrix}$$

is the **consumption matrix**; and

$$\begin{pmatrix} 40{,}000 \\ 90{,}000 \end{pmatrix}$$

is the **availability vector**.

Methods of how to identify the optimal solution for this linear programming problem will be discussed in Chapter 2. Note that the linear programming problem for the PC production has only single criterion (maximizing the total profit) and single resource availability level (from the production manager's point of view). However, in real-world situations, the company may want to simultaneously achieve the maximal production quantities and the maximal profit. The production quantities differ from the profit. Therefore, if these two criteria are under consideration, the linear program has to be extended and reformed. This leads us to think about an MC linear program for the PC problem in the following subsection.

1.3 The MC Linear Programming Problem

In order to see the significance of the extension of linear programming through the PC production problem, we first revise Table 1.1 to reflect the second criterion: the maximal production quantities. Thus, the new Table 1.2 has one more row of data for production quantities than Table 1.1.

By adding the second objective function

$$Max \quad x_1 + x_2 \qquad \text{(Production quantities)}$$

to the LP formulation in Section 1.2, the **MC (or two criteria) linear program** (M1.2) for the PC problem can be expressed as

Table 1.2 The PC Problem for MC

Resource	Type 1	Type 2	Availability Level (units)
Hardware	3	2	40,000
Software	8	3	90,000
Profit ($)	400	200	
Production Quantities	1	1	

Max	$400\, x_1 + 200\, x_2$	(Profit)
Max	$x_1 + x_2$	(Production quantities)
Subject to	$3\, x_1 + 2x_2 \leq 40{,}000$	(Hardware)
	$8\, x_1 + 3x_2 \leq 90{,}000$	(Software)
	$x_1,\, x_2 \geq 0$	

Mathematically speaking, finding the ideal solution that simultaneously represents the maximal production quantities and the maximal profit is almost impossible. However, the theory of MC linear programming allows us to study the tradeoffs of **the criteria space**. In this case, the criteria space is a two dimensional plane consisting of production quantities and profit. The "optimal" solution of MC linear programming is interpreted as an **efficient** or **nondominated solution** that is the acceptable compromise or tradeoff between the multiple criteria. To conduct the tradeoff analysis of multiple criteria, we often put a **weight parameter** on each criterion under consideration. The nondominated solution therefore is a function of weight parameters. Whenever the value of its weights changes, the corresponding nondominated solution may vary. In practice, if the set of nondominated solutions is identified, the decision maker may choose the final solution from one of nondominated solutions for implementation. If the set of nondominated solutions cannot be identified or is difficult to be found, the decision maker may interact with the analyst for identifying a preferred or "near-optimal" solution.

In the PC problem, let λ_1 be the weight parameter for the profit and λ_2 be the weight parameter for production quantities. The MC program of the PC problem (M1.2) becomes a parametric linear program:

Max	$\lambda_1\, (400\, x_1 + 200\, x_2) + \lambda_2\, (x_1 + x_2)$	
Subject to	$3\, x_1 + 2x_2 \leq 40{,}000$	(Hardware)
	$8\, x_1 + 3x_2 \leq 90{,}000$	(Software)
	$x_1,\, x_2 \geq 0$	

The matrix form of this problem is

$$Max\begin{pmatrix}\lambda_1 & \lambda_2\end{pmatrix}\begin{pmatrix}400 & 200\\ 1 & 1\end{pmatrix}\begin{pmatrix}x_1\\ x_2\end{pmatrix}$$

$$Subject\ to\begin{pmatrix}3 & 2\\ 8 & 3\end{pmatrix}\begin{pmatrix}x_1\\ x_2\end{pmatrix} \le \begin{pmatrix}40{,}000\\ 90{,}000\end{pmatrix}$$

$$\begin{pmatrix}x_1\\ x_2\end{pmatrix} \ge 0$$

Solving this MC program requires the examination of all possible values of (λ_1, λ_2). For convenience, we can assume a normalization for (λ_1, λ_2); that is,

let $0 \le \lambda_1, \lambda_2 \le 1$ and $\lambda_1 + \lambda_2 = 1$.

In such a way, we can locate all nondominated solutions when the values of (λ_1, λ_2) change between 0 and 1. Both systematical and interactive methods of solving MC linear programming will be reviewed in Chapter 3.

1.4 The MC^2 Linear Programming Problem

Both linear programming and MC linear programming models for the PC problem implicitly assume that there is only a single person, for example, the production manager in charge of decision making. This may not be true if we consider that a group of decision makers, including the president, chief financial officer, controller, and marketing manager, share the decision making responsibility. Extending a single decision maker to a multiple decision maker setting introduces the multiple resource availability level framework to modeling group decision making problems. The preference values of each involved manager on resource availability can be treated as an individual level of the availability. Then, the modified model has multiple discrete levels of resource availability. It is clear that the multiple decision makers may have different preferences, in terms of their backgrounds and positions in their firm, which result in inevitable conflicts. In addition to the tradeoff between multiple criteria (recall Section 1.3), the "optimal solution" must reflect the compromise or tradeoffs among the multiple discrete levels of resource availability.

For simplicity, suppose that both production manager and financial manager are in charge of the PC production in the example of Section 1.2. We note that the resource availability level of the production manager is different from that of the financial manager. While the production manager views his/her availability level from the perspective of manpower and materials under control, the financial manager may decide the availability level according to the current cash flow and how much the company can borrow from a bank. Suppose the financial manager believes that the Trend Computing Company can produce 65,000 units of hardware for both type 1 and type 2 PC's, and can purchase 45,000 units of software. Then, the new extended data is given in Table 1.3.

Table 1.3 The PC Problem for MC^2

Resource	Type 1	Type 2	Availability Level of Production Manager	Availability Level of Financial Manager
Hardware	3	2	40,000	65,000
Software	8	3	90,000	45,000
Profit ($)	400	200		
Production Quantities	1	1		

In addition to considering the criteria space containing the tradeoffs of multiple criteria, the structure of MC^2 linear programming has a **constraint-level space** that shows all possible tradeoffs of resource availability levels (i.e., the preferences of the production manager and financial manager in the PC problem). To demonstrate the constraint-level space of the PC problem, let γ_1 be the weight parameter for the production manager and γ_2 be the weight parameter for the financial manager. Then, the new parametric constraint levels become:

$$40{,}000\,\gamma_1 + 65{,}000\,\gamma_2 \quad \text{(Hardware)}$$
$$90{,}000\,\gamma_1 + 45{,}000\,\gamma_2 \quad \text{(Software)}$$

An **MC^2 (or two criteria and two constraint-level) linear program** (M1.3) for the PC problem can be formulated as

$$\begin{aligned} &Max && \lambda_1 (400\, x_1 + 200\, x_2) + \lambda_2 (x_1 + x_2) \\ &Subject\ to && 3\, x_1 + 2x_2 \le 40{,}000\, \gamma_1 + 65{,}000\, \gamma_2 \\ & && 8\, x_1 + 3x_2 \le 90{,}000\, \gamma_1 + 45{,}000\, \gamma_2 \\ & && x_1, x_2 \ge 0 \end{aligned}$$

The corresponding matrix form of this MC^2 problem is

$$Max \begin{pmatrix} \lambda_1 & \lambda_2 \end{pmatrix} \begin{pmatrix} 400 & 200 \\ 1 & 1 \end{pmatrix} \begin{pmatrix} x_1 \\ x_2 \end{pmatrix}$$

$$Subject\ to \begin{pmatrix} 3 & 2 \\ 8 & 3 \end{pmatrix} \begin{pmatrix} x_1 \\ x_2 \end{pmatrix} \le \begin{pmatrix} 40{,}000 & 65{,}000 \\ 90{,}000 & 45{,}000 \end{pmatrix} \begin{pmatrix} \gamma_1 \\ \gamma_2 \end{pmatrix}$$

$$\begin{pmatrix} x_1 \\ x_2 \end{pmatrix} \ge 0$$

Let $0 < \gamma_1, \gamma_2 \le 1$ and $\gamma_1 + \gamma_2 = 1$ to normalize the constraint-level parameters. By using the *MC^2 –Simplex Method* that will be studied in Chapter 4, we can systematically locate a set of **potential solutions** to over all possible values of (γ_1, γ_2) and (λ_1, λ_2). For any given value of (γ_1, γ_2) and (λ_1, λ_2), there is a potential solution to optimize the PC problem. The structure of potential solutions provides a comprehensive scenario of all possible tradeoffs or compromises between the profit, the production quantities, the production manager, and the financial manager. The general managerial implications of applications will be discussed in Chapters 16,17, and 21.

1.5 Further Remarks

Based on Dantzig's simplex method, linear programming has been rapidly developed in two directions since 1950's. One is the development of computer software that can solve large-scale linear programming problems. The early focus of the software development was first on card-programmed calculators and then mainframes. Since 1980, the main focus has gradually turned to PC-based solvers. There are more than 30 different commercial software packages claiming the ability to solve large-scale problems. These software include stand-alone solvers and solvers integrated with modeling languages. Many of them are available in Windows environment and some of them have spreadsheet integration. The second direction is the real-world application of linear programming techniques. A recent survey showed that 85% of Fortune 500 companies have used linear programming in business decision making. Applications of linear programming cover transportation, agriculture, economic analysis, finance, energy industries, military, production and public policy.

Because MC linear programming involves the tradeoffs of multiple conflicting criteria, the development of solution algorithms has mainly been in three areas. The first applies the simplex method of linear programming to the MC structure. This results in the MC-simplex method that can *systematically* find all of nondominated solutions. The second uses an interactive approach to achieve one preferred solution that satisfies the decision maker's preference. Tremendous progress has been made on this approach over the last 20 years. There is much software being developed for academic usage and a few for the real-world applications. The third applies goal setting concepts to transform an MC problem into a linear goal programming problem. The latter can then be solved by the linear programming algorithms for minimizing the distance between the ideal solution and the desired solution (see Chapter 13). Like linear programming, the concept of MC linear programming has been attractive to practitioners. Since the 1970's, MC linear programming has been frequently applied in many real-world problems. The application of MC linear programming is parallel to that of linear programming. The reason is that whenever the practitioner uses linear programming and considers multiple criteria, the MC structure provides a good response.

In comparing linear programming and MC linear programming, both software development and application of MC^2 linear programming are currently in an "infancy" stage. The purpose of this book is to emphasize the potential and promise of MC^2 linear programming to academia and practitioners so that they can use this tool to solve the real-world decision problems.

Exercises

1.1 Plateau Com. developed three kinds of memory chips for next generation computers. Chip 1 and Chip 2 are made for PC, while Chip 3 is used for Workstation. The unit prices of Chip 1, Chip 2 and Chip 3 are \$150, \$100, and \$300, respectively. There are two different resources: labor and raw material involving in the production process. To produce a Chip 1, 5 hours of labor and 10 units of raw materials are needed. Each Chip 2 needs 4 hours of labor and 15 units of raw materials. A Chip 3 consumes 10 hours of labor and 8 units of raw materials. Suppose there are 240 labor hours available and 500 units of raw materials available from the production manager point of view.

Formulate a linear program to maximize the total revenue for the production manager.

1.2 In the LP problem 1.1, if the company wants to simultaneously maximize the total revenue and the total production quantities, formulate the MC linear program.

1.3 In addition to the MC problem 1.2, suppose the finance manager believes that 300 labor hours and 450 units of raw materials are available. Incorporate this information into the modeling process and formulate an MC^2 linear program.

1.4 InterWireless Co. is a major cellular phone producer. It has four overseas factories located in Beijing (China), Moscow (Russia), Seoul (Korea), and Tijuana (Mexico). These factories ship cellular phones back to three warehouses in the USA. Warehouses are located at Los Angeles, Omaha, and New York. The unit transportation costs from the factories to warehouses are given in Table 1.4:

Table 1.4

	Los Angeles	Omaha	New York
Beijing	\$10	\$15	\$20
Moscow	\$25	\$20	\$10
Seoul	\$8	\$12	\$18
Tijuana	\$2	\$9	\$12

Suppose the company has to consider the transportation damage rate (percentage) which is estimated as in Table 1.5:

Table 1.5

	Los Angeles	Omaha	New York
Beijing	.1%	.2%	.3%
Moscow	.5%	.3%	.15%
Seoul	.1%	.13%	.2%
Tijuana	.05%	.1%	.2%

The production process is influenced by the global economy. When the economy is in recession, the production amounts of cellular phone are 200 (Beijing), 150 (Moscow), 300 (Seoul), and 250 (Tijuana). The capacities of warehouses are 300 (Los Angeles), 200 (Omaha) and 400 (New York). When the economy is blooming, the production amounts change to 250 (Beijing), 200 (Moscow), 350 (Seoul), and 300 (Tijuana). The demands of warehouses are 350 (Los Angeles), 400 (Omaha) and 350 (New York).

If the prediction of the global economy is in recession, formulate linear programs to minimize the total transportation cost and minimize the total damage rate, respectively.

1.5 For the data of the problem 1.4, if the prediction of the global economy is blooming, formulate linear programs to minimize the total transportation cost and minimize the total damage rate, respectively.

1.6 Given the LP problem 1.4, we assume that the global economy is in recession. If the company wants to simultaneously minimize the total transportation cost and minimize the total damage rate, formulate the corresponding MC linear program.

1.7 Given the LP problem 1.5, if the company wants to simultaneously minimize the total transportation cost and minimize the total damage rate, formulate the MC linear program for the blooming global economy.

1.8 Based on problems 1.6 and 1.7, formulate an MC^2 linear program to study tradeoffs between the total transportation costs, the total damage rate, the bloom economy and the economy recession.

1.9 Midwest Electronics produces a popular VCR. The regular-time production cost of VCR is \$80 and the over-time production cost is \$100. If the products cannot be shipped out, the unit inventory holding cost per 6-month is \$2. However, when demand is higher than supply, the company may potentially lose customers. This is viewed as the shortage cost, which is estimated as $30 per customer. If the future economy is predicted as either "bloom" or "recession," it will affect the production capacity of the company as well as the market demand. Assume the company's aggregate production plan can be made for 6-month base. Then, the data to be used for minimizing the total production cost (regular plus over time) for the next year over fluctuations of the economy is given in Table 1.6. The data of Table 1.7 is used to minimize the total inventory holding cost for the next year. The data for minimizing the total shortage cost is provided in Table 1.8.

Table 1.6

Periods	First 6-month	Second 6-month	Capacity for Bloom Eco.	Capacity for Eco. Recess.
Initial Invent.	20	0	100	90
Regular 1	80	80	120	130
Over-time 1	100	100	50	30
Regular 2	80	80	200	210
Over-time 2	100	100	60	80
Demand for Bloom Eco.	230	300	550	
Demand for Eco. Recess.	340	200		540

Table 1.7

Periods	First 6-month	Second 6-month	Capacity for Bloom Eco.	Capacity for Eco. Recess.
Initial Invent.	0	2	100	90
Regular 1	0	2	120	130
Over-time 1	0	2	50	30
Regular 2	0	0	200	210
Over-time 2	100	100	60	80
Demand for Bloom Eco.	230	300	550	
Demand for Eco. Recess.	340	200		540

Table 1.8

Periods	First 6-month	Second 6-month	Capacity for Bloom Eco.	Capacity for Eco. Recess.
Initial Invent.	0	0	100	90
Regular 1	0	0	120	130
Over-time 1	0	0	50	30
Regular 2	30	0	200	210
Over-time 2	30	0	60	80
Demand for Bloom Eco.	230	300	550	
Demand for Eco. Recess.	340	200		540

If the prediction of the economy is in recession, write three linear programs to minimize the total production cost, minimize the total inventory holding cost, and minimize the total shortage cost, respectively.

1.10 Given the data of the problem 1.9, if the prediction of the global economy is blooming, write three linear programs to minimize the total production cost, minimize the total inventory holding cost, and minimize the total shortage cost, respectively.

1.11 Given the data of the problem 1.9, suppose that the economy will be in recession for the next year, write an MC linear program to simultaneously minimize the total production cost, the total inventory holding cost, and the total shortage cost.

1.12 Given the data of the problem 1.9, if the economy is predicted to be blooming in the next year, write an MC linear program to simultaneously minimize the total production cost, the total inventory holding cost, and the total shortage cost.

1.13 Given the data of the problem 1.9, formulate an MC^2 linear program that can describe the tradeoffs of the total production cost, the total inventory holding cost, and the total shortage cost over fluctuations of the bloom economy and the economy recession.

1.14 United Chemical Corporation has two divisions that process raw materials into intermediate or final products. Division 1 which is located in Kansas City, Missouri, manufactures two kinds of chemicals, called Products 1 and 2. Product 1 is an intermediate product and cannot be sold externally. It can, however, be processed further by Division 2 into a final product. Division 2, located in Atlanta, Georgia, manufactures Product 2 and finalizes Product 1 in Division 1. The executives (president, vice-president for production and vice-president for marketing) agree on three objectives: (1) maximize the overall company's profit, (2) maximize the market share goal of Product 2 in Division 1 and of Products 1 and 2 in Division 2, and (3) maximize the utilized production capacity of the company. The related data is given in Table 1.9. The vice-president for production and vice-president for marketing determine preferences of constraint levels. The president will make the final decision on the basis of the resulting compromises. The constraint levels are shown in Table 1.10.

From the production vice-president's point of view, formulate an MC linear program to simultaneously maximize the overall company's profit, the market share goal, and the utilized production capacity of the company.

1.15 Given the data of the problem 1.14, from the marketing vice-president's point of view, formulate another MC linear program to simultaneously maximize the overall company's profit, the market share goal, and the utilized production capacity.

1.16 Use the data of the problem 1.14 to formulate an MC^2 linear program that can explore compromises of the production vice-president and the marketing vice-president.

Table 1.9

	Division 1		Division 2	
Products	Product 1	Product 2	Product 1	Product 2
Unit Profit ($)	-4	8	13	5
Market Price ($)	0	40	46.2	38
Unit Utilization	4	4	3	2

Table 1.10

Constraints	Production V.P.	Marketing V.P.
Transfer product	0	100
Budget ($)	45,000	40,000
Production capacity of Product 2 in Division 1	38,000	12,000
Production capacity of Product 1 in Division 2	45,000	50,000
Production capacity of Product 2 in Division 2	36,000	10,000

1.17 Global Investment Inc. plans to invest in the construction joint-venture projects. Three are 3 possible projects under consideration. Each project will last 3 years. The decision makers consisting of the president, controller, and engineering manger have identified 3 evaluation criteria for the capital budgeting. The first is maximizing net profit value of each year. The second one is maximizing the growth (net sales increase) of the firm due to each project. The third is maximizing the flexibility of financing (inverse of debt to equity ratio). The data regarding these criteria are given in Tables 1.11, 1.12, and 1.13, respectively. The budget availability level data from the involved decision makers is listed in Table 1.14.

Formulate individual linear programs with integer variables for each evaluation criterion for every decision maker (Hint: there are 9 integer linear programs).

1.18 Use the data of the problem 1.17 to formulate three MC linear programs with integer variables for the president, controller, and engineering manger, respectively.

1.19 Create an MC^2 linear program with appropriate integer variables for solving this capital budgeting problem from the data of problem 1.17 (Details of integer MC^2 formulation can be found in Chapter 5).

Table 1.11

Cash Outlays for Each Year				
Projects	Net Present Value	Year 1	Year 2	Year 3
1	$20	$24	$16	$6
2	16	12	10	2
3	11	8	6	6

Table 1.12

Net Sales Increase for Each Year			
Projects	Year 1	Year 2	Year 3
1	$120	$130	$145
2	100	120	140
3	80	90	95

Table 1.13

Inverse of Debt to Equity Ratio for Each Year			
Projects	Year 1	Year 2	Year 3
1	.85	.90	.98
2	.90	.98	.95
3	.96	.97	.98

Table 1.14

Estimate Budget Availability			
Decision Makers	Year 1	Year 2	Year 3
President	$50	$30	$30
Controller	40	45	30
Engineering Manager	55	40	20

Chapter 2
BASIC TECHNIQUES OF LINEAR PROGRAMMING

This chapter describes some basic techniques of Linear Programming (LP). In Section 2.1, we formally state the mathematical formula of LP, which includes Standard Form and Canonical Form. Section 2.2 demonstrates the graphical method for solving two dimension LP problems. This method motivates the basic thought of developing the well-known Simplex Method as a systematic LP solution procedure. Section 2.3 therefore shows details of the Simplex Method and how it can be implemented by computer program. Section 2.4 deals with a duality approach to LP problems. The economic interpretation of dual model for a given LP problem is also discussed. Section 2.5 focuses on complementary slackness analysis that handles the relationships between primal (decision) variables and dual variables. Section 2.6 reviews sensitivity and parametric analyses when the coefficients of a given LP model are subject to changes. The issues related to computational complexity and various solution algorithms are briefly discussed in Section 2.7. Finally, Section 2.8 provides some further remarks.

2.1 Formulation of Linear Programming

The Standard Form of LP used in this book is given as follows: Identify values of decision variables $x_1, x_2, \ldots, x_n$ from the model

$$
\begin{array}{ll}
\textit{Max} & c_1x_1 + c_2x_2 + \ldots + c_nx_n \\
\textit{Subject to} & a_{11}x_1 + a_{12}x_2 + \ldots + a_{1n}x_n \le d_1 \\
& a_{21}x_1 + a_{22}x_2 + \ldots + a_{2n}x_n \le d_2 \\
& \cdot \\
& \cdot \\
& \cdot \\
& a_{m1}x_1 + a_{m2}x_2 + \ldots + a_{mn}x_n \le d_m \\
& x_j \ge 0, \; j = 1, 2, \ldots n.
\end{array}
$$

Let

$$\mathbf{x} = \begin{pmatrix} x_1 \\ x_2 \\ \vdots \\ x_n \end{pmatrix}$$

be the decision vector;

$$\mathbf{c} = \begin{pmatrix} c_1 & c_2 & \cdots & c_n \end{pmatrix}$$

be the criterion or objective coefficient vector;

$$\mathbf{A} = \begin{pmatrix} a_{11} & a_{12} & \cdots & a_{1n} \\ a_{21} & a_{22} & \cdots & a_{2n} \\ \vdots & \vdots & \vdots & \vdots \\ a_{m1} & a_{m2} & \cdots & a_{mn} \end{pmatrix}$$

be the resource consumption matrix; and

$$\mathbf{d} = \begin{pmatrix} d_1 \\ d_2 \\ \vdots \\ d_m \end{pmatrix}$$

be the constraint (resource availability) level vector.

Then, the matrix representation of the Standard Form (M2.1) can be written as

$$\begin{array}{ll} \textit{Max} & \mathbf{c\,x} \\ \textit{Subject to} & \mathbf{A\,x} \leq \mathbf{d} \\ & \mathbf{x} \geq 0 \end{array}$$

The LP problem in Section 1.2 is a Standard Form. Sometimes a LP problem can be also represented in the following Canonical Form:

Identify values of decision variables $x_1, x_2, \ldots, x_n$ from the model (M2.2)

$$
\begin{array}{ll}
\textit{Max} & c_1x_1 + c_2x_2 + \ldots + c_nx_n \\
\textit{Subject to} & a_{11}x_1 + a_{12}x_2 + \ldots + a_{1n}x_n \le d_1 \\
 & a_{21}x_1 + a_{22}x_2 + \ldots + a_{2n}x_n \le d_2 \\
 & \cdot \\
 & \cdot \\
 & \cdot \\
 & a_{m1}x_1 + a_{m2}x_2 + \ldots + a_{mn}x_n \le d_m \\
 & x_j \ge 0,\ j = 1,2, \ldots n.
\end{array}
$$

Its matrix representation can be coordinately written as

$$
\begin{array}{ll}
\textit{Max} & \mathbf{c\,x} \\
\textit{Subject to} & \mathbf{A\,x} = \mathbf{d} \\
 & \mathbf{x} \ge 0
\end{array}
$$

Both Standard Form and Canonical Form can be easily transformed each other by considering the slack (or unused resource) variables.

Example 2.1 In the PC problem of Section 1.2, let s_1 be the slack variable for the unused units of Hardware and s_2 be the slack variable for the unused units of Software. Since the unused resources do not produce any profit, the Standard Form

$$
\begin{array}{lll}
\textit{Max} & 400\,x_1 + 200\,x_2 & \text{(Profit)} \\
\textit{Subject to} & 3\,x_1 + 2x_2 \le 40{,}000 & \text{(Hardware)} \\
 & 8\,x_1 + 3x_2 \le 90{,}000 & \text{(Software)} \\
 & x_1,\ x_2 \ge 0 &
\end{array}
$$

can be transformed into the Canonical Form:

Max	$400\,x_1 + 200\,x_2 + 0\,s_1 + 0\,s_2$	(Profit)
Subject to	$3\,x_1 + 2x_2 + s_1 + 0\,s_2 = 40{,}000$	(Hardware)
	$8\,x_1 + 3x_2 + 0\,s_1 + s_2 = 90{,}000$	(Software)
	$x_1,\ x_2 \geq 0$	

We also observe that any forms of LP problems can be convert to the Standard Form if some of the following properties are applied:

Property 2.1 *A minimization (Min) problem can be written as a maximization (Max) problem because*

$$\begin{array}{ll} Min & c_1x_1 + c_2x_2 + \ldots + c_nx_n \text{ is equal to} \\ -\,Max & -c_1x_1 - c_2x_2 - \ldots - c_nx_n \end{array}$$

Property 2.2 *An equality constraint can be written as two inequalities*

$$a_1x_1 + a_2x_2 + \ldots + a_nx_n = d \text{ is identical to}$$
$$a_1x_1 + a_2x_2 + \ldots + a_nx_n \leq d \quad \text{and} \quad a_1x_1 + a_2x_2 + \ldots + a_nx_n \geq d$$

Property 2.3 *The sign of inequality constraint can be switched*

$$a_1x_1 + a_2x_2 + \ldots + a_nx_n \geq d \quad \text{is identical to}$$
$$-\,a_1x_1 - a_2x_2 - \ldots - a_nx_n \leq -\,d$$

Property 2.4 *Unconstrained variables can be represented as the combination of two constrained variables*

If x_j is unconstrained, then it can be replaced by two variables u and v, with $u \geq 0$ and $v \geq 0$ such that $x_j = u - v$.

Example 2.2 Consider the LP problem:

Min	$3\,x_1 - 2\,x_2 + 5\,x_3 - 18\,x_4$
Subject to	$-5\,x_1 + 2\,x_2 + 6\,x_3 + 15\,x_4 \leq 25$
	$4\,x_1 + 5x_2 - 8\,x_3 + 9\,x_4 \geq 32$
	$3\,x_1 - 2x_2 + x_3 + 2\,x_4 = 10$
	$x_1,\ x_2,\ x_3 \geq 0$
	x_4 unconstrained.

To transform this problem into the Standard From, we first use Property 2.1 and the objective function becomes

$$-\ Max \qquad -3\ x_1 + 2\ x_2 - 5\ x_3 + 18\ x_4$$

Then, we use Property 2.3 to convert the second constraint

$$4\ x_1 + 5x_2 - 8\ x_3 + 9\ x_4 \geq 32 \quad \text{to} \quad -4\ x_1 - 5x_2 + 8\ x_3 - 9\ x_4 \leq -32$$

By Property 2.2 and Property 2.3, the third constraint $3\ x_1 - 2x_2 + x_3 + 2\ x_4 = 10$ becomes

$$3\ x_1 - 2x_2 + x_3 + 2\ x_4 \leq 10 \quad \text{and} \quad -3\ x_1 + 2x_2 - x_3 - 2\ x_4 \leq -10$$

Finally, using Property 2.4, we replace x_4 by $u - v$ with $u \geq 0$ and $v \geq 0$ in the problem. The Standard Form is

$$\begin{array}{ll} -\ Max & -3\ x_1 + 2\ x_2 - 5\ x_3 + 18\ u - 18\ v \\ Subject\ to & -5\ x_1 + 2\ x_2 + 6\ x_3 + 15\ u - 15\ v \leq 25 \\ & -4\ x_1 - 5x_2 + 8\ x_3 - 9\ u + 9\ v \leq -32 \\ & 3\ x_1 - 2x_2 + x_3 + 2\ u - 2\ v \leq 10 \\ & -3\ x_1 + 2\ x_2 - x_3 - 2\ u + 2\ v \leq -10 \\ & x_1,\ x_2,\ x_3,\ u,\ v \geq 0 \end{array}$$

In order to have consistent discussion, this book will use the Standard Form to represent the LP related problems in most cases.

2.2 Graphical Method for Two Dimension Problems

In this section, we present a simple solution procedure, called the **graphical method** to solve two dimension LP problems. This method will provide not only the basic geometric structure of LP problems, but also important insights and motivations for developing various algorithms in solving large-scale LP problems.

The key of the graphical method is that it identifies the optimal solution of a given LP problem with two decision variables through the meaning and geometry of the objective and constraints, not the computation. The basic steps of this method are given as follows:

1. Graph the feasible solution region

This step first plots each constraint into two-dimensional space on a graph paper and then finds the intersection of all constraints in the graph. Since the intersection is a polygon containing all of the points that satisfy each constraint and therefore are feasible to all constraints, it is called the feasible solution region.

2. Determine the direction of improvement on the objective (criterion) function

To find the optimal solution of the problem, this step plots any two arbitrary value lines of the objective function. The different positions of these lines provide an indication to determine the direction of improvement of the objective function.

3. Read the optimal solution on the direction of improvement of the objective function

This step moves the objective function along the direction of improvement over the feasible solution region. The last corner point of the feasible solution region passed by the objective function is the optimal solution.

Example 2.3 In order to understand the above steps in the graphical method, we demonstrate how to find the optimal solution for the PC problem of Section 1.2 as follows:

Consider

Max	$400\,x_1 + 200\,x_2$	(Profit)
Subject to	$3\,x_1 + 2x_2 \leq 40{,}000$	(Hardware)
	$8x_1 + 3x_2 \leq 90{,}000$	(Software)
	$x_1, x_2 \geq 0$	

Using the graphical method, we have:

Step 1: Because of x_1, $x_2 \geq 0$, we can restrict the graph in a quarter space. The first constraint $3x_1 + 2x_2 \leq 40{,}000$ is a half space. To plot it, we consider the boundary line $3x_1 + 2x_2 = 40{,}000$. Let $x_1 = 0$, then solve $2x_2 = 40{,}000$. This leads to a point (0, 20,000). Let $x_2 = 0$, then solve $3\,x_1 = 40{,}000$. It results in another point (40,000/3, 0). We plot these two point on the quarter space and draw the line $3\,x_1 + 2x_2 =$

40,000 in Figure 2.1, where the half space $3x_1 + 2x_2 \leq 40,000$ is easily identified as the lower -left part of $3x_1 + 2x_2 = 40,000$.

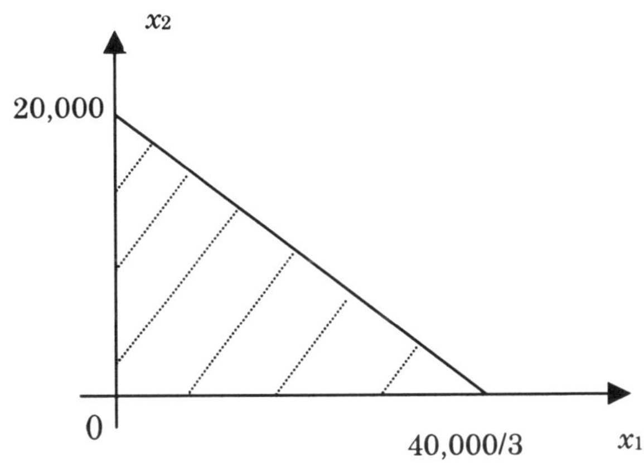

Figure 2.1 The half space $3x_1 + 2x_2 \leq 40,000$ with $x_1, x_2 \geq 0$

Similarly, we graph the half space $8x_1 + 3x_2 \leq 90,000$ with $x_1, x_2 \geq 0$ in Figure 2.2:

We consider the boundary line $8x_1 + 3x_2 = 90,000$. Let $x_1 = 0$, then solve $3x_2 = 90,000$. This leads to a point (0, 30,000). Let $x_2 = 0$, then solve $8x_1 = 90,000$. It follows another point (90,000/8, 0). We first plot these two point on the quarter space and draw the line $8x_1 + 3x_2 = 90,000$, then draw the half space $8x_1 + 3x_2 \leq 90,000$ with $x_1, x_2 \geq 0$.

To find the intersection of all constraints in the graph, we overlap Figure 2.1 on Figure 2.2. The resulting common area in Figure 2.3 is the intersection that contains all of the points that satisfy each constraint. This is called the feasible solution region (X).

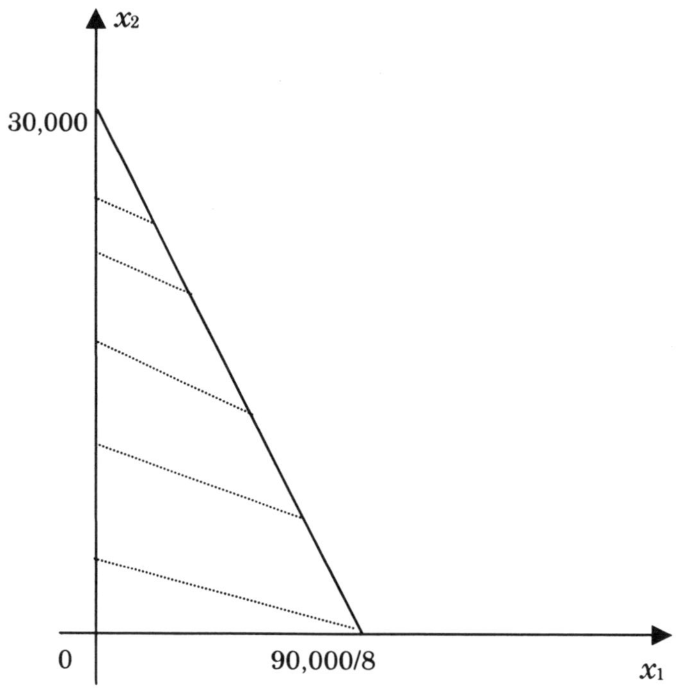

Figure 2.2 The half space $8x_1 + 3x_2 \leq 90{,}000$ with $x_1, x_2 \geq 0$

Step 2: To determine the direction of improvement (here is maximization) on the objective function, we can arbitrarily choose two iso-values as the objective values, say $400x_1 + 200x_2 = 2{,}400{,}000$ (P_1) and $400x_1 + 200x_2 = 1{,}200{,}000$ (P_2). Then, we plot them over the feasible solution region (X). From Figure 2.4, we see that the position P_1 is higher than that of P_2 and if we improve the objective value towards the direction of northeast, the objective value will be continued to increase.

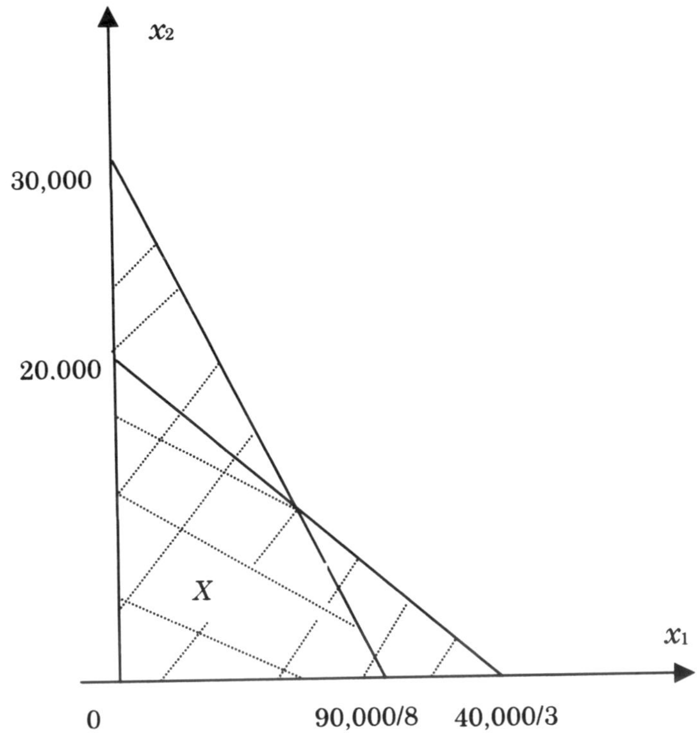

Figure 2.3 The Feasible Solution Region

Step 3: When we move the objective function along the direction of northeast in Figure 2.4 over the X region, the last corner point of the X region passed by the objective function is point (60,000/7, 50,000/7) (see Figure 2.4). It is the optimal solution that we search. If we use the graph paper, then this optimal point is the intersection of equations $3x_1 + 2x_2 = 40,000$ and $8x_1 + 3x_2 = 90,000$. Plugging it into the objective function $400x_1 + 200x_2$, the optimal objective value is \$ 34,000,000/7.

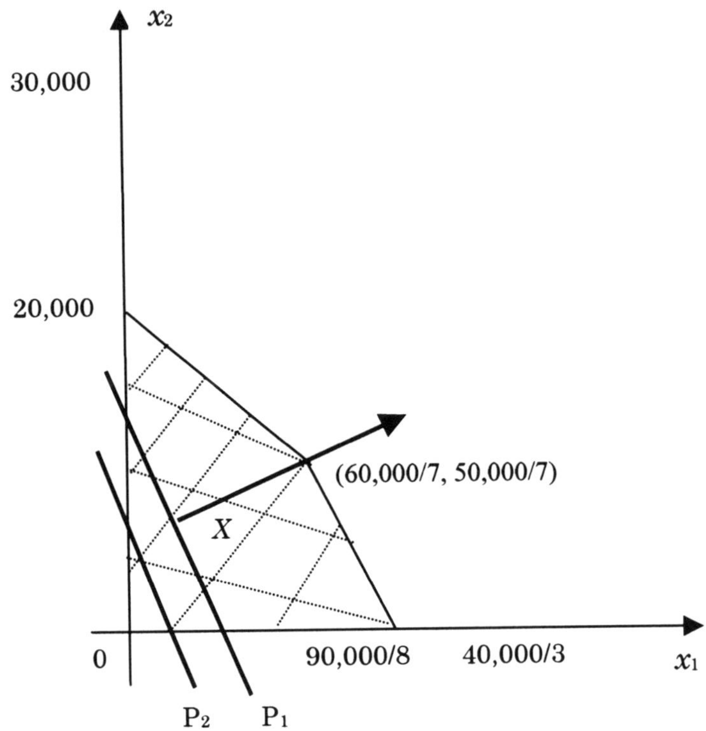

Figure 2.4 The Optimal Solution

2.3 Simplex Method and Computer Solution Procedure

In this section, based on the graphical method, we will use a simple language to describe a systematic solution procedure for LP problems, which is known as the *Simplex Method.* Then, we will have a short discussion on how the simplex method affects the implementation of the computer solution procedure for LP problems.

The graphical method discussed in the last section has motivated the basic idea of developing some sort of systematic procedure to solve LP problems. For example, in the PC problem of Section 1.2, we observe that

(i) The feasible solution region is a polygon (or convex set) which has a number of corner points.

(ii) The optimal solution occurs on a corner point (if there are optimal solutions occurring on two corner points, these solutions must the same or alternative).

Therefore, to solve a LP problem with giving or known corner points, what we can do is to just check or plug these corner points into the objective function to see which one has the maximum or minimum.

In Figure 2.5 (deriving from Figure 2.4), we plug the corner points A, B, C, and O into the objective function $400x_1 + 200x_2$ and obtain:

A: Profit is \$ 2,000,000

B: Profit is \$ 34,000,000/7.

C: Profit is \$ 4,500,000

O: Profit is \$0.

Thus, the optimal solution is the corner point B, which is identical to that we searched in the graphical method.

However, the reality is not simple. When a LP problem has more than two products (variables) and more than two constraints, there may be numerous corner points that cannot be checked one by one. We need a systematic method to effectively identify the optimal solution. The simplex method, initiated by G.B. Dantzig, is the basic one to satisfy the need. This method can reach the effectiveness by two steps. The first is that starting from the origin, say O(0,0), the method provides an efficient way for the direction of the edge (or boundary) to check the next corner point. The second is that at each corner point, the method evaluates the objective and terminates the procedure if the optimal solution is found.

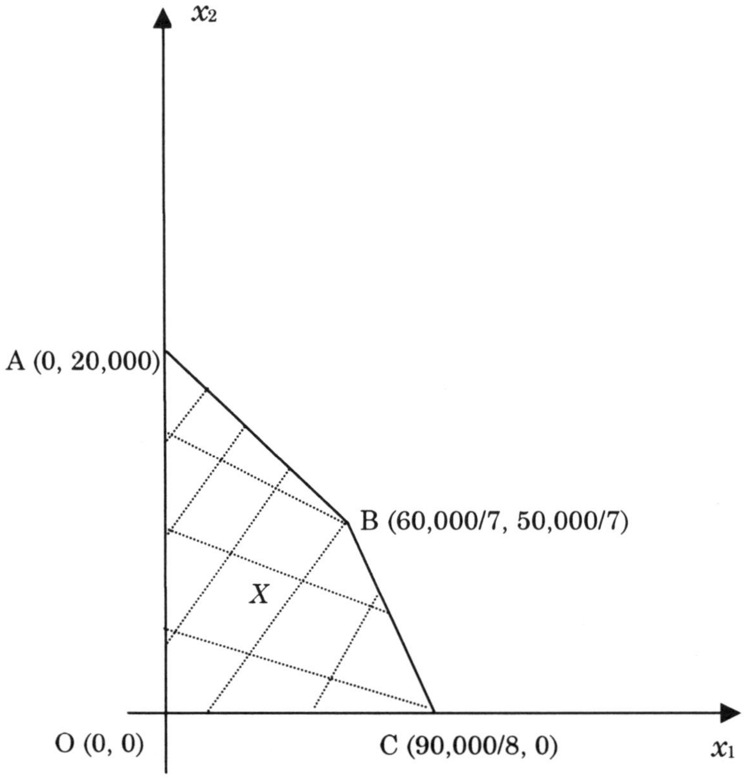

Figure 2.5 The Corner Points and Optimal Solution

Example 2.4 In the following, we use the matrix representation of the Standard Form to show the idea of the simplex method and then walk through the method by examining the PC problem.

Consider the Standard Form (M2.1):

$$\begin{array}{ll} \textit{Max} & \mathbf{c\,x} \\ \textit{Subject to} & \mathbf{A\,x} \le \mathbf{d} \\ & \mathbf{x} \ge 0 \end{array}$$

We add the slack variables, defined as $\mathbf{s} = (s_1, s_2, \ldots, s_m)$ to the constraints, then the problem is

$$\begin{array}{ll} \textit{Max} & \mathbf{c\,x} \\ \textit{Subject to} & \mathbf{A\,x} + \mathbf{I\,s} = \mathbf{d} \\ & \mathbf{x}, \mathbf{s} \ge 0 \end{array}$$

In the above form, **I** is an (mxm)-dimensional identity matrix. We rewrite into the initial simplex tableau:

Table 2.1 The Initial Simplex Tableau

x	**s**	RHS	
A	**I**	**d**	(2.1)
-c	**0**	**0**	(2.2)

In Table 2.1, RHS stands for "Right-Hand-Side." Given a basic variable index set J with basic variables denoted by $\mathbf{x}(J) = (x_{1j}, x_{2j}, \ldots, x_{mj})$, we define $\mathbf{x}(J')$ as the nonbasic variables. We can decompose (or write) the matrix [**A, I**] to [**B, R**], where **B** is the basis matrix with the column index of [**A, I**] in J and **R** is the nonbasis matrix with no column index of [**A, I**] in J. Similarly, we decompose the vector [**-c, 0**] to [$-\mathbf{c}_B$, $-\mathbf{c}_R$], where $-\mathbf{c}_B$ is the subvector of [**-c, 0**] with column index in J and $-\mathbf{c}_R$ is the subvector of [**-c, 0**] with no column index in J. Then, Equation (2.1)-(2.2) can be rewritten as

Table 2.2 The Reform of The Initial Simplex Tableau

$\mathbf{x}(J)$	$\mathbf{x}(J')$	RHS	
$\mathbf{B}$	$\mathbf{R}$	$\mathbf{d}$	(2.3)
$-\mathbf{c}_B$	$-\mathbf{c}_R$	$\mathbf{0}$	(2.4)

We use a row linear transformation on Table 2.2 by:
(2.5) = $\mathbf{B}^{-1}$ (2.3) (i.e., premultiply (2.3) by $\mathbf{B}^{-1}$), and
(2.6) = $\mathbf{c}_B$ (2.5) + (2.4).
This results in Table 2.3 know as the simplex tableau.

Table 2.3 The Simplex Tableau

$\mathbf{x}(J)$	$\mathbf{x}(J')$	RHS	
$\mathbf{I}$	$\mathbf{B}^{-1}\mathbf{R}$	$\mathbf{B}^{-1}\mathbf{d}$	(2.5)
$\mathbf{0}$	$\mathbf{c}_B\mathbf{B}^{-1}\mathbf{R} - \mathbf{c}_R$	$\mathbf{c}_B\mathbf{B}^{-1}\mathbf{d}$	(2.6)

Definition 2.1 Given a basis index set J for a LP problem,
(i) if $\mathbf{x}(J) = \mathbf{B}^{-1}\mathbf{d} \geq 0$, $\mathbf{x}(J)$ is called a feasible solution;
(ii) if the vector $(\mathbf{c}_B \mathbf{B}_J^{-1} \mathbf{R} - \mathbf{c}_R) \geq 0$, the feasible solution $\mathbf{x}(J)$ is called an optimal solution, and its corresponding optimal objective value is $\mathbf{c}_B \mathbf{B}^{-1}\mathbf{d}$.

Generally, (ii) of Definition 2.1 is called the *optimality condition.* The procedure can be summarized as follows.

Simplex Method:

Step 1. Build an initial simplex tableau Table 2.1 and check if the vector $-\mathbf{c} \geq 0$. If not, choose the minimum cell c_k of $-\mathbf{c}$ to improve and the corresponding column k as pivoting column. Then, find minimum d_r/a_{rk} of $\{ d_i/a_{ik} \mid a_{rk} > 0\}$ as the pivoting cell and the corresponding row r as pivoting row.

Step 2. Use the row linear transformation to transform the column $(a_{1k}, a_{2k}, \ldots a_{rk}, \ldots, a_{mk})$ into (0, 0, …1,…, 0) and obtain the simplex tableau Table 2.2.

Step 3. Use Definition 2.1 to see if the optimal solution found. If yes, the procedure terminates. Otherwise, go back to Step 1, and replace $\mathbf{d}$ by $\mathbf{B_J}^{-1}\mathbf{d}$ and $(-\mathbf{c}, \mathbf{0})$ by $(\mathbf{0}, \mathbf{c_B}\,\mathbf{B_J}^{-1}\,\mathbf{R} - \mathbf{c_R})$.

Now we use the simplex method to solve the PC problem of Section 1.2:

Max	$400\,x_1 + 200\,x_2$	(Profit)
Subject to	$3\,x_1 + 2x_2 \leq 40{,}000$	(Hardware)
	$8x_1 + 3x_2 \leq 90{,}000$	(Software)
	$x_1,\ x_2 \geq 0.$	

Step 1. To build an initial simplex tableau, we add the slack variables $\mathbf{s} = (s_1, s_2)$ for the constraints and rewrite the problem as

$$-400\,x_1 - 200\,x_2 = 0$$
$$3\,x_1 + 2x_2 + s_1 = 40{,}000$$
$$8x_1 + 3x_2 \;\; + s_2 = 90{,}000$$

Then, the initial simplex tableau associated with O(0, 0) in Figure 2.5 is:

Table 2.4 The Initial Simplex Tableau of PC Problem

x_1	x_2	s_1	s_2	RHS	d_i/a_{ik}
3 (0)	2	1	0	40,000	13,333
8 (1)	3	0	1	90,000	11,250*
-400 (0)	-200	0	0	0	

Since −400 is smaller than −200, we choose the column x_1 as the pivoting column. Because 40,000/3 = 13,333 and 90,000/8 = 11,250, we choose the second constraint (row) as the pivoting row and the 8 as the pivoting cell.

Step 2. We use the row linear transformation to convert the column (3, 8, -400) into (0, 1, 0). The details are:

(i) Multiplying 1/8 on the second row (8, 3, 0, 1, 90000), we have (1, 3/8, 0, 1/8, 11250).

(ii) We multiply −3 on (1, 3/8, 0, 1/8, 11250) and add the result on the row (3, 2, 1, 0, 40000). Then the new first constraint becomes (0, 0.88, 1, -0.38, 6250).

(iii) We multiply 400 on (1, 3/8, 0, 1/8, 11250) and add the result on the row (-400, -200, 0, 0, 0). Then the new objective row becomes (0, -50, 0, 50, 4500000).

These form the following tableau, associated with the corner point C(90,000/8, 0):

Table 2.5 The Simplex Tableau of the First Pivoting for PC Problem

x_1	x_2	s_1	s_2	RHS	d_i/a_{ik}
0	0.88 (1)	1	-0.38	6250	7142.9*
1	0.38 (0)	0	0.13	11250	29605.3
0	-50 (0)	0	50	4500000	

Step 3. Using Definition 2.1, we have that $\mathbf{x}(C) = \mathbf{x}(s_1, x_1) = (6250, 11250) \geq 0$ is feasible with the objective value 4,500,000. However, -50 on the objective row indicates the solution is not optimal. Then, we go back to Step 1 for the second iteration as follows.

Steps 1′. Since –50 is the smallest cell on the objective row, we choose the column x_2 as the pivoting column. Because 6250/0.88 = 7142.9 and 11250/0.38 = 29605.3, we choose the first constraint (row) as the pivoting row and the 0.88 as the pivoting cell.

Step 2′. We use the row linear transformation to convert the column (0.88, 0.38, -50) of Table 2.5 into (1, 0, 0). The details are:

(i) Multiplying 1/0.88 on the first row (0, 0.88, 1, -0.38, 6250), we have (0, 1, 1.14, -0.43, 7142.9).

(ii) We multiply -0.38 on (0, 1, 1.14, -0.43, 7142.9) and add the result on the row (1, 0.38, 0, 0.13, 11250). Then the new first constraint becomes (1, 0, -0.43, 0.29, 8571.4).

(iii) We multiply 50 on (0, 1, 1.14, -0.43, 7142.9) and add the result on the objective row (0, -50, 0, 50, 4500000). Then the new objective row becomes (0, 0, 57.14, 28.57, 4857142.9).

Then we have Table 2.6, associated with the corner point B(60,000/7, 50,000/7):

Table 2.6 The Optimal Simplex Tableau of PC Problem

x_1	x_2	s_1	s_2	RHS
0	1	1.14	-0.43	7142.9
1	0	-0.43	0.29	8571.4
0	0	57.14	28.57	4857142.9*

Step 3′. By Definition 2.1, we see that in addition to $\mathbf{x}(B) = \mathbf{x}(x_2, x_1) = (7142.9, 8671.4) \geq 0$, the objective function vector (0, 0, 57.14, 28.57) ≥ 0. Thus, $\mathbf{x}(B)$ is the optimal solution with the objective value 4,857,142.9. This result is identical to that of the graphical method.

We note that in the above example, the simplex method only searched three corner points: O, C, and B and terminates at B. This means that the simplex method outperforms the graphical method. The idea of the simplex method provides a manageable framework of computer solution procedure for LP problems. Most of computer packages that use the simplex method have three major algorithms: (i) finding the initial basic solution, (ii) identifying the pivoting cell, and (iii) computing the row linear transformation. Other issues, such as artificial variables, infeasible solution, unbounded solution, degenerate solution, and alternative solution will be found in section 2.8. The related details of these topics are referred to the known LP books listed in the readings of this chapter.

2.4 Duality Theory and Its Economic Interpretation

In this section, we discuss an alternative approach to the original LP problem, which is called the *duality* of LP problems. We first present a simple mathematical procedure to construct the duality, and then, use two examples to illustrate the details of constructing the dualities for maximizing and minimizing problems, respectively.

If we view the Standard Form (M2.1):

$$\begin{array}{ll} \textit{Max} & \mathbf{c\,x} \\ \textit{Subject to} & \mathbf{A\,x} \le \mathbf{d} \\ & \mathbf{x} \ge 0 \end{array}$$

as the original LP problem, its economic interpretation can be stated as given the scarce or limited resource **d**, we want to positively produce the products **x** so that the total profit **cx** will be maximized. However, in the business world, we have to consider other contingency plans in addition to the production plan. For example, alternatively, we may sell all of available resource **d** with the unit price **u**. If the total selling price $\mathbf{u}^t\,\mathbf{d}$ is equal to or better than the total profit **cx**, then we would like to take this opportunity. This idea is captured by the duality and can be stated as instead of producing **x**, we can sell the resource **d** for the unit price **u** and achieve the minimal acceptable **ud**. With respect to the dual problem, the original problem is called the primal problem.

Precisely, let $\mathbf{u} = (u_1, u_2, \ldots, u_m)^t$ be a row vector as the unit price of the column vector of resource $\mathbf{b} = (b_1, b_2, \ldots, b_m)^t$. The dual problem of the original problem (M2.3) with standard form is

$$\begin{array}{ll} \textit{Min} & \mathbf{u}^t\,\mathbf{b} \\ \textit{Subject to} & \mathbf{u}^t\,\mathbf{A} \ge \mathbf{c} \\ & \mathbf{u}^t \ge 0 \end{array}$$

We note that if the primal LP problem is a minimizing problem, its corresponding dual problem will be maximizing problem. The interpretation in production context can be: with the primal problem, we want to minimize the total production cost $\mathbf{u}^t\,\mathbf{b}$ to satisfy the requirement **c**; and with the dual problem, we want to find the maximal acceptable total price **cx** to buy the package **b** without production for the requirement.

Example 2.5 To see the details of constructing the duality, we restate Table 1.1 for the PC problem as follows:

Table 2.7 The Duality of PC Problem

Resource	Type 1(x_1)	Type 2(x_2)	Availability Level (units)
Hardware(u_1)	3	2	40,000
Software(u_2)	8	3	90,000
Profit ($)	400	200	

We define the following **unit price variables**:

Let u_1 = the unit price to sell Hardware parts, and
u_2 = the unit price to sell Software parts.

Then, the *objective (criterion) function* that shows the minimal total selling price can be expressed as

$$\text{Min} \quad 40{,}000\, u_1 + 90{,}000\, u_2$$

The **constraints** of selling hardware and software parts for producing Type 1 and Type 2, respectively can be represented by:

$$3\, u_1 + 8\, u_2 \geq 400 \qquad \text{(Type 1)}$$
$$2\, u_1 + 3\, u_2 \geq 200 \qquad \text{(Type 2)}$$

We assume that variables u_1 and u_2 are nonnegative and the dual problem then is:

$$\begin{array}{lll} \textit{Min} & 40{,}000\, u_1 + 90{,}000\, u_2 & \text{(Total Selling Price)} \\ \textit{Subject to} & 3\, u_1 + 8u_2 \geq 400 & \text{(Type 1)} \\ & 2\, u_1 + 3u_2 \geq 200 & \text{(Type 2)} \\ & u_1,\, u_2 \geq 0. & \end{array}$$

Therefore, when we construct the primal problem for Table 2.7, we look at the variables **x** from the top to down; while we construct the dual problem, we view the variables **u** from left to right. Some main properties of the duality are observed as:

Property 2.5 Given a primal problem with the standard form and its dual problem, if **x**'is a feasible solution of the primal problem and **u**' is a

feasible solution of the dual problem, then the objective value of the primal problem
$\mathbf{cx}' \leq (\mathbf{u}')'\mathbf{b}$, the objective value of the dual problem.

Property 2.6 Given a primal problem with the standard form and its dual problem, if $\mathbf{x}^*$ is the optimal solution of the primal problem and $\mathbf{u}^*$ is the optimal solution of the dual problem, then $\mathbf{cx}^* = (\mathbf{u}^*)'\mathbf{b}$.

Property 2.7 When we use the simplex method to solve the primal problem with the standard form, we can find the optimal solution of the primal problem and the optimal solution of the dual problem simultaneously. This is a one-stone-kill two "birds".

These properties show the significance of considering the duality in the structure of LP problems. The verification of them is pretty straghtforward.

Example 2.6 In Table 2.6, we know the optimal solution of the primal problem is $x_2 = 7142.9$, $x_1 = 8571.4$, and $s_1 = s_2 = 0$. The corresponding dual optimal solution can be found, as u_1 is the value of the cell in the bottom row associated with s_1 column and u_2 is the value of the cell in the bottom row associated with s_2 column. That is, $u_1 = \$57.14$ and $u_2 = \$28.57$. The optimal dual objective value is
$57.14(40{,}000) + 28.57(90{,}000) = \$4857142.9$,
which is equal to the optimal primal objective value
$7142.9(200) + 8571.4(400) = \$4857142.9$.

2.5 Complementary Slackness

This section will briefly discuss a theoretical relationship between the primal and dual LP problems that is called the *complementary slackness*. Given the primal and dual problems we have discussed in the above section, let us define the primal slackness variables as

$\mathbf{s} = \mathbf{d} - \mathbf{Ax} \geq 0$ which is an m-dimensional column vector and the dual surplus variables as

$\mathbf{r} = \mathbf{u}'\mathbf{A} - \mathbf{c} \geq 0$ which is a n-dimensional row vector.

Then, for any feasible solution **x'** of the primal problem and feasible solution **u'** of the dual problem, we use Property 2.5 of the duality and have

$$\begin{aligned}(\mathbf{u}')^t\mathbf{d} - \mathbf{cx}' &= (\mathbf{u}')^t\ \mathbf{d} - (\mathbf{u}')^t\mathbf{Ax}' + (\mathbf{u}')^t\mathbf{Ax}' - \mathbf{cx}'\\ &= (\mathbf{u}')^t\ (\mathbf{d} - \mathbf{Ax}') + ((\mathbf{u}')^t\mathbf{A} - \mathbf{c})\mathbf{x}'\\ &= (\mathbf{u}')^t\mathbf{s} + \mathbf{rx}'\\ &\geq 0.\end{aligned}$$

Thus, the difference between the primal objective value and the dual objective value is classified by the term $(\mathbf{u}')^t\ \boldsymbol{s} + \mathbf{r}^t\mathbf{x}'$. For the optimality condition, **x*** and **u*** are the optimal solutions for the primal and dual problems, respectively, if and only if

$$\mathbf{rx}^* = 0 \text{ and } \mathbf{u}^*\mathbf{s} = 0.$$

This discussion can be stated as

Theorem 2.1 Let **x** be a feasible solution of the primal problem and **u** be a feasible solution of the dual problem. They are the optimal solutions if and only if the complementary conditions

(i) either $r_j = (\mathbf{uA} - \mathbf{c})_j = 0$ or $x_j = 0$ for $j = 1, \ldots, n$ and
(ii) either $s_i = (\mathbf{d} - \mathbf{Ax})_i = 0$ or $u_i = 0$ for $i = 1, \ldots, m$

are satisfied.

Example 2.7 In the example of the PC Problem shown as Table 2.6, we see that

(i) $r_1 = (\mathbf{uA} - \mathbf{c})_1 = 57.14(3) + 28.57(8) - 400 = 0$ and
$r_2 = (\mathbf{uA} - \mathbf{c})_2 = 57.14(2) + 28.57(3) - 200 = 0.$

(ii) $s_1 = s_2 = 0.$

Therefore, by Theorem 2.1, $x_2 = 7142.9$, $x_1 = 8671.4$, and $u_1 =$ \$57.14, $u_2 =$ \$28.57 are the optimal solutions for the primal and dual problems, respectively.

2.6 Sensitivity Analyses

Given a LP problem with the standard form, we note that the problem has been specified by known data of the objective vector **c**, the resource consumption matrix **A** and the constraint level vector **d**. Suppose that **x*** is identified as the optimal solution of the LP problem. We may be interested in what perturbation of known **c, A,** and **d** will allow **x*** to remain optimal. Studying the perturbations is called the ***sensitivity analysis***. This section outlines the basic concepts of the sensitivity analysis regarding the changes in the objective vector **c** and the constraint level vector **d**. The discussion of the change in the resource consumption matrix **A** can found in the LP books listed as the further reading.

Given the optimal solution **x*** with the basis matrix **B** as in Table 2.3, we discuss the following perturbations:

2.6.1 The Change in The Objective Vector **c**

Consider the optimality condition of (ii) of Definition 2.1:

$(\mathbf{c_B} \mathbf{B}^{-1} \mathbf{R} - \mathbf{c_R})_j \geq 0$, for all j not in J.

Let $\mathbf{z} = \mathbf{c_B} \mathbf{B}^{-1}\mathbf{R}$ and the jth component be $z_j = \mathbf{c_B}\, \mathbf{t}_j$, where $\mathbf{t}_j$ is the jth column of $\mathbf{B}^{-1}\mathbf{R}$. Then the optimality condition becomes

$z_j - c_j \geq 0$, for all j not in J.

Suppose that the kth coefficient c_k is changed to $c_k' = c_k + \Delta c_k$. This change may affect the optimality condition, depending on if the corresponding variable x_k is a basic variable or nonbasic variable.

(i) If x_k is a nonbasic variable, $\mathbf{c_B}$ is unchanged, but $z_k - c_k$. Thus,

$z_k - c_k' = (z_k - c_k) - \Delta c_k \geq 0$, or $z_k - c_k \geq \Delta c_k$. This means that in order to remain **x*** to be optimal, the upper bound of the increase of the kth coefficient Δc_k is $z_k - c_k$. However, the objective value will not change since $x_k = 0$ as a nonbasic variable.

(ii) If x_k is a basic variable, let k = pr. The $\mathbf{c_B}$ will change to $\mathbf{c_B}' = \mathbf{c_B} + \Delta c_{pr}\, \mathbf{e}_r$, where $\mathbf{e}_r$ is an $(1 \times m)$ row vector with all zeros but a 1 on the pr th column. For all j not in J (or for all nonbasic columns j),

$z_j' - c_j = \mathbf{c_B}'\mathbf{t}_j - c_j = \mathbf{c_B}\, \mathbf{t}_j + \Delta c_{pr}\, \mathbf{e}_r\, \mathbf{t}_j - c_j = z_j - c_j + \Delta c_{pr}\, t_{rj} \geq 0$, which is

$z_j - c_j \geq - \Delta c_{pr}\, t_{rj}$. Thus, we have the following cases:

(a) If $t_{rj} = 0$, any changes of Δc_{rp} will maintain the optimality condition of $z_j' - c_j$ for all j not in J.

(b) If $t_{rj} > 0$, $\Delta c_{pr} \geq -(z_j - c_j)/t_{rj}$ for all j not in J.

(c) If $t_{rj} < 0$, $\Delta c_{pr} \leq -(z_j - c_j)/t_{rj}$ for all j not in J.

We conclude that the optimal solution $\mathbf{x}^*$ will remain optimal if the change satisfies (a), (b) or (c). If t_{rj} dos not exist, Δc_{pr} in (b) (or (c)) has no lower bound (or upper bound).

Example 2.8 In the final simplex tableau of the PC Problem shown as Table 2.6, since both x_1 and x_2 are basic variables, we have the following discussion for the change in the objective vector c:

(i) For x_1, the lower bound of the change in $c_1 = 400$ is
$-(z_4 - c_4)/t_{24} = -(28.57)/0.29 = -98.52$ for $t_{24} > 0$.
The upper bound of the change in $c_1 = 400$ is
$-(z_3 - c_3)/t_{23} = -(57.14)/(-0.43) = 133$ for $t_{23} < 0$. That is,
$$-98.52 \leq \Delta c_1 \leq 133.$$

(ii) For x_2, the lower bound of the change in $c_2 = 200$ is
$-(z_3 - c_3)/t_{13} = -(57.14)/1.14 = -50.12$ for $t_{13} > 0$.
The upper bound of the change in $c_2 = 200$ is
$-(z_4 - c_4)/t_{14} = -(28.57)/(-0.43) = 66.44$ for $t_{14} < 0$, which is,
$$-50.12 \leq \Delta c_2 \leq 66.44.$$

2.6.2 *The Change in The Constraint Level Vector* **d**

Assume that we change the kth component of **d**, d_k to $d_k' = d_k + \Delta d_k$. In the case, although the objective coefficient of the optimal solution $\mathbf{x}^*$ is independent on the change, the solution $\mathbf{x}^*$ may become infeasible. The new constraint level vector is

$$\mathbf{d}' = \mathbf{d} + \Delta d_k \mathbf{e}_k,$$

where $\mathbf{e}_k$ is an $(m \times 1)$ column vector with all zeros but a 1 on the k th row. The new basic variables

$$\mathbf{x}_B{}' = \mathbf{B}^{-1}\mathbf{d}' = \mathbf{B}^{-1}(\mathbf{d} + \Delta d_k \mathbf{e}_k) = \mathbf{B}^{-1}\mathbf{d} + \Delta d_k \mathbf{B}^{-1}\mathbf{e}_k$$

$$= \mathbf{x}_B + \Delta d_k \mathbf{B}^{-1}\mathbf{e}_k \geq 0,$$ if $\mathbf{x}_B{}'$ is feasible (see (i) of Definition 2.1).

Therefore, we can determine Δd_k by solve $\mathbf{x}_B + \Delta d_k \mathbf{B}^{-1}\mathbf{e}_k \geq 0$ for Δd_k in the given optimal simplex tableau.

Example 2.9 Using Table 2.6 for the PC problem, we can find

$$\mathbf{x_B} = \begin{pmatrix} 7142.86 \\ 8571.43 \end{pmatrix}, \quad \mathbf{B}^{-1} = \begin{pmatrix} 1.14 & -0.43 \\ -0.43 & 0.29 \end{pmatrix}.$$

Then $\mathbf{x_B} + \Delta d_1 \mathbf{B}^{-1}\mathbf{e}_1 \geq 0$ is,

$$7142.86 + 1.14\ \Delta d_1 \geq 0 \text{ and } 8571.43 - 0.43\ \Delta d_1 \geq 0.$$

Solving two inequalities, we get $-6265.7\ 12 \leq \Delta d_1 \leq 19933.6$.

Similarly, $\mathbf{x_B} + \Delta d_2 \mathbf{B}^{-1}\mathbf{e}_2 \geq 0$ implies that

$$7142.86 - 0.43\ \Delta d_2 \geq 0 \text{ and } 8571.43 + 0.29\ \Delta d_2 \geq 0.$$

We solve these two inequalities and get $-29556.7 \leq \Delta d_2 \leq 16610$.

2.7 Complexity Analyses and Other Methods

This section provides a general description about the computational complexity or efficiency of algorithms for solving LP problems. We will start with the basic concepts of computational complexity. Then, we will discuss the complexity of the simplex method, Khachian's ellipsoid method, and Karmarkar's projective method. The detailed theoretical results can be referred to the respective articles listed in the further reading.

2.7.1 Basic Concepts of Computational Complexity

The research issue of computational complexity regarding any mathematical algorithms comes from the Computer Science. The computational complexity measures the efficiency of operating algorithms by using binary coding bits. The operators may include additions, multiplications, and comparisons. The algorithm complexity of LP problems, as a particular example, can be evaluated in terms of the number of operators needed to solve the problems. Consider the Canonical Form (M2.2) of LP problems:

$$\begin{aligned} &\textit{Max} && \mathbf{c\,x} \\ &\textit{Subject to} && \mathbf{A\,x} = \mathbf{d} \\ &&& \mathbf{x} \geq 0 \end{aligned}$$

Without the loss of generality, we assume that **A** is an (mxn) matrix and m, n ≥ 2 and all data in (**c**, **A**, **d**) are integers. Then, the triplet (m, n, L) can represent the size of LP problems, where *L* is the number of binary bits required to story all the data of the LP problem. *L* is also called the input length of the LP problem and is known as a function of logarithm of m, n, c_j, a_{ij}, and d_i. The complexity of an algorithm of the LP problem, thus, is defined as a function of the triplet, *F*(m, n, *L*). If there exists a sufficiently large constant $k > 0$ such that the total number of operators used by the algorithm is less than or equal to *kF*(m, n, *L*), the algorithm is said to be *of order of complexity of O*(*F*(m, n, *L*)) . When *F*(m, n, *L*) is a polynomial function, the algorithm is said to be *of polynomial complexity* or *polynomially bounded.* Otherwise, the algorithm of the LP problem is *of nonpolynomial complexity.*

2.7.2 Complexity of Simplex Method

In practice, the simplex method has been empirically observed to use from 3m/2 to 3m iterations to solve the LP problems. However, when we look at the structure of the simplex method, the situation is different. The Canonical Form of LP problems requires about m(n-m) + n +1 multiplications and m(n – m + 1) additions in an iteration. It is order of *O*(mn). Some simulation results even confirmed that the expected computational effort of the simplex method is of $O(m^2n)$. According to the simplex method, the total number of corner points is

$$\binom{n}{m} \geq 2^m, n \geq m .$$

If a LP problem requires the search of most or all the corner points for finding the optimal solution, it has great potential of exponential (not polynomial) complexity. In fact, V. Klee and G. L. Minty (1972) demonstrated such a example that required traverse all $2^n - 1$ corner points to reach the optimal solution. Therefore, in theory, the simplex method is of exponential complexity.

2.7.3 *Khachian's Ellipsoid Method*

Is there any algorithm of solving LP problems having polynomial complexity? The answer is yes. In 1979, L. G. Khachian proposed a polynomial-time algorithm, now known as the ellipsoid method, of solving the open set of linear inequalities $X = \{\mathbf{x}: \mathbf{Ax} < \mathbf{d}\}$, where **A** is an (mxn) matrix, m, n ≥ 2 and **A** and **d** have all integers elements. Whenever X is not empty and there exists a solution, this method finds the solution. When the method is used to solve the Standard Form of the LP problem, it has been shown that the algorithm requires at most $O((m + n)^2L)$ iterations, which is a polynomial complexity. The basic steps of the method can be outlined as follows.

Ellipsoid Method:

Step 1. If X is not empty, the method defines a sphere in E^n (a n-dimensional Euclidean space), centered at the origin and with a radius covering sufficiently many possible values in the volume of X.

Step 2. If the center of the sphere is in X, the algorithm terminates. Otherwise, the method constructs a sequence of ellipsoids that are monotonically shrinking in the volume of X so that they contain the values of X covered by the initial sphere. Wherever the center of any ellipsoids is identified to lie in the X, the solution is found and the algorithm terminates.

Although in theory, the ellipsoid method is of a polynomial complexity, the computational experience using this method has disappointed. This indeed turns out that this method has little practical power in solving real-life LP problems.

2.7.4 *Karmarkar's Projective Method*

In 1984, N. Karmarkar proposed a better applicable polynomial-time algorithm, which is called the projective method. When this method is used to solve the LP problem: minimize $\{\mathbf{cx}: \mathbf{Ax} = 0, \mathbf{ex} = 1, \mathbf{x} \geq 0\}$, where **A** is an $(m \times n)$ matrix of full row rank and **e** is an $(1 \times m)$ row vector with all 1, it can find an optimal solution and terminate in $O(nL)$ iterations under certain assumptions. It has been also shown that any LP problem can be transformed to this LP format.

In the simplex method, we move from one corner point to another corner point on the boundary of the simplex structure for search the optimal solution. In the projective method, we move through the interior point of the polygon of feasible region by transforming the space at each step so that the current solution (at the beginning, the initial interior solution) is placed at the center domain of the polygon. Since we move from the interior instead of the boundary, the projective method can have a shorter path or less iteration to reach the optimal solution. However, the tradeoff is that we may expect to spend more computation time at each iteration. This computation is consumed by constructing the projective transformation. The main concept of the projective method can be captured as follows without mathematical descriptions.

Projective Method:
Step 1. Given an interior solution in the feasible polygon of the LP problem, if this solution is not close to the optimal solution, we construct a projective transformation $\boldsymbol{T}$ to convert the solution space so that the current solution is near the center of transformed space.
Step 2. We move the current solution in the direction of descent of the objective function to reach the minimum point.
Step 3. We use an inverse transformation $\boldsymbol{T}^{-1}$ to convert the improved solution back to the original solution space and it becomes a new interior solution. Whenever this solution is close enough to the optimal solution, the algorithm terminates. Otherwise, go back to Step 1.

The initial idea of the Karmarkar's projective method has attracted a lot of researchers. So far, numerous similar algorithms have been reported. The application of this method is also very promising. It has great potential to be a strong competitor to the simplex method in solving the real-life LP problems.

2.8 Further Remarks

In addition to the above discussions on LP problems, we outline several related issues in this section. When we use the simplex method to solve the Canonical Form of LP problems, we cannot have a basic solution in the initial simplex tableau because the equalities of the constraints. To overcome this technical difficulty, we can add some *artificial variables* without physical significance to gain a starting basis to pivot other corner points. Given a basis index set J for a LP problem, if for some j, $x_j = (\mathbf{B}^{-1}\mathbf{d})_j < 0$, the solution $\mathbf{x}(J)$ is called an *infeasible solution*. If for some k, $x_k = (\mathbf{B}^{-1}\mathbf{d})_k = 0$, $\mathbf{x}(J)$ is called a *degenerate solution*. The degenerate solution may result in a *cycling*. That is, when a basic variable of degenerate solution whose value is zero is pivoted, the corresponding objective value does not change. After several iterations, we may return to an encountered feasible solution without the termination of an optimal solution. If the pivoting process has no end of increase (for maximizing problems) or decease (for minimizing problems) of the objective value, then the LP problem has an *unbounded solution*. Both infeasible and unbounded solutions are caused by the illness of the LP modeling. In either case, the LP problem should be revisited and reformed. If the objective function is parallel to any of constraints, the problem has two *alternative solutions*, which mean that two different solutions have the same optimal objective value. In this case, although the simplex method terminates at one optimal solution, we can choose the zero cells in the bottom row of the simplex tableau as the pivoting column to achieve another optimal solution.

Exercises

2.1 Solve the following LP problem by using the graphical method:

Max $10\,x_1 + 12\,x_2$

Subject to $3\,x_1 + 2x_2 \leq 24$

$4\,x_1 + 6x_2 \leq 48$

$x_1,\, x_2 \geq 0.$

2.2 Solve the following LP problem by using the graphical method:

Min $5\,x_1 + 3\,x_2$

Subject to $2x_1 + 4x_2 \geq 20$

$x_1 + x_2 \geq 10$

$x_1,\, x_2 \geq 0.$

2.3 Solve the following LP problem by using the graphical method:

Max $x_1 + 2\,x_2$

Subject to $2x_1 + 4x_2 \leq 35$

$4x_1 + 2x_2 \geq 24$

$x_1,\, x_2 \geq 0.$

2.4 Koch Manufacture makes two models of car mufflers: regular and deluxe. The regular model requires 2.5 hours and the deluxe needs 4 hours in the assemble line. Both models use the same steel frame. The production manager has 24 frames and 100 hours available for the production process. The unit profit of the regular model is $40 and that of the deluxe model is $60. The company wants to achieve the maximum profit by using the limited resources to produce quantities of the two models.

(i) Define the decision variables and formulate the LP problem.

(ii) Use the graphical method to solve this LP problem.

2.5 Use the simplex method to solve the problem 2.1.

2.6 Use the simplex method to solve the problem 2.4.

2.7 Define the dual variables for the Koch Manufacture's decision problem, formulate the dual problem, and use the graphical method to find the optimal solution of the dual problem.

2.8 Use Theorem 2.1 to verify the primal optimal solution of the problem 2.4 and the dual optimal solution of the problem 2.7.

2.9 Show the sensitivity analysis of the problem 2.4 on the change in the objective coefficients ($40, $60) (Hint: use the final simplex tableau of the problem 2.6).

2.10 Show the sensitivity analysis of the problem 2.1 on the change in the constraint level (24, 100) (Hint: use the final simplex tableau of the problem 2.6).

Chapter 3
RELATIONSHIPS BETWEEN LP, MC, AND MC^2 PROBLEMS

In this chapter, we discuss mathematical relationships between linear programming (LP), Multiple Criteria (MC) linear programming, and multiple criteria and multiple constraint-level (MC^2) linear programming. This discussion is a preliminary for deriving various MC^2 techniques from Chapter 4 to Chapter 12. Since there are numerous books and papers describing MC problems, we prefer not focusing on the details of MC techniques in the chapter. Instead, we briefly introduce the concepts of MC framework as a bridge to connect LP problems and MC^2 problems in Section 3.1. We show that the decision space (the dimensions of decision variables) is interchangeable with the criteria space (the dimensions of objectives). We also use an example to highlight the ideas of various MC solution algorithms. Section 3.2 traces back the mathematical root of MC^2 problems: matrix optimization that is an extension of LP from a scalar form to vector form, and from vector form to a matrix form. The duality theory of matrix optimization provides a theoretical background for MC and MC^2 problems. Based on the duality theory, Section 3.3 proposes a relationship theory to show the inner connections of MC and MC^2 problems. Because the LP problem is a simple case of MC problems, this theory essentially illustrates the relationship of LP, MC, and MC^2 problems. Section 3.4 elaborates the application perspectives of MC^2 framework, which answer why LP model is extended to MC model, and then MC model is further extended to MC^2 model. Finally, Section 3.5 concludes this chapter with several remarks.

3.1 Framework of MC Linear Programming

In the PC problem of Section 1.3, we have already seen how a LP problem can be extended to an MC problem if a single criterion is replaced by several (or more than one) criteria. This section outlines the solution concepts and algorithms in solving MC problems. For the reader who has no knowledge about MC linear programming, the section serves as preparation to the techniques of MC^2 problems. For those who know MC problems, it could be a highlighted review of MC linear programming.

Consider the Standard Form (M2.1) of LP problems:

$$\begin{array}{ll} \textit{Max} & \mathbf{c\,x} \\ \textit{Subject to} & \mathbf{A\,x} \leq \mathbf{d} \\ & \mathbf{x} \geq 0 \end{array}$$

We note that the single criterion is represented by an $(1 \times n)$ row vector **c**, which may be the unit profit, selling price, customer satisfaction rate, etc. Abstractly, if we want to simultaneously maximize q different criteria represented by a $(q \times n)$ matrix **C**, then, the algebraic form of MC linear programming (M3.1) can be written as

$$\begin{array}{ll} \textit{Max} & \mathbf{C\,x} \\ \textit{Subject to} & \mathbf{A\,x} \leq \mathbf{d} \\ & \mathbf{x} \geq 0 \end{array}$$

As we mentioned in Section 1.3, finding the ideal solution that simultaneously represents all maximal criteria is almost impossible. Instead, we can find a solution that is the best tradeoff between the multiple criteria. This solution is called the nondominated (or efficient, admissible) solution. To show the concept of the solution, we formally bring up the notions of decision space and criteria space as follows:

Definition 3.1 Given an MC problem,

(i) the **decision space** is defined as $\mathbf{X} = \{\mathbf{x} \mid \mathbf{Ax} \leq \mathbf{d},\ \mathbf{x} \geq 0\}$;

(ii) the **criteria space** is defined as $\mathbf{Y} = \{\ \mathbf{y} = \mathbf{Cx} \mid \mathbf{x} \in \mathbf{X}\}$.

We note that the decision space **X** is the feasible solution region that we discussed in Section 2.2 for the graphical method. The LP problem and MC problem have the same decision space. However, the criteria space of the LP problem is a straight line because of the single criterion. The objectives $\mathbf{y} = (y_1, y_2, \ldots, y_q) = (\mathbf{C}^1\mathbf{x}, \mathbf{C}^2\mathbf{x}, \ldots, \mathbf{C}^q\mathbf{x})$ is a vector, where $y_j = \mathbf{C}^j\mathbf{x}$, is the jth objective function, $j = 1, \ldots, q$.

The concept of the nondominated solution is based on Italy economist Pareto's idea for "optimality." We summarize it as the following preference structure:

Pareto Preference:

Given the criteria space $\mathbf{Y} = \{\mathbf{y} = \mathbf{Cx} \mid \mathbf{x} \in \mathbf{X}\}$, for any two objective vectors $\mathbf{y}^1$ and $\mathbf{y}^2$, $\mathbf{y}^1 \geq \mathbf{y}^2$ means that $y_j^1 \geq y_j^2$, and $y_j^1 \neq y_j^2$, for j = 1, . . ., q, then

(i) $\{\succ\} = \{ (\mathbf{y}^1, \mathbf{y}^2) \mid \mathbf{y}^1 \geq \mathbf{y}^2\}$ and

(ii) $\{\sim\} = \{ (\mathbf{y}^1, \mathbf{y}^2) \mid \text{neither } \mathbf{y}^1 \geq \mathbf{y}^2 \text{ nor } \mathbf{y}^2 \geq \mathbf{y}^1\}$.

The notion "$\succ$" refers to "better", while the notion "$\sim$" is called "indefinite."

Definition 3.2 Given an MC problem, the solution $\mathbf{y}^*\in\mathbf{Y}$ is said to be a **nondominated solution** if the set $\{\succ\mathbf{y}^*\}$ is empty.

The set $\{\succ\mathbf{y}^*\}$ means all **y** that are better than $\mathbf{y}^*$. Now we first discuss the conditions of nondominated solution for the criteria space **Y** and then for the decision space **X**. The detailed proofs of these results can be found in Yu (1985).

Theorem 3.1 Given an MC problem, the solution $\mathbf{y}^*\in\mathbf{Y}$ is a **nondominated solution** if and only if for any i $\in$ {1, 2, ..., q}, $\mathbf{y}^*$ uniquely maximizes $\mathbf{y}_i$ for all $\mathbf{y}\in\mathbf{Y}_i(\mathbf{y}^*) = \{\mathbf{y}^*\in\mathbf{Y} \mid y_k \geq y_k^*, k \neq i, \text{ for } k = 1, \ldots, q\}$.

The meaning of Theorem 3.1 says that the necessary and sufficient condition of nondominated solution is if $\mathbf{y}^*$ maximizes a certain dimension on the criteria space **Y** where all y are better than $\mathbf{y}^*$ on the rest dimensions of **Y**. Theorem 3.1 indicates that in order to find a nondominated solution, we can switch some criteria into constraints on **Y**, or vice versa. This is called the *interchangeability* of criteria and constraints, which is a basic idea in the background of goal and compromise programming (see chapter 13). The following result is a sufficient condition of nondominated solution that serves as the basis in the development of MC simplex method (Yu and Zeleny 1975).

Theorem 3.2 Given an MC problem, if the solution $\mathbf{y}^*\in\mathbf{Y}$ maximizes

λ^t **y** over **Y** for some row weight vector λ, where $\lambda^t = (\lambda_1, \lambda_2, \ldots, \lambda_q)$ with $0 < \lambda_i \leq 1$, $i \in \{1, 2, \ldots, q\}$, then **y*** is a **nondominated solution**.

To make a connection between the decision space **X** and the criteria space **Y**, we have:

Definition 3.3 Given an MC problem, the solution **x***∈**X** is a **nondominated solution** if and only if **y***∈**Y** = *f*[**Y**] is a **nondominated solution**.

Then, the conditions of nondominated solution for the decision space **X** can be obtained as:

Theorem 3.3 Given an MC problem,
(i) the solution **x***∈**X** is a **nondominated solution** if and only if for any $i \in \{1, 2, \ldots, q\}$, $f(\mathbf{x}^*)$ uniquely maximizes $f_i(\mathbf{x}^*)$ for all $\mathbf{x} \in \mathbf{X}_i(\mathbf{x}^*) = \{\mathbf{x}^* \in \mathbf{X} \mid f_k(\mathbf{x}^*) \geq f_k(\mathbf{x}^*),\ k \neq i, \text{ for } k = 1, \ldots, q\}$.
(ii) if the solution **x***∈**X** maximizes $\lambda^t f(\mathbf{x}^*)$ over **X** for some row weight vector λ, where $\lambda^t = (\lambda_1, \lambda_2, \ldots, \lambda_q)$ with $0 < \lambda_i \leq 1$, $i \in \{1, 2, \ldots, q\}$, then **x*** is a **nondominated solution**.

Example 3.1 In Section 1.3, the MC program of the PC problem in terms of the decision space **X** is:

Max	$400\,x_1 + 200\,x_2$	(Profit)
Max	$x_1 + x_2$	(Production quantities)
Subject to	$3\,x_1 + 2x_2 \leq 40{,}000$	(Hardware)
	$8\,x_1 + 3x_2 \leq 90{,}000$	(Software)
	$x_1,\ x_2 \geq 0$	

Let $y_1 = 400\,x_1 + 200\,x_2$ and $y_2 = x_1 + x_2$. We solve x_1 and x_2 in terms of y_1 and y_2 and obtain $x_1 = 1/200\ y_1 - y_2$ and $x_2 = -1/200\ y_1 + 2y_2$. Then, we plug them into the constraints and the problem is transformed to the criteria space **Y**:

Max	y_1	(Profit)
Max	y_2	(Production quantities)

Subject to	$1/200\, y_1 + y_2 \leq 40{,}000$ (Hardware)
	$5/200\, y_1 - 2y_2 \leq 90{,}000$ (Software)
	$y_1, y_2 \geq 0$

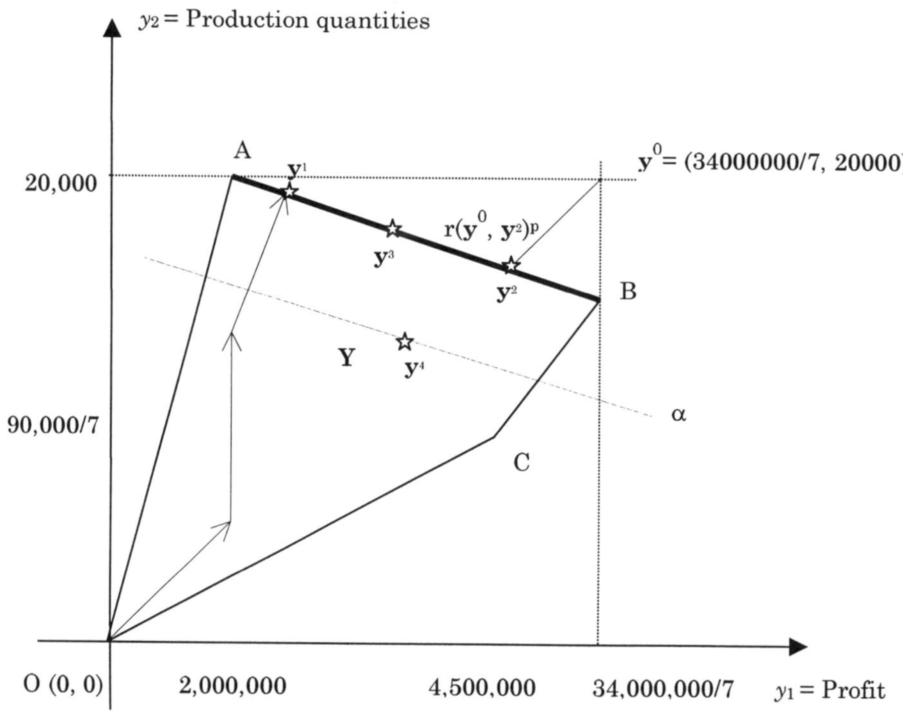

Figure 3.1 The Criteria Space **Y**

The Graph Figure 3.1 shows that the corner points of **Y** are:
A: Profit y_1 is \$ 2,000,000 and Production quantity y_2 is \$ 20,000
B: Profit y_1 is \$ 34,000,000/7 and Production quantity y_2 is \$ 110,000/7
C: Profit y_1 is \$ 4,500,000 and Production quantity y_2 is \$ 90,000/7

O: Profit y_1 is \$ 0 and Production quantity y_2 is 0.

We note that even though the ideal solution is $\mathbf{y}^0 = (y_1^0, y_2^0) =$ (34,000,000/7, 20,000), we cannot achieve it. If we want to only maximize y_1, the optimal solution is $y_1^0 = 34{,}000{,}000/7$. The optimal solution for maximizing y_2 is $y_2^0 = 20{,}000$. However, the nondominated solution $\mathbf{y}^*$ that represents the best tradeoff between Profit y_1 and Production quantity y_2 lies on the segment of the corner points A and B. How to identify $\mathbf{y}^*$ to solve such an MC problem has attracted a lot of researchers who produced numerous MC solving algorithms in the past few decades. Using Figure 3.1 as an illustration, we can summarize those algorithms into the following categories:

MC Solving Algorithms:

(i) *Interactive Algorithms* This class of algorithms start from O(0, 0), the analyst either asks the decision maker to respond some favorable points in **Y** or constructs an improvement direction over **Y** to gradually $\mathbf{y}^1$ on the segment of the corner points A and B as the preferred nondominated solution (see the further reading section).

(ii) *Goal and Compromise Algorithms* These approached construct various distance functions $\mathbf{r}(\mathbf{y}^0, \mathbf{y})^p$ from the ideal solution $\mathbf{y}^0$ to any point $\mathbf{y}$ on the segment of the corner points A and B by using different norms p, $1 \le p \le \infty$. A nondominated solution $\mathbf{y}^2$ on the segment of the corner points A and B is identified through minimizing the distance function (see chapter 13).

(iii) *MC Simplex Algorithm* This approach is systematic. The nondominated solution is represented by $\mathbf{y}^3 = \lambda_1 y_1 + \lambda_2 y_2$, with $0 < \lambda_i \le 1$, i =1, 2 (see Theorem 3.1). When λ_i value is changed, we can locate all nondominated solutions on the segment of the corner points A and B (Yu and Zeleny 1975).

(iv) *Fuzzy Satisficing Algorithm* Instead of finding the nonodominated solution on the segment of the corner points A and B, this method seeks a number of satisficing solution $\mathbf{y}^1$ according to the line of the aspiration level α, $0 \le \alpha \le 1$, which is acceptable by the decision maker, over the criteria space **Y**. Therefore, $\mathbf{y}^1$ is near (or sub) nondominated solution.

3.2 Matrix Optimization

This section traces back a mathematical root of MC programming, which is called the linear matrix programming problem. According to Gale, Kuhn and Tucker (1951), a matrix **B** having a certain property is said to be *maximal* or *minimal* if there is no other matrix $\mathbf{B}^0$ having the property such that $\mathbf{B}^0 \geq \mathbf{B}$ or $\mathbf{B}^0 \leq \mathbf{B}$, respectively.

A linear matrix programming is written as:

$$Max\ \{\mathbf{B} \mid \mathbf{Cx} \geq \mathbf{B}\gamma,\ \mathbf{Ax} \leq \mathbf{D}\gamma,\ \mathbf{x} \geq 0,\ \gamma > 0\}, \tag{3.1}$$

where **A** is an $(m \times n)$ matrix, **D** is an $(m \times p)$ matrix, **C** is a $(q \times n)$ matrix, and **B** is a $(q \times p)$ matrix of variables, and **x** is a n dimensional column vector and γ is p dimensional column vector, respectively.

Definition 3.4 A feasible solution $(\mathbf{x}^*, \gamma^*, \mathbf{B}^*)$ for (3.1) is said to be a nondominated solution for (3.1) if and only if there is no such other feasible solution $(\mathbf{x}^0, \gamma^0, \mathbf{B}^0)$ of (3.1) for which $\mathbf{B}^0 \geq \mathbf{B}^*$.

Note that if $(\mathbf{x}^*, \gamma^*, \mathbf{B}^*)$ is a nondominated solution for (3.1), then $\mathbf{B}^*$ is said to be maximal for (3.1). Gale, Kuhn and Tucker (1951) also showed the following existence theorem:

Theorem 3.4 The maximal solution $\mathbf{B}^*$ exists for the matrix problem (3.1) if and only if the following condition both hold:

(i) $\mathbf{Ax} \leq \mathbf{D}\gamma$ for some $\mathbf{x} \geq 0$, $\gamma > 0$, and

(ii) $\mathbf{Cx} \geq 0$, for no $\mathbf{x} \geq 0$ such that $\mathbf{Ax} \leq 0$.

The importance of the Gale, Kuhn and Tucker's work can be summarized as follows:

(i) It shows that mathematically, scalar optimization (or linear programming) can be extended to vector optimization (or MC programming), that then can be further extended to matrix optimization (or MC^2 linear programming).

(ii) The conditions of solution existence in linear programming can be parallely applied to that of vector optimization and matrix optimization.

To see the insight of the importance, we have to elaborate the duality of the matrix optimization as follows:

A duality of Gale, Kuhn and Tucker's matrix maximization (3.1) can be stated as

$$Min\ \{\mathbf{B} \mid \mathbf{u}^t\mathbf{D} \leq \lambda^t\mathbf{B},\ \mathbf{u}^t\mathbf{A} \geq \lambda^t\mathbf{C},\ \mathbf{u} \geq 0,\ \lambda > 0\}, \qquad (3.2)$$

where **u** is an m dimensional row vector and λ is q dimensional row vector, respectively.

Definition 3.5 A dual feasible solution ($\mathbf{u}^*$, λ^*, $\mathbf{B}^*$) for (3.2) is said to be a nondominated solution for (3.2) if and only if there is no such other dual feasible solution ($\mathbf{u}^0$, λ^0, $\mathbf{B}^0$) of (3.2) for which $\mathbf{B}^0 \leq \mathbf{B}^*$.

Based on Definitions 3.4 and 3.5, the duality theorem of Gale, Kuhn and Tucker (1951) is stated as:

Theorem 3.5

(i) (3.1) has a nondominated solution if and only if (3.2) has a nondominated solution.

(ii) Let ($\mathbf{x}^*$, γ^*, $\mathbf{B}^*$) be a nondomonated solution for (3.1). Then there exists a nondominated solution ($\mathbf{u}^0$, λ^0, $\mathbf{B}^0$) for (3.2) such that $\mathbf{B}^* = \mathbf{B}^0$. An analogous statement holds for (3.2).

(iii) If ($\mathbf{x}^*$, γ^*, $\mathbf{B}^*$) is a nondominated solution for (3.1), then $\mathbf{C}\mathbf{x}^* = \mathbf{B}^*\gamma^*$. An analogous statement holds for (3.2).

(iv) Let ($\mathbf{x}^*$, γ^*, $\mathbf{B}^*$) and ($\mathbf{u}^*$, λ^*, $\mathbf{B}^*$) be feasible solutions for (3.1) and (3.2) respectively, with $\mathbf{C}\mathbf{x}^* = \mathbf{B}^*\gamma^*$ and $\mathbf{u}^*\mathbf{D} = \lambda^*\mathbf{B}^*$. Then ($\mathbf{x}^*$, γ^*, $\mathbf{B}^*$) and ($\mathbf{u}^*$, λ^*, $\mathbf{B}^*$) are nondominated solutions for (3.1) and (3.2), respectively.

Gyetvàn and Shi (1992) presented a weak duality theorem of matrix optimization:

Theorem 3.5 Let ($\mathbf{x}$, γ, $\mathbf{B}_1$) and ($\mathbf{u}$, λ, $\mathbf{B}_2$) be two feasible solutions for (3.1) and (3.2), respectively. Then $\mathbf{B}_1 \geq \mathbf{B}_2$ may not hold.

Proof. By contradiction, assume that $\mathbf{B}_1 \geq \mathbf{B}_2$. As ($\mathbf{x}$, γ, $\mathbf{B}_1$) and ($\mathbf{u}$, λ, $\mathbf{B}_2$) are feasible solutions,

$$\mathbf{C}\mathbf{x} \geq \mathbf{B}\gamma,\ \mathbf{A}\mathbf{x} \leq \mathbf{D}\gamma,\ \mathbf{x} \geq 0,\ \gamma > 0, \qquad (3.3)$$

$$\mathbf{u}^t\mathbf{D} \leq \lambda^t\mathbf{B},\ \mathbf{u}^t\mathbf{A} \geq \lambda^t\mathbf{C},\ \mathbf{u} \geq 0,\ \lambda > 0. \qquad (3.4)$$

Multiplying both sides of the inequalities in (3.3) and (3.4) by appropriate vectors **u**, λ, and **x**, γ, respectively, we get

$$\lambda^t \mathbf{C}\mathbf{x} \geq \lambda^t \mathbf{B}_1 \gamma, \quad \mathbf{u}^t \mathbf{A}\mathbf{x} \leq \mathbf{u}^t \mathbf{D}\gamma, \text{ and}$$
$$\mathbf{u}^t \mathbf{D}\gamma \leq \lambda^t \mathbf{B}_2 \gamma, \quad \mathbf{u}^t \mathbf{A}\mathbf{x} \geq \lambda^t \mathbf{C}\mathbf{x}.$$

Given $\mathbf{B}_1 \geq \mathbf{B}_2$, $\gamma > 0$ and $\lambda > 0$, then $\lambda^t \mathbf{B}_1 \gamma > \lambda^t \mathbf{B}_2 \gamma$. After rearranging we get the following series of inequalities:

$$\mathbf{u}^t \mathbf{A}\mathbf{x} \leq \mathbf{u}^t \mathbf{D}\gamma \leq \lambda^t \mathbf{B}_2 \gamma < \lambda^t \mathbf{B}_1 \gamma \leq \lambda^t \mathbf{C}\mathbf{x} \leq \mathbf{u}^t \mathbf{A}\mathbf{x},$$

namely $\mathbf{u}^t \mathbf{A}\mathbf{x} < \mathbf{u}^t \mathbf{A}\mathbf{x}$, which is obviously a contradiction.

Now, let us introduce the notion of a nondominated pair.

Definition 3.6 A pair of solutions $\{(\mathbf{x}^*, \gamma^*, \mathbf{B}^*), (\mathbf{u}^*, \lambda^*, \mathbf{B}^*)\}$ is called an **nondominated pair** of problems (3.1) and (3.2) if and only if

$$\mathbf{C}\mathbf{x}^* = \mathbf{B}^* \gamma^* \text{ and } (\mathbf{u}^*)^t \mathbf{D} = (\lambda^*)^t \mathbf{B}^*,$$

where $(\mathbf{x}^*, \gamma^*, \mathbf{B}^*)$ and $(\mathbf{u}^*, \lambda^*, \mathbf{B}^*)$ are nondominated solutions for (3.1) and (3.2), respectively.

The following theorem is also not included in the work of Gale, Kuhn and Tucker (1951), but it can be interpreted as a generalization of the complementary slackness theorem of Park (1982) MC programming.

Theorem 3.6 (Complementary Slackness Theorem)

Let $(\mathbf{x}^*, \gamma^*, \mathbf{B}^*)$ and $(\mathbf{u}^*, \lambda^*, \mathbf{B}^*)$ be feasible solutions for (3.1) and (3.2), respectively. A pair of solutions $\{(\mathbf{x}^*, \gamma^*, \mathbf{B}^*), (\mathbf{u}^*, \lambda^*, \mathbf{B}^*)\}$ is a nondominated pair if and only if

$$((\mathbf{u}^*)^t \mathbf{A} - (\lambda^*)^t \mathbf{C})\mathbf{x}^* = 0 \text{ and } (\mathbf{u}^*)^t (\mathbf{A}\mathbf{x}^* - \mathbf{D}\gamma^*) = 0.$$

In other words,

$x_i^* > 0$ implies $(\mathbf{u}^*)^t \mathbf{a}^i = (\lambda^*)^t \mathbf{c}^i$,
$u_j^* > 0$ implies $\mathbf{a}_j \mathbf{x}^* = \mathbf{d}_j \gamma^*$,
$(\mathbf{u}^*)^t \mathbf{a}^i > (\lambda^*)^t \mathbf{c}^i$ implies $x_i^* = 0$, and
$\mathbf{a}_j \mathbf{x}^* < \mathbf{d}_j \gamma^*$ implies $u_j^* = 0$,

where $\mathbf{a}^i$ and $\mathbf{c}^i$ are the column vectors of the matrices **A** and **C**, $\mathbf{a}_j$ and $\mathbf{d}_j$ are the row vectors of the matrices **A** and **D**, x_i^* and u_j^* are the components of the vectors $\mathbf{x}^*$ and $\mathbf{u}^*$, respectively.

Proof. Since $(\mathbf{x}^*, \gamma^*, \mathbf{B}^*)$ and $(\mathbf{u}^*, \lambda^*, \mathbf{B}^*)$ are feasible solutions for (3.1) and (3.2), we have

$\mathbf{Cx}^* \geq \mathbf{B}^*\gamma^*$, $\mathbf{Ax}^* \leq \mathbf{D}\gamma^*$, $\mathbf{x}^* \geq 0$, $\gamma^* > 0$, and
$(\mathbf{u}^*)^t\mathbf{D} \leq (\lambda^*)^t\mathbf{B}^*$, $(\mathbf{u}^*)^t\mathbf{A} \geq (\lambda^*)^t\mathbf{C}$, $\mathbf{u}^* \geq 0$, $\lambda^* > 0$.

Similar to the proof of Theorem 3.5, multiplying both sides of the first two inequalities by λ^*, $\mathbf{u}^*$ and the second two inequalities by γ^*, $\mathbf{x}^*$, we have

$$(\lambda^*)^t\mathbf{Cx}^* \leq (\mathbf{u}^*)^t\mathbf{Ax}^* \leq (\mathbf{u}^*)^t\mathbf{D}\gamma^* \leq (\lambda^*)^t\mathbf{B}^*\gamma^* \leq (\lambda^*)^t\mathbf{Cx}^*. \quad (3.5)$$

Let us suppose that $((\mathbf{u}^*)^t\mathbf{A} - (\lambda^*)^t\mathbf{C})\mathbf{x}^* = 0$ and $(\mathbf{u}^*)^t(\mathbf{Ax}^* - \mathbf{D}\gamma^*) = 0$, namely that $(\mathbf{u}^*)^t\mathbf{Ax}^* = (\lambda^*)^t\mathbf{Cx}^*$ and $(\mathbf{u}^*)^t\mathbf{Ax}^* = (\mathbf{u}^*)^t\mathbf{D}\gamma^*$. Then, (3.5) may be written in the following form:

$$(\lambda^*)^t\mathbf{Cx}^* = (\mathbf{u}^*)^t\mathbf{Ax}^* = (\mathbf{u}^*)^t\mathbf{D}\gamma^* \leq (\lambda^*)^t\mathbf{B}^*\gamma^* \leq (\lambda^*)^t\mathbf{Cx}^*. \quad (3.6)$$

Note that the first and the last part of the inequality series (3.6) are identical. Thus, we have $(\mathbf{u}^*)^t\mathbf{D}\gamma^* = (\lambda^*)^t\mathbf{B}^*\gamma^*$ and $(\lambda^*)^t\mathbf{B}^*\gamma^* = (\lambda^*)^t\mathbf{Cx}^*$. As $\gamma^* > 0$ and $\lambda^* > 0$, $\mathbf{Cx}^* = \mathbf{B}^*\gamma^*$ and $(\mathbf{u}^*)^t\mathbf{D} = (\lambda^*)^t\mathbf{B}^*$. By Definition 3.6, $(\mathbf{x}^*, \gamma^*, \mathbf{B}^*)$ and $(\mathbf{u}^*, \lambda^*, \mathbf{B}^*)$ are a nondominated pair.

Conversely, suppose that $\{(\mathbf{x}^*, \gamma^*, \mathbf{B}^*), (\mathbf{u}^*, \lambda^*, \mathbf{B}^*)\}$ is a nondominated pair. $\mathbf{Cx}^* = \mathbf{B}^*\gamma^*$ and $(\mathbf{u}^*)^t\mathbf{D} = (\lambda^*)^t\mathbf{B}^*$. So, in this case (3.5) has the following form:

$$(\lambda^*)^t\mathbf{Cx}^* \leq (\mathbf{u}^*)^t\mathbf{Ax}^* \leq (\mathbf{u}^*)^t\mathbf{D}\gamma^* = (\lambda^*)^t\mathbf{B}^*\gamma^* = (\lambda^*)^t\mathbf{Cx}^*. \quad (3.7)$$

The first and the last part of the inequality series (3.7) are identical. Consequently, $(\mathbf{u}^*)^t\mathbf{Ax}^* = (\lambda^*)^t\mathbf{Cx}^*$ and $(\mathbf{u}^*)^t\mathbf{Ax}^* = (\mathbf{u}^*)^t\mathbf{D}\gamma^*$.

3.3 Relationship Theory

This section first discusses the relationship between the duality theory of linear matrix programming and that of MC^2 programming, and then the relationship between MC^2 programming and MC programming. Finally, we use a theorem to integrate the relationships of MC^2 problem, MC problem and LP problem.

Using the terminology of Gale, Kuhn and Tucker, the initial formulation of MC²-programming (Seiford and Yu 1979) can be transferred as the following primal-dual pair:

$$Max\ \{\lambda^t\mathbf{C}\mathbf{x}\ |\ \mathbf{A}\mathbf{x} \le \mathbf{D}\gamma,\ \mathbf{x} \ge 0,\ \gamma > 0,\ \lambda > 0\}, \quad (3.8)$$

$$Min\ \{\mathbf{u}^t\mathbf{D}\gamma\ |\ \mathbf{u}^t\mathbf{A} \ge \lambda^t\mathbf{C},\ \mathbf{u} \ge 0,\ \gamma > 0,\ \lambda > 0\}. \quad (3.9)$$

Definition 3.7
(i) A feasible solution $(\mathbf{x}^*, \gamma^*)$ for (3.8) is said to be a **nondominated solution** for (3.8) if and only if there is no such other feasible solution $(\mathbf{x}^0, \gamma^0)$ of (3.8) for which $\lambda^t\mathbf{C}\mathbf{x}^0 > \lambda^t\mathbf{C}\mathbf{x}^*$, for some $\lambda > 0$.
(ii) A feasible solution $(\mathbf{u}^*, \lambda^*)$ for (3.9) is said to be a **nondominated solution** for (3.9) if and only if there is no such other solution $(\mathbf{u}^0, \lambda^0)$ of (3.9) for which $(\mathbf{u}^0)^t\mathbf{D}\gamma < (\mathbf{u}^*)^t\mathbf{D}\gamma$, for some $\gamma > 0$.

The concept of nondominated pair for the MC² programming problem is given by:

Definition 3.8 A pair of nondominated solutions $\{(\mathbf{x}^*, \gamma^*), (\mathbf{u}^*, \lambda^*)\}$ is called a **nondominated pair** of problems (3.8) and (3.9) if and only if $(\lambda^*)^t\mathbf{C}\mathbf{x}^* = (\mathbf{u}^*)^t\mathbf{D}\gamma^*$.

The following two theorems, adapted from Seiford and Yu (1979) and Chapter 8 of Yu (1985), correspond to the weak duality theorem and the complementary slackness theorem in the linear matrix optimization.

Theorem 3.7 If $(\mathbf{x}, \gamma)$ and $(\mathbf{u}, \lambda)$ are feasible solutions for (3.8) and (3.9), respectively, then $(\lambda^*)^t\mathbf{C}\mathbf{x}^* \le (\mathbf{u}^*)^t\mathbf{D}\gamma^*$.

Theorem 3.8 Let $(\mathbf{x}, \gamma)$ and $(\mathbf{u}, \lambda)$ be feasible solutions for (3.8) and (3.9), respectively. They are a nondominated pair for (3.8) and (3.9) if and only if

$$(\mathbf{u}^*)^t(\mathbf{A}\mathbf{x}^* - \mathbf{D}\gamma^*) = 0 \text{ and } ((\mathbf{u}^*)^t\mathbf{A} - (\lambda^*)^t\mathbf{C})\mathbf{x}^* = 0.$$

We state the following theorem to show the close relation between the MC² programming (3.8) and (3.9) and the linear matrix programming problems (3.1) and (3.2).

Theorem 3.9

(i) If $(\mathbf{x}^0, \gamma^0, \mathbf{B}^0)$ and $(\mathbf{u}^0, \lambda^0, \mathbf{B}^0)$ are feasible solutions for (3.1) and (3.2), respectively, then $(\mathbf{x}^0, \gamma^0)$ and $(\mathbf{u}^0, \lambda^0)$ are feasible solutions for (3.8) and (3.9), respectively.

(iii) Conversely, if $(\mathbf{x}^0, \gamma^0)$ and $(\mathbf{u}^0, \lambda^0)$ are feasible solutions for (3.8) and (3.9), respectively, then there is a $\mathbf{B}^0$ such that $(\mathbf{x}^0, \gamma^0, \mathbf{B}^0)$ and $(\mathbf{u}^0, \lambda^0, \mathbf{B}^0)$ are feasible solutions for (3.1) and (3.2), respectively.

Proof. Suppose that $(\mathbf{x}^0, \gamma^0, \mathbf{B}^0)$ and $(\mathbf{u}^0, \lambda^0, \mathbf{B}^0)$ are two feasible solutions for (3.1) and (3.2), respectively. Then,

$$\mathbf{Cx}^0 \geq \mathbf{B}^0\gamma^0,\ \mathbf{Ax}^0 \leq \mathbf{D}\gamma^0,\ \mathbf{x}^0 \geq 0,\ \gamma^0 > 0, \text{ and}$$
$$(\mathbf{u}^0)^t\mathbf{D} \leq (\lambda^0)^t\mathbf{B}^0,\ (\mathbf{u}^0)^t\mathbf{A} \geq (\lambda^0)^t\mathbf{C},\ \mathbf{u}^0 \geq 0,\ \lambda^0 > 0.$$

This implies that $\mathbf{Ax}^0 \leq \mathbf{D}\gamma^0$, $\mathbf{x}^0 \geq 0$, $\gamma^0 > 0$, and $(\mathbf{u}^0)^t\mathbf{A} \geq (\lambda^0)^t\mathbf{C}$, $\mathbf{u}^0 \geq 0$, $\lambda^0 > 0$. Thus, $(\mathbf{x}^0, \gamma^0)$ and $(\mathbf{u}^0, \lambda^0)$ are two feasible solutions for (3.8) and (3.9), respectively.

Conversely, if $(\mathbf{x}^0, \gamma^0)$ and $(\mathbf{u}^0, \lambda^0)$ are two feasible solutions for (3.8) and (3.9) respectively, then both

$\mathbf{Ax}^0 \leq \mathbf{D}\gamma^0$, $\mathbf{x}^0 \geq 0$, $\gamma^0 > 0$, and $(\mathbf{u}^0)^t\mathbf{A} \geq (\lambda^0)^t\mathbf{C}$, $\mathbf{u}^0 \geq 0$, $\lambda^0 > 0$ hold.

By applying Theorem 4 of Gale, Kuhn and Tucker (1951) there exists $\mathbf{B}^0$ such $(\mathbf{x}^0, \gamma^0, \mathbf{B}^0)$ and $(\mathbf{u}^0, \lambda^0, \mathbf{B}^0)$ become feasible solutions for (3.1) and (3.2), respectively.

Corollary 3.1

(i) If $\{(\mathbf{x}^*, \gamma^*, \mathbf{B}^*), (\mathbf{u}^*, \lambda^*, \mathbf{B}^*)\}$ is a nondominated pair of problems (3.1) and (3.2), then $\{(\mathbf{x}^*, \gamma^*), (\mathbf{u}^*, \lambda^*)\}$ is a nondominated pair of problems (3.8) and (3.9).

(ii) Conversely, if $\{(\mathbf{x}^*, \gamma^*), (\mathbf{u}^*, \lambda^*)\}$ is a nondominated pair of problems (3.8) and (3.9), then there is a $\mathbf{B}^0$ such that $\{(\mathbf{x}^*, \gamma^*, \mathbf{B}^*), (\mathbf{u}^*, \lambda^*, \mathbf{B}^*)\}$ is a nondominated pair of problems (3.1) and (3.2).

Proof. To show (i), let $\{(\mathbf{x}^*, \gamma^*, \mathbf{B}^*), (\mathbf{u}^*, \lambda^*, \mathbf{B}^*)\}$ be a nondominated pair of problems (3.1) and (3.2). Then, by Definition 3.6, we have

$$\mathbf{Cx}^* = \mathbf{B}^*\gamma^* \text{ and } (\mathbf{u}^*)^t\mathbf{D} = (\lambda^*)^t\mathbf{B}^*,$$

which imply that $(\lambda^*)^t\mathbf{C}\mathbf{x}^* = (\lambda^*)^t\mathbf{B}^*\gamma^*$ and $(\mathbf{u}^*)^t\mathbf{D}\gamma^* = (\lambda^*)^t\mathbf{B}^*\gamma^*$, respectively. Thus,
$(\lambda^*)^t\mathbf{C}\mathbf{x}^* = (\mathbf{u}^*)^t\mathbf{D}\gamma^*$ which means $\{(\mathbf{x}^*, \gamma^*), (\mathbf{u}^*, \lambda^*)\}$ is a nondominated pair of problems (3.8) and (3.9) by Definition 3.8.

To show (ii), let $\{(\mathbf{x}^*, \gamma^*), (\mathbf{u}^*, \lambda^*)\}$ be a nondominated pair of problems (3.8) and (3.9). Then, by Definition 3.8, we have $(\lambda^*)^t\mathbf{C}\mathbf{x}^* = (\mathbf{u}^*)^t\mathbf{D}\gamma^*$.

Applying Theorem 4 of Gale, Kuhn and Tucker (1951), we can find $\mathbf{B}^*$ such that

$$(\lambda^*)^t\mathbf{C}\mathbf{x}^* = (\lambda^*)^t\mathbf{B}^*\gamma^* = (\mathbf{u}^*)^t\mathbf{D}\gamma^*.$$

Note that $\lambda^* > 0$ and $\gamma^* > 0$, thus the above implies that

$$\mathbf{C}\mathbf{x}^* = \mathbf{B}^*\gamma^* \text{ and } (\mathbf{u}^*)^t\mathbf{D} = (\lambda^*)^t\mathbf{B}^*.$$

By Definition 3.6, we conclude that $\{(\mathbf{x}^*, \gamma^*, \mathbf{B}^*), (\mathbf{u}^*, \lambda^*, \mathbf{B}^*)\}$ is a nondominated pair of problems (3.1) and (3.2).

From Theorem 3.9 and Corollary 3.1 we see that the duality of linear matrix programming and that of MC^2 programming are closely related. Indeed, they are equivalent in the sense that the respective feasible solutions are in “one to one correspondence”.

Now we describe the connection between the MC programming and the MC^2 programming. Kornbluth (1974) used the following pair of linear systems to construct a duality for the MC programming problem:

$$Max\ \{\mathbf{C}\mathbf{x} \mid \mathbf{A}\mathbf{x} \le \mathbf{D}\gamma,\ \mathbf{x} \ge 0,\ \gamma > 0\}, \tag{3.10}$$

$$Min\ \{\mathbf{u}^t\mathbf{D} \mid \mathbf{u}^t\mathbf{A} \ge \lambda^t\mathbf{C},\ \mathbf{u} \ge 0,\ \lambda > 0\}. \tag{3.11}$$

From Kornbluth's dual pair (3.10) and (3.11), it is easy to get the form of MC^2 programming (3.8) and (3.9). Let $\gamma^0 > 0$ be a fixed vector, then (3.10) can be written as follows:

$$Max\ \{\mathbf{C}\mathbf{x} \mid \mathbf{A}\mathbf{x} \le \mathbf{D}\gamma^0,\ \mathbf{x} \ge 0\}. \tag{3.12}$$

The following theorem shows the relationship between MC programming and LP (see Philip 1972, Yu and Zeleny 1975).

Theorem 3.10 The vector **x*** is a nondominated solution for (3.12) if and only if there is a vector $\lambda^0 > 0$ for which **x*** is an optimal solution for the following scalar LP problem:

$$Max\ \{(\lambda^0)^t\mathbf{Cx}\ |\ \mathbf{Ax} \le \mathbf{D}\gamma^0,\ \mathbf{x} \ge 0\}. \tag{3.13}$$

We note that according to the duality theorem of LP problems (Recall Property 2.6), (3.13) has an optimal solution if and only if there is a vector **u***, which is an optimal solution for the following dual problem of (3.13):

$$Min\ \{\mathbf{u}^t\mathbf{D}\gamma^0\ |\ \mathbf{u}^t\mathbf{A} \ge \lambda^0\mathbf{C},\ \mathbf{u} \ge 0\}. \tag{3.14}$$

Thus, if **x*** is a nondominated solution for (3.12), then there exist $\lambda^0 > 0$ and **u*** so that **u*** is an optimal solution to (3.14). As $\gamma^0 > 0$, Theorem 3.10 also implies that **u*** is a nondominated solution for

$$Min\ \{\mathbf{u}^t\mathbf{D}\ |\ \mathbf{u}^t\mathbf{A} \ge (\lambda^0)^t\mathbf{C},\ \mathbf{u} \ge 0\},$$

which is the dual problem of (3.12).

Note that in the above analysis, vectors $\gamma^0 > 0$ and $\lambda^0 > 0$ are arbitrary. Thus the pair of (3.13) and (3.14) are equivalent to that of (3.8) and (3.9). Therefore, the MC duality and the MC^2 duality are closely related. Furthermore, the duality of linear matrix programming, MC^2 duality, and MC duality are equivalent in the sense that their respective feasible solutions are in "one to one correspondence". Finally, we use the following theorem as a conclusion of this section:

Theorem 3.10 (Relationship of LP, MC, and MC^2 problems)
If **x*** is a nondominated solution for the MC^2 problem (3.8), then $(\mathbf{x}^*, \gamma^0)$, $\gamma^0 > 0$, is a nondominated solution for the MC problem (3.12). Furthermore, $(\mathbf{x}^*, \gamma^0, \lambda^0)$, $\gamma^0 > 0$, $\lambda^0 > 0$, is an optimal solution for the LP problem (3.13).

3.4 Extension Based on Application Perspectives

In addition to the theoretical reasons of extensions from the LP model to the MC model, and then from the MC model to the MC^2 model, this section provides two descriptive roots of the MC^2 model as its application perspectives. Then some potential application fields which can be formulated as an MC^2 model will be outlined for the reader who is interested in practical problem-solving approaches.

3.4.1 Two Roots

A first MC^2 root is economic in nature. Supply and demand relationships in economic theory have been extensively explored by economists (Henderson 1922, Schultz 1938, Phelps 1947). Both supply and demand can be viewed as the resource availability. In a market-oriented economy, a famous fundamental theorem of Keynes (1936) recognized that firms faced the formidable task of determining simultaneously what and how much should be produced, what techniques should be used and what input mix should be utilized to grow the most profit in the least costly way. If the resource availability is a function of employment, then it fluctuates depending on economic situations or levels of economic activity. Those situations are commonly known as economic booms, recessions, or recoveries. In microeconomic theory, both supply and demand of Keynesian thought are often represented as functions of price (Henderson and Quandt 1980). In either the short or long run, supply and demand (resource availabilities) fluctuate as changes occur in economic situations. An equilibrium solution can then be identified in fluctuations. Suppose that resource availability functions of Keynesian theory are linear. Then, the values of resource availability on certain economic situations can be considered as some extreme (or discrete) points of linear resource availability functions. In this case, a linear combination of the availability levels can be used to recover the linear resource availability functions. Thus, with assessment on the resource availability levels for the economic situations, the traditional LP with a single resource availability level can be expanded into multiple resource availability levels. Economic theory supports this essential idea.

A second MC^2 root is a group decision making situation (Hwang and Lin 1987). It is common knowledge that in many real-world decision problems a group of involved managers, such as the president, production, finance, accounting, and marketing managers, share the decision making responsibility. However, LP models implicitly assume that there is only a single person in charge of decision making. Extending a single decision maker to a multiple decision maker setting introduces the multiple resource availability level framework to modeling group decision making problems. The preference values of each involved manager on resource availability can be treated as an individual level of the availability. Then, the modified model has multiple discrete levels of resource availability. It is clear that the multiple decision makers may have different preferences, in terms of their backgrounds and positions in their firm, which result in inevitable conflicts. The optimal solution must reflect the compromise or trade-offs among the multiple discrete levels of resource availability.

3.4.2 Potential Application Fields

The flexibility feature of the MC^2 model fosters a great potential of applying it in solving many real-world problems. We list some application areas where we have been applying the MC^2 model as follows. The details of the related research findings will be discussed in Chapters 16-21.

Transfer Pricing and Capital Budgeting

In Business Accounting, a transfer pricing problem occurs when a firm processes raw materials into intermediate or final products from one division to another division. This problem involves multiple objective, such as maximizing the overall company's profit, maximizing the market share goal of products in each division, and maximizing the utilized production capacity of the company so that each division manager can avoid any underutilization of normal production capacity. However, existing transfer pricing models formulated by linear programming and goal programming cannot provide a comprehensive scenario of all possible optimal trade-offs between multiple objectives under consideration for a given transfer pricing problem. Using the MC^2 model (3.8), Shi, Kwak and Lee (1998) studied these optimal trade-offs.

Similarly, the MC^2 model (3.8) can be used to formulate a capital budget problem that not only incorporates multiple criteria over future periods of investment, but also allows multiple (a group of) decision makers to involve in the decision process (Kwak *et al.* 1996).

Aggregate Production Planning

Production planning is one of the most important functions in the process of production and operations management. In the production planning, managers of a manufacture firm need to make a crucial decision on which specific aggregate levels of production, inventory and work force have to be produced to meet possible demands. Such a decision is traditionally interpreted as to find a best combination of the production, inventory and work force quantities that yields a minimum overall cost.

Production planning has drawn a great deal of attention from both practitioners and academia. Historically, economists initiated research on production planning and product development. Since then, production planning has been extensively studied in the literature. By realizing fluctuations in demand, production and inventory on a seasonal basis, linear programming approach regards production periods that consist of inventory, regular and overtime production rates as "resources," and the corresponding demand periods as "destinations." Then, a transportation method of linear programming is applied to identify an optimal solution that yields the minimum total costs of production and inventory in supplying the product units to the demands. The framework of this model has a great impact on many modeling developments, because it provided a useful channel of not only using mathematical modeling approaches to production planning, but also solving a real aggregate production planning problem by any available commercial linear programming software. A number of authors adopted techniques of goal programming and MC programming to solve aggregate production planning problems with multiple objectives, for instance, treating production, inventory, work force and shortage as individual objectives.

These current models, however, have explicitly or implicitly assumed that both capacity (supply) and demand levels are fixed. That is, capacity and demand levels in the models are given for a certain situation. It is well known that in reality, capacity and demand fluctuate on both seasonal and situational bases. Using the MC^2 transportation model of (Shi 1995), Shi and Haase (1996) proposed an MC^2 model of aggregate production planning. This production planning model provides manufacture decision makers with a systematic and comprehensive scenario about all possible optimal trade-offs of aggregate production planning depending on the multiple factors.

Telecommunication Management and Data Files Allocation

The MC^2 model can be used to select telecommunication network systems (Lee, Nazem and Shi 1994, Nazem *et al.* 1996, Shi, Nazem and Lee 1996). Given a set of candidate cities for a telecommunication network, we want to select some subsets of candidate cities as hub city designs for the telecommunication system, such as an information super highway. The selected hub city design (a subset of candidate cities) maximizes the multiple criteria: population, economy, education, health care, and transportation, subject to the location constraints with multiple resource availability levels. By applying the MC^2 model (3.8), we can construct optimal hub city designs to reflect policy (decision) makers' goal-seeking and compromise behavior. Similarly, the MC^2 model (3.8) can be applied to the problem of allocating data files over a wide area network (Lee, Shi and Stolen 1994, Shi, Nazem and Lee 1996).

3.5 Further Remarks

The demonstration of both mathematical and practical relationships between the LP, MC and MC^2 problems in this chapter has further clarified the significance of MC^2 techniques that we will discuss in the following chapters. We want to address a few more comments before closing this chapter. First, Section 3.1 is a very brief introduction on MC problems. As we mentioned in Section 1.1, since 1970's, the MC model has been recognized as one of the most efficient techniques dealing with the tradeoffs of conflicting criteria in real-world problems. Some well-known MC books are cited in the reading of this chapter for the interested reader. However, Figure 3.1 shows the highlight summary of various MC algorithms. The reader may think over the importance of the purpose using this figure. Second, the work of Gale, Kuhn and Tucker (1951) lay on the heart of MC and MC^2 theory. A careful reader can see that starting from their matrix optimization (3.1), we can perform a deduction process to obtain the MC^2 formulation (3.8), then the MC formulation (3.12), and finally the LP formulation (3.13) by adding the criteria weight vector λ and/or the constraint level vector γ. In addition, the primal analysis of this deduction is parallel to the dual analysis. Inversely, an induction process can be conducted from LP to MC, and then from MC to MC^2 models. Third, two application roots (the economic tradeoff and decision-making tradeoff) support the applicability of MC^2 models in the real world. All of these three comments together tell that the MC^2 approach is mathematically natural and practically useful for business decision making.

Exercises

3.1 Identify the best tradeoff segment of the criteria space for the following MC problem by using the graphical method:

Max	$10\, x_1 + 12\, x_2$
Max	$15\, x_1 + 8\, x_2$
Subject to	$3\, x_1 + 2x_2 \leq 24$
	$4x_1 + 6x_2 \leq 48$
	$x_1,\ x_2 \geq 0.$

3.2 Identify the best tradeoff segment of the criteria space for the following MC problem by using the graphical method:

Min	$5\, x_1 + 3\, x_2$
Min	$2\, x_1 + 7\, x_2$
Subject to	$2x_1 + 4x_2 \geq 20$
	$x_1 + x_2 \geq 10$
	$x_1,\ x_2 \geq 0.$

3.3 Identify the best tradeoff segment of the criteria space for the following MC problem by using the graphical method:

Min	$4\, x_1 - 3\, x_2$
Max	$5\, x_1 + 9\, x_2$
Subject to	$2x_1 + 4x_2 \leq 20$
	$x_1 + x_2 \leq 10$
	$x_1,\ x_2 \geq 0.$

3.4 Intuitively find a nondominated or near nondominated solution for the problem 3.1 by
(i) Interaction with the decision maker,
(ii) Using the ideal solution as the goal setting,
(iii) Combing the objectives,

(iv) Setting the aspiration level.

3.5 Intuitively find a nondominated or near nondominated solution for the problem 3.2 by

(v) Interaction with the decision maker,
(vi) Using the ideal solution as the goal setting,
(vii) Combing the objectives,
(viii) Setting the aspiration level.

3.6 Prove Theorem 4 of Gale, Kuhn and Tucker (1951):

There exists a solution, $\mathbf{B}^0$ for (3.1) and (3.2), if and only if the following existence conditions are both satisfied:

(i) $\mathbf{Ax} \leq \mathbf{D}\gamma$ for some $\mathbf{x} \geq 0$, $\gamma > 0$,
(ii) $\mathbf{u}^t\mathbf{A} \geq \lambda^t\mathbf{C}$, for some $\mathbf{u} \geq 0$, $\lambda > 0$.

3.7 Propose three different examples that illustrate the possible real-life applications of MC^2 framework.

Chapter 4
MC^2 LINEAR PROGRAMMING

In this chapter, we formally describe fundamental issues about multiple criteria and multiple constraint-level (MC^2) linear programming. Most of the contents are adapted from Chapter 8 of Yu (1985) in a slightly different fashion. Section 4.1 elaborates the formulation of MC^2 linear programming. Although Section 1.4 showed the example of MC^2 format of PC problem and Section 3.3 used MC^2 linear programming in the relationship theory, this section studies the basis of the MC^2 formulation for the simplex method. Section 4.2 builds the concepts of potential solution for MC^2 problems, which is an extension of nondominated solutions in Definition 3.4. A potential solution of an MC^2 problem is optimal for a certain range of (γ, λ). Section 4.3 shows a pivoting procedure for a basis of MC^2 linear programming. Section 4.4 provides a potential solution structure of MC^2 problems. This structure contains a set of potential solutions that cover all possible values of (γ, λ). The MC^2 simplex method is proposed with illustrative examples in Section 4.5. Section 4.6 discusses the concepts of MC^2 duality, which will be used in Chapter 8 for Fuzzy MC^2 duality and Chapter 11 for dual optimal system designs. Section 4.7 outlines the developed two MC^2 computer software that is part of MC^2 software packages attached with this book. Finally, Section 4.8 discusses several remarks.

4.1 Formulation of MC^2 Linear Programming

Recall from Section 3.1, an algebraic form of the MC linear programming (M3.1) can be represented by

$$\begin{array}{ll} \textit{Max} & \mathbf{C\,x} \\ \textit{Subject to} & \mathbf{A\,x} \le \mathbf{d} \\ & \mathbf{x} \ge 0 \end{array}$$

We note that the objective function means that we want to simultaneously maximize q different criteria written by the $(q \times n)$

matrix **C**, while the (n × m) matrix **A** is the unit consumption of resources and the (m × 1) vector **d** is the constraint (resource availability) level.

To form MC² linear programming, we replace **d** by an (m × p) matrix **D** and the problem (M3.1) becomes

$$\begin{aligned} &\mathit{Max} && \mathbf{C\,x} \\ &\mathit{Subject\ to} && \mathbf{A\,x} \le \mathbf{D} \\ & && \mathbf{x} \ge 0 \end{aligned}$$

The portion "**Ax** ≤ **D**" is not inequalities. We may view the constraint as a given set of "feasible" solutions, **x**, whenever **Ax** is contained in the linear combination generated by the columns of **D**. Thus, the columns of **D** can be interpreted as possible initial resource levels or extreme points of the set that contains the random constraint levels. In other words, this MC² problem implies that "**x** is feasible if **Ax** lies in a convex set generated by **D**". In order to see how this interpretation works, we introduce two **weight parameters**: the constraint parameter $\gamma > 0$, and the criteria parameter $\lambda > \mathbf{0}$ into the formulation. Note that we refer γ as a (p × 1) column vector and λ^t as a (1 × q) row vector. Thus, the MC² linear programming (M4.1) can be written as

$$\begin{aligned} &\mathit{Max} && \lambda^t \mathbf{C\,x} \\ &\mathit{Subject\ to} && \mathbf{A\,x} \le \mathbf{D}\gamma \\ & && \mathbf{x} \ge 0 \end{aligned}$$

From now on, we will refer this form as the MC² problem without confusion. Based on this form, a solution concept, called potential solution will be formally defined in the next section as an extension of nondominated solution for the MC problem. Parallel to the form of MC² problem, the dual MC² problem (M4.2) can be introduced as

$$\begin{aligned} &\mathit{Min} && \mathbf{u}^t\,\mathbf{D}\gamma \\ &\mathit{Subject\ to} && \mathbf{u}^t\,\mathbf{A} \ge \lambda^t\,\mathbf{C} \\ & && \mathbf{u} \ge 0 \end{aligned}$$

Here, the dual variable $\mathbf{u}^t$ is a $(1 \times m)$ row vector. The significance of MC^2 duality will be shown in Section 4.5.

4.2 Concepts of Potential Solutions

Before building the solution concept for the MC^2 problem, we recall the linear basis discussed in Section 2.3. Given a basic variable index set J in a linear system, the variable $\mathbf{x} = (x_1, x_2, \ldots, x_n)$ can be divided as two parts: the basic variables denoted by $\mathbf{x}(J) = (x_{1j}, x_{2j}, \ldots, x_{mj})$ and the nonbasic variables $\mathbf{x}(J')$ which has n-m components. Since the MC^2 problem is a linear system, we have:

Definition 4.1 A basis (index set) J is called a potential basis a potential solution for the MC^2 problem if and only if there exist $\gamma^0 > 0$ and $\lambda^0 > 0$ such that J is an optimal basis for

$$\begin{array}{ll} \textit{Max} & (\lambda^0)^t \mathbf{C}\mathbf{x} \\ \textit{Subject to} & \mathbf{A}\mathbf{x} \leq \mathbf{D}\gamma^0 \\ & \mathbf{x} \geq 0 \end{array}$$

Remark 4.1 If $\mathbf{D}$ is replaced by the vector $\mathbf{d}$, then the MC^2 problem reduces to the MC problem. Furthermore, if $\mathbf{C}$ is replaced by the vector $\mathbf{c}$, the MC problem becomes the LP problem. This is another way to explain Theorem 3.10.

Based on Definition 4.1, we can further elaborate the following:

Definition 4.2 A basis J is a primal potential solution for the MC^2 problem if and only if J is an optimal basis for some $\gamma > 0$ and $\lambda = 0$

.

Definition 4.3 A basis J is a dual potential solution for the MC^2 problem if and only if J is an optimal basis for $\gamma = 0$ and some $\lambda > 0$.

Remark 4.2 We label Definition 4.2 for a primal potential solution because the basic solution is primal infeasible for a negative right-hand side. A similar remark holds for Definition 4.3. Clearly, a basis J is a potential solution if and only if J is both a primal and a dual potential solution.

4.3 MC² Simplex Pivoting

This section is motivated by the simplex method of linear program (Dantzig 1963). As we see from the MC² problem, given a potential solution J, its the basic variables should be denoted by $\mathbf{x}(J, \gamma)$ because they are the function of γ. In addition, the objective of the MC² problem is a function of (γ, λ). Therefore, finding a systematic way to identify all nondominated solutions over various values of (γ, λ) is not trivial task. Fortunately, with the help of Definition 4.1, we can utilize the LP simplex method to do the job. The creativity of **MC² simplex method** is that (i) it uses the LP simplex routine to compute the data (**C**, **A**, **D**) for a number of bases; and (ii) given a basis, it applies Definitions 4.1–4.3 to recover the certain range of (γ, λ) so that the basis satisfies the feasible and then the optimal conditions. Now we proceed the details of MC² simplex method (Seiford and Yu 1979):

In order to utilize the advantage of LP simplex method, we first assume the parameters (γ, λ) are fixed. Then, after adding the column slack variables, defined as $\mathbf{s} = (s_1, s_2, \ldots, s_m)$ to the constraints, then the MC² problem (M4.1) is

$$\begin{aligned} &\textit{Max} && \lambda^t \mathbf{C}\,\mathbf{x} \\ &\textit{Subject to} && \mathbf{A}\,\mathbf{x} + \mathbf{I}\,\mathbf{s} = \mathbf{D}\gamma \\ & && \mathbf{x}, \mathbf{s} \geq 0 \end{aligned}$$

As mentioned in Section 2.3, **I** presents the (mxm)-dimensional identity matrix, where the diagonal cell of the matrix is one and the rest is zero. Then, the initial MC² simplex tableau can be written as:

Table 4.1 The Initial Simplex Tableau

x	s	RHS	
A	I	Dγ	(4.1)
$-\lambda^t\mathbf{C}$	0	0	(4.2)

Given a basis J with the basic variables $\mathbf{x}(J)$ and the nonbasic variables $\mathbf{x}(J')$, we decompose the matrix $[\mathbf{A}, \mathbf{I}]$ to $[\mathbf{B}, \mathbf{R}]$, where $\mathbf{B}$ is the basis matrix with the column index of $[\mathbf{A}, \mathbf{I}]$ in J and $\mathbf{R}$ is the nonbasis matrix with no column index of $[\mathbf{A}, \mathbf{I}]$ in J. Similarly, we decompose the vector $[-\lambda^t\mathbf{C}, \mathbf{0}]$ to $[-\lambda^t\mathbf{C}_B, -\lambda^t\mathbf{C}_R]$, where $-\lambda^t\mathbf{C}_B$ is the subvector of $[-\lambda^t\mathbf{C}, \mathbf{0}]$ with column index in J and $-\lambda^t\mathbf{C}_R$ is the subvector of $[-\lambda^t\mathbf{C}, \mathbf{0}]$ with no column index in J. Then, Equation (4.1)-(4.2) can be rewritten as

Table 4.2 The Reform of the Initial Simplex Tableau

$\mathbf{x}(J)$	$\mathbf{x}(J')$	RHS	
$\mathbf{B}$	$\mathbf{R}$	$\mathbf{D}\gamma$	(4.3)
$-\lambda^t\mathbf{C}_B$	$-\lambda^t\mathbf{C}_R$	$\mathbf{0}$	(4.4)

We use a row linear transformation on Table 4.2 by:
(4.5) = $\mathbf{B}^{-1}$ (4.3) (i.e., premultiply (4.3) by $\mathbf{B}^{-1}$), and
(4.6) = $-\lambda^t\mathbf{C}_B$ (4.5) + (4.4).
This results in Table 4.3 known as the simplex tableau.

Table 4.3 The Simplex Tableau

$\mathbf{x}(J)$	$\mathbf{x}(J')$	RHS	
$\mathbf{I}$	$\mathbf{B}^{-1}\mathbf{R}$	$\mathbf{B}^{-1}\mathbf{D}\gamma$	(4.5)
$\mathbf{0}$	$\lambda^t\mathbf{C}_B\mathbf{B}^{-1}\mathbf{R} - \lambda^t\mathbf{C}_R$	$\lambda^t\mathbf{C}_B\mathbf{B}^{-1}\mathbf{D}\gamma$	(4.6)

Note that Tables 4.1-4.3 follow the computation of LP simplex method since the parameters (γ, λ) are fixed. If (γ, λ) are not fixed, the routine cannot be carried out. However, we can drop (γ, λ) from Table 4.3 and obtain Table 4.4 that has the computing result of the data $(\mathbf{C}, \mathbf{A}, \mathbf{D})$. For this reason, we should call it the MC^2 Simplex Tableau for the basis J.

Table 4.4 The MC² Simplex Tableau

$\mathbf{x}(J)$	$\mathbf{x}(J')$	RHS	
$\mathbf{I}$	$\mathbf{B}^{-1}\mathbf{R}$	$\mathbf{B}^{-1}\mathbf{D}$	(4.7)
$\mathbf{0}$	$\mathbf{C}_B\mathbf{B}^{-1}\mathbf{R}-\mathbf{C}_R$	$\mathbf{C}_B\mathbf{B}^{-1}\mathbf{D}$	(4.8)

To recover the certain range of (γ, λ) for the basis J, we restate the potential solution in terms of Table 4.3 as follows:

Definition 4.4 Given a basis J for an MC² problem,

(i) $\Gamma(J) = \{\gamma > 0 \mid \mathbf{B}^{-1}\mathbf{D}\gamma \geq 0\}$ is called the primal parameter set; and

(ii) $\Lambda(J) = \{\lambda > 0 \mid \lambda^t(\mathbf{C}_B\mathbf{B}^{-1}\mathbf{R} - \mathbf{C}_R) \geq 0\}$ is called the dual parameter set.

Definition 4.5 Given a basis J for an MC² problem,

(i) The basic solution $\mathbf{x}(J, \gamma) = \mathbf{B}^{-1}\mathbf{D}\gamma$ is feasible if and only if $\gamma \in \Gamma(J)$;

(ii) The solution $\mathbf{x}(J, \gamma)$ is optimal if and only if $\gamma \in \Gamma(J)$ and $\lambda \in \Lambda(J)$;

(iii) The objective "value" of the problem is denoted by $V(J, \lambda, \gamma) = \lambda^t\mathbf{C}_B\mathbf{B}^{-1}\mathbf{D}\gamma$, which is a function of (γ, λ).

Generally, (ii) of Definition 4.5 is called the **optimality condition**. Then, we have:

Theorem 4.1

(i) J is a primal potential basis if and only if $\Gamma(J) \neq \varnothing$;

(ii) J is a dual potential basis if and only if $\Lambda(J) \neq \varnothing$; and

(iii) J is a potential basis if and only if the Cartesian product $\Gamma(J) \times \Lambda(J) \neq \varnothing$.

This theorem can be easily shown by Definition 4.1-4.5. For the convenience of our following discussion, we define the matrices of Table 4.4 as:

$\mathbf{Y}(J) = [\mathbf{I}, \mathbf{B}^{-1}\mathbf{R}]$, $\mathbf{W}(J) = [\mathbf{B}^{-1}\mathbf{D}]$,
$\mathbf{Z}(J) = [\mathbf{0}, \mathbf{C}_B\mathbf{B}^{-1}\mathbf{R} - \mathbf{C}_R]$, and $\mathbf{V}(J) = [\mathbf{C}_B\mathbf{B}^{-1}\mathbf{D}]$.

Remark 4.3 We notice that if the kth column of $\mathbf{Z}(J)$, $\mathbf{Z}_k(J) \leq \mathbf{0}$ (thus $\mathbf{Z}_k(J) \neq 0$), then $\Lambda(J) = \varnothing$. Similarly, if the ith row of $\mathbf{W}(J)$, denoted by $\mathbf{W}^i(J) \leq 0$, then it implies $\Gamma(J) = \varnothing$.

4.4 Potential Solution Structure

Given a basis J, we can use J' to represent the set of nonbasic vectors. For any element of $\mathbf{Y}$, $y_{pk} \neq 0$ and $k \in J'$, we can pivot to a new tableau associated with the basis $K = J \cup\{k\}\backslash\{j_p\}$, where j_p is the basic vector associated with the pth row of the tableau. We classify such a pivot as **a primal pivot** if $y_{pk} > 0$ and **a dual pivot** if $y_{pk} < 0$. Thus, if a particular basis J is a dual potential but not a primal potential we could dual pivot to remove the primal infeasibility. More specifically, because J is a dual potential solution, the corresponding tableau is dual feasible for any $\lambda \in \Lambda(J)$. If the constraint set is consistent for $\gamma^0 > 0$ we can use the dual simplex method to pivot and remove primal infeasibility. This pivot selection rule requires $y_{pk} < 0$, which is classified as the "dual pivot". Similar remarks can be made concerning primal pivots and a primal (but not dual) potential solution.

With the computation routine described as the above, we now address some key theoretical issues, such as separation, adjacency, and connectedness of the pivoting process.

In separation and adjacency, we study the relationship between the potential sets of two adjacent bases. This relationship will insure the pivoting steps go from a given potential basis to another adjacency potential basis and the separation hyperplane of those two bases can be detected. We introduce the concepts of null constraint and effective constraint:

Definition 4.6 Suppose J is a dual potential solution. If for some $k \in J'$, $\mathbf{Z}_k(J) \neq 0$ and there exists $\lambda^0 \in \Lambda(J)$ such that $(\lambda^0)^t\mathbf{Z}_k(J) = 0$, then $\mathbf{Z}_k(J)$ is called an **effective constraint** of $\Lambda(J)$. If for some $k \in J'$, $\mathbf{Z}_k(J) = 0$, then $\mathbf{Z}_k(J)$ is called a **null constraint** of $\Lambda(J)$.

Definition 4.7 Suppose J is a primal potential solution. If there exists $\gamma^0 \in \Gamma(J)$ such that $\mathbf{W}^p(J)\gamma^0 = 0$, $\mathbf{W}^p(J) \neq 0$, then $\mathbf{W}^p(J)$ is called an **effective constraint** of $\Gamma(J)$. If for some row p, $\mathbf{W}^p(J) = 0$, then $\mathbf{W}^p(J)$ is called an **null constraint** of $\Gamma(J)$.

Now we are able to elaborate on the details of the pivoting process. Let $y_{pk} \neq 0$ be the pivot element and suppose that we pivot from the basis J to a new basis $K = J \cup \{k\} \setminus \{j_p\}$, the adjacent basis of J. This means that the Cartesian space $\Gamma(J) \times \Lambda(J)$ is transformed to the Cartesian space $\Gamma(K) \times \Lambda(K)$. This is captured by the following pivoting procedure.

MC² Pivoting Procedure:

For the primal pivoting of $\mathbf{Z}(K)$,

$\mathbf{Z}_j(K) = 0$ if $j \in K$,

$\mathbf{Z}_j(K) = -\mathbf{Z}_k(J)/\,y_{pk}$ if $j = j_p \in K'$,

$\mathbf{Z}_j(K) = \mathbf{Z}_j(J) - \mathbf{Z}_k(J)\,y_{pj}\,/\,y_{pk}$ if $j \in K' \setminus \{j_p\}$,

For the dual pivoting of $\mathbf{W}(K)$,

$\mathbf{W}^i(K) = \mathbf{W}^p(J)/\,y_{pk}$ if $i = p$,

$W^i(K) = W^i(J) - y_{ik}\,/y_{pk}\,W^p(J)$ if $i \neq p$.

We define the hyperplanes for $Z_k(J)$ and $W^p(J)$, respectively as:

$\mathbf{H}_k \equiv \{\lambda > 0 \mid \lambda^t\mathbf{Z}_k(J) = 0\}$, and

$\mathbf{G}_p \equiv \{\gamma > 0 \mid \mathbf{W}^p(J)\gamma = 0\}$.

Then, the following two cases have to be considered:

Case 1. $\mathbf{Z}_k(J)$ is an effective constraint for $\Lambda(J)$ and $y_{pk} > 0$ is the primal pivot element.

Because $\Lambda(J) \subseteq \{\lambda > 0 \mid \lambda^t\mathbf{Z}_k(J) \geq 0\}$, and $\Lambda(K) \subseteq \{\lambda > 0 \mid \lambda^t\mathbf{Z}_k(J) \leq 0\}$, the hyperplane separates $\Lambda(J)$ and $\Lambda(K)$ and the following relationship hold:

$$\mathbf{H}_k \cap \Lambda(J) = \mathbf{H}_k \cap \Lambda(K) = \Lambda(J) \cap \Lambda(K). \qquad (4.9)$$

In addition, under the pivot on y_{pk}, the value γ^0 is contained in the set

$$\Gamma(J) \cap \Gamma(K) = \{\gamma > 0 \mid \mathbf{W}(J)\gamma \geq 0;\ \mathbf{W}^p(J)\gamma \geq y_{ik}\,/y_{pk}\,W^p(J)\,\gamma, \text{ for } y_{ik} > 0\}. \qquad (4.10)$$

Definition 4.8 Whenever (4.9) and (4.10) hold, we say that the potential sets J and K are **primal adjacent**. In this case, the sets $\Lambda(J)$ and $\Lambda(K)$ are "abut", while $\Gamma(J)$ and $\Gamma(K)$ are "overlapped".

Case 2. $\mathbf{W}^p(J)$ is an effective constraint for $\Gamma(J)$ and $y_{pk} < 0$ is the dual pivot element.
Since $\Gamma(J) \subseteq \{\gamma > 0 \mid W^p(J)\gamma \geq 0\}$, and $\Gamma(K) \subseteq \{\gamma > 0 \mid W^p(J)\gamma \leq 0\}$, we see that $\mathbf{G}_p$ is a hyperplane which separates $\Gamma(J)$ and $\Gamma(K)$ and furthermore

$$\mathbf{G}_p \cap \Gamma(J) = \mathbf{G}_p \cap \Gamma(K) = \Gamma(J) \cap \Gamma(K). \qquad (4.11)$$

Similar to (4.10), under the pivot on y_{pk}, the value λ^0 is contained in the set $\Lambda(J) \cap \Lambda(K) = \{\lambda > 0 \mid \lambda^t\mathbf{Z}(J) \geq 0;\ \lambda^t\mathbf{Z}_j(J) \leq \lambda^t\mathbf{Z}_k(J)\ y_{pj}/y_{pk}, \text{ for } y_{pj} < 0\}$.

$$(4.12)$$

Definition 4.9 Whenever (4.11) and (4.12) hold, we say that the potential sets J and K are **dual adjacent**. If the potential sets are either primal or dual adjacent, we refer to that they are adjacent.

Form the above results, we summarize the connection between two adjacent potential bases as follows:

Theorem 4.2 Let J be a potential solution and $y_{pk} \neq 0$ be the pivoting element which yields an adjacent basis K.

(i) if $y_{pk} > 0$ and $\mathbf{Z}_k(J)$ is an effective constraint, then $\mathbf{H}_k$ as defined above is a hyperplane which separates $\Lambda(J)$ and $\Lambda(K)$ and (4.9) and (4.10) hold.

(ii) if $y_{pk} < 0$ and $\mathbf{W}^p(J)$ is an effective constraint, then $\mathbf{G}_p$ as defined above is a hyperplane which separates $\Gamma(J)$ and $\Gamma(K)$ and (4.11) and (4.12) hold.

Note that if $\mathbf{Z}_k(J)$ is a null constraint then $\Lambda(J) = \Lambda(K)$ and if $\mathbf{W}^p$ is a null constraint then $\Gamma(J) = \Gamma(K)$. We can use the following theorem to locate the effective and null constraints.

Theorem 4.3

(i) $\mathbf{Z}_k(J)$ is an effective or null constraint for $\Lambda(J)$ if and only if there exists vector $(\mathbf{v}, \mathbf{w}) \geq 0$ such as $\mathbf{vZ}(J) + \mathbf{eZ}(J) = \mathbf{w}$ with $\mathbf{w}_k = 0$;

(ii) $\mathbf{W}^p(J)$ is an effective or null constraint for $\Gamma(J)$ if and only if there exists vector $(\mathbf{v}, \mathbf{w}) \geq 0$ such as $\mathbf{W}(J)\mathbf{v}^t + \mathbf{W}(J)\mathbf{e}^t = \mathbf{w}^t$ with $\mathbf{w}_p = 0$;

where $\mathbf{e} = (1, 1, \ldots, 1)$ is the vector of ones.

In the study of connectedness, we want to show that the set of potential bases is potential bases J and K, we can reach K from J by a sequence of pivots through adjacent connected. That is, given any two potential bases. Two supporting theorems are:

Theorem 4.4 Suppose (γ^1, λ^1) and (γ^2, λ^2) are points in the potential sets of potential solutions L and J, respectively. Then there is a potential solution K whose potential set contains the points (γ^1, λ^2).

Theorem 4.5 Let (γ^0, λ^1) and (γ^0, λ^2) be points in the potential sets of potential solutions J and K, respectively. Then there is a finite sequence of potential solutions $\{J_i\}^m_{i=1}$ with $J_1 = J$ and $J_m = K$. Furthermore, J_i and J_{i+1} are adjacent potential solutions with J_{i+1} obtained by introducing either an effective or null constraint for $\Lambda(J_i)$. The $\{J_i\}$ also satisfy

(i) $\gamma^0 \in \cap_i \Gamma(J_i)$,

(ii) $[\lambda^1, \lambda^2] \subseteq \cup_i \Lambda(J_i)$,

(iii) $\Lambda(J_i) \cap [\lambda^1, \lambda^2] \neq \varnothing$, for any i, where $[\lambda^1, \lambda^2]$ is the line segment connecting point λ^1 and λ^2.

Remark 4.4 We observe that if we switch (γ^0, λ^1) and (γ^0, λ^2) by (γ^1, λ^0) and (γ^2, λ^0), a similar "dual" result can be obtained for the connection of J and K.

The proofs of both Theorem 4.4 and Theorem 4.5 can be found in the Chapter 8 of Yu (1985). Using these results, we have:

Theorem 4.6 A set of potential solutions, denoted by ϑ, is connected.

Proof. Let L and J be potential solutions with (γ^1, λ^1) and (γ^2, λ^2) in their respective potential sets. By Theorem 4.4, there exists a potential solution K whose potential set contains (γ^1, λ^2). By Theorem 4.5, there exists a sequence of adjacent potential solutions $\{J_i\}^m_{i=1}$, such that $J_1 = J$ and $J_m = K$. By the "dual" result of Theorem 4.5 (see Remark 4.4), there exists a sequence of adjacent potential solutions $\{K_i\}^r_{i=1}$ such that $K_1 = K$ and $K_r = J$. Because L and J were arbitrary, ϑ is connected.

4.5 MC² Simplex Method

Based on Sections 4.3and 4.4, this section first summarizes the MC² simplex method for locating the set of potential solutions, ϑ , and then provides an illustrative example.

MC² Simplex Method:

Step 1. Given an MC² problem with the data (**C**, **A**, **D**), we use Tables 4.1-4.4 to compute the basis J without (γ, λ).

Step 2. For J, we use Definitions 4.4-4.5 and Theorem 4.1 to find the primal parameter set $\Gamma(J)$ and the dual parameter set $\Lambda(J)$. If J is a potential solution, we record J as an element of ϑ.

Step 3. We apply Theorem 4.3 to detect the effective or null constraint $\mathbf{Z}_k(J)$ for $\Lambda(J)$ and the effective or null constraint $\mathbf{W}^p(J)$ for $\Gamma(J)$.

Step 4. We use the MC² Pivoting Procedure to pivot an adjacent basis K from J. The pivoting is guaranteed by Theorems 4.4-4.5.

Step 5. We return to Step 2 for K until the space (γ, λ) is covered by the number of parameter Cartesian spaces of the identified potential solutions, which the set ϑ that we want to find.

Example 4.1

We use the above MC² simplex method to solve the PC MC² problem in Section 1.4:

$$Max \quad (\lambda_1 \quad \lambda_2)\begin{pmatrix} 400 & 200 \\ 1 & 1 \end{pmatrix}\begin{pmatrix} x_1 \\ x_2 \end{pmatrix}$$

$$Subject\ to \begin{pmatrix} 3 & 2 \\ 8 & 3 \end{pmatrix}\begin{pmatrix} x_1 \\ x_2 \end{pmatrix} \le \begin{pmatrix} 40{,}000 & 65{,}000 \\ 90{,}000 & 45{,}000 \end{pmatrix}\begin{pmatrix} \gamma_1 \\ \gamma_2 \end{pmatrix}$$

$$\begin{pmatrix} x_1 \\ x_2 \end{pmatrix} \ge 0$$

In order to show a two-dimensional picture of (γ, λ), we use $0 < \gamma_1, \gamma_2 \le 1$ and $\gamma_1 + \gamma_2 = 1$ to normalize the constraint-level parameters $\gamma = (\gamma_1, \gamma_2)$ and $0 < \lambda_1, \lambda_2 \le 1$ and $\lambda_1 + \lambda_2 = 1$ to normalize the criteria parameters $\lambda = (\lambda_1, \lambda_2)$. Then, we use the MC² simplex method as follows:

Step 1. Using the data (**C**, **A**, **D**), we form the MC^2 Simplex tableau (see Table 4.5):

Table 4.5

	x_1	x_2	x_3	x_4	RHS	
x_3	(3)	2	1	0	40000	65000
x_4	8	3	0	1	90000	45000
	-400	-200	0	0	0	0
	-1	-1	0	0	0	0

Note that the variables x_3 and x_4 are the slack variables.

Step 2. Since $\mathbf{Z}_1 < 0$, we immediately see that $J = \{3, 4\}$ is not a dual potential solution. However, $J = \{3, 4\}$ is a primal potential solution (e.g., $\gamma = (0.5, 0.5)$).

We now pivot (on the circled element in Tableau 4.5) to the basis $J = \{1, 4\}$. The resulting Tableau is Table 4.6. Because of $\mathbf{W}^2 < 0$, $J = \{1, 4\}$ is not a primal potential solution. We continue pivoting on the circled element of Table 4.6 from $J = \{1, 4\}$ to $J = \{1, 2\}$ as listed in Table 4.7. Applying Definition 4.4, we solve

$8671.43\gamma_1 - 15000\gamma_2 \geq 0$, $7142.86\gamma_1 + 55000\gamma_2 \geq 0$, and $\gamma_1 + \gamma_2 = 1$.

This results in $\Gamma(J\{1, 2\}) = \{0.64 \leq \gamma_1 \leq 1\}$.

Table 4.6

	x_1	x_2	x_3	x_4	RHS	
x_1	1	0.67	0.33	0	13333.33	21666.67
x_4	0	(-2.33)	-2.67	1	-16666.67	-128333.33
	0	66.67	133.33	0	5333333.33	8666666.67
	0	-0.33	0.33	0	13333.33	21666.67

Table 4.7

	x_1	x_2	x_3	x_4	RHS	
x_1	1	0	-0.43	0.29	8671.43	-15000
x_2	0	1	1.14	-0.43	7142.86	55000
	0	0	57.14	28.57	4857142.86	5000000
	0	0	0.71	-0.14	15714.29	40000

Similarly, we solve
$57.14\lambda_1 + 0.71\lambda_2 \geq 0$, $28.57\lambda_1 - 0.14\lambda_2 \geq 0$, and $\lambda_1 + \lambda_2 = 1$ and obtain
$\Lambda(J\{1, 2\}) = \{0.0049 \leq \lambda_1 \leq 1\}$.

We see that $\Gamma(J\{1, 2\}) \times \Lambda(J\{1, 2\}) \neq \varnothing$. By Theorem 4.1, $J = \{1, 2\}$ is a potential solution. We name $J = J_1$.

Step 3. To find the effective or null constraint $\mathbf{Z}_k(J_1)$ for $\Lambda(J_1)$, we use Theorem 4.3 (i) form the subroutine Table 4.8 based on Table 4.7. Without any pivoting, we see that $v_1 = v_2 = 0$ (nonbasic variables) and $w_1 = 57.86$ and $w_2 = 28.43$. After all possible pivoting, we find that there are no zero values of w_1 and w_2. Thus, $\Lambda(J_1)$ has no effective constraints.

Table 4.8

	v_1	v_2	w_1	w_2	Sum $\mathbf{eZ}(J_1)$
w_1	-57.14	-0.71	1.00	0.00	57.86
w_2	-28.57	0.14	0.00	1.00	28.43

To detect the effective or null constraint $\mathbf{W}^p(J_1)$ for $\Gamma(J_1)$, we form Table 4.9 from Table 4.7 (see Theorem 4.3 (ii)).

Table 4.9

	v_1	v_2	w_1	w_2	Sum $\mathbf{W}^p(J_1)\mathbf{e}^t$
w_1	-8571.43	15000	1.00	0.00	-6428.57
w_2	-7142.86	-55000	0.00	1.00	62142.86

We pivot the circled element in Tableau 4.9 and obtain Table 4.10.

Table 4.10

	v_1	v_2	w_1	w_2	Sum $\mathbf{W}^p(J_1)\mathbf{e}^t$
v_1	1.00	-1.75	0.00	0.00	0.75
w_2	0	-67500	-0.83	1.00	67500

In Table 4.10, constraint (row) x_1 in Table 4.7, corresponding to w_1=0. we see that v_1= 0.75, v_2= 0 (nonbasic variable), w_1=0 (nonbasic variable) and w_2= 67500. Thus, $\Gamma(J_1)$ has one effective

Step 4. We use the dual pivoting of MC^2 Pivoting Procedure to pivot J_1 an adjacent basis J_2. Note that the pivoting element is -0.43, which is only negative cell in the first row (corresponding to the basic variable x_1) in Table 4.7. Hence, we introduce x_3 and remove x_1 from Table 4.7. The result is Table 4.11.

Table 4.11

	x_1	x_2	x_3	x_4	RHS	
x_3	-2.33	0	1	-0.67	-20000	35000
x_2	2.67	1	0	0.33	30000	15000
	1.33.33	0	0	66.67	6000000	3000000
	1.67	0	0	0.33	30000	15000

Step 5. For J_2 = {3, 2}, we apply Definitions 4.4-4.5 and Theorem 4.1 and find: $\Gamma(J\{3, 2\}) = \{0 < \gamma_1 \leq 0.64\}$ and $\Lambda(J\{3, 2\}) = \{0 < \lambda_1 \leq 1\}$.

Since $\Gamma(J\{3, 2\}) \times \Lambda(J\{3, 2\}) \neq \emptyset$, by Theorem 4.1, $J_2 = \{3, 2\}$ is another potential solution.

To find the effective or null constraint $\mathbf{W}^p(J_2)$ for $\Gamma(J_2)$, we form the subroutine Table 4.12 from Table 4.11. We pivot the circled element in Table 4.12 and obtain Table 4.13. We see that $v_1 = 0.75$, $v_2 = 0$ (nonbasic variable), $w_1 = 0$ (nonbasic variable) and $w_2 = 67500$. Thus, $\Gamma(J_2)$ has one effective constraint x_3 (corresponding to $w_1 = 0$). That is, the first column of Table 4.11 with respect to x_3 is subject to the dual pivoting for the next adjacent basis.

Table 4.12

	v_1	v_2	w_1	w_2	Sum $\mathbf{W}^p(J_2)\mathbf{e}^t$
w_1	20000	-35000	1.00	0.00	15000
w_2	-30000	-15000	0.00	1.00	45000

Table 4.13

	v_1	v_2	w_1	w_2	Sum $\mathbf{W}^p(J_2)\mathbf{e}^t$
v_1	1	-1.75	0	0.00	0.75
w_2	0	-67500	1.50	1.00	67500

To find the effective or null constraint $\mathbf{Z}_k(J_2)$ for $\Lambda(J_2)$, Table 4.14 is formed from Table 4.11. Without any pivoting, we see that $v_1 = v_2 = 0$ (nonbasic variable), $w_1 = 135$ and $w_2 = 67$. After all possible pivoting, we conclude that $\Lambda(J_2)$ has no effective constraint.

Table 4.14

	v_1	v_2	w_1	w_2	Sum $\mathbf{e}\mathbf{Z}_k(J_2)$
w_1	-133.33	-1.67	1.00	0	135
w_2	-66.67	-0.33	0	1.00	67

Step 4′. Table 4.11 has two negative cells: -2.33 with respect to column x_1 and –0.67 with respect to x_4. We do not need to x_1 because this will return to the basis $J_1 = \{1, 2\}$. We now use -0.67 of Table 4.11 to pivot J_2

an adjacent basis $J_3 = \{4, 2\}$. That is, we introduce x_4 and remove x_3 from Table 4.11. The result is Table 4.15.

Table 4.15

	x_1	x_2	x_3	x_4	RHS	
x_4	3.50	0	-1.50	1.00	30000	-52500
x_2	1.50	1.00	0.50	0	20000	32500
	-100	0	100	0	4000000	6500000
	0.50	0	0.50	0	20000	52500

Step 5′. For $J_3 = \{4, 2\}$, we apply Definitions 4.4-4.5 and Theorem 4.1 and find: $\Gamma(J\{4, 2\}) = \{0.64 < \gamma_1 \leq 1\}$ and $\Lambda(J\{3, 2\}) = \{0 < \lambda_1 \leq 0.0049\}$. Because $\Gamma(J\{4, 2\}) \times \Lambda(J\{4, 2\}) \neq \varnothing$, by Theorem 4.1, $J_3 = \{4, 2\}$ is the third potential solution.

From Figure 4.1, we see that the set of $\Gamma(J\{1, 2\}) \times \Lambda(J\{1, 2\})$, $\Gamma(J\{3, 2\}) \times \Lambda(J\{3, 2\})$, and $\Gamma(J\{4, 2\}) \times \Lambda(J\{4, 2\})$ covers the space (γ, λ). Therefore, the set of potential solutions for PC MC^2 problem is $\vartheta = \{J_1, J_2, J_3\}$.

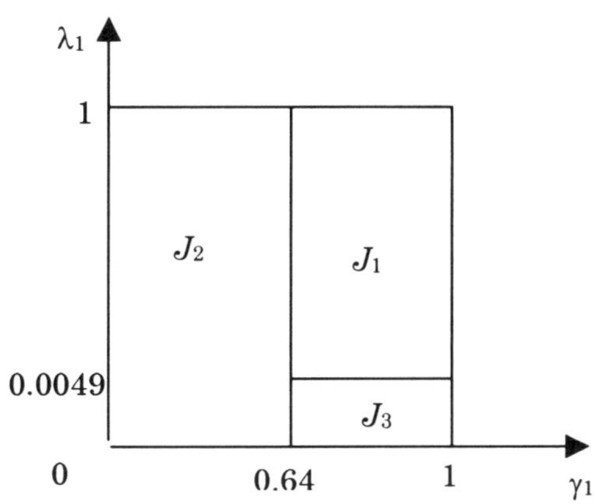

Figure 4.1

4.6 MC² Duality

As discussed in Section 3.2 for the duality of matrix optimization (3.2), this section formally states the MC² duality that is analogous to the duality of linear programming (LP). The LP duality has the following fundamental properties:

(i) The dual of the dual is the primal.

(ii) The duality inequality is always satisfied.

(iii) The duality is perfect. That is, if one program has a finite optimal solution the both have an optimal solution and the two values are equal.

The following discussion shows that the MC² duality shares similar properties mentioned in the above.

Recalling from Sectionn4.1, if we call the original MC² problem is the primal (P) (or M4.1) problem:

$$\begin{array}{ll} \textit{Max} & \boldsymbol{\lambda}^t \mathbf{C}\,\mathbf{x} \\ \textit{Subject to} & \mathbf{A}\,\mathbf{x} \leq \mathbf{D}\gamma \\ & \mathbf{x} \geq 0 \end{array}$$

Then, the dual (D) (or M4.2) MC² problem is written as

$$\begin{array}{ll} \textit{Min} & \mathbf{u}^t\,\mathbf{D}\gamma \\ \textit{Subject to} & \mathbf{u}^t\,\mathbf{A} \geq \boldsymbol{\lambda}^t\,\mathbf{C} \\ & \mathbf{u} \geq 0 \end{array}$$

Definition 4.10 ($\mathbf{x}$; γ) is a (P)-feasible solution if and only if $\gamma > \mathbf{0}$ and $\mathbf{x}$ satisfies

$$\begin{array}{c} \mathbf{Ax} \leq \mathbf{D}\gamma \\ \mathbf{x} \geq 0. \end{array}$$

Definition 4.11 ($\mathbf{u}$; λ) is a (D)-feasible solution if and only if $\lambda > \mathbf{0}$ and $\mathbf{u}$ satisfies

$$\begin{array}{c} \mathbf{u}^t\,\mathbf{A} \geq \boldsymbol{\lambda}^t\mathbf{C} \\ \mathbf{u} \geq 0. \end{array}$$

Definition 4.12 ($\mathbf{x}$; γ) is a nondominated solution to (P) if and only if $\mathbf{x}$ is a nondominated solution to

$$\begin{array}{ll} \textit{Max} & \mathbf{C\,x} \\ \textit{Subject to} & \mathbf{A\,x} \leq \mathbf{D}\gamma \\ & \mathbf{x} \geq 0 \end{array}$$

Definition 4.13 ($\mathbf{u}$; λ) is a nondominated solution to (D) if and only if $\mathbf{u}$ is a nondominated solution to

$$\begin{array}{ll} \textit{Min} & \mathbf{u}^t\,\mathbf{D} \\ \textit{Subject to} & \mathbf{u}^t\,\mathbf{A} \geq \lambda^t\,\mathbf{C} \\ & \mathbf{u} \geq 0 \end{array}$$

In Definitions 4.4-4.5 and Theorem 4.1, we defined potential solutions in terms of bases because some values of γ in (P) may cause degeneracy and hence there will not be a one-to-one correspondence between bases and corner points. With each basis, the following remark shows the identity between the potentiality of the basis and the feasibility of the "solutions" for (P) and (D).

Remark 4.5 From (i) of Definition 4.4 and (i) of Theorem 4.1, we see that $\mathcal{J}$ is a primal potential solution if and only if for some $\gamma > 0$, the basic solution for $\mathcal{J}$, $\mathbf{x}(\mathcal{J}, \gamma)$, satisfies

$$\begin{array}{c} \mathbf{Ax} \leq \mathbf{D}\gamma \\ \mathbf{x} \geq 0. \end{array}$$

Thus from (i) of Definition 4.5, the basic solution $\mathbf{x}(\mathcal{J}, \gamma) = \mathbf{B}^{-1}\mathbf{D}\gamma$, associated with a primal potential basis $\mathcal{J}$ is (P)-feasible for all $\gamma \in \Gamma(\mathcal{J})$;

Similarly, from (ii) of Definition 4.4 and (ii) of Theorem 4.1, $\mathcal{J}$ is a dual potential solution if and only if for some $\lambda > \mathbf{0}$, $\lambda^t\mathbf{Z}(\mathcal{J}) \geq \mathbf{0}$, where $\mathbf{Z}(\mathcal{J}) = [\mathbf{0},\ \mathbf{C}_B\mathbf{B}^{-1}\mathbf{R} - \mathbf{C}_R]$. That is, for every $\lambda \in \Lambda(\mathcal{J})$, the dual basic variables, defined as $\mathbf{u}(\mathcal{J}, \lambda) = \lambda^t\mathbf{C}_B\mathbf{B}^{-1}$ are (D)-feasible.

Note that using Remark 4.5, it is easy to show the fundamental property (i): The dual of the dual is the primal.

In the MC^2 simplex method, because the computation is accomplished via the tableau, we state the following theorems in terms of both bases and solutions. Since the results for solutions follow immediately from the results for bases, we only prove the latter.

We first establish the MC^2 duality inequality. Recalling from Section 4.2, in the MC^2 simplex tableau, for a basis J, we have

$$\mathbf{V}(J) = [\mathbf{C}_B \mathbf{B}^{-1} \mathbf{D}].$$

This is stated in (iii) of Definition 4.5 as the objective "value" of the MC^2 problem $V(J, \lambda, \gamma) = \lambda^t \mathbf{C}_B \mathbf{B}^{-1} \mathbf{D}\gamma$, which is a function of (γ, λ), for the basic solution $\mathbf{x}(J, \gamma)$. Then, the relationship of the primal and dual bases is given as

Theorem 4.7 (MC^2 Weak duality)
(i) If J is a primal potential basis and K is a dual potential basis then

$$V(J, \lambda, \gamma) \le V(K, \lambda, \gamma) \text{ for all } (\gamma, \lambda) \in \Gamma(J) \times \Lambda(J).$$

(ii) If $(\mathbf{x}; \gamma)$ and $(\mathbf{u}; \lambda)$ are feasible solutions for (P) and (D), respectively, then

$$\lambda^t \mathbf{C}\mathbf{x} \le \mathbf{u}^t \mathbf{D}\gamma.$$

The proof of Theorem 4.7 can be found in Chapter 8 of Yu (1985). Basically, Theorem 4.7 tells the fundamental property (ii): The duality inequality is always satisfied.

To show the fundamental property (iii): The duality is perfect, we have the following theorem:

Theorem 4.8
(i) Let J be a primal potential basis and K be a dual potential basis. Then for each $(\gamma, \lambda) \in \Gamma(J) \times \Lambda(J)$, there is a potential basis L such that

$$V(J, \lambda, \gamma) \le V(L, \lambda, \gamma) \le V(K, \lambda, \gamma).$$

(ii) Let $(\mathbf{x}; \gamma)$ and $(\mathbf{u}; \lambda)$ be feasible solutions for (P) and (D), respectively. Then there exists feasible solutions $\mathbf{x}^*(J, \gamma)$ and $\mathbf{u}^*(J, \lambda)$ for (P) and (D), respectively, such that

$$\lambda^t \mathbf{C}\mathbf{x} \le \lambda^t \mathbf{C}\mathbf{x}^* = (\mathbf{u}^*)^t \mathbf{D}\gamma \le \mathbf{u}^t \mathbf{D}\gamma.$$

The proof of Theorem 4.8 can be also found in Chapter 8 of Yu (1985). This result immediately brings up the following extension:

Corollary 4.1
(i) Suppose J is a primal potential basis and K is a dual potential basis. If, for some $(\gamma, \lambda) \in \Gamma(J)\times\Lambda(J)$, $V(J, \lambda, \gamma) = V(K, \lambda, \gamma)$, then J and K are both potential solutions.

(ii) Suppose($\mathbf{x}$; γ) and ($\mathbf{u}$; λ) are both feasible solutions for (P) and (D), respectively. If $\lambda\mathbf{C}\mathbf{x} \doteq \mathbf{u}\mathbf{D}\gamma$, then J and K are both nondominated solutions.

Similar to the complementary slackness for matrix optimization (see Theorem 3.6 of Section 3.2), the MC² complementary slackness becomes the following theorem.

Theorem 4.9 (MC² Complementary Slackness) Let ($\mathbf{x}$; γ) and ($\mathbf{u}$; λ) be feasible solutions for (P) and (D), respectively. They are nondominated solutions if and only if $\mathbf{u}^t(\mathbf{Ax} - \mathbf{D}\gamma) = \mathbf{0}$ and $(\lambda^t\mathbf{C} - \mathbf{u}^t\mathbf{A})\ \mathbf{x} = \mathbf{0}$.

Example 4.2 Given the PC MC² problem in Section 1.4, we can construct its (D) problem as:

$$Min\quad (u_1 \quad u_2)\begin{pmatrix} 40{,}000 & 65{,}000 \\ 90{,}000 & 45{,}000 \end{pmatrix}\begin{pmatrix} \gamma_1 \\ \gamma_2 \end{pmatrix}$$

$$Subject\ to\quad (u_1 \quad u_2)\begin{pmatrix} 3 & 2 \\ 8 & 3 \end{pmatrix} \ge (\lambda_1 \quad \lambda_2)\begin{pmatrix} 400 & 200 \\ 1 & 1 \end{pmatrix}$$

$$(u_1 \quad u_2) \ge 0$$

Let us observe Table 4.11 of Example 4.1 which is associated with the potential solution $J_2 = \{3, 2\}$. This table contains the submatrices:

$$\mathbf{B}^{-1}\mathbf{D} = \begin{pmatrix} -20{,}000 & 35{,}000 \\ 30{,}000 & 15{,}000 \end{pmatrix} \quad \text{and} \quad \mathbf{C_B}\mathbf{B}^{-1} = \begin{pmatrix} 0 & 66.67 \\ 0 & 0.33 \end{pmatrix}.$$

If we choose $\gamma = (1/2,\ 1/2)$, where $\gamma_1 = 1/2 \in \Gamma(J\{3, 2\}) = \{0 < \gamma_1 \le 0.64\}$ and $\lambda = (1/3,\ 2/3)$, where $\lambda_1 = 1/3 \in \Lambda(J\{3, 2\}) = \{0 < \lambda_1 \le 1\}$. Then

$$\mathbf{x}(J_2, \gamma) = \mathbf{B}^{-1}\mathbf{D}\gamma = \begin{pmatrix} 7500 \\ 22500 \end{pmatrix}, \text{ and}$$

$$\mathbf{u}(J_2, \lambda) = \lambda \mathbf{C}_B \mathbf{B}^{-1} = (44.47,\ 0.22)$$

Thus, $(x_1, x_2) = (7500, 22500)$ and $(u_1, u_2) = (44.47, 0.22)$ are (P) and (D) feasible for $\gamma = (1/2, 1/2)$ and $\lambda = (1/3, 2/3)$, respectively.

Furthermore, the functional values $\lambda\mathbf{Cx}$ are equal to $\mathbf{u}\mathbf{D}\gamma$ (Theorem 4.9). This will be true for any $(\gamma, \lambda) \in \Gamma(J_2) \times \Lambda(J_2)$ since

$$\lambda^t \mathbf{Cx} = \lambda^t \mathbf{C}_B\, \mathbf{x}(J_2, \gamma) = \lambda^t \mathbf{C}_B \mathbf{B}^{-1}\mathbf{D}\gamma = \mathbf{u}^t \mathbf{D}\gamma.$$

In the case of $\gamma = (1/2, 1/2)$ and $\lambda = (1/3, 2/3)$, $\lambda\mathbf{Cx} = \mathbf{u}\mathbf{D}\gamma = 1{,}515{,}000$. The verification of further relationships is left to the reader.

4.7 MC² Computer Software

As we see from the above discussion, locating a set of potential solutions, ϑ, by using the MC² simplex method involves the computation on both MC² simplex tableau and the unknown parameter sets $\Gamma(J)$ and $\Lambda(J)$. Therefore, the MC² simplex method can consist of five subroutines in each iteration: pivoting, determining primal feasibility and dual feasibility, and determining the effective constraints for the parameter sets $\Gamma(J)$ and $\Lambda(J)$ of a known potential basis J. This section provides the development and introduction of two MC² computer software based on the MC² simplex method. The first developed by Chien, Shi and Yu (1989) is a man-machine interactive software that is identical to Steps of the MC² simplex method in Section 4.4. The second one developed by Hao and Shi (1996) which intends to solve large-scale MC² problems.

4.7.1 A Man-machine Interactive Software

This PC based MC² program was designed to assist the user in solving MC² problems. The data (**C**, **A**, **D**) of original MC² problem are first entered interactively into the screen of the PC. Then, the set of all potential solutions, ϑ, is identified by a man-machine interactive process. Note that the (γ, λ) ranges of some potential solutions may overlap and the (γ, λ) space is completely covered. Through the screen representation, the user can observe any MC² simplex tableau that he/she wants to pivot. Because of the displaying the MC² simplex tableau, the program is only effective to solve a small-scale problem (e.g., twenty variables and ten constraints). The code of software was written in Pascal according to the flow chart represented in Figure 4.2.

4.7.2 A File Input-output Software

The file input-output software was designed to save the computation time so that a large-scale MC^2 problem (e.g., 500 variables and 200 constraints) can be solved. The user can freely choose an input format for the data (**C**, **A**, **D**). Without showing MC^2 simplex tableaus on the screen, this program reports a set of potential solutions, ϑ, by printing an output file containing (γ, λ) ranges, the objective payoff $V(J, \lambda, \gamma)$ and the basic variables $\mathbf{x}(J, \gamma)$. This set of the (γ, λ) ranges for ϑ covers the (γ, λ) space without overlap. Therefore, the software has three items: the input file, execution program, and output file. The code of execution program is written in C++ with the application of probability simulation. A simple flow chart is given in Figure 4.3. Both software including menu are included in the attached PC diskette on the back of the book for the interested readers.

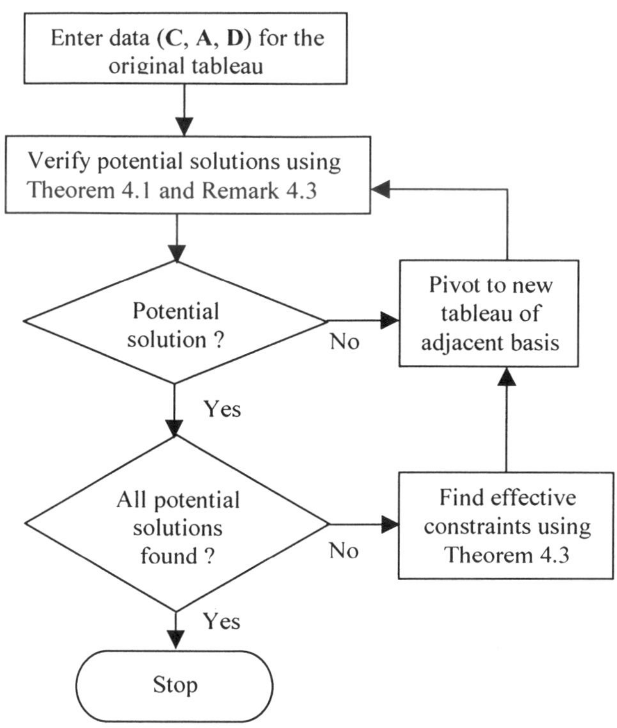

Figure 4.2

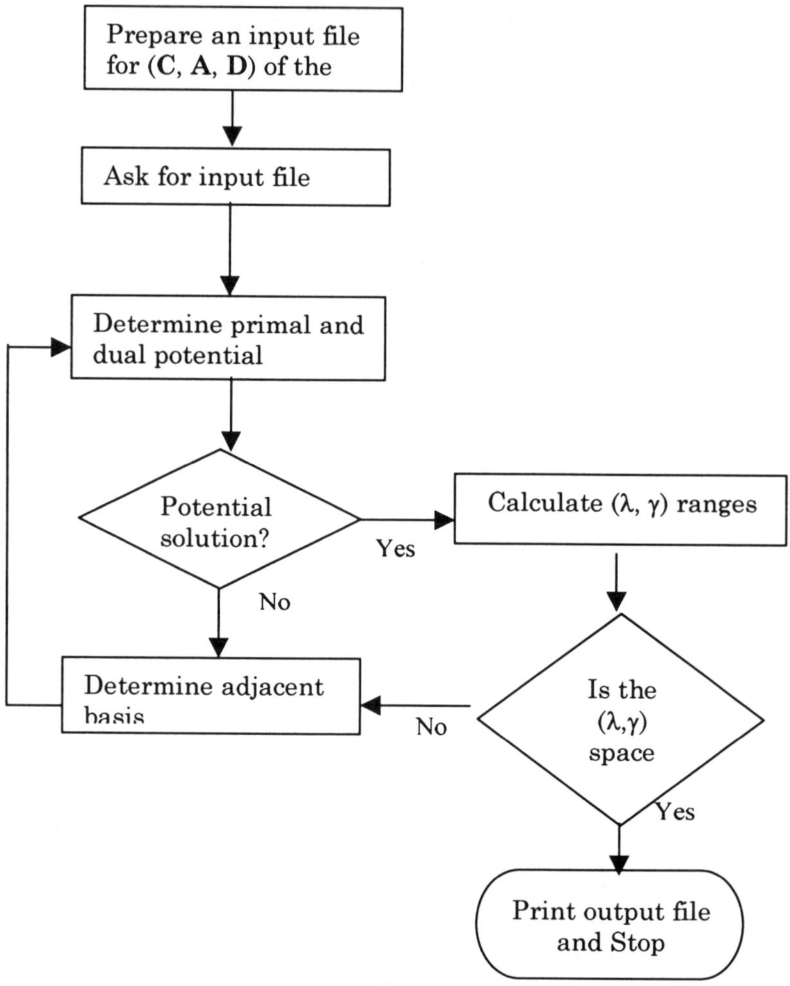
Prepare an input file
for (C, A, D) of the
Ask for input file
Determine primal and
dual potential
Potential
solution?
Yes
Calculate (λ, γ) ranges
No
Determine adjacent
basis
No
Is the
(λ,γ)
space
Yes
Print output file
and Stop

Figure 4.3

4.8 Further Remarks

Linear functions not only have many unique mathematical properties, but also have been frequently used to estimate real-world phenomenon that has non-linear characteristics. The MC^2 problems, like MC problems enjoy a number of unique linear features. To facilitate computation, we have introduced the MC^2 simplex method as natural extension of the ordinary LP simplex method. Our description in this chapter is heavily based on Seiford and Yu (1979) and the Chapter 8 of Yu (1985). Shi (1998) provides a comprehensive review on MC^2 theory and applications. The extensive study of MC^2 problems in optimal system designs can be found in Shi (1991). In addition to the MC^2 software developed by Chien, Shi and Yu (1989) and Hao and Shi (1996), Ge and Shi (1997) has proposed a Window-based MC^2 program. Due to its immaturity, this program needs to be further advanced. Other MC^2 related computer software for solving MC^2 integer programming and MC^2 transportation problems will be discussed in Chapter 5 and Chapter 6, respectively. There are many interesting and related articles to the field of MC linear programming with the changes of parameters. For example, Gal (1979) discussed extended simplex methods from a parametric programming point of view.

Exercises

4.1 Solve the following MC^2 problem by using the MC^2 simplex method:

$$\begin{array}{ll} \textit{Max} & 10\,x_1 + 12\,x_2 \\ \textit{Max} & 15\,x_1 + 8\,x_2 \\ \textit{Subject to} & 3\,x_1 + 2x_2 \le (24, -35) \\ & 4x_1 + 6x_2 \le (24, 48) \\ & x_1, x_2 \ge 0. \end{array}$$

4.2 In the MC^2 simplex tableau of each potential solutions of problem 4.1, find the matrices:

(i) $\mathbf{Y}(J) = [\mathbf{I}, \mathbf{B}^{-1}\mathbf{R}]$,

(ii) $\mathbf{W}(J) = [\mathbf{B}^{-1}\mathbf{D}]$,

(iii) $\mathbf{Z}(J) = [\mathbf{0}, \mathbf{C}_B\mathbf{B}^{-1}\mathbf{R} - \mathbf{C}_R]$, and

(iv) $\mathbf{V}(J) = [\mathbf{C}_B\mathbf{B}^{-1}\mathbf{D}]$.

Applying Theorem 4.3, identify primal and dual effective constraints for each potential solutions of problem 4.1 in subroutines.

4.3 Solve problem 4.1 by (i) computer program of Chien, Shi and Yu (1989) and (ii) Hao and Shi (1996), respectively.

4.4 Transfer the "Min" objective function to a "Max" function and solve the following MC^2 problem by using the MC^2 simplex method:

$$\begin{array}{ll} \textit{Min} & 4\,x_1 - 3\,x_2 \\ \textit{Max} & 5\,x_1 + 9\,x_2 \\ \textit{Subject to} & 2x_1 + 4x_2 \leq (20,\ 30) \\ & x_1 + \ x_2 \leq (10,\ -50) \\ & x_1,\ x_2 \geq 0. \end{array}$$

4.5 Construct an MC^2 problem (e.g., two or more criteria and constraint levels) with 20 variables and 10 constraints. Then use the computer program of Chien, Shi and Yu (1989) to solve it.

4.6 Construct an MC^2 problem (e.g., two criteria and two constraint levels or three criteria and three constraint levels) with 500 variables and 60 constraints. Then use the computer program of Hao and Shi (1996) to solve it.

4.7 Prove Theorem 4.3.

4.8 Prove Theorem 4.4.

4.9 Prove Theorem 4.5.

Part Two: Special MC^2 Problems

Chapter 5
MC² INTEGER PROBLEMS

For the last four decades, traditional integer programming problems have been extensively studied by many researchers (for example, see Hu 1969, Garfinkel and Nemhauser 1972, Zionts 1974, Salkin 1975, Taha 1975, Nemhauser and Wolsey 1988, and Parker and Rardin 1988). Various algorithms using the branch-and-bound technique have been developed to solve these integer programming problems (Land and Doig 1960, Dakin 1965, Balas 1965, Lawler and Wood 1966, Mitten 1970, Geoffrion and Marsten 1972, Crowder *et al* 1983, Johnson *et al* 1985, and Van Roy and Wolsey 1987). However, because the mathematical model of traditional integer programming has only a single criterion (objective) and a single resource availability level (right hand side), it may have limitations in dealing with real-world problems. For example, the problem of selecting hub cities for rural telecommunication systems can be formulated as "integer programming," but the problem must consider multiple criteria, such as population, economy, education, health care, and transportation (Lee *et al* 1994 and Nazem *et al* 1995). In addition, resource availability levels of the problem can be differently represented by the telecommunication company and the local community.

It has been well recognized that many real-world problems should have multiple criteria (MC) (Zeleny 1974, Gal 1977, Goicoechea *et al* 1982, Yu 1985, Steuer 1986, and Dyer *et al* 1992). Formulations of MC linear integer programming problems and interactive algorithms for solving such problems have been explored by Villarreal and Karwan 1981, Zionts 1981, Korhonen *et al* 1984, Ramesh 1985, Ramesh *et al* 1985, Gabbani and Magazine 1986, and Ramesh *et al* 1989.

This chapter is to discuss mathematical formulations of MC² integer problems that extend integer linear programming and MC linear programming. Section 5.1 explores the concepts of integer potential solutions, which include zero-one (or binary) potential solution and general integer potential solution. Section 5.2 illustrates MC² zero-one linear programming (MC²BIP) and develops a branch-and-bound procedure to solve such integer programming problems (Shi and Lee 1997). This approach first relaxes the original zero-one MC² problem into a regular MC² problem, and then solve it for zero-one potential solutions. If zero-one potential solution is not found, it branches the decision variable as MC^2 zero-one integer so that each subproblem in the branches is an MC^2 problem. The upper bound of the subproblem is defined as the expected objective value of its relaxation problem with a probability distribution over parameters of multiple criteria λ and multiple constraint levels γ. Section 5.3 builds a general MC^2 integer linear programming (MC^2IP)(Li and Shi 1999). Instead of using the branch-and-bound approach, we create a branch-and-partition algorithm, in which the decision variable is branched according to the range of parameter λ for the potential solution and then a general MC^2 integer solution is identified from partitions of the range of parameter γ. Section 5.4 introduces computer software to solve the MC^2 BIP problems that is also part of MC^2 software packages attached with this book. Finally, Section 5.5 provides further comments.

5.1 Integer Potential Solutions

Given a basis J for MC^2 linear programming (M4.1):

$$\begin{array}{ll} \textit{Max} & \lambda^t \mathbf{C}\,\mathbf{x} \\ \textit{Subject to} & \mathbf{A}\,\mathbf{x} \leq \mathbf{D}\gamma \\ & \mathbf{x} \geq 0 \end{array}$$

if there is $\gamma > 0$ such that $\mathbf{x}(J, \gamma) = \mathbf{B}^{-1}\mathbf{D}\gamma > 0$, we call J a *nondegenerate feasible basis*; and if at least one component of $\mathbf{x}(J, \gamma)$ is zero, we call J a *degenerate feasible basis.*

Many real-world decision problems, for example, selecting telecommunication hub cities, can be classified as integer programming problems under a multiple criteria environment. The variable of hub city selection is zero-one (i.e., selected is one and not selected is zero). Similarly, if the decision variable represents a tangible product, such as a number of ships to be purchased, then the variable should be integer. In the former case, it is necessary and important to develop an MC^2 BIP model; while in the latter, a general MC^2 IP model needs to be built. In this section, we first discuss the definitions of "0-1 integer numbers" and "integer number" for the MC^2 problem (M4.1), then describe the concept of integer potential solutions.

In order to do this, we recall that given a basis J, its basic solution $\mathbf{x}(J, \gamma) = \mathbf{B}^{-1}\mathbf{D}\gamma$ is a function of parameter γ. If the jth component of $\mathbf{x}(J, \gamma)$ is a decision variable, then it is called a basic decision variable, denoted by $x_j(J, \gamma)$; otherwise, it is a basic slack variable, denoted by $s_j(J, \gamma)$. We also recall that the parameter γ is normalized, i.e., $0 \le \gamma_k \le 1$, k = 1, . . ., p and $\gamma_1 + \gamma_2 + \ldots + \gamma_p = 1$. We now define "0-1 integer numbers" for the MC^2 problem (M4.1) as follows:

Definition 5.1 Given a basis J, (i) for $j \in J$, the decision variable $x_j(J, \gamma)$ is said to be an " MC^2-integer number 1" if and only if $x_j(J, \gamma) = \gamma_1 + \gamma_2 + \ldots + \gamma_p = 1$; (ii) for $j \in J$, $x_j(J, \gamma)$ is said to be an "MC^2-integer number 0" if and only if $x_j(J, \gamma) = 0$; and (iii) for $j \notin J$, $x_j(J, \gamma)$ is an " MC^2-integer number 0," which is denoted by $x_j(J, \gamma) = 0$.

Note that we do not put the restriction of "0-1 integer numbers" in Definition 5.1 on any basic slack variable $s_j(J, \gamma)$.

The condition if $x_j(J, \gamma) = \gamma_1 + \gamma_2 + \ldots + \gamma_p = 1$ in (i) of Definition 5.1 implies that when all of the decision situation forces mutually agree to select $x_j(J, \gamma)$ (i.e., d_{jk} of D is 1 for all k), then the summation of resulting utility value (weights) is 1, i.e., $x_j(J, \gamma) = \Sigma d_{jk}\gamma_k = \Sigma\gamma_k = 1$. In this case, $x_j(J, \gamma)$ is defined as an "MC^2-integer number 1" for J (For the utility function analysis, see Keeney and Raiffa 1976). Otherwise, if $x_j(J, \gamma)$ is not in J or degenerate, then $x_j(J, \gamma) = 0$ (i.e., $x_j(J, \gamma)$ has the utility value of 0).

A definition of the general MC^2-integer can be similarly given by:

Definition 5.2 Given a basis J, the decision variable $x_j(J, \gamma)$ is said to be an "MC^2-integer" if and only if $x_j(J, \gamma) = d_{j1}\gamma_1 + d_{j2}\gamma_2 + \ldots + d_{jp}\gamma_p$ is an integer.

The 0-1 potential solution and integer potential solution for the MC^2 problem (M4.1) can be described as

Definition 5.3 (i) A given potential solution $\mathbf{x}(J, \gamma)$ is said to be an "MC^2-0-1 potential solution" if and only if all decision variables $x_j(J, \gamma)$ are "MC^2-integer number 1's or 0's." If $\mathbf{x}(J, \gamma)$ is an "MC^2-0-1 potential solution", then the corresponding J is called "MC^2-0-1 potential basis."

(ii) The potential solution $\mathbf{x}(J, \gamma)$ is said to be an "MC^2-integer potential solution" if and only if all decision variables $x_j(J, \gamma)$ are "MC^2-integer." If $\mathbf{x}(J, \gamma)$ is an "MC^2-integer potential solution", then the corresponding J is called " MC^2-integer potential basis."

5.2 MC² Zero-One Problem and A Branch-and-Bound Procedure

According to (i) of Definition 5.3, we call φ a set of all "MC²-0-1 integer potential bases" if every J of φ is an "MC²-0-1 integer potential basis." Based on the above discussion, we can present a mathematical model of an MC² BIP problem (M5.1) as follows:

$$\begin{aligned} &\mathit{Max} && \lambda^t \mathbf{C}\,\mathbf{x} \\ &\mathit{Subject\ to} && \mathbf{A}\,\mathbf{x} \le \mathbf{D}\gamma \\ & && x_j(\gamma) = \sum_{k=1}^{p} \gamma_k = 1, \mathit{or}\ 0, \mathit{for\ all}\ j. \end{aligned}$$

A "relaxation problem" of (M5.1) can be proposed as the problem (M5.2):

$$\begin{aligned} &\mathit{Max} && \lambda^t \mathbf{C}\,\mathbf{x} \\ &\mathit{Subject\ to} && \begin{pmatrix} \mathbf{A} \\ \mathbf{I}_n \end{pmatrix} \mathbf{x} \le \begin{pmatrix} \mathbf{D} \\ \mathbf{1}_{nxp} \end{pmatrix} \gamma \\ & && \mathbf{x} \ge 0, \end{aligned}$$

Where $\mathbf{I}_n$ is an identity matrix and $\mathbf{1}_{nxp}$ is a $(n \times p)$ matrix with 1 in all cells.

Applying Definitions 4.4-4.5 and Theorem 4.1, the relationship between (M5.1) and (M5.2) can be stated by the following theorem:

Theorem 5.1

(i) The set of all feasible solutions for (M5.1) is a subset of all the feasible solutions for (M5.2).

(ii) Furthermore, if every potential basis for Problem (M5.2) is an MC²-integer potential basis, then the set of all MC²-0-1 integer potential bases for Problem (M5.2) is also the set of all MC²-0-1 integer potential bases for Problem (M5.1).

By Theorem 5.1 and Definitions 5.1 and 5.3(i), we may solve the relaxation problem (M5.2) to identify the " MC²-0-1 integer potential bases" for a given MC²BIP problem (M5.1).

The following example is used to illustrate how to solve an MC^2BIP problem (M5.1) by solving the relaxation problem (M5.2).

Example 5.1

Consider the following MC^2 BIP problem:

$$Max\ (\lambda_1 \quad \lambda_2)\begin{pmatrix} 3 & 2 \\ -1 & 4 \end{pmatrix}\begin{pmatrix} x_1 \\ x_2 \end{pmatrix}$$

$$Subject\ to \begin{pmatrix} 1 & 3 \\ 2 & 1 \end{pmatrix}\begin{pmatrix} x_1 \\ x_2 \end{pmatrix} \le \begin{pmatrix} 5 & 8 \\ 6 & 3 \end{pmatrix}\begin{pmatrix} \gamma_1 \\ \gamma_2 \end{pmatrix}$$

$$x_j = \gamma_1 + \gamma_2 = 1, \text{or } 0, \text{j} = 1, 2.$$

Its relaxation is an MC^2 problem:

$$Max\ (\lambda_1 \quad \lambda_2)\begin{pmatrix} 3 & 2 \\ -1 & 4 \end{pmatrix}\begin{pmatrix} x_1 \\ x_2 \end{pmatrix}$$

$$Subject\ to \begin{pmatrix} 1 & 3 \\ 2 & 1 \\ 1 & 0 \\ 0 & 1 \end{pmatrix}\begin{pmatrix} x_1 \\ x_2 \end{pmatrix} \le \begin{pmatrix} 5 & 8 \\ 6 & 3 \\ 1 & 1 \\ 1 & 1 \end{pmatrix}\begin{pmatrix} \gamma_1 \\ \gamma_2 \end{pmatrix}$$

$$\begin{pmatrix} x_1 \\ x_2 \end{pmatrix} \ge 0.$$

We solve the above MC^2 problem and then obtain two " MC^2-0-1 integer potential bases," which are denoted by $\varphi = \{K_1, K_2\}$ as in Table 5.1. We see that from Table 5.1, the decision variables $x_j \ge 0$, j = 1, 2 in K_1 and x_1 in K_2 are " MC^2 integer number 1's" for (M5.2) by Definition 5.1. Thus, by Definition 5.3(i), K_1 and K_2 are " MC^2-0-1 integer potential bases" for (M5.2) (note that there is no restriction of MC^2-integer numbers on any slack variables in K_1 and K_2). By Theorem 5.1, they are also "MC^2-0-1 integer potential bases" for (M5.1).

Table 5.1 MC²-0-1 Integer Potential Bases

Bases	$\Gamma(K_j)$	$\Lambda(K_j)$	Value of $\mathbf{x}(K_j)$
$\mathbf{x}(K_1)$ $= (x_1, x_2, s_1, s_2)$	$0 < \gamma_1 \leq 1$	$1/4 \leq \lambda_1 \leq 1$	$x_1 = x_2 = \gamma_1 + \gamma_2$ $s_1 = \gamma_1 + 4\gamma_2$ $s_2 = 3\gamma_1$
$\mathbf{x}(K_2)$ $= (x_1, s_1, s_2, s_3)$	$0 < \gamma_1 \leq 1$	$0 < \lambda_1 \leq 1/4$	$x_1 = \gamma_1 + \gamma_2$ $s_1 = 2\gamma_1 + 5\gamma_2$ $s_2 = 5\gamma_1 + 2\gamma_2$ $s_3 = \gamma_1 + \gamma_2$

Unfortunately, most MC²BIP problems of (M5.1) cannot be solved only through its relaxation problem (M5.2) since the solutions for (M5.2) may not be all MC²-0-1 integer potential solutions (recall (ii) of Theorem 5.1). To avoid this pitfall, we can divide a given MC²BIP (M5.1) into a series of subproblems, and then solve relaxations of those subproblems for their MC²-0-1 integer potential solutions. For this purpose, we adopt the branch-and-bound method of binary integer linear programming to develop a branch-and-bound solution procedure to solve MC²BIP problems.

We note that in the branch-and-bound method for solving linear programming problems (Hillier and Lieberman 1990), once the branching is made by partitioning the original feasible set into smaller subsets, each subproblem needs to be solved to obtain the "bound." Such a bound can be determined by the optimal objective value of the subproblem. Thus, when adopting the branch-and-bound method for solving an MC²BIP problem (M5.1), we have to deal with the "bound" of an MC² problem. Because the objective payoff of each potential basis for the MC² problem is a function of (γ, λ), finding the value of the total objective payoffs of all potential bases for the MC² problem involves the assessment of the likelihood of (γ, λ) over the various points of its range. With proper assumptions, the uncertainty may be represented by random variables with some known probability distributions, such as a normal, exponential, or uniform distribution.

In this section, for an illustrative purpose, we assume that there is a *joint uniform distribution* over parameters (γ, λ). The value of the "expected total objective payoff" of the potential bases for the relaxation problem of a subproblem under a uniform distribution is used to determine the "bound" of the subproblem. This is different from simply computing expectations for the objective coefficients and the constraint level (or right-hand-side) values and then solving the MC^2BIP problem as a standard "0-1" integer program. Our procedure has the advantage of presenting all possible potential bases associated with ranges of parameters (γ, λ) which offer the decision makers a comprehensive scenario to make the final decision. However, the latter will shrink many decision choices into one basis with the given expected values.

Definition 5.4 Let $\mathbf{EV}(\varphi)$ be the expected total objective payoff of $\varphi = \{K_1, \ldots, K_r\}$ for a given MC^2BIP problem, then

$$\mathbf{EV}(\varphi) = \sum_{i=1}^{r} \int_{S_i} \mathbf{V}(K_i)\, dF(\gamma, \lambda)$$

where $\mathbf{V}(K_i)$ is the objective payoff (a function of (γ, λ) of the MC^2-0-1 integer potential solution K_i, for i = 1, . . ., r; $F(\gamma, \lambda)$ is the joint cumulative uniform distribution which is determined by a given estimation method; and $S_i = \Gamma(K_i) \times \Lambda(K_i)$ is the optimal situation set of (γ, λ) of K_i, for i = 1, . . ., r.

By integrating the above results with the well-known framework of branch-and-bound methods, we propose the following algorithm to solve the MC^2BIP problem (M5.1):

Algorithm 5.1 (An MC^2BIP Branch-and-Bound Algorithm):

Step 0. (Initialization) Set Bound $Z^* = -\infty$, where Z^* is denoted as the value of the expected total objective payoff $\mathbf{EV}(\varphi^*)$ for the current incumbent (see Definition 5.4). Apply Step 2, Step 3, and Step 4 described below to the given MC^2BIP problem (M5.1). If not fathomed (dismissed), classify this problem as one "remaining subproblem" and perform the following steps for each iteration.

Steps for Each iteration:

Step 1. (Branching) Select the problem which has been created most recently from the remaining (unfathomed) subproblems (break ties according to which has the larger bound). Choose the first one in the natural ordering of the MC^2 binary integer variables to be the "branching variable." Let x_j be this decision variable (not slack variable), and then branch from the problem node of the solution tree to create two new subproblems by setting $x_j(J, \gamma) = \gamma_1 + \gamma_2 + \ldots + \gamma_p = 1$ or $= 0$.

Step 2. (Bounding) For each new subproblem, its "bound" is obtained as follows. First use the MC^2-simplex method to solve its relaxation problem (which is an MC^2 problem). And then find the expected total objective payoff **EV**(φ^*) (see Definition 5.4) for the relaxation problem. Lastly, set the bound of the subproblem to be **EV**(φ^*).

Step 3. (Fathoming) For each new subproblem, apply the three fathoming tests given below, and eliminate the subproblems which are fathomed by any of the tests.

Test 1: Its bound $\leq Z^*$.

Test 2: Its relaxation problem has no feasible solutions for some values of γ.

Test 3: Each potential solution for its relaxation problem has MC^2-integer values (MC^2-integer number 1 or 0) for all of the MC^2 binary integer variables in the basis (see (i) Definitions 5.1 and 5.3). If the expected total objective payoff of the relaxation problem is larger than the incumbent Z^*, it becomes the new incumbent and Test 1 is reapplied to all unfathomed subproblems with the new incumbent Z^{**}.

Step 4. (Optimality Test) Stop when there are no remaining subproblems; the current incumbent is "optimal." Otherwise, go to another iteration to continue.

The computer software of Chien, Shi and Yu (1989) or Hao and Shi (1996) can be used to identify a set of all potential bases for each relaxation problem in the above procedure.

The following adopted real-world example is used to demonstrate how Algorithm 5.1 can be employed to solve a class of real-world MC^2 binary integer problems.

Example 5.2

Nazem *et al* (1995) and Lee *et al* (1994) suggest that the problem of selecting hub cities for building rural telecommunication networks be formulated using an integer linear programming with multiple criteria. Two decision forces are involved in the decision-making process. One force is the telecommunications company that submits its proposal to a local legislature (community) for approval. Another force is the local legislature that has the authority to approve or disapprove the proposal. Both forces estimate the budget and various costs, including operating costs, with their own preferences. If a compromise solution is reached, i.e, the proposal is passed, the company starts building the fiber optics networks and incurs the costs. In designing the network, both decision forces consider population, economy, education, health care, transportation, and intertown distances as criteria for local community development.

Let N = {1, 2, 3, 4} be the set of possible candidate hub cities in a rural area where telecommunication networks might be built. Consider two important criteria: population and economy, where the population in a candidate city has 1,000 people as a unit and the economy, measured by the number of employees of a candidate city, has 1,000 employees as a unit. There are two constraints, the budget for building the fiber optics networks and the available operation capital in a given planning horizon (both have $ 1,000,000 as a unit). The resource availability levels are estimated by the local business community and the telecommunication company, respectively (see Table 5.2). The goal is to decide which of the N candidate cities should be selected as the "optimal" hubs so that the expected total payoff can be maximized.

Table 5.2 Resource consumption, availability, and contribution

	Resources required				Resource levels ($1 million per unit)	
City	1	2	3	4	Community	Company
Budget	4	3	2	6	10	8
Operation cost	1	2	2	3	4	6
Contribution Coefficients	3	6	2	4	(Population: 1000 per unit)	
	1	3	1	3	(Economy: 1000 per unit)	

Since there are two resource availability levels, let $x_j(J, \gamma) = \gamma_1 + \gamma_2 = 1$, if both the local business community and the telecommunication company agree with choosing City j as a hub city; otherwise, let $x_j(J, \gamma) = 0$, j = 1, 2, 3, 4.

Based on information provided by Table 5.2, the corresponding MC^2 BIP problem (M5.1) is formulated as follows:

$$Max\begin{pmatrix}\lambda_1 & \lambda_2\end{pmatrix}\begin{pmatrix}3 & 6 & 2 & 4\\ 1 & 3 & 1 & 3\end{pmatrix}\begin{pmatrix}x_1\\ x_2\\ x_3\\ x_4\end{pmatrix}$$

$$Subject\ to\begin{pmatrix}4 & 3 & 2 & 6\\ 1 & 2 & 2 & 3\end{pmatrix}\begin{pmatrix}x_1\\ x_2\\ x_3\\ x_4\end{pmatrix}\leq\begin{pmatrix}10 & 8\\ 4 & 6\end{pmatrix}\begin{pmatrix}\gamma_1\\ \gamma_2\end{pmatrix}$$

$$x_j = \gamma_1 + \gamma_2 = 1, \text{or } 0, \text{j} = 1, 2, 3, 4.$$

By using Algorithm 5.1 to solve the above problem, we have:

Step 0. (Initialization) Let $Z^* = -\infty$. By replacing $x_j(J, \gamma) = \gamma_1 + \gamma_2 = 1$ or 0 by $0 \leq x_j \leq \gamma_1 + \gamma_2 = 1$ or 0, for j = 1, 2, 3, 4, the relaxation of the original MC^2 BIP problem is formed as:

$$Max\begin{pmatrix}\lambda_1 & \lambda_2\end{pmatrix}\begin{pmatrix}3 & 6 & 2 & 4\\ 1 & 3 & 1 & 3\end{pmatrix}\begin{pmatrix}x_1\\ x_2\\ x_3\\ x_4\end{pmatrix}$$

$$Subject\ to\begin{pmatrix}4 & 3 & 2 & 6\\ 1 & 2 & 2 & 3\\ 1 & 0 & 0 & 0\\ 0 & 1 & 0 & 0\\ 0 & 0 & 1 & 0\\ 0 & 0 & 0 & 1\end{pmatrix}\begin{pmatrix}x_1\\ x_2\\ x_3\\ x_4\end{pmatrix}\leq\begin{pmatrix}10 & 8\\ 4 & 6\\ 1 & 1\\ 1 & 1\\ 1 & 1\\ 1 & 1\end{pmatrix}\begin{pmatrix}\gamma_1\\ \gamma_2\end{pmatrix}$$

$$x_j \geq 0,\ j = 1, 2, 3, 4.$$

By using computer software of the MC^2-simplex method, we solve this problem and obtain the set of all potential solutions, denoted by $\varphi^* = \{K_1, K_2, K_3, K_4, K_5, K_6, K_7\}$ as in Table 5.3. By Definition 5.4, from the MC^2 -simplex tableau, we find that the expected total objective payoff $\mathbf{EV}(\varphi^*)$ is 10.92 (see Figure 5.1). From Table 5.3, we see that none of potential solution of φ^* are the MC^2-integer potential solutions by Definition 5.3 (i). For example, both x_1 and x_2 in K_2 are "MC^2-integer number 1's," but x_3 and x_4 are not. Thus, K_2 is not an MC^2 integer potential solution. To find all MC^2-integer potential solutions for the given problem, we have to go to the first iteration.

Steps for iteration 1:

Step 1. (Branching) We partition the set of all primal potential (feasible) solutions for the original problem into two subsets by setting $x_j = \gamma_1 + \gamma_2 = 1$ for one subset and $x_j = 0$ for the other subset. Then we have the following two subproblems:

Subproblem 1: the original problem plus an additional constraint $x_j = 0$; and

Subproblem 2: the original problem plus an additional constraint $x_j = \gamma_1 + \gamma_2 = 1$.

In Step 0, we form the relaxations of subproblems 1 and 2, respectively, by replacing $x_j = \gamma_1 + \gamma_2 = 1$ or 0 by $0 \le x_j \le \gamma_1 + \gamma_2 = 1$ or 0, for j = 2, 3, 4.

Step 2. (Bounding) We solve the relaxation of subproblem 1 and obtain the set of all potential solutions, denoted by $\varphi^{*^1} = \{K^1{}_1, K^1{}_2, K^1{}_3\}$ as in Table 5.4. The expected total objective payoff $\mathbf{EV}(\varphi^{*^1})$ is 10.86. Solving the relaxation of subproblem 2 results in $\varphi^{*^2} = \{K^2{}_1, K^2{}_2, K^2{}_3\}$ as in Table 5.5. The expected total objective payoff $\mathbf{EV}(\varphi^{*^2})$ is 7.78. Thus, the relaxations of these two subproblems yield the following results (see the solution tree in Figure 1):

The relaxation of subproblem 1: none of φ^{*^1} are the MC^2-integer potential solutions.

Bound for subproblem 1: $\mathbf{EV}(\varphi^{*^1}) \le 10.86$.

The relaxation of subproblem 2: none of φ^{*^2} are the MC^2-integer potential solutions.

Bound for subproblem 2: $\mathbf{EV}(\varphi^{*^2}) \le 7.78$.

Step 3. (Fathoming) Both Subproblems 1 and 2 fail all fathoming tests.

Step 4. (Optimality Test) Both Subproblems 1 and 2 do not provide "optimal integer solution." Thus we go to the second iteration.

Table 5.3 Potential Bases of Relaxation of the Original MC² BIP Problem

Bases	$\Gamma(K_j)$	$\Lambda(K_j)$	Value of $\mathbf{x}(K_j)$
$\mathbf{x}(K_1) =$ $(x_1, x_2, x_4, s_3, s_5, s_6)$	$1/2 \le \gamma_1 \le 5/6$	$0 < \lambda_1 \le 1/2$	$x_1 = 1.5\gamma_1 - 1.5\gamma_2$ $x_2 = \gamma_1 + \gamma_2$ $x_4 = .17\gamma_1 + 1.83\gamma_2$
$\mathbf{x}(K_2) =$ $(x_1, x_2, x_3, x_4, s_5, s_6)$	$1/2 \le \gamma_1 \le 5/6$	$1/2 \le \lambda_1 \le 1$	$x_1 = x_2 = \gamma_1 + \gamma_2$ $x_3 = -.5\gamma_1 + 2.5\gamma_2$ $x_4 = .67\gamma_1 - .67\gamma_2$
$\mathbf{x}(K_3) =$ $(x_1, x_2, x_3, x_4, s_3, s_6)$	$1/6 \le \gamma_1 \le 1/2$	$1/2 \le \lambda_1 \le 3/4$	$x_1 = 2.5\gamma_1 - .5\gamma_2$ $x_2 = x_3 = \gamma_1 + \gamma_2$ $x_4 = -.83\gamma_1 + .83\gamma_2$
$\mathbf{x}(K_4) =$ $(x_2, x_3, x_4, s_3, s_5, s_6)$	$1/6 \le \gamma_1 \le 1/2$	$0 < \lambda_1 \le 1/2$	$x_2 = \gamma_1 + \gamma_2$ $x_3 = -1.5\gamma_1 + 1.5\gamma_2$ $x_4 = 1.67\gamma_1 + .33\gamma_2$
$\mathbf{x}(K_5) =$ $(x_2, x_3, x_4, s_2, s_3, s_6)$	$0 < \gamma_1 \le 1/6$	$0 < \lambda_1 \le 3/4$	$x_2 = x_3 = \gamma_1 + \gamma_2$ $x_4 = .83\gamma_1 + .5\gamma_2$
$\mathbf{x}(K_6) =$ $(x_1, x_2, x_3, s_2, s_3, s_6)$	$0 < \gamma_1 \le 1/2$	$3/4 \le \lambda_1 \le 1$	$x_1 = 1.25\gamma_1 + .75\gamma_2$ $x_2 = x_3 = \gamma_1 + \gamma_2$
$\mathbf{x}(K_7) =$ $(x_1, x_2, x_4, s_1, s_5, s_6)$	$5/6 \le \gamma_1 \le 1$	$0 < \lambda_1 \le 1$	$x_1 = x_2 = \gamma_1 + \gamma_2$ $x_4 = .33\gamma_1 + \gamma_2$

Table 5.4 Potential Bases of Relaxation of Subproblem 1

Bases	$\Gamma(K^1_j)$	$\Lambda(K^1_j)$	Value of $\mathbf{x}(K^1_j)$
$\mathbf{x}(K^1_1) =$ $(x_2, x_3, x_4, s_4, s_5)$	$1/6 \le \gamma_1 \le 1/2$	$0 < \lambda_1 \le 1$	$x_2 = \gamma_1 + \gamma_2$ $x_3 = -1.5\gamma_1 + 1.5\gamma_2$ $x_4 = 1.67\gamma_1 + .33\gamma_2$
$\mathbf{x}(K^1_2) =$ $(x_2, x_3, x_4, s_2, s_6)$	$0 < \gamma_1 \le 1/6$	$0 < \lambda_1 \le 1$	$x_2 = x_3 = \gamma_1 + \gamma_2$ $x_4 = .83\gamma_1 + .5\gamma_2$
$\mathbf{x}(K^1_3) =$ $(x_2, x_4, s_1, s_4, s_5)$	$1/6 \le \gamma_1 \le 1/2$	$1/2 \le \lambda_1 \le 3/4$	$x_2 = \gamma_1 + \gamma_2$ $x_4 = .67\gamma_1 + 1.33\gamma_2$

Table 5.5 Potential Bases of Relaxation of Subproblem 2

Bases	$\Gamma(K^2_j)$	$\Lambda(K^2_j)$	Value of $\mathbf{x}(K^2_j)$
$\mathbf{x}(K^2_1) =$ $(x_2, x_3, x_4, s_4, s_5)$	$1/2 \le \gamma_1 \le 5/6$	$0 < \lambda_1 \le 1$	$x_2 = \gamma_1 + \gamma_2$ $x_3 = -.5\gamma_1 + 2.5\gamma_2$ $x_4 = .67\gamma_1 - .67\gamma_2$
$\mathbf{x}(K^2_2) =$ $(x_2, x_4, s_2, s_4, s_5)$	$0 < \gamma_1 \le 1/2$	$0 \le \lambda_1 \le 1$	$x_2 = \gamma_1 + \gamma_2$ $x_3 = 1.5\gamma_1 + .5\gamma_2$
$\mathbf{x}(K^2_3) =$ $(x_2, x_4, s_1, s_4, s_5)$	$5/6 \le \gamma_1 \le 1$	$0 < \lambda_1 \le 1$	$x_2 = \gamma_1 + \gamma_2$ $x_4 = .33\gamma_1 + \gamma_2$

For the sake of space, we skip the details of the third to seventh iterations. However, the interested reader can refer to the detail through Shi and Lee (1997). The results of all of eight iterations is represented by a tree structure in Figure 5.1. Now we conclude this example by the eighth iteration as follows:

Steps for iteration 8:

Step 1. (Branching) We partition the set of all primal potential solutions for subproblem 13 into two subsets by setting replacing $x_4 = \gamma_1 + \gamma_2 = 1$ for one subset and $x_4 = 0$ for the other subset. Thus, we have the following new subproblems:

Subproblem 15: subproblem 13 plus an additional constraint $x_4 = 0$; and

Subproblem 16: subproblem 13 plus an additional constraint $x_4 = \gamma_1 + \gamma_2 = 1$.

Step 2. (Bounding) The resulting branching variable x_4 is the last decision variable. Subproblem 15 has the corresponding $(\gamma_1 + \gamma_2, \gamma_1 + \gamma_2, 0, 0)$ and subproblem 16 has the corresponding $(\gamma_1 + \gamma_2, \gamma_1 + \gamma_2, 0, \gamma_1 + \gamma_2)$ as the values of the decision variables. There is no relaxation needed for these two subproblems. The results are (see Figure 1):

Subproblem 15: $(\gamma_1 + \gamma_2, \gamma_1 + \gamma_2, 0, 0)$ is an MC^2-integer potential solution with $\mathbf{EV}(\varphi^{*15}) = 6.5$.

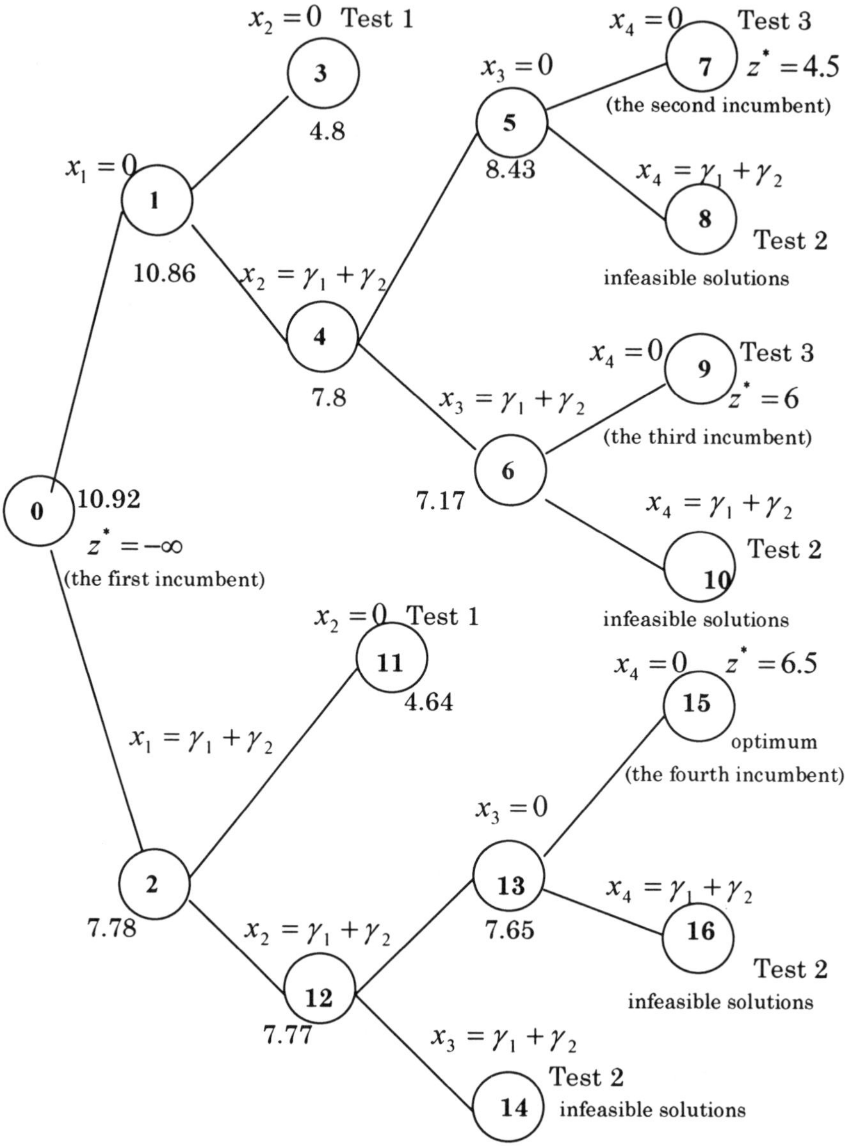

Figure 5.1 The Solution Tree for Example 5.2

Subproblem 16: $(\gamma_1 + \gamma_2, \gamma_1 + \gamma_2, 0, \gamma_1 + \gamma_2)$ is infeasible for some values of γ.

Step 3. (Fathoming) Subproblem 15 passes Test 3 and subproblem 16 passes Test 2. Because $\mathbf{EV}(\varphi^{*15}) = 6.5 > Z^* = 6$, 6.5 becomes the value for the new incumbent, i.e., $Z^* = 6.5$. Subproblem 16 is fathomed.

Step 4. (Optimality Test) Since there is no remaining (unfathomed) subproblem, we conclude that: $(\gamma_1 + \gamma_2, \gamma_1 + \gamma_2, 0, 0)$ is "optimal," and we have solved the original problem. Therefore, the conclusion is that we should select Cities 1 and 2 as the hub cities for building the rural telecommunication networks. The solution tree of the original MC^2 BIP problem is shown in Figure 5.1.

5.3 General MC² Integer Problem and A Branch-and-Partition Procedure

This section, based on Li and Shi (1999), first builds a general MC^2 integer linear programming (MC^2IP). Then, a branch-and-partition algorithm is created for identifying a set of integer potential solutions that covers the (γ, λ) space. This approach differs from the branch-and-bound for the MC^2BIP problem (M5.1). Given an MC^2IP problem, we use the MC^2-simplex method to find a set of potential solutions for its relaxation problem. For each solution with its range of parameter λ, we branch the decision variable $x_j(J, \gamma)$ and produce some subproblems. These subproblems are associated with a partition of the range of parameter γ, because the decision variable $x_j(J, \gamma)$ is a function of γ. Repeating such a way, a set of integer potential solutions will be identified for the (γ, λ) space.

Because some real-world decision problems require an integer solution (for example, purchasing a number of giant oil tankers or ocean cruisers), integer programming problems under a multi-criteria and multi-constraint level environment becomes necessary and important to be explored. For this purpose, we propose a general MC^2 integer linear programming (MC^2IP) model (M5.3) as follows:

$$\begin{aligned} &\textit{Max} && \lambda^t \mathbf{C}\,\mathbf{x} \\ &\textit{Subject to} && \mathbf{A}\,\mathbf{x} \le \mathbf{D}\gamma \\ & && \mathbf{x} \textit{ are } \text{nonnegative integers.} \end{aligned}$$

We see that model (M5.3) is defined on general integer, which contains binary (or 0-1) integers in Shi and Lee (1997)'s MC^2BIP formulation. We also note that in the branch-and-bound method for solving integer linear programming problems (Dakin 1965, Hu 1969, Lawler and Wood 1966, and Mitten 1970), once the branching is made by dividing the original feasible set into smaller subsets, each sub-problem needs to be solved to obtain the "bound." Such a bound can be determined by the optimal objective value of the sub-problem. The branch-and-bound algorithm of Shi and Lee (1997) for solving MC^2BIP is an application of LP branch-and-bound method, in which the bound is represented by the expected objective pay-off over a certain (γ, λ) range. Instead of using a "bound" to compute the sub-problems, this paper proposes an algorithm that first branches decision variables $x_j(J, \gamma)$ according to the parameter γ (since $x_j(J, \gamma)$ is a function of γ) on the known λ range, and then searches for an integer potential solution on each (γ, λ) partition. Because we identify a set of integer potential solutions over the "partitions" of the (γ, λ) space through a sequence without calculating a "bound", this method is called "a branch-and-partition" algorithm so as to distinguish from the known "branch-and-bound" method. In addition, the set of integer potential solutions identified by this method is not unique since the searching sequence can vary.

Definition 5.4. A given potential solution $\mathbf{x}(J, \gamma)$ is said to be an "MC^2-integer potential solution" if and only if all decision variables $x_j(J, \gamma)$ in $\mathbf{x}(J, \gamma)$ are integer. If $\mathbf{x}(J, \gamma)$ is an "MC^2-integer potential solution", then the corresponding J is called "MC^2-integer potential basis."

Definition 5.5. A problem corresponding to model (M4.1) is said to be the relaxation problem of model (M5.3).

Theorem 5.2. The set of all feasible solutions for an MC^2IP problem (M5.3) is a sub-set of all the feasible solutions for its relaxation problem (M4.1). Furthermore, if every potential basis for the relaxation problem (M4.1) is an MC^2-integer potential basis, then the set of all MC^2-integer potential bases for the relaxation problem (M4.1) is also the set of all MC^2-integer potential bases for the MC^2IP problem (M5.3).

Based on the above results, we propose the following algorithm to solve the MC^2IP problem. Its flow chart can be expressed as Figure 5.2.

Algorithm 5.2 (An MC^2IP Branch-and-partition Algorithm)**:**

Step 1. Ignoring the integer restrictions, solve the relaxation problem as an MC^2 linear programming in order to obtain its potential basic solutions.

Step 2. Select the first potential basic solution T_1 in nature order to solve the equations of all the basic decision variables for $x_j(T_1, \gamma) \in [I]$ (where [I] is a integer set) in $\Gamma(T_1) \times \Lambda(T_1)$. We can obtain a solution set $\Omega(\gamma)$ of the equations for all $x_j(T_1, \gamma)$. Set $\Gamma(T_1) = \Gamma(T_1) \backslash \Omega(\gamma)$.

Step 3. Select a basic decision variable x_j as the branching variable; take the solutions of the equation $x_j(T_1, \gamma) \in [I]$ in $\Gamma(T_1)$ as the interface to divide $\Gamma(T_1)$. The purpose of doing this step is to make sure x_j has one integer value at most in any sub-set of $\Gamma(T_1)$.

Step 4. Select a sub-set $\Gamma_1(T_1)$ to find the branching points b_1 and b_2. Where b_1 is the largest integer not exceeding any value of $x_j(T_1, \gamma)$ in $\Gamma_1(T_1)$, and b_2 is the next larger integer from its sibling b_1.

Step 5. Construct the descendant MC^2 linear programming problems and solve them with MC^2-simplex method. Each descendant has just one additional restriction $x_j \le b_1$ or $x_j \ge b_2$.

Step 6. Take the descendant MC^2 linear programming problems as the original problems repeating the steps from Step 2 to Step 5 until both of descendant problems have MC^2-integer potential solutions for any γ in $\Gamma_1(T_1)$.

Step 7. Rearrange the MC^2-integer potential solutions by taking the solution that has maximum objective value in any sub-set of $\Gamma_1(T_1)$.

Step 8. Select another sub-set of $\Gamma(T_1)$ repeating the steps from Step 4 to Step 7 until all the sub-sets of $\Gamma(T_1)$ have been handled.

Step 9. Select another potential solution of the relaxation problem and repeat Step 2 to Step 8 until all the potential solutions have been selected.

Remark 5.1 In general, the integer solutions of $x_j(T_1, \gamma)$ in Step 3 result from (p-1) dimensional equations about $\gamma_1, \gamma_2, \ldots, \gamma_p$ because of the normalization $\gamma_1 + \gamma_2 + \ldots + \gamma_p = 1$. In order to determine an interface of γ to partition $\Gamma_1(T_1)$ for Step 4, we have to know (p-2) components of γ for solving the (p-1) components.

The following example is used to illustrate how the above algorithm can be employed to solve the MC²IP problems.

Example 5.3. Given an MC²IP problem:

$$Max \begin{pmatrix} \lambda_1 & \lambda_2 \end{pmatrix} \begin{pmatrix} 2 & 5 \\ 4 & 2 \end{pmatrix} \begin{pmatrix} x_1 \\ x_2 \end{pmatrix}$$

$$Subject\ to \begin{pmatrix} 1 & 2 \\ 2 & 1 \end{pmatrix} \begin{pmatrix} x_1 \\ x_2 \end{pmatrix} \leq \begin{pmatrix} 4 & 3 \\ 2 & 5 \end{pmatrix} \begin{pmatrix} \gamma_1 \\ \gamma_2 \end{pmatrix}$$

$$\begin{pmatrix} x_1 \\ x_2 \end{pmatrix} \text{are nonnegative integers}$$

Using Algorithm 5.2, we solve the problem as follows:

Step 1. Ignoring the integer restrictions, solve the relaxation problem as an MC² linear programming. We can obtain the potential solutions that are shown in Table 5.6. We can express the parameter spaces corresponding to the potential solutions as Figure 5.3.

Step 2. Select the potential solution T_1 shown in Table 5.6.
For x_1: $7/3 - 7/3\gamma_1 \in [I]$, then $\gamma_1 = 1/7, 4/7$ and 1.
For x_2: $1/3 + 5/3\gamma_1 \in [I]$, then $\gamma_1 = 2/5$, and 1.

Table 5.6 Potential Bases of Relaxation of the Original MC^2 IP Problem

Bases	$\Gamma(T_j)$	$\Lambda(T_j)$	Value of $\mathbf{x}(T_j)$
$\mathbf{x}(T_1) = (x_1, x_2)$	$0 < \gamma_1 \le 1$	$0 < \lambda_1 \le 6/7$	$x_1 = 7/3 - 7/3\gamma_1$ $x_2 = 1/3 + 5/3\gamma_1$
$\mathbf{x}(T_2) = (x_2, s_2)$	$0 < \gamma_1 \le 1$	$6/7 \le \lambda_1 \le 1$	$x_2 = 3/2 + 1/2\gamma_1$ $s_2 = 7/2 - 7/2\gamma_1$

Because $\Omega(\gamma) = \{\gamma_1 = 1\} \neq \varnothing$, we get an MC^2-integer potential solution at $\gamma_1 = 1$, that is $\mathbf{x}^* = (0, 2)$ at $\gamma_1 = 1$, and $\lambda_1 \in (0, 6/7]$. Set $\Gamma(T_1) = \Gamma(T_1)\backslash\Omega(\gamma) = \{0 < \gamma_1 \le 1\}$ and go to Step 3.

Step 3. Select a basic decision variable x_2 as the branching variable and take $\gamma_1 = 2/5$ as the interface to divide $\Gamma(T_1)$. It can be divided into $\Gamma_1(T_1) = \{0 < \gamma_1 \le 2/5\}$ and $\Gamma_2(T_1) = \{2/5 < \gamma_1 < 1\}$.

Step 4. Select a sub-set $\Gamma_1(T_1)$ to find the branching points b_1 and b_2, where $b_1 = 0$, and $b_2 = 1$.

Step 5. Construct the descendant MC^2 linear programming problems MC^2-LP1 and MC^2-LP2 (they are shown as bellow). Solve MC^2-LP1 and MC^2-LP2 with MC^2-simplex method we can obtain the results that are shown in Table 5.7. Note that both $\Gamma_1(T_{11})\times\Lambda_1(T_{11})$ for MC^2-LP1 and $\Gamma_1(T_{11})\times\Lambda_1(T_{11})$ for MC^2-LP2 cover $\Gamma_1(T_1)\times\Lambda_1(T_1) = \{0 < \gamma_1 \le 2/5\}\times\{0 < \lambda_1 \le 6/7\}$.

Step 6. Select Table 5.7 for repeating the steps from Step 2 to Step 5, we can obtain following integer potential solutions:

$\mathbf{x}^* = (0, 2)$ at $\gamma_1 \in (0, 1/3)$ and $\lambda_1 \in (0, 6/7]$;
$\mathbf{x}^* = (2, 0)$ at $\gamma_1 = 1/3$ and $\lambda_1 \in (0, 6/7]$;
$\mathbf{x}^* = (1, 0)$ at $\gamma_1 \in (1/3, 2/5]$ and $\lambda_1 \in (0, 6/7]$.

Select Table 5.8 for repeating the steps from Step 2 to Step 5, we can obtain following integer potential solutions:

$\mathbf{x}^* = (1, 1)$ at $\gamma_1 = 0$ and $\lambda_1 \in (0, 6/7]$;
$\mathbf{x}^* = (1, 1)$ at $\gamma_1 \in (0, 2/5]$ and $\lambda_1 \in (0, 6/7]$.

Step 7. Rearrange the integer potential solutions and the result is shown in Figure 5.4.

MC²-LP1:

$$\text{Max} \quad \begin{pmatrix} \lambda_1 & \lambda_2 \end{pmatrix} \begin{pmatrix} 2 & 5 \\ 4 & 2 \end{pmatrix} \begin{pmatrix} x_1 \\ x_2 \end{pmatrix}$$

$$\text{Subject to} \begin{pmatrix} 1 & 2 \\ 2 & 1 \end{pmatrix} \begin{pmatrix} x_1 \\ x_2 \end{pmatrix} \le \begin{pmatrix} 4 & 3 \\ 2 & 5 \end{pmatrix} \begin{pmatrix} \gamma_1 \\ \gamma_2 \end{pmatrix}$$

$$x_2 \le 0 \, (\textit{Additional restriction})$$

$$\begin{pmatrix} x_1 \\ x_2 \end{pmatrix} \ge 0.$$

MC²-LP2:

$$\text{Max} \quad \begin{pmatrix} \lambda_1 & \lambda_2 \end{pmatrix} \begin{pmatrix} 2 & 5 \\ 4 & 2 \end{pmatrix} \begin{pmatrix} x_1 \\ x_2 \end{pmatrix}$$

$$\text{Subject to} \begin{pmatrix} 1 & 2 \\ 2 & 1 \end{pmatrix} \begin{pmatrix} x_1 \\ x_2 \end{pmatrix} \le \begin{pmatrix} 4 & 3 \\ 2 & 5 \end{pmatrix} \begin{pmatrix} \gamma_1 \\ \gamma_2 \end{pmatrix}$$

$$x_2 \ge 1 \, (\textit{Additional restriction})$$

$$\begin{pmatrix} x_1 \\ x_2 \end{pmatrix} \ge 0.$$

Table 5.7 Potential Bases from MC²–LP1 and MC²-LP2 for $\Gamma_1(T_1)$

Bases	$\Gamma_1(T_{1j})$	$\Lambda_1(T_{1j})$	Value of $\mathbf{x}(T_j)$
$\mathbf{x}(T_{11}) = (x_1, x_2, s_1)$	$0 < \gamma_1 \le 1$	$0 < \lambda_1 \le 1$	$x_2 = 0$ $x_1 = 5 - 3/\gamma_1$ $s_1 = 1 + 5\gamma_1$
$\mathbf{x}(T_{12}) = (x_1, x_2, s_2)$	$0 < \gamma_1 \le 2/5$	$0 \le \lambda_1 \le 6/7$	$x_2 = 1$ $x_1 = 1 + \gamma_1$ $s_2 = 2 - 5\gamma_1$

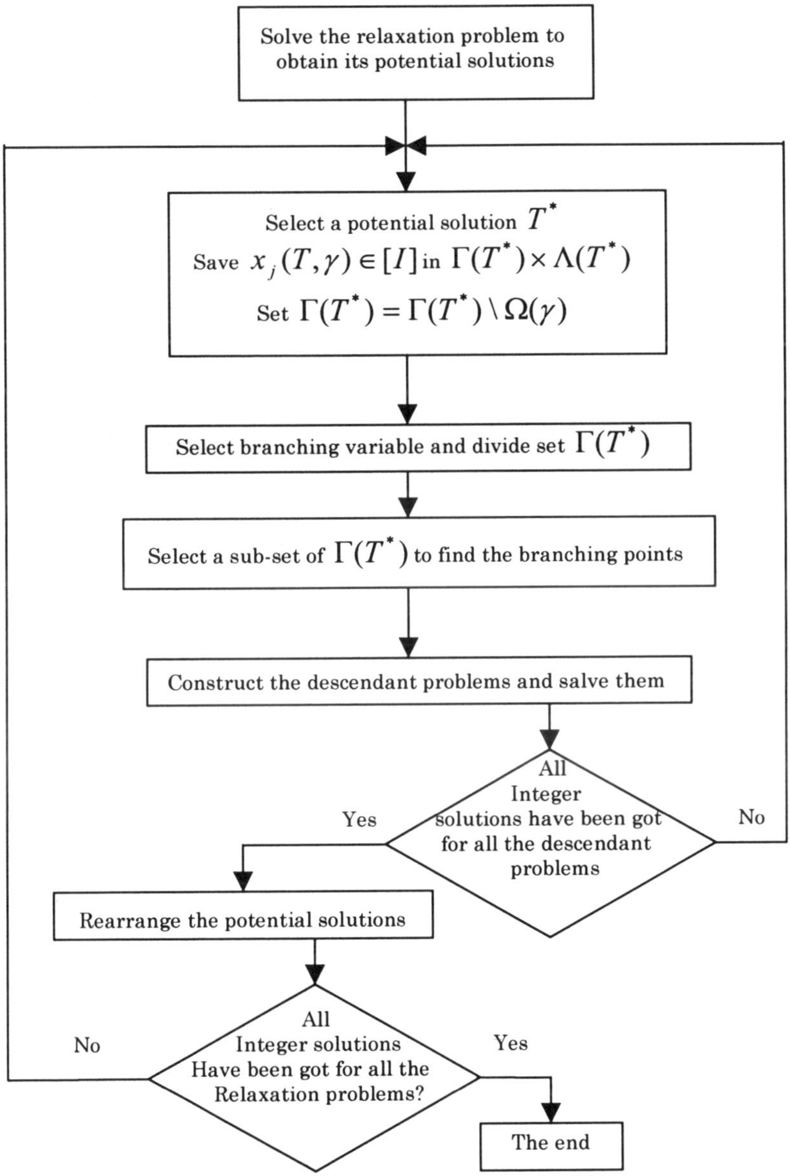

Figure 5.2 Flow Chart of Branch-and Partition Algorithm

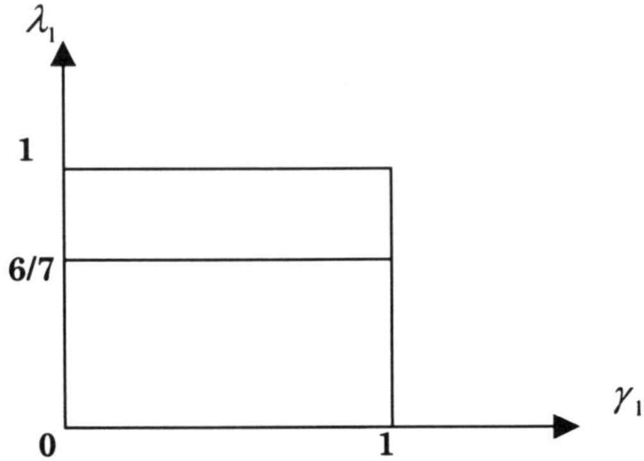

Figure 5.3 Parameter Spaces Corresponding to The Potential Solutions

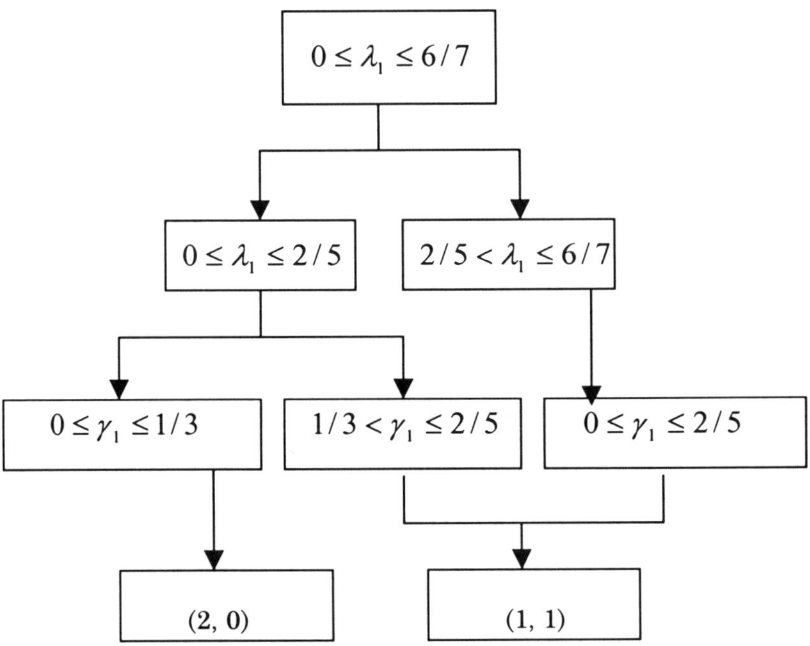

Figure 5. 4 Parameter Partition $\Gamma_1(T_1)$ for Some Integer Potential Solutions

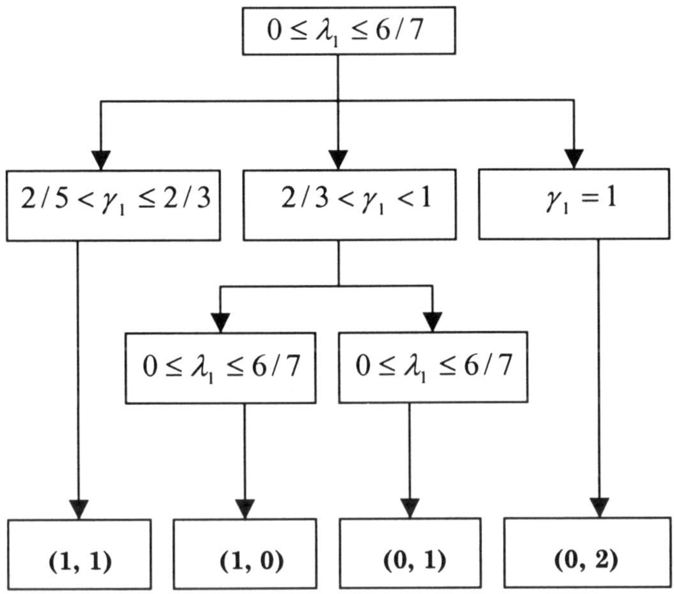

Figure 5.5 Parameter Partition $\Gamma_1(T_2)$ for Some Integer Potential Solutions

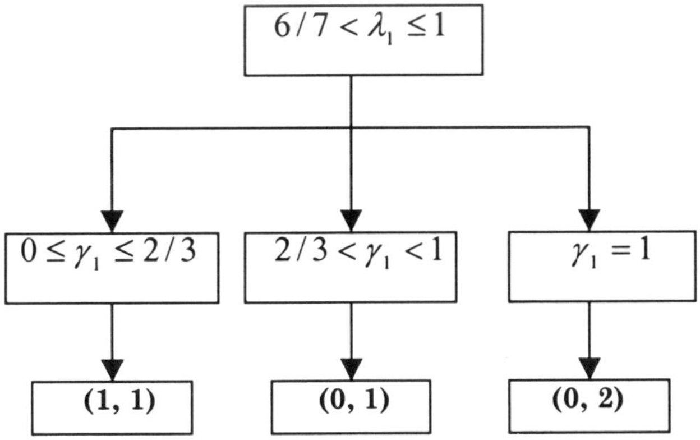

Figure 5.6 Parameter Partition $\Gamma_2(T_1)$ for Some Integer Potential Solutions

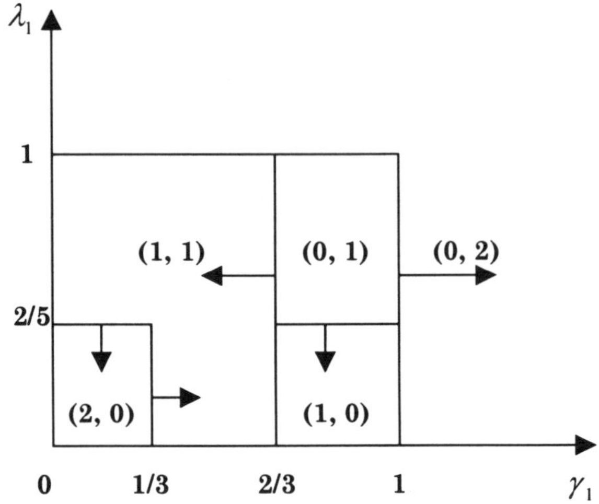

Figure 5.7 Integer Potential Solutions of Example 2

Step 8. Select another sub-set $\Gamma_2(T_1)$ for repeating the steps from Step 4 to Step 7. The result is shown in Figure 5.5.

Step 9. Select another potential solution of the original relaxation problem T_2 shown in Table 5.6 to repeat the steps from Step 2 to Step 8. The result is shown in Figure 5.6. Therefore, a set of five MC^2-integer potential bases (solutions) on the (γ, λ) space is identified as shown in Figure 5.7.

5.4 Computer Software for MC^2 Integer Problem

Hao and Shi (1997) developed a computer procedure for solving MC^2BIP problems. Instead of implementing Algorithm 5.1 completely, the computer code adopted a numeration method to solve binary integer linear problems with two criteria and two constraint levels. Similar to the large-scale MC^2 program of Hao and Shi (1996), the software consists of an input file, execution program, and output file. First, the user uses any format to create the input file for the data (**C**, **A**, **D**). Then, the user opens the execution program that was written by Visual C++5.0 in Window 98 or Window NT. Whenever, the input file is called by the execution program, the computing results of a set of MC^2-0-1 integer potential solutions is created in a output file. The output file includes all

(γ, λ) ranges, the objective payoff $V(J, \lambda, \gamma)$ and the basic variables $\mathbf{x}(J, \gamma)$. In addition, Fan and Shi (2000) used a branch-and-bound algorithm of linear programming to find an optimal solution of an MC^2IP problem with a fixed (γ, λ). Then it generates the integer potential solution by simulating the (γ, λ) ranges. These programs are part of the MC^2 software package attached in the back of the book.

We observe that the software of Hao and Shi (1996) can be applied to perform both Algorithms 5.1 and 5.2 for solving the relaxation problems. Therefore, we would be interested in further developing computer software that can systematically and effectively perform each step of Algorithms 5.1 and 5.2, respectively. In the software, we want to realize not only some functions such as decision variables branching, bounding and/or parameters partitioning, but also the post-optimality analysis under the circumstance of adding additional restriction. We are currently working on such a software development.

5.5 Further Remarks

In order to assess the performance of Algorithms 5.1 and 5.2, we have tested more than 28 MC^2BIP and MC^2IP problems with different number of criteria, constraint levels, variables and constraints, including the problem in Examples 5.2 and 5.3. When the criteria and constraint levels are two dimensions (like Example 5.1: $q = p = 2$), we can easily obtain the integer potential solutions with the different values of λ_1 and γ_1, where $0 \leq \lambda_1, \lambda_2 \leq 1$, $\lambda_1 + \lambda_2 = 1$ and $0 \leq \gamma_1, \gamma_2 \leq 1$, $\gamma_1 + \gamma_2 = 1$. When the criteria and constraint levels are more than two dimensions, we can obtain the integer potential solutions on different combination planes about (γ, λ) using normalized equations ($\gamma_1 + \gamma_2 + \ldots + \gamma_p = 1$ and $\lambda_1 + \lambda_2 + \ldots + \lambda_q = 1$). The normalized equations can decrease the dimensions of (γ, λ) from p and q to p-1 and q-1, respectively. Thus, if we want to obtain a definite answer we must know the information about (γ, λ) in p-2 and q-2 dimensions, respectively.

Integer programming, known as a broadly practical field, is an extension of linear programming with the additional restriction that the decision variable must be integers. The MC^2 model is proposed to deal with the problem involving fluctuations of resources in multiple criteria space. It is natural that both MC^2BIP and MC^2IP models, which are developed based on integer programming and MC^2 theory, will certainly empower the tools in the real-world problem solving.

Exercises

5.1 Solve the following MC^2 integer problem by using Algorithm 5.1:

Max	$10\,x_1 + 12\,x_2$
Max	$15\,x_1 + 8\,x_2$
Subject to	$3\,x_1 + 2\,x_2 \leq (24, -35)$
	$4\,x_1 + 6\,x_2 \leq (24, 42)$
	x_1, x_2 are 0 or 1.

5.2 Solve the following MC^2 integer problem by using Algorithm 5.1:

Max	$10\,x_1 + 12\,x_2 + 15\,x_3 + 11\,x_4$
Max	$15\,x_1 + 8\,x_2 + 20\,x_3 + 6\,x_4$
Subject to	$3\,x_1 + 2\,x_2 + 4\,x_3 + 6\,x_4 \leq (30, 50)$
	$4\,x_1 + 6\,x_2 + 15\,x_3 + 3\,x_4 \leq (90, 70)$
	x_1, x_2, x_3 and x_4 are 0 or 1.

5.3 Solve problems 5.1 and 5.2 by using the computer program of Hao and Shi (1997).

5.4 Solve the following MC^2 integer problem by using Algorithm 5.2:

Max	$10\,x_1 + 12\,x_2$
Max	$15\,x_1 + 8\,x_2$
Subject to	$3\,x_1 + 2\,x_2 \leq (24, -35)$
	$4\,x_1 + 6\,x_2 \leq (24, 42)$
	x_1, x_2 are integers.

5.5 Solve the following MC^2 integer problem by using Algorithm 5.2:

Max	$10\,x_1 + 12\,x_2 + 15\,x_3 + 11\,x_4$
Max	$15\,x_1 + 8\,x_2 + 20\,x_3 + 6\,x_4$
Subject to	$3\,x_1 + 2\,x_2 + 4\,x_3 + 6\,x_4 \leq (30, 50)$
	$4\,x_1 + 6\,x_2 + 15\,x_3 + 3\,x_4 \leq (90, 70)$
	x_1, x_2, x_3 and x_4 are integers.

5.6 Construct an MC^2 binary integer problem (e.g., two or more criteria and constraint levels) with 5 variables and 4 constraints. Then use the computer program of Hao and Shi (1997) to solve it.

5.7 Based on the data of the problem 1.17 (see Table 1.11 - 1.14), build and solve an MC^2 linear program with appropriate integer variables for the capital budgeting problem.

Chapter 6
MC^2 TRANSPORTATION PROBLEMS

The traditional transportation problem was initiated by Hichcock (1941) and Koopmans (1947). Dantzig (1951) applied his simplex method to solve the transportation problems. Charnes and Cooper (1954) developed the stepping stone method and Reinfeld and Vogel (1958) proposed the modified distribution method as an alternative way to find an optimal solution. The northwest corner rule is often used to find an initial basic feasible solution in the process of solving the transportation problems. Houthakker (1955), Vogel (1958), and Russell (1969) also developed methods of finding an initial basic feasible solution. The traditional transportation model, however, may have limitations in dealing with the real world problems because it has only a single criterion (objective) and a single supply and demand level. For example, a transportation problem may involve three objectives: the minimization of total energy cost consumed in transportation time, and the minimization of total product deterioration during transportation; and two supply and demand levels: the supply and demand quantities when the economy is in recession and the supply and demand quantities when the economy is booming. The traditional transportation model certainly fails to handle this case.

A transportation problem with two criteria was proposed by Aneja and Nair (1979). By using the techniques in multiple criteria (MC) linear programming (Zeleny 1974 and Yu 1985), they developed an algorithm to identify the set of nondominated extreme points in the criteria space. Other formulations and solution methods for MC transportation problems were studied by Lee and Moore (1973), Diaz (1979), Isermann (1979), Current and Min (1986), Ringuest and Rinks (1987) and the citations therein. But, all of those approaches have assumed there is only a fixed supply and demand level in the transportation problem. Thus, they cannot be used to deal with a transportation problem involving fluctuations of supply and demand.

This chapter is to propose two mathematical formulations of MC^2 transportation problems within the framework of MC^2 linear programming. One is called the steady MC^2 transportation problem and anther one is the dynamic MC^2 transportation problem. The steadiness means that the distribution or shipment of the amount of goods from a factory to a warehouse is to be done in a certain stage (season). The related optimal solution thus is identified for one steady stage. If the distribution crosses several stages, the problem is called a dynamic MC^2 transportation problem. Its optimal solution must be the best plan for all stages under consideration. In this chapter, we shall develop two different algorithms to solve such transportation problem, respectively.

In Section 6.1, we will discuss how to expand the known a steady MC transportation model to a steady MC^2 transportation model. The basic theoretical issues regarding the steady MC^2 transportation model will also be explored. In Section 6.2, an algorithm is developed to solve the steady MC^2 transportation model. In this algorithm, the traditional northwest corner rule is adopted to find an initial basic feasible solution for a given steady MC^2 transportation problem. Then the MC^2-simplex method is applied to locate the set of all potential solutions over possible changes of the criteria coefficient parameter and the supply and demand parameter for the MC^2 transportation problem. A numerical example is used to perform the algorithm. Section 6.3 describes a dynamic MC^2 transportation model that allows the distribution to be carried out through several stages. In Section 6.4, another algorithm that solves the dynamic MC^2 transportation problems is provided based on the modifications of the algorithm of the steady MC^2 transportation model. We will use an example with three stages to test this algorithm. Section 6.5 reviews two computer software developed to solve the steady MC^2 transportation model. Finally, some remarks are given in Section 6.6.

6.1 A Steady MC^2 Transportation Problem

Suppose the goods or products are transported from all factories to all warehouses in a given stage or season. Then, a steady MC^2 transportation problem (M6.1) can be proposed as follows (Shi 1995):

$$
\begin{aligned}
Min \quad & \sum_{i=1}^{m}\sum_{j=1}^{n} c_{ij}^{1} x_{ij} \\
& \sum_{i=1}^{m}\sum_{j=1}^{n} c_{ij}^{2} x_{ij} \\
& \vdots \\
& \sum_{i=1}^{m}\sum_{j=1}^{n} c_{ij}^{q} x_{ij} \\
Subject\ to \quad & \sum_{j=1}^{n} x_{ij} = (a_{i1}, \quad \cdots, \quad a_{ip}), \; i=1,\cdots,m, \\
& \sum_{i=1}^{m} x_{ij} = (b_{j1}, \quad \cdots, \quad b_{jp}), \; j=1,\cdots,n, \\
& x_{ij} \geq 0 \ for\ all\ (i, j),
\end{aligned}
$$

where c_{ij}^{k} = the "cost " of transporting a unit product from source i to destination j when the kth criterion is under consideration, k = 1, . . ., q;

a_{is} = the sth supply availability level at i, s = 1, . . ,p, which is assumed to be positive;

b_{js} = the sth demand availability level at j, s = 1, . . ,p, which is assumed to be positive;

x_{ij} = the amount of the products transported from i to j.

As mentioned before, the "cost" of transporting a unit product in problem (M6.1) could be the energy cost consumed in transportation, the transportation time, or the product deterioration during transportation. The supply and demand levels could be the desirability levels believed by the different decision makers or the economic situations. The interpretation of the criteria and supply and demand levels in the MC² transportation problem (M6.1) should depend on the features of the real world problems.

Mathematically speaking, problem (M6.1) is a special case of the MC² linear programming problem (M4.1) (because the maximizing objective is replaced by the minimizing objective and the "≤" constraint is changed to the "="constraint). Thus, all of the results presented in Chapter 4 for (M4.1) can be applied to problem (M6.1).

By putting the criteria parameter γ and the constraint level parameter λ on problem (M4.1), it becomes (M6.2):

$$Min \quad (\lambda_1, \ \cdots, \ \lambda_q) \begin{pmatrix} \sum_{i=1}^{m}\sum_{j=1}^{n} c_{ij}^{1} x_{ij} \\ \sum_{i=1}^{m}\sum_{j=1}^{n} c_{ij}^{2} x_{ij} \\ \vdots \\ \sum_{i=1}^{m}\sum_{j=1}^{n} c_{ij}^{q} x_{ij} \end{pmatrix}$$

$$Subject\ to \quad \sum_{j=1}^{n} x_{ij} = (a_{i1}, \ \cdots, \ a_{ip}) \begin{pmatrix} \gamma_1 \\ \vdots \\ \gamma_p \end{pmatrix}, i = 1, \cdots, m,$$

$$\sum_{i=1}^{m} x_{ij} = (b_{j1}, \ \cdots, \ b_{jp}) \begin{pmatrix} \gamma_1 \\ \vdots \\ \gamma_p \end{pmatrix}, j = 1, \cdots, n,$$

$$x_{ij} \geq 0 \ for\ all\ (i, j).$$

The MC² transportation problem (M6.2) is a parametric linear programming problem. If the values of (γ, λ) were known, it could be solved by a traditional transportation algorithm (Houthakker 1955). If γ was known, problem (M6.2) could be solved by some MC transportation algorithm (For example, Aneja and Nair 1979). However, when parameters (γ, λ) cannot be known ahead of decision time, these approaches would not be effective because there are infinitely many possible combinations of (γ, λ). In addition, the changes of γ may make the original choice infeasible, while that of λ can render the choice not optimal. With unknown (γ, λ), this paper applies the MC²-simplex method to develop an algorithm for solving problem (M6.2). In order to do so, some theoretical properties of problem (M6.2) are studied as follows.

Lemma 6.1 For any $\gamma^0 > 0$, $\sum_{i=1}^{m}\sum_{s=1}^{p} a_{is}\gamma_s^0 = \sum_{j=1}^{n}\sum_{s=1}^{p} b_{js}\gamma_s^0$ if and only if $\sum_{i=1}^{m} a_{is} = \sum_{j=1}^{n} b_{js}$ for any s = 1, . . ., p.

Proof: The proof of sufficiency is trivial. We only show the necessity:

Suppose that $\sum_{i=1}^{m} a_{is_0} \neq \sum_{j=1}^{n} b_{js_0}$, for some s_0. Without loss of generality, we assume that $\sum_{i=1}^{m} a_{is_0} > \sum_{j=1}^{n} b_{js_0}$. We choose $\gamma_s > 0$, such that γ_s approaches zero for all $s \neq s_0$. Then, we see that $\gamma_{s_0} = 1 - \sum_{s=s_0} \gamma_s$ approaches 1 as $\sum_{s=s_0} \gamma_s$ approaches zero. This implies that we can properly choose γ^0 so that $\sum_{i=1}^{m}\sum_{s=1}^{p} a_{is}\gamma_s^0 > \sum_{j=1}^{n}\sum_{s=1}^{p} b_{js}\gamma_s^0$.

Theorem 6.1 Given $\gamma^0 > 0$, the MC² transportation problem (M6.2) has a feasible solution at γ^0 if $\sum_{i=1}^{m} a_{is} = \sum_{j=1}^{n} b_{js}$, for any s = 1, . . ., p.

Proof: Given a $\gamma^0 > 0$, we suppose $\sum_{i=1}^{m} a_{is} = \sum_{j=1}^{n} b_{js}$, for any s = 1, . . ., p.

By Lemma 6.1, we have $\sum_{i=1}^{m}\sum_{s=1}^{p} a_{is}\gamma_s^0 = \sum_{j=1}^{n}\sum_{s=1}^{p} b_{js}\gamma_s^0$ = H, for any $\gamma_s^0 > 0$.

This implies the existence of a feasible solution, namely x_{ij} = $\left(\sum_{s=1}^{p} a_{is}\gamma_s^0\right)\left(\sum_{s=1}^{p} b_{js}\gamma_s^0\right)$/H, for all (i, j) (see page 321 of Gass 1985).

The condition, for any s = 1, . . ., p, of Theorem 6.1 is imposed in the following discussion to guarantee that problem (M6.2) has a feasible solution for a γ. This condition means that when the sth supply and demand level is under consideration, the total amount of products transported is equal to the total amount received.

With the above condition, it is observed that there are at most m+n-1 linearly independent constraints in problem (M6.2). This leads to the following property:

Theorem 6.2 Given a primal potential basis J for problem (M6.2), the basic variables **x**(*J*) contain at most m+n-1 components x_{ij}.

The proof of Theorem 6.2 can be readily shown by integrating the results in Section 4.2 and the corresponding theoretical result in the traditional transportation problem (Cooper and Steinberg 1974 and Gass 1985). If the number of the basic variables with positive in **x**(*T*) is exactly equal to m+n-1, then the solution *T* is called a *nondegenerate solution* for all $\gamma \in \Gamma(T)$. If the number of the basic variables with positive in **x**(*T*) is less than m+n-1, then the solution T is called a *degenerate solution* for all $\gamma \in \Gamma(T)$. In this chapter, the MC^2 transportation problems having nondegenerate solution are examined. It should be noted that the degenerate case can be similarly studied.

6.2 An Algorithm for the Steady MC² Transportation Model

This section develops an algorithm for solving the MC² transportation problem (M6.2) (or (M6.1)) based on the framework of the MC²-simplex method in Section 4.3.

Let $c_{ij} = (c_{ij}^1, \ldots, c_{ij}^q)^t$ be the vector of q criteria coefficients when x_{ij} is transported from supply source i to demand destination j. Let $u_i(\lambda)$ be the simplex multiplier (implicit price) corresponding to supply i, and $v_j(\lambda)$ be the simplex multiplier corresponding to demand j (Dantzig 1951, Hillier and Lieberman 1990). Both $u_i(\lambda)$ and $v_j(\lambda)$ are functions of in the situation. Apply (ii) of Definition 4.4, the dual parameter set $\Lambda(T)$ of a basis T for problem (M6.2) can be identified by:

Theorem 6.3 A basis T is a dual potential basis for problem (M6.2) if

(i) $u_i(\lambda) + v_j(\lambda) - \lambda^t c_{ij} = 0$, for every (i, j) such that x_{ij} is a basic variable; and

(ii) $u_i(\lambda) + v_j(\lambda) - \lambda^t c_{ij} \leq 0$, for every (i, j) such that x_{ij} is a nonbasic variable.

Note that (i) of Theorem 6.3 provides a way to calculate $u_i(\lambda)$ and $v_j(\lambda)$ corresponding to a basis T for problem (M6.2), while (ii) of Theorem 6.3, which is used to find the dual parameter set $\Lambda(T)$, yields a *potentiality test* for the basis T if the primal parameter set $\Gamma(T) \neq \varnothing$.

There are three important tasks for developing such an MC² algorithm: (i) find an initial basic feasible solution; (ii) select the "entering basic variable" and the "leaving basic variable" at an iteration where the solution is not a potential solution; and (iii) select the "entering basic variable" and the "leaving basic variable" at an iteration where the solution is a potential solution. The third task is not needed for the traditional transportation problem because the traditional algorithm usually terminates if an optimal solution is found. However, for a given MC² transportation problem (M6.2), this task is definitely needed to pivot from current potential solution to other adjacent potential solutions (see Chapter 8 of Yu 1985).

In this algorithm, for convenience, the traditional northwest corner rule (Charnes and Cooper 1954) is adopted to find an initial basic feasible solution for problem (M6.2). Other alternative methods, such as Vogel's approximation method and Russell's approximation method could also be used to find a better initial basic feasible solution for problem (M6.2). With the initial basic feasible solution, the MC²-simplex method then can be applied to locate the set of all potential solutions over possible changes of (γ, λ) for problem (M6.2). According to Section 4.2, a variable x_{ij} for problem (M6.2) is a function of γ and the objective coefficient of the variable is a function of λ. Thus, finding a way for determining the entering basic variable and the leaving basic variable at each iteration requires an estimate of the "upper bound" of the value of the chosen variable or the value of the objective coefficient (If the MC² transportation problem has a maximizing criterion, the "lower bounds" of the values of the objective coefficients should be considered). The following lemma can be used for this purpose.

Lemma 6.2 (i) $-\sqrt{p\left(\sum_{s=1}^{p} d_s^2\right)} \le \sum_{s=1}^{p} d_s \gamma_s \le \sqrt{p\left(\sum_{s=1}^{p} d_s^2\right)}$,

where d_s is a_{is} or b_{js} for all (i, j) and $\gamma = (\gamma_1, \cdots, \gamma_p)$ is the supply and demand parameters;

(ii) $-\sqrt{q\left(\sum_{k=1}^{q} (c_{ij}^k)^2\right)} \le \sum_{k=1}^{q} \lambda_k c_{ij}^k \le q\left(\sum_{k=1}^{q} (c_{ij}^k)^2\right)$, for all (i, j) where $\lambda = (\lambda_1, \cdots, \lambda_q)$ is the objective parameter.

Proof: (i) By Chauchy inequality,

$$\left(\sum_{s=1}^{p} d_s \gamma_s\right)^2 \le \left(\sum_{s=1}^{p} d_s^2\right)\left(\sum_{s=1}^{p} \gamma_s^2\right).$$

Since $0 \le \gamma_s \le 1$, $0 \le \gamma_s^2 \le 1$. Then, $\sum_{s=1}^{p} \gamma_s^2 \le \sum_{s=1}^{p} 1 = p$. Thus,

$\left(\sum_{s=1}^{p} d_s \gamma_s\right)^2 \le \left(\sum_{s=1}^{p} d_s^2\right)\left(\sum_{s=1}^{p} \gamma_s^2\right) \le p\left(\sum_{s=1}^{p} d_s^2\right)$. Taking the square root on both sides of $\left(\sum_{s=1}^{p} d_s \gamma_s\right)^2 \le p\left(\sum_{s=1}^{p} d_s^2\right)$, (i) is derived.

Similarly, the proof of (ii) can be shown.

Let $\psi = \{T_1, \ldots, T_k\}$ be the set of *all* potential solutions over all possible changes of (γ, λ) for problem (M6.2). The following lemma allows one to choose an entering variable for pivoting from a known potential solution T_i to an adjacent feasible (primal potential) or potential solution T_j.

Lemma 6.3 Given a potential solution T, its nonbasic variable x_{ij} is an entering variable for the next pivoting if there is a $\lambda \in \Lambda(T)$ so that $u_i(\lambda) + v_j(\lambda) - \lambda^t c_{ij} = 0$.

Lemma 6.3 is a special case of the definition of the "effective constraint" in Definition 4.6 for the MC² linear programming problem. By integrating the above discussion, an MC² algorithm for solving the steady MC² transportation problems is proposed as follows:

Algorithm 6.1

Step 0 (*Initialization Step*). Set $\psi^0 = \varnothing$. Find an initial basic feasible (primal potential) solution by using the northwest corner rule Charnes and Cooper (1954). This means that select x_{11}, which corresponds to the *northwest corner* (allocation cell) (1,1) of the MC² transportation simplex tableau (see Example 6.1), as the first basic variable. Either the p availability levels of supply 1 are exhausted simultaneously or demand 1 is completely satisfied over the p requirement levels. Note that the sth availability level of supply 1 is provided to only the sth requirement level of demand 1, s = 1, . . ., p. If the p availability levels of supply 1 are exhausted, then x_{21} will be processed; otherwise, if the p requirement levels of demand 1 are satisfied, then x_{12} will be chosen. Therefore, if x_{ij}

is the last selected variable, next select $x_{i+1,j}$ if the p availability levels of supply i are exhausted. Otherwise, next select $x_{i,j+1}$. In such a way, an initial basic feasible solution can be found. In the case of the nondegenerate solution, the number of the basic variables with positive values is exactly equal to m+n-1. Let K be the initial basic feasible solution (basis). By using Definition 4.4, find the primal parameter set $\Gamma(K)$ and the dual parameter set $\Lambda(K)$. Calculate $u_i(\lambda)$ and $v_j(\lambda)$ by setting its $u_1(\lambda) = 0$. Then, solve the set of equations $u_i(\lambda) + v_j(\lambda) = \lambda^t c_{ij}$, for every (i, j) such that x_{ij} is the basic variable (see (i) of Theorem 6.3). Go to the steps of the following iterations.

Steps of the ith iteration:

Step 1. By using (iii) of Theorem 4.1, if a basis K is not a potential solution, then go to Step 2. If a basis K is a potential solution, then set $K = T_j$, $\psi^j = \psi^{j-1} \cup \{T_j\}$. For every (γ, λ), if there is a T of ψ^j such that $(\gamma, \lambda) \in \Gamma(J) \times \Lambda(J)$, then $\psi^j = \psi$ and go to Step 7; otherwise, go to Step 3.

Step 2. When a basis K is not a potential solution, determine the entering basic variable: Select the nonbasic variable x_{ij} having the *largest upper bound* of $u_i(\lambda) + v_j(\lambda) - \lambda^t c_{ij} \geq 0$ with $\lambda_1 + \lambda_2 + \ldots + \lambda_q = 1$ for $0 \leq \lambda_k \leq 1$, k = 1, . . ., q, by (ii) of Lemma 6.2.

Step 3. When a basis K is a potential solution, select the nonbasic variable x_{ij} which satisfies Lemma 6.3 as the entering basic variable. If there are several x_{ij} satisfying Lemma 6.3, choose one arbitrarily.

Step 4. Given a basis K, determine the leaving variable: Identify the loop formed by the nonbasic cell (i, j) corresponding to the entering basic variable x_{ij}, and the basic cells in the MC^2 transportation simplex tableau. Number the cells in this loop consecutively: starting with the (i, j), set "+" on the odd-numbered cells, and "-" on the even-numbered cells. Among the cells with "-", select the basic variable having the *smallest upper bound* of the value by using (i) of Lemma 6.2.

Step 5. For an adjacent basis Q of K, determine the new basic variables: Add the values of the leaving basic variable to the allocation cell with "+"

in the loop, respectively, for the p supply or demand levels. Subtract these values from the allocation cell with "-" in the loop, respectively, for the p supply or demand levels. The result will be a new MC^2 transportation simplex tableau for the basis Q.

Step 6. For the basis Q, first find the primal parameter set $\Gamma(Q) \neq \varnothing$ by using (i) of Definition 4.4. Second calculate $u_i(\lambda)$ and $v_j(\lambda)$ by selecting the row having the largest number of allocations and setting its $u_i(\lambda) = 0$. The, solve the set of equations $u_i(\lambda) + v_j(\lambda) = \lambda^t c_{ij}$, for all (i, j) such that x_{ij} is a basic variable (see (i) of Theorem 6.3). If $u_i(\lambda) + v_j(\lambda) - \lambda^t c_{ij} \leq 0$, for every (i, j) such that x_{ij} is a nonbasic variable (see (ii) of Theorem 6.3), then Q is a dual potential solution; that is, $\Lambda(Q) \neq \varnothing$, according to (ii) of Theorem 4.1. Go back to Step 1 of the i+1th iteration.

Step 7. The algorithm terminates since ψ^j is the set of all potential solutions over all possible changes of (γ, λ) for problem (M6.2).

The following example adopted from Reinfeld and Vogel (1958) is used to show how the above algorithm can be employed to solve the steady MC^2 transportation problems.

Example 6.1 Consider a transportation problem involving three factories, denoted by F_1, F_2, and F_3, and four warehouses, denoted by W_1, W_2, W_3, and W_4. A particular product is transported from the ith factory to the jth warehouse, i = 1, 2, 3, and j = 1, 2, 3, 4. Both the supply level and demand level vary with the economic situations. Suppose that if the economy is in recession, the supply capacity level of F_1 is 30 tons, F_2 is 40 tons, and F_3 is 53 tons; while the demand requirement level of W_1 is 22 tons, W_2 is 35 tons, W_3 is 25 tons, and W_4 is 41 tons. If the economy is booming, the supply capacity level of F_1 is 20 tons, F_2 is 55 tons, and F_3 is 60 tons; while the demand requirement level of W_1 is 30 tons, W_2 is 30 tons, W_3 is 22 tons, and W_4 is 53 tons. Assume there are two criteria under consideration: (i) the minimization of total energy cost consumed in transportation, and (ii) the minimization of total product deterioration during transportation. The energy cost for transporting one ton products and the deterioration percentage of one ton products during transportation are given in Table 6.1 and Table 6.2, respectively. The goal is to locate the set of all potential solutions for the above transportation problem.

Table 6.1 The Energy Cost Consumed for Transporting One Ton Products ($10 per unit cost)

	W_1	W_2	W_3	W_4
F_1	23	27	16	18
F_2	12	17	20	51
F_3	22	28	12	32

Table 6.2 The Deterioration Percentage of One Ton Products During Transportation

	W_1	W_2	W_3	W_4
F_1	4	2	3	1
F_2	2	3	4	2
F_3	1	5	2	3

Note that in the above problem when the sth supply and demand level is under consideration, s = 1, 2, the total amount of products transported is equal to the total amount received. That is, when the economy is in recession, 30 +40 + 53 = 123 = 22 +35 + 25 +41; and when the economy is booming, 20 +55 + 60 = 135 = 30 +30 + 22 +53. By Theorem 3.1, the problem has a feasible solution. Let $\gamma = (\gamma_1 \quad \gamma_2)$ be the supply and demand parameter over the economic situations such that $0 < \gamma_1, \gamma_2 \leq 1$ and $\gamma_1 + \gamma_2 = 1$, and $\lambda = (\lambda_1 \quad \lambda_2)$ be the criteria parameter over two criteria such that $0 < \lambda_1, \lambda_2 \leq 1$ and $\lambda_1 + \lambda_2 = 1$. Let ψ be the set of all potential solutions over all possible changes of (γ, λ) for the transportation problem. In the following, Algorithm 6.1 is used to locate ψ.

Step 0 (Initialization Step). Set $\psi^0 = \varnothing$. The initial MC² transportation simplex tableau of the problem is presented by Table 6.3. In the allocation cell (i, j) of Table 6.3, the figure in the left upper small square is the coefficient of the first criterion and the figure in the right upper small square is the coefficient of the second criterion for the variable x_{ij}. The circled figures of Table 6.3 are related allocations. For the northwest corner (allocation cell) of Table 6.3, (1, 1), when the economy is in recession, 22 tons are transported to W_1 from F_1; and when the economy is booming, 30 tons are transported to W_1 from F_1. Since W_1 is satisfied, the allocation cell (1, 2) proceeds. In (1, 2), when the economy is in recession, 8 tons are transported to W_2 from F_1; and when the economy is booming, -10 tons are transported to W_2 from F_1, which means that F_1 owes 10 tons to W_2 because in the cell (1, 1) 10 more tons are transported from F_1 to W_1. Since F_1 is exhausted, the allocation cell (2, 2) proceeds. In (2, 2), when the economy is in recession, 27 tons are transported to W_2 from F_2; and when the economy is booming, 40 tons are transported to W_2 from F_2. This leads to proceed the allocation cell (2, 3), because W_2 is satisfied. In (2, 3), when the economy is in recession, 13 tons are transported to W_3 from F_2; and when the economy is booming, 15 tons are transported to W_3 from F_2. Because F_2 is exhausted, the allocation cell (3, 3) proceeds. In (3, 3), when the economy is in recession, 12 tons are transported to W_3 from F_3; and when the economy is booming, 7 tons are transported to W_3 from F_3. Now, the allocation cell (3, 4) proceeds, because W_3 is satisfied. In (3, 4), when the economy is in recession, 41 tons are transported to W_4 from F_3; and when the economy is booming, 53 tons are transported to W_4 from F_3. Let K_1 be the resulting initial feasible solution (basis). Then, $\mathbf{x}(K_1)$ = $(x_{11}, x_{12}, x_{22}, x_{23}, x_{33}, x_{34})$, where the number of basic variable is 6 (= 3 + 4 - 1). The values of x_{ij} in $\mathbf{x}(K_1)$, the primal parameter set $\Gamma(K_1)$ and the dual parameter set $\Lambda(K_1)$ are given in Table 6.4. Since the basic variables x_{ij} of K_1 satisfy (i) of Theorem 6.3, the following is obtained:

For x_{11}: $\lambda^t c_{11} = 23\lambda_1 + 4\lambda_2 = u_1(\lambda) + v_1(\lambda)$.

For x_{12}: $\lambda^t c_{12} = 27\lambda_1 + 2\lambda_2 = u_1(\lambda) + v_2(\lambda)$.

For x_{22}: $\lambda^t c_{22} = 17\lambda_1 + 3\lambda_2 = u_1(\lambda) + v_2(\lambda)$.

For x_{23}: $\lambda^t c_{23} = 20\lambda_1 + 4\lambda_2 = u_2(\lambda) + v_3(\lambda)$.

For x_{33}: $\lambda^t c_{33} = 12\lambda_1 + 2\lambda_2 = u_3(\lambda) + v_3(\lambda)$.

For x_{34}: $\lambda^t c_{34} = 32\lambda_1 + 3\lambda_2 = u_3(\lambda) + v_4(\lambda)$.

Set $u_1(\lambda) = 0$, then $v_1(\lambda) = 23\lambda_1 + 4\lambda_2$, $v_2(\lambda) = 27\lambda_1 + 2\lambda_2$, $u_2(\lambda) = -10\lambda_1 + \lambda_2$, $v_3(\lambda) = 30\lambda_1 + 3\lambda_2$, $u_3(\lambda) = -18\lambda_1 - \lambda_2$, and $v_4(\lambda) = 50\lambda_1 + 4\lambda_2$.
Go to the steps of the first iteration.

Table 6.3 The Initial MC^2 Transportation Simplex Tableau of K_1

	W_1		W_2		W_3		W_4		a_{11}	a_{11}
F_1	23	4	27	2	16	3	18	1	30 +	20
	(22)	(30)	(8)	(10)						
F_2	12	2	17	3	20	4	51	2	40	55
		+	(27)	(40)	(13)	(15)				
F_3	22	1	28	5	12	2	32	3	53	60
				+	(12)	(7)	(41)	(53)	–	
b_{11}	22		35		25		41		123	
b_{12}	30		30		22		53			135

Steps of the first iteration:

Step 1: By (iii) of Theorem 4.1, the basis K_1 is not a potential solution since $\Lambda(K_1) = \varnothing$. Go to Step 2.

Step 2: Determine the entering baisc variable corresponding to K_1 based on the results of Step 0:

By (ii) of Lemma 6.2, the corresponding nonbasic variable x_{ij} are checked and the upper bounds of $u_i(\lambda) + v_j(\lambda) - \lambda^t c_{ij} \geq 0$ with $0 \leq \lambda_1, \lambda_2 \leq 1$ and $\lambda_1 + \lambda_2 = 1$ are obtained as follows.

Table 6.4 The Initial Feasible (Primal Potential) Solution of K_1

$\mathbf{x}(K_1)$	$\Gamma(K_1)$	$\Lambda(K_1)$	Value of x_{ij} (K_1)
$(x_{11}, x_{12}, x_{22}, x_{23}, x_{33}, x_{34})$	$5/9 \le \gamma_1 \le 1$	$\varnothing$	$x_{11}= 22\gamma_1 + 30\gamma_2$ $x_{12}= 8\gamma_1 + 10\gamma_2$ $x_{22}= 27\gamma_1 + 40\gamma_2$ $x_{23}= 13\gamma_1 + 15\gamma_2$ $x_{33}= 12\gamma_1 + 7\gamma_2$ $x_{34}= 41\gamma_1 + 53\gamma_2$

For x_{13}: $u_1(\lambda) + v_3(\lambda) - \lambda^t c_{13} = 14\lambda_1 + 2\lambda_2 \le 20.08$;

For x_{14}: $u_1(\lambda) + v_4(\lambda) - \lambda^t c_{14} = 32\lambda_1 + 3\lambda_2 \le 45.64$; and

For x_{21}: $u_2(\lambda) + v_4(\lambda) - \lambda^t c_{21} = \lambda_1 + 3\lambda_2 \le 4.49$ ($u_i(\lambda)$ and $v_j(\lambda)$ are referred to Step 0).

Therefore, select x_{14} as the entering variable since its upper bound 45.64 is the largest one. Go to Step 4.

Step 4: Determine the leaving variable corresponding to K_1 based on Table 6.3:

Select a loop formed by the cells: (1, 4), (1, 2), (2, 2), (2, 3) (3, 3), and (3, 4), corresponding to the variables x_{14}, x_{12}, x_{22}, x_{23}, x_{33}, and x_{34}, respectively (see Table 6.3). Note that x_{14} is the entering variable and x_{12}, x_{22}, x_{23}, x_{33}, and x_{34} are the basic variables in Table 6.3. Then, number (1, 4) as "+", (1, 2) as "-", (2, 2) as "+", (2, 3) as "-", (3, 3) as "+", and (3, 4) as "-". From the cells with "-", the upper bounds of the value of the basic variables are obtained by using Lemma 6.2 (see Table 6.4):

$x_{12} = 8\gamma_1 - 10\gamma_2 \le 18.19$; $x_{23} = 13\gamma_1 + 15\gamma_2 \le 28.19$; and $x_{34} = 41\gamma_1 + 53\gamma_2 \le 95.15$.

Select x_{12} as the leaving variable since its upper bound 18.19 is the smallest one. Go to Step 5.

Step 5: For an adjacent basis K_2 of K_1, determine the new basic variables: Add $8\gamma_1 - 10\gamma_2$, the value of the leaving basic variable x_{12}, to the allocations of $(1, 4)^+$, $(2, 2)^+$, $(3, 3)^+$ in the loop, respectively. Then, subtract $8\gamma_1 - 10\gamma_2$ from the allocations of $(1, 2)^-$, $(2, 3)^-$, and $(3, 4)^-$ in the loop, respectively. This results in the MC² transportation simplex tableau for the basis K_2.

Step 6: For the basis K_2, $\mathbf{x}(K_2) = (x_{11}, x_{14}, x_{22}, x_{23}, x_{33}, x_{34})$, the primal parameter set $\Gamma(K_1) \neq \emptyset$, and the dual parameter set $\Lambda(K_1) = \emptyset$ can be identified (see Shi 1992). Since the basic variables x_{ij} of K_2 satisfy (i) of Theorem 6.3, the following are obtained:

For x_{11}: $\lambda^t c_{11} = 23\lambda_1 + 4\lambda_2 = u_{11}(\lambda) + v_1(\lambda)$.

For x_{14}: $\lambda^t c_{14} = 18\lambda_1 + \lambda_2 = u_1(\lambda) + v_4(\lambda)$.

For x_{22}: $\lambda^t c_{22} = 17\lambda_1 + 3\lambda_2 = u_2(\lambda) + v_2(\lambda)$.

For x_{23}: $\lambda^t c_{23} = 20\lambda_1 + 4\lambda_2 = u_2(\lambda) + v_3(\lambda)$.

For x_{33}: $\lambda^t c_{33} = 12\lambda_1 + 2\lambda_2 = u_3(\lambda) + v_3(\lambda)$.

For x_{34}: $\lambda^t c_{34} = 32\lambda_1 + 3\lambda_2 = u_3(\lambda) + v_4(\lambda)$.

Set $u_1(\lambda) = 0$, then $v_1(\lambda) = 23\lambda_1 + 4\lambda_2$, $v_4(\lambda) = 18\lambda_1 + \lambda_2$, $u_3(\lambda) = 14\lambda_1 + 2\lambda_2$, $v_3(\lambda) = -2\lambda_1$, $u_2(\lambda) = 22\lambda_1 + 4\lambda_2$, and $v_2(\lambda) = -5\lambda_1 - \lambda_2$.

Go to the steps of the second iteration. Through the second and third iterations, we can identify those two adjacent bases $\{K_3, K_4\}$. Here, K_3 is a primal potential solution; while K_4 is a potential solution with $\mathbf{x}(K_4) = (x_{14}, x_{21}, x_{22}, x_{31}, x_{33}, x_{34})$, $\Gamma(K_4) = \{\gamma_1 > 0 \mid 0 < \gamma_1 \le 1, \gamma_1 + \gamma_2 = 1\}$, $\Lambda(K_4) = \{\lambda_1 > 0 \mid 2/31 \le \lambda_1 \le 1, \lambda_1 + \lambda_2 = 1\}$. The objective payoff of K_4 is:

$$\lambda' C(K_4)x(K_4) = (\lambda_1, \ \lambda_2)\begin{pmatrix} 18 & 12 & 17 & 22 & 12 & 32 \\ 1 & 2 & 3 & 1 & 2 & 3 \end{pmatrix}\begin{pmatrix} 30 & 20 \\ 5 & 25 \\ 35 & 30 \\ 17 & 5 \\ 25 & 22 \\ 11 & 33 \end{pmatrix}\begin{pmatrix} \gamma_1 \\ 1-\gamma_1 \end{pmatrix}$$

$$= (\lambda_1, \ 1-\lambda_1)\begin{pmatrix} 2221 & 2600 \\ 245 & 308 \end{pmatrix}\begin{pmatrix} \gamma_1 \\ 1-\gamma_1 \end{pmatrix}$$

which is a function of $(\gamma_1, \lambda_1) \in \Gamma(K_4) \times \Lambda(K_4)$. The $u_i(\lambda)$ and $v_j(\lambda)$ are obtained:

For x_{14} : $\lambda^t c_{14} = 18\lambda_1 + \lambda_2 = u_1(\lambda) + v_4(\lambda)$.

For x_{21} : $\lambda^t c_{21} = 12\lambda_1 + 2\lambda_2 = u_2(\lambda) + v_1(\lambda)$.

For x_{22} : $\lambda^t c_{22} = 17\lambda_1 + 3\lambda_2 = u_2(\lambda) + v_2(\lambda)$.

For x_{31} : $\lambda^t c_{31} = 22\lambda_1 + \lambda_2 = u_3(\lambda) + v_1(\lambda)$.

For x_{33} : $\lambda^t c_{33} = 12\lambda_1 + 2\lambda_2 = u_3(\lambda) + v_3(\lambda)$.

For x_{34} : $\lambda^t c_{34} = 32\lambda_1 + 3\lambda_2 = u_3(\lambda) + v_4(\lambda)$.

Set $u_3(\lambda) = 0$ (corresponding to the row having the largest number of allocations in the simplex tableau of K_4), then $v_4(\lambda) = 32\lambda_1 + 3\lambda_2$, $v_3(\lambda) = 12\lambda_1 + 2\lambda_2$, $v_1(\lambda) = 22\lambda_1 + \lambda_2$, $u_1(\lambda) = -14\lambda_1 - 2\lambda_2$, $u_2(\lambda) = -10\lambda_1 + \lambda_2$, and $v_2(\lambda) = 27\lambda_1 + 2\lambda_2$.

For the sake of space, the details of the process are not elaborated here, but can be found in Shi (1992). In the following, we show the fourth and further iterations:

Steps of the fourth iteration:

Step 1: Since K_4 is a potential solution, set $K_4 = T_1$ and $\psi^1 = \psi^0 \cup \{T_1\} = \{T_1\}$. Note that there exists some $(\gamma_1, \lambda_1) \notin \Gamma(T_1)\times\Lambda(T_1)$. Go to Step 3.

Step 3: From the simplex tableau of K_4, check the nonbasic variable x_{ij} for $u_i(\lambda) + v_j(\lambda) - \lambda c_{ij} = 0$ with $\lambda_1 + \lambda_2 = 1$ for $0 < \lambda_k \le 1$, k = 1, 2:

For x_{11}: $u_1(\lambda) + v_1(\lambda) - \lambda^t c_{11} = 15\lambda_1 + 5\lambda_2 = 0$; then $\lambda_1 = -1/2$.

For x_{12}: $u_1(\lambda) + v_2(\lambda) - \lambda^t c_{12} = 14\lambda_1 + 2\lambda_2 = 0$; then $\lambda_1 = -1/6$.

For x_{13}: $u_1(\lambda) + v_3(\lambda) - \lambda^t c_{13} = 18\lambda_1 + 3\lambda_2 = 0$; then $\lambda_1 = -1/5$.

For x_{23}: $u_2(\lambda) + v_3(\lambda) - \lambda^t c_{23} = 18\lambda_1 + \lambda_2 = 0$; then $\lambda_1 = -1/17$.

For x_{24}: $u_2(\lambda) + v_4(\lambda) - \lambda^t c_{24} = 29\lambda_1 - 2\lambda_2 = 0$; then $\lambda_1 = 2/31$.

For x_{32}: $u_3(\lambda) + v_2(\lambda) - \lambda^t c_{32} = \lambda_1 + 3\lambda_2 = 0$; then $\lambda_1 = 2/3$.

Thus, both x_{24} and x_{32} satisfy Lemma 6.3. Choose x_{24} arbitrarily as entering variable. Go to Step 4.

Step 4: Determine the leaving variable corresponding to K_4 based on its simplex tableau:

Select a loop formed by the cells: $(2, 4)^+$, $(2, 1)^-$, $(3, 1)^+$, and $(3, 4)^-$, corresponding to the nonbasic variables x_{24}, and the basic variables x_{21}, x_{31}, and x_{34}, respectively, in the simplex tableau of K_4. From the cells with "-", the upper bounds of the value of the basic variables are obtained by using Lemma 6.2 (Shi 1992):

$$x_{21} = 5\gamma_1 + 25\gamma_2 \leq 36.20 \text{ and } x_{34} = 11\gamma_1 + 33\gamma_2 \leq 49.39.$$

Select x_{21} as the leaving variable since its upper bound 36.20 is the smallest one. Go to Step 5.

Step 5: For an adjacent basis K_5 of T_1, determine the new basic variables: Add $5\gamma_1 + 25\gamma_2$, the value of the leaving basic variable x_{21}, to the allocations of $(2, 4)^+$ and $(3, 1)^+$ in the loop, respectively. Then, subtract $5\gamma_1 + 25\gamma_2$ from the allocations of $(2, 1)^-$ and $(3, 4)^-$ in the loop, respectively. The result is the MC^2 transportation simplex tableau for the basis K_5 given in Table 6.5.

Step 6: For the basis K_5, $\Gamma(K_5) = (x_{14}, x_{22}, x_{24}, x_{31}, x_{33}, x_{34})$, the primal parameter set $\Gamma(K_5) \neq \emptyset$, and the dual parameter set $\Lambda(K_5) \neq \emptyset$ are identified as in Table 6.6. By (iii) of Theorem 4.1, K_5 is a potential solution since $\Gamma(T_5) \times \Lambda(T_5) \neq \emptyset$. From Table 6.5, the objective payoff of K_5 is:

$$\lambda' C(K_5)x(K_5) = (\lambda_1, \ \lambda_2) \begin{pmatrix} 18 & 12 & 17 & 22 & 12 & 32 \\ 1 & 2 & 3 & 1 & 2 & 3 \end{pmatrix} \begin{pmatrix} 30 & 20 \\ 35 & 30 \\ 5 & 25 \\ 22 & 30 \\ 25 & 22 \\ 6 & 8 \end{pmatrix} \begin{pmatrix} \gamma_1 \\ 1-\gamma_1 \end{pmatrix}$$

$$= (\lambda_1, \ 1-\lambda_1) \begin{pmatrix} 2366 & 3325 \\ 235 & 258 \end{pmatrix} \begin{pmatrix} \gamma_1 \\ 1-\gamma_1 \end{pmatrix}$$

The $u_i(\lambda)$ and $v_j(\lambda)$ are obtained as:

For x_{14}: $\lambda^t c_{14} = 18\lambda_1 + \lambda_2 = u_1(\lambda) + v_4(\lambda)$.

For x_{22}: $\lambda^t c_{22} = 17\lambda_1 + 3\lambda_2 = u_2(\lambda) + v_2(\lambda)$.

For x_{24}: $\lambda^t c_{24} = 51\lambda_1 + 2\lambda_2 = u_2(\lambda) + v_4(\lambda)$.

For x_{31}: $\lambda^t c_{31} = 22\lambda_1 + \lambda_2 = u_3(\lambda) + v_1(\lambda)$.

For x_{33}: $\lambda^t c_{33} = 12\lambda_1 + 2\lambda_2 = u_3(\lambda) + v_3(\lambda)$.

For x_{34}: $\lambda^t c_{34} = 32\lambda_1 + 3\lambda_2 = u_3(\lambda) + v_4(\lambda)$.

Set $u_3(\lambda) = 0$ (corresponding to the row having the largest number of allocations in the simplex tableau of K_5), then $v_4(\lambda) = 32\lambda_1 + 3\lambda_2$, $v_3(\lambda) = 12\lambda_1 + 2\lambda_2$, $v_1(\lambda) = 22\lambda_1 + \lambda_2$, $u_1(\lambda) = -14\lambda_1 - 2\lambda_2$, $u_2(\lambda) = 19\lambda_1 - \lambda_2$, and $v_2(\lambda) = -2\lambda_1 + 4\lambda_2$.

Go to the steps of the fifth iteration.

Table 6.5 The MC² Transportation Simplex Tableau of K_5

	W_1		W_2		W_3		W_4		a_{11}	a_{12}
F_1	23	4	27	2	16	3	18	1	30	20
							(30)	(20)		
F_2	12	2	17	3	20	4	51	2	40	55
			(35)	(30)			(5)	(25)		
F_3	22	1	28	5	12	2	32	3	53	60
	(22)	(30)			(25)	(22)	(6)	(8)		
b_{11}	22		35		25		41		123	
b_{12}	30		30		22		53			135

Table 6.6 The initial feasible (primal potential) solution of K_5

$\mathbf{x}(K_5)$	$\Gamma(K_5)$	$\Lambda(K_5)$	Value of x_{ij} (K_5)
$(x_{14}, x_{22}, x_{24}, x_{31}, x_{33}, x_{34})$	$0 < \gamma_1 \leq 1$	$0 < \lambda_1 \leq 2/31$	$x_{14} = 30\gamma_1 + 20\gamma_2$ $x_{22} = 35\gamma_1 + 30\gamma_2$ $x_{24} = 5\gamma_1 + 25\gamma_2$ $x_{31} = 22\gamma_1 + 30\gamma_2$ $x_{33} = 25\gamma_1 + 22\gamma_2$ $x_{34} = 6\gamma_1 + 8\gamma_2$

Steps of the fifth iteration:

Step 1: Since K_5 is a potential solution, set $K_5 = T_2$ and $\psi^2 = \psi^1 \cup \{T_2\} = \{T_1, T_2\}$. For every (γ_1, λ_1), there is a T_j of ψ^2 such that $(\gamma_1, \lambda_1) \in \Gamma(T_j) \times \Lambda(T_j)$. Go to Step 7.

Step 7: The process terminates since $\psi^2 = \psi$ is the set of all potential solutions over all possible changes of (γ_1, λ_1) for the given transportation problem.

6.3 A Dynamic MC² Transportation Problem

This section extends the steady MC² transportation model discussed in Sections 6.1 and 6.2 to the context in which we make a sequence of interrelated decisions. Suppose that an MC² transportation problem can be divided into stages, with a policy decision required at each stage. The solution procedure is designed to find an optimal policy for the overall problem, i.e., a prescription of the optimal sub-policy decision at each stage for each of the possible states. To solve this problem, we employ a dynamic programming ideology to find the optimal sub-policies and the optimal policy for MC² transportation problems.

The dynamic MC² (DMC²) transportation model associated with several stages can be formulated as (M6.3) (Li and Shi 2000):

$$Min \quad (\lambda_1, \quad \cdots, \quad \lambda_q) \begin{pmatrix} \sum_{k=1}^{t}\sum_{i=1}^{m}\sum_{j=1}^{n} c_{ijk}^{1} x_{ijk} \\ \sum_{k=1}^{t}\sum_{i=1}^{m}\sum_{j=1}^{n} c_{ijk}^{2} x_{ijk} \\ \vdots \\ \sum_{k=1}^{t}\sum_{i=1}^{m}\sum_{j=1}^{n} c_{ijk}^{q} x_{ijk} \end{pmatrix}$$

$$Subject\ to \quad \sum_{k=1}^{t}\sum_{j=1}^{n} x_{ijk} = (a_{i1}, \quad \cdots, \quad a_{ip}) \begin{pmatrix} \gamma_1 \\ \vdots \\ \gamma_p \end{pmatrix}, i = 1, \cdots, m,$$

$$\sum_{k=1}^{t}\sum_{i=1}^{m} x_{ijk} = (b_{j1}, \quad \cdots, \quad b_{jp}) \begin{pmatrix} \gamma_1 \\ \vdots \\ \gamma_p \end{pmatrix}, j = 1, \cdots, n,$$

$$a_{ir} \geq 0,\ b_{jr} \geq 0\ for\ all\ (i, j, r),$$

$$x_{ijk} \geq 0\ for\ all\ (i, j, k),$$

where x_{ijk} = the amount of products transported from source i to destination j at the kth stage, k = 1, . . ., t;

c_{ijk}^{w} = the "cost " of transporting a unit product from source i to destination j when the wth criterion, w = 1, . . ., q, is under consideration at the kth stage, k = 1, . . ., t;

a_{is} = the sth supply availability level at i, s = 1, . . ,p, which is assumed to be positive;

b_{js} = the sth demand availability level at j, s = 1, . . ,p, which is assumed to be positive.

We observe that in the above DMC^2 transportation model (M6.3), the total supply $\sum_{i=1}^{m} a_{is}$ is equal to the total demand $\sum_{j=1}^{n} b_{js}$, which does not change over the different stages. Thus, the statement of Theorem 6.1 also holds for (M6.3). In other words, the following is true:

Theorem 6.4 Given $\gamma^0 > 0$, the MC² transportation problem (M6.3) has a feasible solution at γ^0 if $\sum_{i=1}^{m} a_{is} = \sum_{j=1}^{n} b_{js}$, for any s = 1, . . ., p.

Because x_{ijk} is determined across the different stages, we have:

Theorem 6.5 Given a primal potential basis J for problem (M6.3), the basic variables **x**(*J*) contain at most m+n-1 components x_{ijk}, where k is some of {1, . . ., t}.

To address the relationship between the sub-process on one stage and the whole process on all stages, we need to state a number of definitions as follows:

Definition 6.1 A solving procedure for (M6.3) involves moving backward stage by stage. The sub-process of problem (M6.3) that contains the first *h* stages is called *h-DMC² problem*. An optimal sequence of interrelated decisions associated with the *h-DMC²* problem is called *an optimal sub-policy* if $h < t$, and *an optimal policy* overall problem (M6.3) if $h = t$.

Definition 6.2 For h-DMC² problem, $cell(i, j)$ is *a non-basic cell* if any x_{ijk} ($k = 1,..., h$) is non-basic variable; otherwise, $cell(i, j)$ is *a basic cell*.

Definition 6.3 As boundary condition, we assume all the cost parameters at Stage 0 are the same with at Stage 1. It means $c^w_{ij0} = c^w_{ij1}$.

Definition 6.4 $\sigma_{ijk} = \lambda^t c_{ij*} - (u_i(\lambda) + v_j(\lambda))$ is called the *improvement difference* of non-basic $cell(i, j)$ at the kth stage where $\lambda^t c_{ij*} = \min\{\lambda^t c_{ijl} \mid l = 1,..., k\}$.

Definition 6.5 $\Omega(T)$ is called the *premise condition set* about (λ, γ) that makes the basis *T* exist. Because a_{ir} and b_{jr} are nonnegative for all r, i, j in the DMC² transportation (M6.3), $\gamma \in (0,1]$ keep the same all the time. Setting $\Omega(T) = \{0 < \lambda \le 1\}$ when *T* is an initial basis of stage-0, otherwise setting $\Omega(T) = \Lambda(T)$.

In the case of Stage h= 0, it means there is only one stage (stage-0) under considering. A DMC² transportation problem (M6.3) is reduced to an MC² transportation problem in this case. Set $\eta^0 = \varnothing$, $\xi^0 = \varnothing$. Find an initial basic solution T by using the Vogel's rule or the other rules that are available. $\Omega(T) = \{0 < \lambda \leq 1\}$ at this time. Either the sth availability level of supply i are exhausted simultaneously or demand j is completely satisfied over the sth requirement level if x_{ij0} is a basic variable (see Example 6.2). The number of the basic variables should be exactly equal to $m+n-1$. Let B be the initial basic solution (basis). The following theorem is similar to Theorem 6.3 for (M6.3):

Theorem 6.6 A basis B is a dual potential basis for h-DMC² problem if

(i) $u_i(\lambda) + v_j(\lambda) - \lambda^t c_{ijh} = 0$, for all (i, j) such that $cell(i, j)$ is a basic cell and x_{ijh} is a basic variable; and

(ii) $u_i(\lambda) + v_j(\lambda) - \lambda^t c_{ijh} \leq 0$, for all (i, j) such that $cell(i, j)$ is a non-basic cell and x_{ijh} is a nonbasic variable; where $\lambda^t c_{ijh} = \min\{\lambda^t c_{ijl} \mid l = 1, \ldots, h\}$.

Note that (i) of Theorem 6.6 provides a way to calculate $u_i(\lambda) + v_j(\lambda)$ corresponding to a basis B for problem (M6.3), while (ii) of Theorem 6.3, which is used to find the dual parameter set $\Lambda(B)$, yields a potentiality test for the basis B if the primal parameter set $\Gamma(B) \neq \varnothing$.

Lemma 6.4 Given a potential solution B (at Stage h-1), its non-basic variable x_{ijh} is an entering variable for the next pivoting (at Stage h) if there is a $\lambda \in \Lambda(B)$ so that $u_i(\lambda) + v_j(\lambda) - \lambda^t c_{ijh} = 0$, where $\lambda^t c_{ijh} = \min\{\lambda^t c_{ijl} \mid l = 1, \ldots, k\}$.

6.4 An Algorithm for The Dynamic MC² Transportation Model

Based on the above discussion, we describe an algorithm to solve the DMC² transportation problem (M6.3) as follows:

Algorithm 6.2

Step 0 For Stage $h = 0$, set $\eta^0 = \varnothing$, $\xi^0 = \varnothing$. Find an initial basic solution T_0 by using the Vogel's rule or the other rules that are available. $\Omega(T_0) = \{0 < \lambda \leq 1\}$ if such a basis T_0 exist. Otherwise, find T_0 and set $\Omega(T_0) = \Lambda(T_0)$. From the result simplex table, calculate $u_i(\lambda)$ and $v_j(\lambda)$ by setting $u_1(\lambda) = 0$ arbitrarily. Then solve the set of equations in (i) of Theorem 6.6. Set $h = 1$ and go to the steps of the following iterations.

Step 1. For Stage h, from a basic solution T that was been built in the last stage h-1, check the basic $cell(i, j)$ for $\lambda^t c_{ijh-1} = \lambda^t c_{ijh}$ with $\lambda_1 + \lambda_2 + \ldots + \lambda_q = 1$ for $0 \leq \lambda_k \leq 1$, $k = 1, \ldots, q$, where $\lambda^t c_{ijh-1} = \min\{\lambda^t c_{ijl} \mid l = 1, \ldots, h-1\}$ is described in (i) of Theorem 6.6. There are $(m+n-1)$ sets regarding λ to be checked out. Take all the sets satisfying $\Omega(T)$ as the interface to divide $\Omega(T)$ and $\Omega(T)$ can be divided at most into 2^{m+n-1} sub-sets. The purpose doing this step is to make sure that $\lambda^t c_{ijh-1} < \lambda^t c_{ijh}$ or $\lambda^t c_{ijh-1} \geq \lambda^t c_{ijh}$ in any sub-set of $\Omega(T)$.

Step 2. Check the basic $cell(i, j)$ in a sub-set of $\Omega(T)$. Move the allocations to Stage h for the basic cells $\lambda^t c_{ijh-1} \geq \lambda^t c_{ijh}$ and keep the allocations in Stage h-1 for the basic cells $\lambda^t c_{ijh-1} < \lambda^t c_{ijh}$. The goal doing Step 2 is to obtain a basic solution of the h-DMC² transportation problem. One sub-set of $\Omega(T)$ corresponds to a basic solution of the h-DMC² problem. Choose a basic solution of the h-DMC² problem K and go to the next step.

Step 3. By using (iii) of Theorem 4.1, if K is not a potential solution go to Step 4, otherwise set $K = B_j$ and $\eta^j = \eta^{j-1} \cup \{B_j\}$. For every given (λ, γ) in the sub-set of $\Omega(T)$, if there is a B of η^j such that $(\lambda, \gamma) \in$

$\Gamma(B)\times\Lambda(B)$, then $\eta^j = \eta$ and go to Step 9; otherwise go to Step 5, where (λ, γ) is the parameter space associated with the initial basic solution.

Step 4. Determine the entering basic variable: Select the non-basic variable x_{ijh} associated with non-basic $cell(i, j)$ having the largest upper bound of $u_i(\lambda) + v_j(\lambda) > \lambda^t c_{ijh}$ by applying (ii) of Lemma 6.2, where $\lambda^t c_{ijh} = \min\{\lambda^t c_{ijl} \mid l = 1,\ldots,h\}$. If r is not unique, we select the largest one. Go to Step 6.

Step 5. Select the non-basic variable x_{ijh} that satisfies Lemma 6.4 as the entering basic variable. If x_{ijh} is not unique, choose one arbitrarily.

Step 6. Determine the leaving variable: Identify the loop formed by the non-basic $cell(i, j)$ corresponding to the entering basic variable x_{ijh}, and the basic cells in the h-DMC^2 transportation simplex tableau. Number the cells in this loop consecutively: Starting with $cell(i, j)$, set "+" on the odd-numbered cells and "–" on the even-numbered cells. Among the cells with "–", select the basic variable having the smallest upper bound of the value by using (i) of Lemma 6.2.

Step 7. For an adjacent basis Q of K, determine the new basic variables: Add the values of the leaving basic variable to the entering basic variable x_{ijh} and all the other basic variables allocated in the cell with "+" in the loop, respectively, for the sth supply or demand levels. Subtract these values from the leaving basic variable and all the other basic variables allocated in the cell with "–" in the loop, respectively, for the sth supply or demand levels. The result will be a new simplex tableau for the basis Q.

Step 8. For the basis Q, first find the primal parameter set $\Gamma(Q) \neq \varnothing$. Second, calculate $u_i(\lambda)$ and $v_j(\lambda)$ by setting $u_i(\lambda) = 0$. Then, solve the equality set (i) of Theorem 6.6. If the inequality set (ii) of Theorem 6.6 is tenable, then Q is a dual potential solution, that is $\Lambda(Q) \neq \varnothing$. Otherwise $\Lambda(Q) = \varnothing$. Go to Step 3.

Step 9 Set $\xi^i = \xi^{i-1} \cup \eta$. If there is a B of ξ^i such that $(\lambda, \gamma) \in \Gamma(B) \times \Lambda(B)$, for every $0 < \lambda \le 1$ and $0 < \gamma \le 1$, then go to Step **10.** Otherwise, if the initial basis that in the course has been finished, there is a B of ξ^i such that $(\lambda, \gamma) \in \Gamma(B) \times \Lambda(B)$ for every given (λ, γ) in $\Omega(T)$, go to Step 1; if the initial basis has not been finished, go to Step 2 from another sub-set of $\Omega(T)$ entering the next iteration.

Step 10 If $h < t$, let $h = h+1$ and go to Step 1. If $h = t$, the whole procedure terminates.

The following simple example is used to show how the above algorithm can be employed to solve the DMC^2 transportation problem (M6.3).

Example 6.2 Consider a transportation problem involving three suppliers, denoted by P_1, P_2, P_3, and three demanders, denoted by D_1, D_2, and D_3. A particular product is transported from the ith supplier to the jth demander, $i = 1,2,3$ and $j = 1,2,3$. Both the supply level and demand level change with the economic situations. Suppose that if the economy is in recession, the supply capacity level of P_1 is 7units, P_2 is 5 units, and P_3 is 8 units; while the demand requirement level of D_1 is 6 units, D_2 is 10 units, and D_3 is 4 units. If the economy is booming, the supply capacity level of P_1 is 8 units, P_2 is 10 units, and P_3 is 12 units; while the demand requirement level of D_1 is 10 units, D_2 is 15 units, and D_3 is 5 units. For simplicity, we assume there are two criteria under consideration:

(i) The minimization of total transportation expenses consumed in transportation, and

(ii) The minimization of total product deterioration during transportation.

Suppose the transportation can take place in every month of a Quarter (A month will be a stage and the Quarter has three stages). The transportation expense and the deterioration for transportation one unit products at every month are given in Table 6.7, respectively. For example, in Month 1, [3, 6] means that the unit transportation expense from P_1 to D_1 is \$3, while its unit product deterioration is 6. Similarly, in Month 3, [4, 7] represents that the unit transportation expense from P_3 to D_3 is \$4, while the unit product deterioration is 7. The goal is to locate the set of all potential solutions for this three-stage transportation problem.

Table 6.7 Transportation Expense and Deterioration in Three Months

	D_1			D_2			D_3		
	1	2	3	1	2	3	1	2	3
P_1	[3, 6]	[6, 9]	[7, 10]	[5, 5]	[4, 3]	[6, 8]	[2, 4]	[3, 6]	[10, 6]
P_2	[8, 10]	[4, 6]	[6, 8]	[4, 8]	[5, 5]	[8, 10]	[6, 12]	[4, 5]	[9, 11]
P_3	[9, 10]	[7, 12]	[12, 10]	[12, 11]	[10, 8]	[8, 9]	[10, 8]	[6, 8]	[4, 7]

Note that in the above problem the total amount of products transported is equal to the total amount received in the both levels. By Theorem 6.4, the problem has a feasible solution. Let $\boldsymbol{\gamma}$ be the supply and demand parameter over the economic situations, such that $\gamma_1 + \gamma_2 = 1$ and $\gamma_1, \gamma_2 > 0$. Let $\boldsymbol{\lambda}$ be the criteria parameter over the two criteria, such that $\lambda_1 + \lambda_2 = 1$ and $\lambda_1, \lambda_2 > 0$. Let η be the set of all potential solutions overall possible changes of $(\boldsymbol{\lambda}, \boldsymbol{\gamma})$ for the three stages transportation problem. Applying Algorithm 6.2, we have:

Step 0. Set $\eta^0 = \varnothing$. Let us just consider Stage h=0 for the DMC^2 transportation problem. Using the Vogel's rule according to Definition 6.3, we can obtain the initial basis T_0 of Stage 0 (see Table 6.8) such that $\Omega(T_0) = \{0 < \lambda_1 < 1\}$ (see Definition 6.5). In the allocation *cell*(i, j) of Table 6.8, the figure in the left upper corner is the related allocations when the economy is in recession and the figure in the right lower corner is the related allocations when the economy is in booming. For example, the allocations in *cell*(1,1) of Table 6.8 are $6\gamma_1 + 10\gamma_2$ units and the zeros that follow 6 and 10 in brackets express the allocations take place in Stage 0. The figures in middle brackets are the improvement differences of non-basic cells.

For non-basic $cell(i,j)$ the improvement differences at Stage **0** is (see Definition 6.4) $\sigma_{ij0} = \boldsymbol{\lambda}' c_{ij0} - (u_i(\lambda) + v_j(\lambda))$

Since the basic variables x_{ij0} of T_0 satisfy (i) of Theorem 6.6, the following equalities are obtained:

For x_{110}: $3\lambda_1 + 6\lambda_2 = u_1(\lambda) + v_1(\lambda)$;

For x_{130}: $2\lambda_1 + 4\lambda_2 = u_1(\lambda) + v_3(\lambda)$;

For x_{220}: $4\lambda_1 + 8\lambda_2 = u_2(\lambda) + v_2(\lambda)$;

For x_{320}: $12\lambda_1 + 11\lambda_2 = u_3(\lambda) + v_2(\lambda)$; and

For x_{330}: $10\lambda_1 + 8\lambda_2 = u_3(\lambda) + v_3(\lambda)$.

Set $u_1(\lambda) = 0$, then $v_1(\lambda) = 3\lambda_1 + 6\lambda_2$, $v_3(\lambda) = 2\lambda_1 + 4\lambda_2$, $u_3(\lambda) = 8\lambda_1 + 4\lambda_2$, $v_2(\lambda) = 4\lambda_1 + 7\lambda_2$, and $u_2(\lambda) = \lambda_2$. For example, the number $(5,3)$ in cell $(2,1)$ of Table 6.8 represents $\sigma_{210} = 5\lambda_1 + 3\lambda_2$. Set h=1 and go to the steps of the first iteration.

Table 6.8 The Initial Basis T_0 of Stage **0** for Example 6.2

	D_1	D_2	D_3
P_1	6(0) 10(0)	(1, -2)	1(0) -2(0)
P_2	(5, 3)	5(0) 10(0)	(4, 7)
P_3	(-2, 0)	5(0) 5(0)	3(0) 7(0)

Steps of The Iteration in Stage 1:

Step 1. From the initial basic solution T_0 that was been built in Stage 0, check the basic cells for $\boldsymbol{\lambda}' c_{ij1} = \boldsymbol{\lambda}' c_{ij0}$ with $\lambda_1 + \lambda_2 = 1$ and $0 < \lambda_i < 1, i = 1,2$. We can get the same set $\{0 < \lambda_1 < 1\}$ as the boundary points to divide $\Omega(T_0)$ because $c_{ij1} = c_{ij0}$ (Definition 6.3). Thus $\Omega(T_0)$ keep the same.

Step 2. Check the basic *cell*(i, j) in $\Omega(T_0)$. Because $\boldsymbol{\lambda}' c_{ij1} = \boldsymbol{\lambda}' c_{ij0}$ for every (i, j), move the allocations to stage-1 for all the basic cells. Now we get an initial basic solution in Stage 1 (see Table 6.9). For non-basic cells $\sigma_{ij1} = \boldsymbol{\lambda}' c_{ij*} - (u_i(\lambda) + v_j(\lambda))$, there is $\sigma_{ij1} = \sigma_{ij0}$ for all non-basic cells because $c_{ij0} = c_{ij1}$ for all i, j. Let K_1 be the resulting initial basic solution, then $\mathbf{x}(K_1) = (x_{111}, x_{131}, x_{221}, x_{321}, x_{331})$.

Set $\mathbf{x}(K_1) \geq 0$, obviously the primal parameter set $\boldsymbol{\Gamma}(K_1) = \{2/3 \leq \gamma_1 < 1\}$. Set $\sigma_{ij1} \geq 0$ for all i, j, then the dual parameter set $\boldsymbol{\Lambda}(K_1) = \varnothing$. Go to the next step.

Table 6.9 A Basic Solution K_1 in Stage 1

	D_1	D_2	D_3
P_1	6(1) – 10(1)	(1, -2)	1(1) + -2(1)
P_2	(5, 3)	5(1) 10(1)	(4, 7)
P_3	(-2, 0) +	5(1) 5(1)	3(1) – 7(1)

Step 3. By (ii) of Theorem 4.1, the basis K_1 is not a potential solution since $\boldsymbol{\Lambda}(K_1) = \varnothing$. Go to Step 4.

Step 4. Determine the entering basic variable corresponding to K_1. By applying (ii) of Lemma 6.2, the upper bounds for $\sigma_{ij1} < 0$ can be obtained as $\sigma_{311} = -2\lambda_1 + 0\lambda_2 \leq \sqrt{2((-2)^2 + 0^2)} \approx 2.83$. Therefore, select

x_{311} as the entering variable since its upper bound 2.83 is the largest one (in fact there is only σ_{311} satisfying $\sigma_{ij1} < 0$). Go to Step 6.

Step 6. Determine the leaving variable corresponding to K_1. Select a loop formed by the cells(3,1), (3,3), (1,3) and (1,1), corresponding to the variables $x_{311}, x_{331}, x_{131}, x_{111}$, respectively (see Table 6.9). Then *cell*(3,1), (1,3) are marked with "+" and (3,3), (1,1) with "–". From the cells with "–", the upper bounds of value of the basic variables are obtained by using (i) of Lemma 6.2 (see Table 6.9):

$$x_{111} = 6\gamma_1 + 10\gamma_2 \leq \sqrt{2(6^2 + 10^2)} \approx 16.49$$

$$x_{331} = 3\gamma_1 + 7\gamma_2 \leq \sqrt{2(3^2 + 7^2)} \approx 10.77$$

Select x_{331} as the leaving variable since its upper bound 10.77 is the smallest. Continue to Step 7.

Step 7. For an adjacent basis K_2, determine the new basic variables: Add $3\gamma_1 + 7\gamma_2$, the value of the leaving basic variable x_{331}, to the allocations of (3,1) and (1,3) in the loop, and then subtract $3\gamma_1 + 7\gamma_2$ from the allocations of (3,3) and (1,1) in the loop. This result is shown in Table 6.10.

Table 6.10 The Basic Solution K_2 in Stage 1

	D_1	D_2	D_3
P_1	3(1) / 3(1)	(-1, -2)	4(1) / 5(1)
P_2	(7, 3)	5(1) / 10(1)	(6, 7)
P_3	3(1) / 7(1)	5(1) / 5(1)	(2, 0)

Step 8. The primal parameter set $\Gamma(K_2) \neq \varnothing$ and the dual parameter set $\Lambda(K_2) = \varnothing$ can be identified by Theorem 4.1 (see Table 6.10). Go to Step 3.

Through the second iteration, we can identify the adjacent base K_3. By using Definition 4.3, we have $\Gamma(K_3) = \{0 < \gamma_1 < 1\}$ and $\Lambda(K_3) = \{2/3 \leq \lambda_1 < 1\}$ (see Table 6.11).

Thus, K_3 is a potential solution because $\Gamma(K_3) \neq \varnothing, \Lambda(K_3) \neq \varnothing$ (by Theorem 4.1).

Set $K_3 = B_1$ and $\eta^1 = \eta^0 \cup \{B_1\} = \{B_1\}$. There is only B_1 in η^1 and for $\Omega(T_0) = \{0 < \lambda_1 < 1\}$, there exists some $(\lambda_1, \gamma_1) \notin \Gamma(B_1) \times \Lambda(B_1)$. Therefore, go to Step 5.

Table 6.11 The Potential Solution K_3 in Stage 1

	D_1	D_2	D_3
P_1	(1, 2)	3(1) 3(1)	4(1) 5(1)
P_2	(7, 3)	5(1) 10(1)	(5, 5)
P_3	6(1) 10(1)	2(1) 2(1)	(1, -2)

Step 5. From the simplex tableau (Table 6.11) of K_3, check the non-basic variable x_{ij1} for $u_i(\lambda) + v_j(\lambda) - \lambda' c_{ij1} = 0$

Obviously, only x_{331} satisfy the above condition, so choose x_{331} as entering variable. Go to the next step. Through step 6, 7, 8 we can obtain Table 6.12.

Table 6.12 The Potential Solution K_4 in Stage 1

	D_1	D_2	D_3
P_1	(2, 0)	5(1) 5(1)	2(1) 3(1)
P_2	(8, 1)	5(1) 10(1)	(5, 5)
P_3	6(1) 10(1)	(-1, 2)	2(1) 2(1)

We can identify the basis K_4 is a potential solution with $\Gamma(K_4)=\{0<\gamma_1<1\}$ and $\Lambda(K_4)=\{0<\lambda_1\le 2/3\}$ (see Table 6.12). Set $K_4=B_2$ and $\eta^2=\eta^1\cup\{B_2\}=\{B_1,B_2\}$. For every (λ_1,γ_1) in $\Omega(T_0)=\{0<\lambda_1<1\}$, there is a B_j of η^2 such that $(\lambda_1,\gamma_1)\in\Gamma(B_j)\times\Lambda(B_j)$. Set $\eta^2=\eta_1$ and go to Step 9.

Step 9. Set $\xi^1=\xi^0\cup\eta_1=\eta_1$. For every $0<\lambda_1<1$ and $0<\gamma_1<1$, there is a B_j of ξ^1 such that $(\lambda_1,\gamma_1)\in\Gamma(B_j)\times\Lambda(B_j)$, go to Step10.

Step 10. Because h = 1 < 3, set h = h+1 = 2 and go to the Steps of the iteration in Stage 2. For the sake of space, the details of the process in Stage 2 are not elaborated here. The results six potential solutions of Stage 2 are shown from Table 6.13 to Table 6.18. These (λ,γ) ranges are:

$\Gamma(L_1)=\{0<\gamma_1<1\}$ and $\Lambda(L_1)=\{0<\lambda_1\le 1/2\}$ (Table 6.13).

$\Gamma(L_2)=\{0<\gamma_1<1\}$ and $\Lambda(L_2)=\{1/2\le\lambda_1\le 3/5\}$ (Table 6.14).

$\Gamma(L_3)=\{0<\gamma_1<1\}$ and $\Lambda(L_3)=\{3/5\le\lambda_1\le 2/3\}$ (Table 6.15).

$\Gamma(L_4)=\{0<\gamma_1<1\}$ and $\Lambda(L_4)=\{2/3\le\lambda_1\le 3/4\}$ (Table 6.16).

$\Gamma(L_5)=\{0<\gamma_1<1\}$ and $\Lambda(L_5)=\{3/4<\lambda_1\le 5/6\}$ (Table 6.17).

$\Gamma(L_6)=\{0<\gamma_1<1\}$ and $\Lambda(L_6)=\{5/6\le\lambda_1<1\}$ (Table 6.18).

Table 6.13 The Potential Solution L_1 in Stage 2

	D_1	D_2	D_3
P_1	(0, 1)	7(2) 8(2)	(2, 1)
P_2	5(2) 10(2)	(0, 1)	(3, 1)
P_3	1 (1) 0(1)	3(2) 7(2)	4(2) 5(2)

Table 6.14 The Potential Solution L_2 in Stage 2

	D_1	D_2	D_3
P_1	(2, -1)	7(2) 8(2)	(-1, 2)
P_2	5(2) 10(2)	(-2, 3)	(1, 3)
P_3	1(2) 0(2)	3(2) 7(2)	4(2) 5(2)

Table 6.15 The Potential Solution L_2 in Stage 2

	D_1	D_2	D_3
P_1	(0, 2)	7(2) 8(2)	(0, 4)
P_2	2(2) 3(2)	3(2) 7(2)	(1, 3)
P_3	4(2) 7(2)	(2, -3)	4(2) 5(2)

Table 6.16 The Potential Solution L_4 in Stage 2

	D_1	D_2	D_3
P_1	(0, 2)	7(2) 8(2)	(0, 4)
P_2	2(2) 3(2)	3 (2) 7(2)	(1, 3)
P_3	4(2) 7(2)	(2, -3)	4(2) 5(2)

Table 6.17 The Potential Solution L_5 in Stage 2

	D_1	D_2	D_3
P_1	(-1, 5)	7(2) 8(2)	(-1, 7)
P_2	2(2) 3(2)	3 (1) 7 (1)	(1, 3)
P_3	4(2) 7(2)	(3, -6)	4(2) 5(2)

Table 6.18 The Potential Solution L_6 in Stage 2

	D_1	D_2	D_3
P_1	2 (1) 3 (1)	5(2) 5(2)	(0, 2)
P_2	(1, -5)	5 (1) 10 (1)	(2, -2)
P_3	4(2) 7(2)	(2, -1)	4(2) 5(2)

Steps of The Iteration In Stage-3:

Step 1. Let $T = L_1$ in Table 6.13, so $\mathbf{\Omega}(T) = \mathbf{\Omega}(L_1) = \{0 < \lambda_1 \le 1/2\}$. Check basic $cell(i,j)$ in Table 6.13 for $\boldsymbol{\lambda}' c_{ij3} = \boldsymbol{\lambda}' c_{ij2}$ with $\lambda_1 + \lambda_2 = 1$ and $\lambda_1, \lambda_2 \in (0,1)$, where e = 1, 2. For example, it is $c_{122} = (c_{122}{}^1, c_{122}{}^2)$, but $c_{312} = (c_{311}{}^1, c_{311}{}^2)$.

For x_{122}: $6\lambda_1 + 8\lambda_2 = 4\lambda_1 + 3\lambda_2$; then $\lambda_1 = 3/5$.
For x_{212}: $6\lambda_1 + 8\lambda_2 = 4\lambda_1 + 6\lambda_2$; then no solution for λ_1.
For x_{311}: $12\lambda_1 + 10\lambda_2 = 9\lambda_1 + 10\lambda_2$; then $\lambda_1 = 0$.
For x_{322}: $8\lambda_1 + 9\lambda_2 = 10\lambda_1 + 8\lambda_2$; then $\lambda_1 = 1/3$.
For x_{332}: $4\lambda_1 + 7\lambda_2 = 6\lambda_1 + 8\lambda_2$; then $\lambda_1 = -1$.

Thus, we can only obtain $\lambda_1 = 1/3$ ($cell(3,2)$) satisfy $\mathbf{\Omega}(T) = \{0 < \lambda_1 \le 1/2\}$, so we divide $\mathbf{\Omega}(T)$ into $\mathbf{\Omega}(T^*)$ and $\mathbf{\Omega}(T^{**})$, where

$$\mathbf{\Omega}(T^*) = \{0 < \lambda_1 \le 1/3\},\ \mathbf{\Omega}(T^{**}) = \{1/3 \le \lambda_1 \le 1/2\}.$$

Step 2. Check the basic $cell(i,j)$ of Table 6.13 in $\mathbf{\Omega}(T^*)$. Through the step we can obtain Table 6.19. In Table 6.19, the allocations that are followed by "(1)", "(2)", or "(3)" express the transportation take place in Stage 1, Stage 2, or Stage 3. Table 6.19 gives a potential solution Q_1 and it can be express as: transport $7\gamma_1 + 8\gamma_2$ units from P_1 to D_2 at Stage 2, $5\gamma_1 + 10\gamma_2$ units from P_2 to D_1 at Stage 2, γ_1 units from P_3 to D_1 at Stage 1, $3\gamma_1 + 7\gamma_2$ units from P_3 to D_2 at Stage 2, $4\gamma_1 + 5\gamma_2$ units from P_3 to D_3 at Stage 3. The other tables give the rest potential solutions have the same meaning. Where $\mathbf{\Gamma}(Q_1) = \{0 < \gamma_1 < 1\}$ and $\mathbf{\Lambda}(Q_1) = \{0 < \lambda_1 \le 1/3\}$.

The improvement difference σ_{ij3} must be paid more attention. Where $\lambda^{t} c_{ijh} = \min\{\lambda' c_{ijl} \mid l = 1,2,3\}$ and $u_i(\lambda), v_j(\lambda)$ must be connected with the stage that basic variable take place (see Definition 6.4). For instance (where $\mathbf{\Lambda}(Q_1) = \{0 < \lambda_1 \le 1/3\}$),

$$c_{22h} = \min\{(4,8),(5,5),(8,10)\} = (5,5) = c_{222}.$$

For x_{122}: $\boldsymbol{\lambda}' c_{122} = 4\lambda_1 + 3\lambda_2 = u_1(\lambda) + v_2(\lambda)$

For x_{212}: $\boldsymbol{\lambda}' c_{212} = 4\lambda_1 + 6\lambda_2 = u_2(\lambda) + v_1(\lambda)$

For x_{311} : $\boldsymbol{\lambda}' c_{311} = 9\lambda_1 + 10\lambda_2 = u_3(\lambda) + v_1(\lambda)$

For x_{322} : $\boldsymbol{\lambda}' c_{322} = 10\lambda_1 + 8\lambda_2 = u_3(\lambda) + v_2(\lambda)$

For x_{333} : $\boldsymbol{\lambda}' c_{333} = 4\lambda_1 + 7\lambda_2 = u_3(\lambda) + v_3(\lambda)$.

Set $u_1(\lambda) = 0$, then $v_2(\lambda) = 4\lambda_1 + 3\lambda_2$, $u_3(\lambda) = 6\lambda_1 + 5\lambda_2$, $v_1(\lambda) = 3\lambda_1 + 5\lambda_2$, $u_2(\lambda) = \lambda_1 + \lambda_2$, and $v_3(\lambda) = -2\lambda_1 + 2\lambda_2$, so $\sigma_{223} = (5,5) - [(1,1) + (4,3)] = (0,1)$.

Step 3. Set $Q_1 = B_1$ and $\eta^1 = \eta^0 \cup \{B_1\} = \{B_1\}$. For every given $(\boldsymbol{\lambda}, \boldsymbol{\gamma})$ in $\boldsymbol{\Omega}(T^*) = \{0 < \lambda_1 \leq 1/3\}$, then is a B_1 of η^1 such that $(\lambda_1, \gamma_1) \in \boldsymbol{\Gamma}(B_1) \times \boldsymbol{\Lambda}(B_1)$. Then set $\eta^1 = \eta_1$ and go to Step 9.

Table 6.19 The Potential Solution Q_1 in Stage 3

	D_1	D_2	D_3
P_1	(0, 1)	7 (2) / 8 (2)	(4, 2)
P_2	5 (2) / 10 (2)	(0, 1)	(5, 2)
P_3	1 (1) / 0 (1)	3 (2) / 7 (2)	4(3) / 5(3)

Step 9. Set $\xi^1 = \xi^0 \cup \eta_1 = \{B_1\}$. There is only B_1 in ξ^1 and there exists some $(\lambda_1, \gamma_1) \notin \boldsymbol{\Gamma}(B_1) \times \boldsymbol{\Lambda}(B_1)$ for $0 < \lambda_1 < 1$ and $0 < \gamma_1 < 1$. For a given $(\boldsymbol{\lambda}, \boldsymbol{\gamma})$ in $\boldsymbol{\Omega}(L_1) = \{0 < \lambda_1 < 1/2\}$, there exists some $(\lambda_1, \gamma_1) \notin \boldsymbol{\Gamma}(B_1) \times \boldsymbol{\Lambda}(B_1)$ either. Thus the initial $T = L_1$ in Table 6.13 has not been finished, so go to Step 2.

Step 2. Check the basic $cell(i, j)$ of Table 6.13 in $\boldsymbol{\Omega}(T^{**})$. Through the step we can obtain Table 6.20. Table 6.20 gives a potential solution Q_2 . $\boldsymbol{\Gamma}(Q_2) = \{0 < \gamma_1 < 1\}$, $\boldsymbol{\Lambda}(Q_2) = \{1/3 \leq \lambda_1 \leq 1/2\}$.

Step 3. Set $Q_2 = B_2$ and $\eta^2 = \eta^1 \cup \{B_2\} = \{B_1, B_2\}$. For every given (λ, γ) in $\Omega(T^{**}) = \{1/3 \leq \lambda_1 \leq 1/2\}$, there is a basis B_2 of η^2 such that $(\lambda_1, \gamma_1) \in \Gamma(B_2) \times \Lambda(B_2)$. Set $\eta^2 = \eta_2$ and go to Step 9.

Step 9. Set $\xi^2 = \xi^1 \cup \eta_2 = \eta_1 \cup \eta_2 = \{B_1, B_2\}$, for $0 < \lambda_1 < 1$ and $0 < \gamma_1 < 1$ there exists some $(\lambda_1, \gamma_1) \notin \{\Gamma(B_1) \cup \Gamma(B_2)\} \times \{\Lambda(B_1) \cup \Lambda(B_2)\}$. For all (λ, γ) in $\Omega(L_1) = \{0 < \lambda_1 \leq 1/2\}$, there is a basis B_j of η^2 such that $(\lambda_1, \gamma_1) \in \Gamma(B_j) \times \Lambda(B_j)$, so go back to Step 1 at Stage 3.

Step 1. Let $T = L_2$ in Table 6.14, we can obtain Table 6.21. It gives a potential solution Q_3,

where $\Gamma(Q_3) = \{0 < \gamma_1 < 1\}$ and $\Lambda(Q_3) = \{1/2 \leq \lambda_1 \leq 3/5\}$.

From the initial solution in Table 6.15, Table 6.16, and Table 6.17, we can obtain the potential solutions same as that one in Table 6.21. Set $\xi^6 = \xi^5 \cup \eta_6 = \{B_1, ..., B_6\} = \{0 < \lambda_1 \leq 5/6\}$. Table 6.21 gives a potential solution when $\{1/2 \leq \lambda_1 \leq 5/6\}$.

Table 6.20 The Potential Solution Q_2 in Stage 3

	D_1	D_2	D_3
P_1	(-2, 2)	7 (2) 8 (2)	(2, 3)
P_2	5 (2) 10 (2)	(1, 0)	(5, 2)
P_3	1 (1) 0 (1)	3(3) 7(3)	4(3) 5(3)

Table 6.21 The Potential Solution Q_3 in Stage 3

	D1	D2	D3
P1	(0, 0)	7 (2) 8(2)	(2, 3)
P2	5 (2) 10 (2)	(0, 2)	(3, 4)
P3	1 (2) 0 (2)	3(3) 7(3)	4(3) 5(3)

Let $T = L_6$ in Table 6.18, we can obtain Table 6.22. It gives a potential solution Q_4. Here $\Gamma(Q_4) = \{0 < \gamma_1 < 1\}$, and $\Lambda(Q_4) = \{5/6 \le \lambda_1 < 1\}$. Set $Q_4 = B_7$ and $\eta^7 = \eta^6 \cup \{B_7\} = \{B_1,..., B_7\}$.

For every given (λ, γ) in $\Omega(T) = \{5/6 \le \lambda_1 < 1\}$, there exists B_7 of η^7 such that $(\lambda_1, \gamma_1) \in \Gamma(B_7) \times \Lambda(B_7)$. Set $\eta^7 = \eta_7$ and go to Step 9.

Step 9. Set $\xi^7 = \xi^6 \cup \eta_7 = \{B_1,..., B_7\}$, for $0 < \lambda_1 < 1$ and $0 < \gamma_1 < 1$ there is a basis B_j of ξ^7 such that $(\lambda_1, \gamma_1) \in \Gamma(B_j) \times \Lambda(B_j)$. Go to Step 10.

Table 6.22. The Potential Solution Q_4 in Stage 3

	D1	D2	D3
P1	2 (1) 3 (1)	5 (2) 5 (2)	(2, 3)
P2	(1, -5)	5 (1) 10 (1)	(4, -1)
P3	4 (2) 7 (2)	(0, 0)	4(3) 5(3)

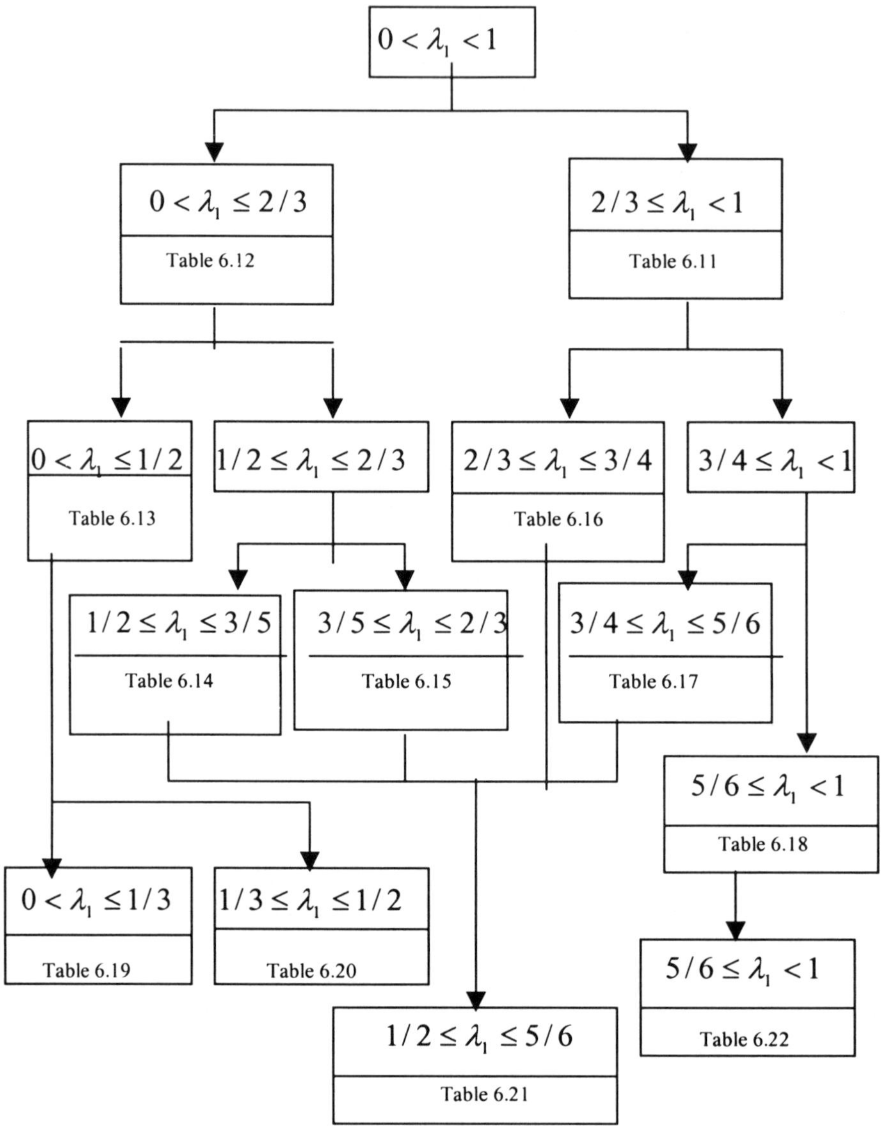

$0 < \lambda_1 < 1$
$0 < \lambda_1 \le 2/3$
Table 6.12
$2/3 \le \lambda_1 < 1$
Table 6.11
$0 < \lambda_1 \le 1/2$
Table 6.13
$1/2 \le \lambda_1 \le 2/3$
$2/3 \le \lambda_1 \le 3/4$
Table 6.16
$3/4 \le \lambda_1 < 1$
$1/2 \le \lambda_1 \le 3/5$
Table 6.14
$3/5 \le \lambda_1 \le 2/3$
Table 6.15
$3/4 \le \lambda_1 \le 5/6$
Table 6.17
$5/6 \le \lambda_1 < 1$
Table 6.18
$0 < \lambda_1 \le 1/3$
Table 6.19
$1/3 \le \lambda_1 \le 1/2$
Table 6.20
$5/6 \le \lambda_1 < 1$
Table 6.22
$1/2 \le \lambda_1 \le 5/6$
Table 6.21

Figure 6.1

Step 10. The whole procedure terminates because $h = 3 = t$.

The potential solutions with respect to λ in every stage can be shown in Figure 6.1.

6.5 Computer Software for MC² Transportation Problem

Hasse and Shi (1994) and Huang and Shi (1997) developed two computer software for solving the steady MC² transportation problems. There are two differences in the development of the software. The first one is that in the coding technology, Hasse and Shi (1994) used ANSI C and complied with the Borland C++ 4.0 Compiler, while Huang and Shi (1997) was written and compiled on Unix C++. The second difference is the implementation of Algorithm 6.1. Because Hasse and Shi (1994) directly followed the steps of Algorithm 6.1, the extensive coding works have been used on the determination of the (γ, λ) ranges for each feasible or potential solutions in the iterations. In Huang and Shi (1997), a steady MC² transportation problem is treated as a linear transportation problem with a fixed (γ^0, λ^0). Whenever an optimal solution of the linear problem is identified, a simulation technique is used to calculate the related (γ, λ) values. In such a way, we can find a set of potential solutions that covers the (γ, λ) space. The coding of Huang and Shi (1997) is much simpler than Haase and Shi (1994). As a result, given a certain large-size steady MC² transportation problem, the former is far more efficient (less CPU time) than the latter. In the following, we introduce the outline of Huang and Shi (1997) as an interesting comparison to Algorithm 6.1.

Algorithm 6.3 (Alternative to Perform Algorithm 6.1 by Computer):

Step 1. Select (γ^0, λ^0) randomly from [0,1], respectively.

Step 2. Convert a steady MC² transportation problem (M6.2) into a linear transportation problem with the fixed (γ^0, λ^0), then

Step 2.1 Apply Northwest corner rule to get an initial basic feasible solution;

Step 2.2 Check to see if this feasible solution is a potential solution for (γ^0, λ^0);

Step 2.3 If yes, go to Step 3; if no, continue;

Step 2.4 Determine the entering variable;

Step 2.5 Find the loop and determine the leaving variable;
Step 2.6 Pivot the loop, go to Step 2.2;

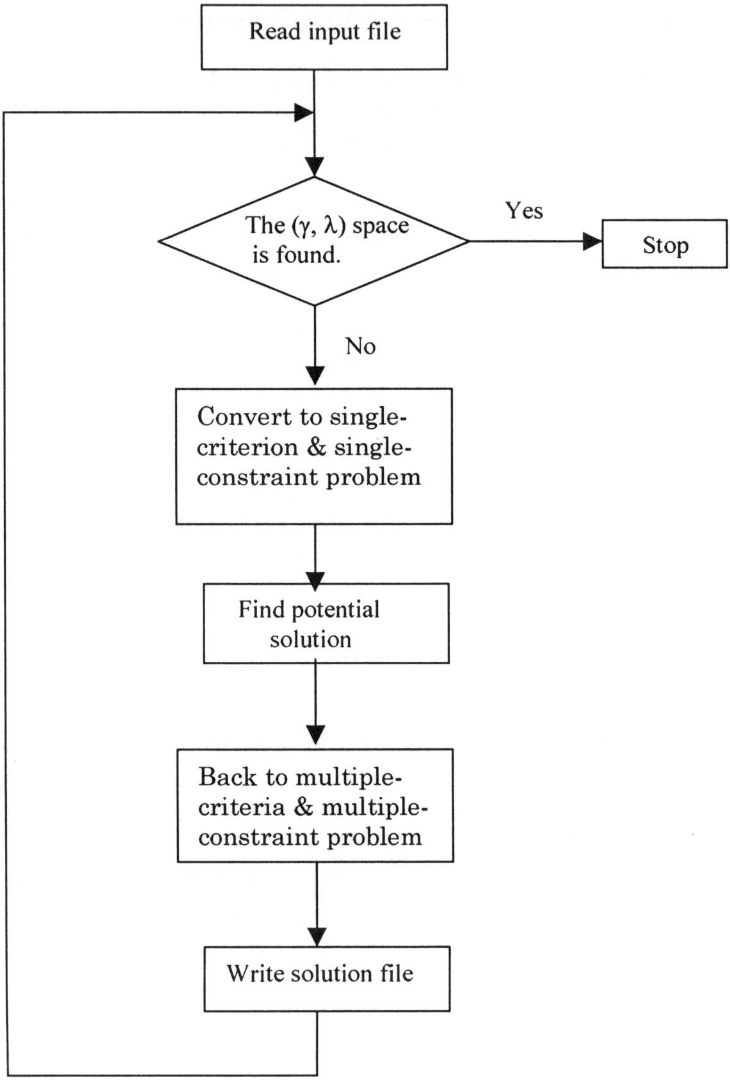

Figure 6. 2

Step 3. Convert the linear transportation problem with the fixed (γ^0, λ^0) back to the steady MC^2 transportation problem (M6.2) by using the basic variables found in the correspondent the MC^2 transportation problem.

Step 4. Calculate the (γ, λ) values for the resulting potential solution by simulation.

Step 5. Check to see if there exist (γ, λ) values from $[0,1]$ that are excluded from the calculated (γ, λ) range.

Step 6. If yes, go to Step 2. Otherwise, the set of potential solutions is found and the algorithm terminates.

A simple flowchart of Algorithm 6.3 is given as Figure 6.2. The design of both Hasse and Shi (1994) and Huang and Shi (1997) is similar to that of MC^2 program of Hao and Shi (1996). That is, the software consists of an input file, execution program, and output file. For the computational efficiency, we put Huang and Shi (1997) as part of the MC^2 software package attached in the back of the book. To the date, a research project is undertaken towards the implementation of Algorithm 6.2 to solve the dynamic MC^2 transportation problem (M6.3).

6.6 Further Remarks

This chapter has extended the known transportation model with single criterion and single constraint level to two different MC^2 transportation models. One is the steady MC^2 transportation model in which the distribution or shipment of the amount of goods from a factory to a warehouse is to be done in a given stage (season). Another is the dynamic MC^2 transportation model in which the distribution crosses several stages. The optimal solution of the second model is the best plan for all stages under consideration. This chapter has also discussed two computer software developed to solve the steady MC^2 transportation model. Theoretically speaking, similar software for solving the dynamic

MC^2 transportation model can be developed. This is an ongoing research project. The framework of the MC^2 transportation problem can be immediately applied to many research areas, such as aggregate production planning in production/operations management (Shi and Haase 1996), which will be explored in Chapter 21.

Another interesting research problem deals with constructing contingency plans for each potential solution of the MC^2 transportation problem. Recall that for a given potential solution, if there is some γ does not belong to the primal parameter set for the solution, then the solution becomes infeasible; and if there is some λ does not belong to the dual parameter set for the solution, then the solution is not optimal. In either case, contingency plans for the potential solution of the MC^2 transportation problem need to be constructed to recover the feasibility and maintain the optimality. Various approaches to constructing the contingency plans will be discussed in Chapter 9-12.

Exercises

6.1 Plateau Fruits supplies two kinds of Apples (Red Delicious and Golden Delicious) for three supermarkets. The unit shipping cost and unit shipping damage cost are given in Table 6.23 and Table 6.24, respectively. Assume that the supply level and demand level vary with the weather situations. If the weather is good for the Apple growth, the supply level to S_1 is 20 tons, S_2 is 20 tons, and S_3 is 20 tons; while the demand requirement level of RD is 50 tons and GD is 50 tons. If the weather is not good, the supply level to S_1 is 10 tons, S_2 is 15 tons, and S_3 is 20 tons; while the demand requirement level for RD is 30 tons and GD is 25 tons.

(i) Formulate a steady MC^2 transportation problem that minimizes both total shipping cost and total product damage cost during transportation simultaneously.

(ii) Solve this problem by hands through Algorithm 6.1.

(iii) Solve this problem by the software of Huang and Shi (1997).

Table 6.23 The Shipping Cost One Ton Apples ($10 per unit)

	S_1	S_2	S_3
RD	8	7	4
GD	6	8	5

Table 6.24 The Damage Cost of Apples during Transportation ($10 per unit)

	S_1	S_2	S_3
RD	4	2	3
GD	2	3	4

6.2 In the problem of 6.1, suppose the shipping can occur in three different seasons. The shipping cost and the damage cost of the first season are Table 6.23 and 6.24, respectively. The costs for the second and third seasons are given in Table 6.25 to Table 6.28.

(i) Formulate a dynamic MC^2 transportation problem with three seasons (stages) to minimize both total shipping cost and total product damage cost during transportation simultaneously.

(ii) Solve this problem by hands through Algorithm 6.2.

(iii) Compare the result with that of the problem 6.1. What is different?

Table 6.25 The Shipping Cost One Ton Apples for Season 2 ($10 per unit)

	S_1	S_2	S_3
RD	6	5	9
GD	7	4	6

Table 6.26 The Damage Cost of Apples during Transportation for Season 2 ($10 per unit)

	S_1	S_2	S_3
RD	2	3	1
GD	3	1	5

Table 6.27. The Shipping Cost One Ton Apples for Season 3 ($10 per unit)

	S_1	S_2	S_3
RD	2	10	5
GD	4	7	3

Table 6.28. The Damage Cost of Apples during Transportation for Season 3 ($10 per unit)

	S_1	S_2	S_3
RD	3	5	2
GD	4	5	3

6.3 Consider a steady transportation problem (Table 6.29) with two criteria (the upper values in the cell) and two constraint levels (a_{ij} for supplies and b_{ij} for demands).

(i) Formulate a steady MC^2 transportation problem that minimizes both criteria simultaneously.

(ii) Solve this problem by hands through Algorithm 6.1.

(iii) Solve this problem by the software of Huang and Shi (1997).

6.4 Use the data of Table 1.4 to Table 1.5 as well as the data described in the problem 1.4 to formulate a steady MC^2 transportation model for studying tradeoffs between the total transportation cost, the total damage rate, the bloom economy and the economy recession. Then, solve this problem by the software of Huang and Shi (1997).

6.5 Refer to the problem 1.9 and the data of Table 1.6 and Table 1.7, Midwest Electronics produces a popular VCR. The regular-time production cost of VCR is \$80 and the over-time production cost is \$100. If the products cannot be shipped out, the unit inventory holding cost per 6-month is \$2. When demand is higher than supply, the company will not consider the shortage. If the future economy is predicted as either "bloom" or "recession," it will affect the production capacity of the company as well as the market demand. Assume the company's aggregate production plan can be made for 6-month base. Then, the data to be used for minimizing the total production cost (regular plus over time) for the next year over fluctuations of the economy is given in Table 1.6. The data of Table 1.7 is used to minimize the total inventory holding cost for the next year.

(i) Formulate a steady MC^2 transportation problem for the company's aggregate production plan that minimizes both total production cost (regular plus over time) and inventory holding cost for the next year.

(ii) Solve this problem by the software of Huang and Shi (1997).

(iii) Interpret the meaning of the potential solutions.

Table 6.29 The Data MC Transportation Problem for Problem 6.3

	W_1		W_2		W_3		W_4		a_{11}	a_{12}
F_1	25	3	30	4	15	2	F_1	25	3	30
F_2	15	3	12	3	22	4	F_2	15	3	12
F_3	22	2	30	4	15	3	F_3	22	2	30
b_{11}	100		200		250		150		700	
b_{12}	300		200		100		200			800

6.6 Based on the information of the problem 6.5, create a dynamic aggregate production plan model and solve it by Algorithm 6.2.

Part Three: Fuzzy MC^2 Solutions

Chapter 7
FUZZY MC^2 PROBLEMS

It has been recognized that in many decision making problems, instead of finding the exist "optimal solution" (a goal value), decision makers often approach a "satisfying solution" between upper and lower aspiration levels that can be represented by the upper and lower bounds of acceptability for objective payoffs, respectively (Charnes and Cooper 1961, Lee 1972, Yu 1985, and Shi and Yu 1989). This behavior, which is has an important and pervasive impact on human decision making (Lindsay and Norman 1972), is called the decision makers' goal-seeking and compromise behavior. Zimmermann (1978) applied it as the basis of his pioneering work on fuzzy linear programming. The reader can recall this fuzzy satisficing algorithm from Section 3.1. As we notice that the fuzzy satisficing solution illustrated in Figure 3.1 is not a nondominated solution, but a near (or sub) nondominated solution.

In this chapter, we extend the fuzzy approach to tackle the MC^2 problems. Section 7.1 layouts a fundamental approach to fuzzy MC^2 problems. It explores a fuzzy potential solution for a given fuzzy MC^2 problem. We first treat the MC^2 problem as a problem with fuzzy optimal objectives. Next, we use the MC^2-simplex method to locate two sets of potential solutions for a maximum MC^2 problem and a minimum MC^2 problem, respectively. With these two sets of potential solutions, we reform the ranges of the decision parameters (γ, λ) so that each reformed range results in both the upper and lower bounds of the objective payoff over the changes of (γ, λ) within the range. Then, we construct a membership function for each of the ranges. Finally, for every reformed range, a fuzzy potential solution is determined in terms of the membership function and a primal potential basis. Section 7.2 is a special of the approach described in Section 7.1. We can view the basic variable $\mathbf{x}(J, \gamma) = \mathbf{B}^{-1}\mathbf{D}\gamma$ as a parameter form with respect the parameter γ. Then, we set the condition $\gamma_1 + \gamma_2 + \ldots + \gamma_p = 1$ as a constraint in the process of finding the aspiration level for the fuzzy MC^2 objective value. As a result, the fuzzy potential solution will be identified along with a specific γ^0 value, which must be included in one of fuzzy potential solutions discussed in Section 7.1. After our discussion for the fuzzy MC^2 problems, Section 7.3 concludes the chapter with further comments and reading.

7.1 Fuzzy Potential Solutions

Given a MC^2 linear programming (M4.1):

$$\begin{array}{ll} \textit{Max} & \lambda^t \mathbf{C}\,\mathbf{x} \\ \textit{Subject to} & \mathbf{A}\,\mathbf{x} \leq \mathbf{D}\gamma \\ & \mathbf{x} \geq 0 \end{array}$$

we now assume that the decision makers have a goal seeking and compromise behavior for a satisfying solution. In this section, the satisfying solution is represented by a fuzzy potential solution with a satisfying level between an upper aspiration bound and a lower aspiration bound of the objective payoff. To describe such a fuzzy potential solution, we first rewrite (M4.1) as the following fuzzy-objective MC^2 problem (M7.1):

$$\begin{array}{ll} \textit{Fuzzy-Max} & \lambda^t \mathbf{C}\,\mathbf{x} \\ \textit{Subject to} & \mathbf{A}\,\mathbf{x} \leq \mathbf{D}\gamma \\ & \mathbf{x} \geq 0 \end{array}$$

Then, we split (M7.1) into two crisp optimization problems: one is (M4.1) and another is given by (M7.2):

$$\begin{array}{ll} \textit{Min} & \lambda^t \mathbf{C}\,\mathbf{x} \\ \textit{Subject to} & \mathbf{A}\,\mathbf{x} \leq \mathbf{D}\gamma \\ & \mathbf{x} \geq 0 \end{array}$$

We note that (M7.2) is a minimum MC^2 problem, while (M4.1) is a maximum MC^2 problem. By using the MC^2-simplex method (see Section 4.3), we can identify two sets of potential solutions over all possible changes of (γ, λ) for (M4.1) and (M7.2), respectively. Let $\Phi = \{U_1, U_2, \ldots, U_r\}$ be the set of potential solutions over all possible changes of (γ, λ) for (M4.1) satisfying

$$(\Gamma(U_i) \times \Lambda(U_i)) \cap (\Gamma(U_g) \times \Lambda(U_g)) = \varnothing \text{ for } i \neq g;\ i, g = 1, \ldots, r,$$

and let $\Psi = \{L_1, L_2, \ldots, L_k\}$ be the set of potential solutions over all possible changes of (γ, λ) for (M7.2) satisfying

$$(\Gamma(L_j)\times\Lambda(L_j)) \cap (\Gamma(L_h)\times\Lambda(L_h)) = \emptyset \text{ for } j \neq h;\ j, h = 1, \ldots, k.$$

Then, given (M7.1), we may view U_i, i = 1, ..., r, as the "upper" potential solution for $\Gamma(U_i)\times\Lambda(U_i)$ and L_j, j = 1, ..., k, as the "lower" potential solution for $\Gamma(L_j)\times\Lambda(L_j)$. This means that for each $(\gamma, \lambda) \in \Gamma(U_i)\times\Lambda(U_i)$, $\mathbf{x}(U_i, \gamma)$ maximizes (M4.1), while for $(\gamma, \lambda) \in \Gamma(L_j)\times\Lambda(L_j)$, $\mathbf{x}(L_j, \gamma)$ minimizes (M7.2).

Let $\Gamma\times\Lambda$ be the Cartesian product space of parameters of (γ, λ). Then we have the following definition:

Definition 7.1 The space $\Gamma\times\Lambda$ is said to be covered by the set of potential solutions, $\vartheta = \{J_1, J_2, \ldots, J_r\}$ for (M4.1) if

$$\Gamma\times\Lambda \subseteq (\Gamma(J_1)\times\Lambda(J_1)) \cup \ldots \cup (\Gamma(J_r)\times\Lambda(J_r)).$$

Given $\Phi = \{U_1, U_2, \ldots, U_r\}$ and $\Psi = \{L_1, L_2, \ldots, L_k\}$, by Definition 7.1, we see that the space $\Gamma\times\Lambda$ is covered by both Φ and Ψ. To establish the upper and lower bounds of the objective payoff for a certain range of (γ, λ), we can reform the (γ, λ) ranges determined by Φ and Ψ by taking intersections over the parameter sets of elements of Φ and Ψ.

Let $R(i, j) = (\Gamma(U_i)\times\Lambda(U_i)) \cap (\Gamma(L_j)\times\Lambda(L_j)) \neq \emptyset$, i = 1, ..., r; j, = 1, ..., k, be the reformed (γ, λ) ranges and $\Re = \cup\ \{R(i, j) \mid \text{for all } (i, j)\}$ be a reformed the (γ, λ) space. Then, we derive the following results:

Theorem 7.1

(i) $R(i, j) \subseteq \Gamma(U_i)\times\Lambda(U_i)$ and $R(i, j) \subseteq \Gamma(L_j)\times\Lambda(L_j)$, for all i = 1, ..., r; j, = 1, ..., k.

(ii) $R(i, j) \cap R(g, h) = \emptyset$ for $i \neq g$; $j \neq h$; i, g = 1, ..., r; j, h = 1, ..., k.

(iii) $\Gamma\times\Lambda \subseteq \Re$.

Proof. (i) This follows from the definition of $R(i, j)$.

(ii) Note that $R(i, j) \cap R(g, h) = ((\Gamma(U_i)\times\Lambda(U_i)) \cap (\Gamma(L_j)\times\Lambda(L_j))) \cap ((\Gamma(U_g)\times\Lambda(U_g)) \cap (\Gamma(L_h)\times\Lambda(L_h)))$. By the associative law of the ordinary set theory, the right hand side of the equality is:

$((\Gamma(U_i)\times\Lambda(U_i)) \cap (\Gamma(U_g)\times\Lambda(U_g))) \cap ((\Gamma(L_j)\times\Lambda(L_j)) \cap (\Gamma(L_h)\times\Lambda(L_h))) = \varnothing\cap\varnothing = \varnothing$, for $i \neq g$; $j \neq h$; i, g = 1, ..., r; j, h = 1, ..., k. Thus, $R(i, j) \cap R(g, h) = \varnothing$.

(iii) Since the space $\Gamma\times\Lambda$ is covered by both Φ and Ψ,
$\Gamma\times\Lambda \subseteq \cup\{\Gamma(U_i)\times\Lambda(U_i) \mid i = 1, ..., r\}$ and $\Gamma\times\Lambda \subseteq \cup\{\Gamma(L_j)\times\Lambda(L_j) \mid j = 1, ..., k\}$. This implies that
$\Gamma\times\Lambda \subseteq \{\cup\{\Gamma(U_i)\times\Lambda(U_i) \mid i = 1, ..., r\}\}\cap\{\cup\{\Gamma(L_j)\times\Lambda(L_j) \mid j = 1, ..., k\}\}$.
By the distributive law of the ordinary set theory, we see that
$\{\cup\{\Gamma(U_i)\times\Lambda(U_i) \mid i = 1, ..., r\}\}\cap\{\cup\{\Gamma(L_j)\times\Lambda(L_j) \mid j = 1, ..., k\}\}$
$= \cup\{\Gamma(U_i)\times\Lambda(U_i) \cap \Gamma(L_j)\times\Lambda(L_j) \mid \text{for all } (i, j)\}$.
For $\Gamma(U_i)\times\Lambda(U_i) \cap \Gamma(L_j)\times\Lambda(L_j) \neq \varnothing$, i = 1, ..., r; j, = 1, ..., k, the right hand side of the last equality is $\Re = \cup \{R(i, j) \mid \text{for all } (i, j)\}$. Thus, $\Gamma\times\Lambda \subseteq \Re$.

Remark 7.1 Since $\Gamma\times\Lambda$ is the space of all (γ, λ), we see that $\Gamma\times\Lambda \supseteq \Re$. By (iii) of Theorem 7.1, $\Gamma\times\Lambda = \Re$.

Theorem 7.1 shows characteristics of the reformed (γ, λ) ranges on $\Re$. (i) of Theorem 7.1 says that such a range on $\Re$ is contained in both a (γ, λ) range determined by an element of Φ and a (γ, λ) range determined by an element of Ψ. This guarantees that given a range R(i, j), there is a corresponding upper potential solution from an element of Ψ. (ii) of Theorem 7.1 says that any two reformed (γ, λ) ranges on $\Re$ are not overlapped, which allows us to construct a fuzzy membership function for each (γ, λ) range on $\Re$ without worrying about overlapping situations. (iii) of Theorem 7.1 states that the resulting $\Re$ is covered by the space $\Gamma\times\Lambda$. This leads us to find a set of fuzzy potential solutions for all possible changes of (γ, λ). Thus, from Theorem 7.1, we have the following theorem.

Theorem 7.2

(i) For a given R(i, j) on $\Re$, $\mathbf{x}(U_i, \gamma)$ is its upper potential solution and $\mathbf{x}(L_j, \gamma)$ is its lower potential solution.

(ii) For any $(\gamma, \lambda) \in \Gamma\times\Lambda$, there is a corresponding maximum solution resulting from an element of Φ and a corresponding minimum solution resulting from an element of Ψ.

Proof.
(i) This follows from the definition of $R(\mathrm{i}, \mathrm{j})$ and (i) of Theorem 7.1.
(ii) By (iii) of Theorem 7.1, $\Gamma \times \Lambda \subseteq \Re$. Hence, for any $(\gamma, \lambda) \in \Gamma \times \Lambda$, there is a (γ, λ) range on $\Re$, say $R(\mathrm{i}, \mathrm{j})$, such that $(\gamma, \lambda) \in R(\mathrm{i}, \mathrm{j})$. By (i), $\mathbf{x}(U_{\mathrm{i}}, \gamma)$ is the maximum solution and $\mathbf{x}(L_{\mathrm{j}}, \gamma)$ is the minimum solution with respect to (γ, λ).

Now we discuss how to construct a fuzzy membership function for each reformed (γ, λ) range on $\Re$. Given a reformed (γ, λ) range on $\Re$, both its upper potential solution $\mathbf{x}(U_{\mathrm{i}}, \gamma)$ and lower potential solution $\mathbf{x}(L_{\mathrm{j}}, \gamma)$ are specified. That is, let $\mathbf{B}_U$ be the basis for $\mathbf{x}(U_{\mathrm{i}}, \gamma)$ and the corresponding objective payoff be denoted by $\mathrm{V}(U_{\mathrm{i}}; \lambda, \gamma) = \lambda^t \mathbf{C_B} \mathbf{B}_U^{-1} \mathbf{D}\gamma$ and let $\mathbf{B}_L$ be the basis for $\mathbf{x}(L_{\mathrm{j}}, \gamma)$ and the corresponding objective payoff be denoted by $\mathrm{V}(L_{\mathrm{j}}; \lambda, \gamma) = \lambda^t \mathbf{C_B} \mathbf{B}_L^{-1} \mathbf{D}\gamma$. Then, whenever any $(\gamma, \lambda) \in R(\mathrm{i}, \mathrm{j})$ are specified, $\mathrm{V}(U_{\mathrm{i}}; \lambda, \gamma)$ is the upper aspiration bound with respect U_{i}, while $\mathrm{V}(L_{\mathrm{j}}; \lambda, \gamma)$ is the lower aspiration bound with respect to L_{j}. By applying the fuzzy approach methodologies in linear programming, a linear membership function of any $R(\mathrm{i}, \mathrm{j})$ on $\Re$ for (M7.1) can be construct as (M7.3):

$$\mu_{R(i,j)}(\mathbf{x}) = \begin{cases} 1, & \lambda^t \mathbf{C}\mathbf{x} \geq \mathrm{V}(U_i;\gamma,\lambda) \\ \dfrac{\lambda^t \mathbf{C}\mathbf{x} - \mathrm{V}(L_j;\gamma,\lambda)}{\mathrm{V}(U_i;\gamma,\lambda) - \mathrm{V}(L_j;\gamma,\lambda)}, & \mathrm{V}(L_j;\gamma,\lambda) < \lambda^t \mathbf{C}\mathbf{x} < \mathrm{V}(U_i;\gamma,\lambda) \\ 0, & \mathrm{V}(L_j;\gamma,\lambda) \geq \lambda^t \mathbf{C}\mathbf{x} \end{cases}$$

From (M7.3), we see that if $\lambda^t \mathbf{C}\mathbf{x} = \mathrm{V}(U_i;\gamma,\lambda)$, then $\mu_{R(i,j)}(\mathbf{x}) = 1$ and $\mathbf{x}(U_{\mathrm{i}}, \gamma)$ is the maximum solution for (M7.1); if $\lambda^t \mathbf{C}\mathbf{x} = \mathrm{V}(L_j;\gamma,\lambda)$, then $\mu_{R(i,j)}(\mathbf{x}) = 0$ and $\mathbf{x}(L_{\mathrm{j}}, \gamma)$ is the minimum solution for (M7.1). The fuzzy potential solutions can be defined in terms of the membership function (M7.3).

Definition 7.2 Let K be a primal potential basis of (M7.1) with the basic variables $\mathbf{x}(K, \gamma)$ and the corresponding primal parameter set $\Gamma(K)$.

$\mathbf{x}(K, \gamma)$ is said to be a fuzzy potential solution of R(i, j) for (M7.1) if $\Gamma(U_i) \cap \Gamma(L_j) \subseteq \Gamma(K)$ and $\mathbf{x}(K, \gamma)$ has a membership function (M7.3). In this case, K is called a fuzzy potential basis of R(i, j).

Remark 7.2 By Definition 7.2, a primal potential basis K of (M7.1) could be as a fuzzy potential basis for several reformed (γ, λ) ranges on $\Re$.

Definition 7.3 Let ($\mathbf{x}(K, \gamma)$; $\mu_{R(i,j)}(\mathbf{x})$) be a fuzzy potential solution of R(i, j) for (M7.1). The set $\Im = \{(\mathbf{x}(K, \gamma);\ \mu_{R(i,j)}(\mathbf{x})\ ) \mid$ for all R(i, j) and $K\}$ is called a set of fuzzy potential solutions for all possible changes of $(\gamma, \lambda) \in \Gamma \times \Lambda$.

Remark 7.3 Given of $(\gamma^0, \lambda^0) \in R$(i, j), the solution ($\mathbf{x}(K, \gamma^0)$; $\mu_{R(i,j)}(\mathbf{x})$) is a fuzzy optimal solution for the linear programming with (γ^0, λ^0) resulting from (M7.1).

Based on the above discussion, the relationship between fuzzy potential solution and potential solution is summarized as follows.

Theorem 7.3

(i) Given a fuzzy potential solution ($\mathbf{x}(K, \gamma)$; $\mu_{R(i,j)}(\mathbf{x})$), there exists an upper and a lower potential solution for R(i, j).

(ii) Conversely, with an upper and a lower potential solution of R(i, j), a fuzzy potential solution ($\mathbf{x}(K, \gamma)$; $\mu_{R(i,j)}(\mathbf{x})$) can be constructed.

Proof. (i) Given a fuzzy potential solution ($\mathbf{x}(K, \gamma)$; $\mu_{R(i,j)}(\mathbf{x})$), if K is U_i or L_j, then the proof is complete. Otherwise, we solve (M4.1) to obtain the upper potential solution for R(i, j) and solve (M7.2) to obtain the lower potential solution for R(i, j). (ii) This follows from (I) of Theorem 7.2, the membership function (M7.3) and Definition 7.2.

(i) of Theorem 7.3 basically shows that both the upper and lower potential solutions are extreme cases of a fuzzy potential solutions for R(i, j) on $\Re$. (ii) of Theorem 7.3 allows us to construct a fuzzy potential solution from given upper and lower potential solutions for a $R(\mathrm{i}, \mathrm{j})$. A heuristic algorithm for effectively and systematically constructing all fuzzy potential solutions for all possible changes of (γ, λ) of (M7.1) is given as follows:

Algorithm 7.1

Step 1. Given a fuzzy-objective MC^2 problem (M7.1), identify the set of all primal potential solutions (feasible) bases, denoted by $\varsigma = \{K_1, K_2, \ldots, K_v\}$ using the MC^2-simplex method (see Section 4.3).

Step 2. Break the problem (M7.1) into the maximum problem (M4.1) and the minimum problem (M7.2). Then use the MC^2-simplex method to obtain $\Phi = \{U_1, U_2, \ldots, U_r\}$ for (M4.1) and $\Psi = \{L_1, L_2, \ldots, L_k\}$ for (M7.2). Since the constraints of (M7.1), (M4.1) and (M7.2) are identical, $\Phi \subseteq \varsigma$ and $\Psi \subseteq \varsigma$. Thus, Φ and Ψ can be readily found from ς.

Step 3. Based on Φ and Ψ, form the (γ, λ) range:
$R(\mathrm{i}, \mathrm{j}) = (\Gamma(U_\mathrm{i}) \times \Lambda(U_\mathrm{i})) \cap (\Gamma(L_\mathrm{j}) \times \Lambda(L_\mathrm{j})) \neq \emptyset$, i = 1, ..., r; j, = 1, ..., k, and the (γ, λ) space $\Re = \cup \{R(\mathrm{i}, \mathrm{j}) \mid \text{for all } (\mathrm{i}, \mathrm{j})\}$.

Step 4. For each $R(\mathrm{i}, \mathrm{j})$ on $\Re$, construct the fuzzy membership function (M7.3).

Step 5. For each $R(\mathrm{i}, \mathrm{j})$ on $\Re$, identify the corresponding fuzzy potential solution with the membership function. This can be readily done by checking $K \in \varsigma \backslash (\Phi \cup \Psi)$ for $\Gamma(U_\mathrm{i}) \cap \Gamma(L_\mathrm{j}) \subseteq \Gamma(K)$. By Definition 7.2, such a $\mathbf{x}(K, \gamma)$ is a fuzzy potential solution with a membership function $\mu_{R(i,j)}(\mathbf{x})$. Note that the fuzzy potential solution may not be unique for $R(\mathrm{i}, \mathrm{j})$.

Step 6. The algorithm terminates when the set of $\Im$ of Definition 7.3 is identified.

Example 7.1 Consider the following fuzzy-objective MC² problem:

$$Fuzzy-Max \begin{pmatrix} \lambda_1 & \lambda_2 \end{pmatrix} \begin{pmatrix} 2 & 1 & 1 & 0 & -1 \\ 1 & 0 & 1 & 2 & 2 \end{pmatrix} \begin{pmatrix} x_1 \\ x_2 \\ x_3 \\ x_4 \\ x_5 \end{pmatrix}$$

$$Subject\ to \begin{pmatrix} 1 & 0 & 2 & 1 & 0 \\ 0 & 1 & 1 & 2 & 2 \\ 1 & 1 & 0 & 0 & 1 \end{pmatrix} \begin{pmatrix} x_1 \\ x_2 \\ x_3 \\ x_4 \\ x_5 \end{pmatrix} \leq \begin{pmatrix} 20 & 50 \\ 40 & 20 \\ 30 & 40 \end{pmatrix} \begin{pmatrix} \gamma_1 \\ \gamma_2 \end{pmatrix}$$

$$x_i \geq 0, \mathrm{i} = 1,\ 2,\ 3,\ 4, 5.$$

By using Algorithm 7.1, we have:

Step 1. Let $x_i \geq 0$, I = 6, 7, 8 be slack variables. Then, by using the computer software of Chien, Shi and Yu (1989), we identified a set of all primal potential bases of the problem, denoted by and x_2 By using computer software of the MC²-simplex method, we solve this problem and obtain the set of all potential solutions, denoted by $\varsigma = \{K_1, K_2, \ldots, K_{27}\}$ shown as in Table 7.1.

Step 2. Break the given problem into the following maximum and minimum problems:

$$Max \quad \begin{pmatrix} \lambda_1 & \lambda_2 \end{pmatrix} \begin{pmatrix} 2 & 1 & 1 & 0 & -1 \\ 1 & 0 & 1 & 2 & 2 \end{pmatrix} \begin{pmatrix} x_1 \\ x_2 \\ x_3 \\ x_4 \\ x_5 \end{pmatrix}$$

$$Subject\ to \begin{pmatrix} 1 & 0 & 2 & 1 & 0 \\ 0 & 1 & 1 & 2 & 2 \\ 1 & 1 & 0 & 0 & 1 \end{pmatrix} \begin{pmatrix} x_1 \\ x_2 \\ x_3 \\ x_4 \\ x_5 \end{pmatrix} \le \begin{pmatrix} 20 & 50 \\ 40 & 20 \\ 30 & 40 \end{pmatrix} \begin{pmatrix} \gamma_1 \\ \gamma_2 \end{pmatrix}$$

$$x_i \ge 0, \mathrm{i} = 1,\ 2,\ 3,\ 4, 5.$$

$$Min \quad \begin{pmatrix} \lambda_1 & \lambda_2 \end{pmatrix} \begin{pmatrix} 2 & 1 & 1 & 0 & -1 \\ 1 & 0 & 1 & 2 & 2 \end{pmatrix} \begin{pmatrix} x_1 \\ x_2 \\ x_3 \\ x_4 \\ x_5 \end{pmatrix}$$

$$Subject\ to \begin{pmatrix} 1 & 0 & 2 & 1 & 0 \\ 0 & 1 & 1 & 2 & 2 \\ 1 & 1 & 0 & 0 & 1 \end{pmatrix} \begin{pmatrix} x_1 \\ x_2 \\ x_3 \\ x_4 \\ x_5 \end{pmatrix} \le \begin{pmatrix} 20 & 50 \\ 40 & 20 \\ 30 & 40 \end{pmatrix} \begin{pmatrix} \gamma_1 \\ \gamma_2 \end{pmatrix}$$

$$x_i \ge 0, \mathrm{i} = 1,\ 2,\ 3,\ 4, 5.$$

Let U_{i} be an upper potential basis for the above maximum problem and $\Gamma(U_{\mathrm{i}})$ and $\Lambda(U_{\mathrm{i}})$ be its corresponding primal and dual parameter sets, respectively, and $V(U_i; \lambda, \gamma)$ be its objective payoff, which is a function of (γ, λ). Let L_j be a lower potential basis for the above minimum problem, $\Gamma(L_j)$ and $\Lambda(L_j)$) be its corresponding primal and dual parameter bases, respectively, and $V(L_j; \lambda, \gamma)$ be its objective payoff. We obtain the set of upper potential bases, denoted by $\Phi = \{U_1, U_2, U_3, U_4\}$ (see Table 7.2) for the maximum problem and the set of lower potential bases, denoted by $\Psi = \{L_1, L_2\}$ (see Table 7.3) for the minimum problem. In Table 7.2-7.3, when γ_1 and λ_1 are given, γ_2 and λ_2 are uniquely specified due to $\gamma_1 + \gamma_2 = 1$ and $\lambda_1 + \lambda_2 = 1$. The set Φ is found by using (ii) of Definition 4.4 to check the maximality of K in ς, while Ψ is found by applying (ii) of Definition 4.4 to check the minimality of K in ς. Note that $\Phi \subseteq \varsigma$ and $\Psi \subseteq \varsigma$. For example, $U_1 = K_{18}$ and $L_1 = K_1$. The graphical representation of Φ is given in Figure 7.1 and that of Ψ is given in Figure 7.2.

Table 7.1 Primal Potential Bases ς

Bases	$\mathbf{x}(K_i, \gamma)$	Bases	$\mathbf{x}(K_i, \gamma)$
$K_1 = L_1$	(x_6, x_7, x_8)	K_{15}	(x_2, x_3, x_6)
K_2	(x_1, x_7, x_8)	K_{16}	(x_1, x_2, x_4)
K_3	(x_1, x_2, x_3)	K_{17}	(x_1, x_2, x_6)
$K_4 = L_2$	(x_5, x_6, x_8)	$K_{18} = U_1$	(x_1, x_2, x_7)
$K_5 = L_1$	(x_4, x_6, x_8)	K_{19}	(x_1, x_3, x_9)
K_6	(x_3, x_6, x_8)	$K_{20} = U_2$	(x_1, x_3, x_7)
K_7	(x_3, x_7, x_8)	K_{21}	(x_1, x_3, x_5)
K_8	(x_3, x_5, x_8)	$K_{22} = U_3$	(x_1, x_4, x_5)
K_9	(x_3, x_4, x_8)	$K_{23} = U_4$	(x_1, x_4, x_7)
K_{10}	(x_2, x_6, x_8)	K_{24}	(x_1, x_4, x_8)
K_{11}	(x_2, x_6, x_7)	K_{25}	(x_1, x_5, x_7)
K_{12}	(x_2, x_5, x_6)	K_{26}	(x_1, x_5, x_6)
K_{13}	(x_2, x_4, x_6)	K_{27}	(x_1, x_6, x_7)
K_{14}	(x_2, x_3, x_9)		

Table 7.2 Upper Potential Bases Φ

Bases	$\Gamma(U_i)$	$\Lambda(U_i)$	$V(U_i; \lambda, \gamma)$
U_1	$1/2 \le \gamma_1 \le 1$	$1/2 \le \lambda_1 \le 1$	$(\lambda_1 \quad \lambda_2)\begin{pmatrix} 50 & 90 \\ 20 & 50 \end{pmatrix}\begin{pmatrix} \gamma_1 \\ \gamma_2 \end{pmatrix}$
U_2	$0 < \gamma_1 \le 1/2$	$3/4 \le \lambda_1 \le 1$	$(\lambda_1 \quad \lambda_2)\begin{pmatrix} 55 & 85 \\ 25 & 45 \end{pmatrix}\begin{pmatrix} \gamma_1 \\ \gamma_2 \end{pmatrix}$
U_3	$0 < \gamma_1 \le 1$	$0 < \lambda_1 \le 1/2$	$(\lambda_1 \quad \lambda_2)\begin{pmatrix} 15 & 80 \\ 55 & 60 \end{pmatrix}\begin{pmatrix} \gamma_1 \\ \gamma_2 \end{pmatrix}$
U_4	$0 < \gamma_1 \le 1/2$	$1/2 \le \lambda_1 \le 3/4$	$(\lambda_1 \quad \lambda_2)\begin{pmatrix} 60 & 80 \\ 10 & 60 \end{pmatrix}\begin{pmatrix} \gamma_1 \\ \gamma_2 \end{pmatrix}$

Table 7.3 Lower Potential Bases Ψ

Bases	$\Gamma(L_j)$	$\Lambda(L_j)$	$V(L_j; \lambda, \gamma)$
L_1	$1/2 \le \gamma_1 \le 1$	$1/2 \le \lambda_1 \le 1$	$(\lambda_1 \quad \lambda_2)\begin{pmatrix} 0 & 0 \\ 0 & 0 \end{pmatrix}\begin{pmatrix} \gamma_1 \\ \gamma_2 \end{pmatrix}$
L_2	$0 < \gamma_1 \le 1/2$	$3/4 \le \lambda_1 \le 1$	$(\lambda_1 \quad \lambda_2)\begin{pmatrix} -20 & -10 \\ 40 & 20 \end{pmatrix}\begin{pmatrix} \gamma_1 \\ \gamma_2 \end{pmatrix}$

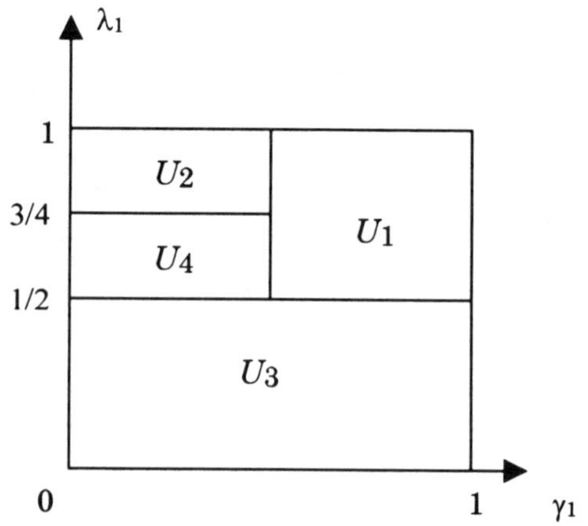

Figure 7.1 All Upper Potential Bases Φ

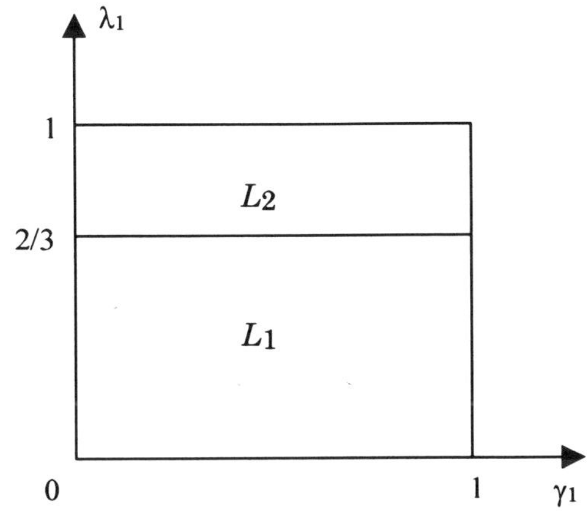

Figure 7.2 All Lower Potential Bases ψ

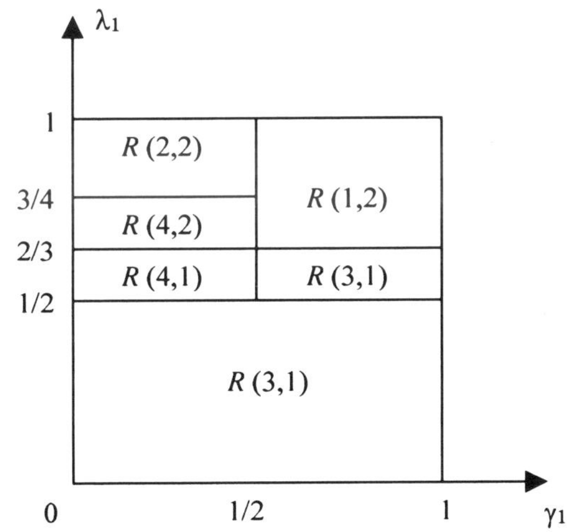

Figure 7.3 All Reformed Ranges $R(\mathrm{i}, \mathrm{j})$

Step 3. Based on Φ and Ψ, we form the (γ, λ) range:
$R(\mathrm{i, j}) = (\Gamma(U_i)\times\Lambda(U_i)) \cap (\Gamma(L_j)\times\Lambda(L_j)) \neq \varnothing$, i = 1, 2, 3, 4; j, = 1, 2, and the (γ, λ) space $\Re = \cup \{R(\mathrm{i, j}) \mid \text{for all (i, j)}\}$ as follows:

$R(1, 1) = \{ (\gamma_1, \lambda_1) \mid 1/2 \le \gamma_1 \le 1,\ 1/2 \le \lambda_1 \le 2/3\}$,
$R(1, 2) = \{ (\gamma_1, \lambda_1) \mid 1/2 \le \gamma_1 \le 1,\ 2/3 \le \lambda_1 \le 1\}$,
$R(2, 2) = \{ (\gamma_1, \lambda_1) \mid 0 < \gamma_1 \le 1/2,\ 3/4 \le \lambda_1 \le 1\}$,
$R(3, 1) = \{ (\gamma_1, \lambda_1) \mid 0 < \gamma_1 \le 1,\ 0 < \lambda_1 \le 1/2\}$,
$R(4, 1) = \{ (\gamma_1, \lambda_1) \mid 0 < \gamma_1 \le 1/2,\ 1/2 \le \lambda_1 \le 2/3\}$, and
$R(4, 2) = \{ (\gamma_1, \lambda_1) \mid 0 < \gamma_1 \le 1/2,\ 2/3 \le \lambda_1 \le 3/4\}$; and

$\Re = \{R(1, 1), R(1, 2), R(2, 2), R(3, 1), R(4, 1), R(4, 2)\}$, which is graphically shown as Figure 7.3.

Step 4. For each $R(\mathrm{i, j})$ on $\Re$, we construct the fuzzy membership function (M7.3) as follows:

$$\mu_{R(1,1)}(\mathbf{x}) = \begin{cases} 1, & \lambda^t\mathbf{C}\,\mathbf{x} \ge \mathrm{V}(U_1;\gamma,\lambda) \\ \dfrac{\lambda^t\mathbf{C}\,\mathbf{x} - \mathrm{V}(L_1;\gamma,\lambda)}{\mathrm{V}(U_1;\gamma,\lambda) - \mathrm{V}(L_1;\gamma,\lambda)}, & \mathrm{V}(L_1;\gamma,\lambda) < \lambda^t\mathbf{C}\,\mathbf{x} < \mathrm{V}(U_1;\gamma,\lambda) \\ 0, & \mathrm{V}(L_1;\gamma,\lambda) \ge \lambda^t\mathbf{C}\,\mathbf{x} \end{cases}$$

$$\mu_{R(1,2)}(\mathbf{x}) = \begin{cases} 1, & \lambda^t\mathbf{C}\,\mathbf{x} \ge \mathrm{V}(U_1;\gamma,\lambda) \\ \dfrac{\lambda^t\mathbf{C}\,\mathbf{x} - \mathrm{V}(L_2;\gamma,\lambda)}{\mathrm{V}(U_1;\gamma,\lambda) - \mathrm{V}(L_2;\gamma,\lambda)}, & \mathrm{V}(L_2;\gamma,\lambda) < \lambda^t\mathbf{C}\,\mathbf{x} < \mathrm{V}(U_1;\gamma,\lambda) \\ 0, & \mathrm{V}(L_2;\gamma,\lambda) \ge \lambda^t\mathbf{C}\,\mathbf{x} \end{cases}$$

$$\mu_{R(2,2)}(\mathbf{x}) = \begin{cases} 1, & \lambda^t\mathbf{C}\,\mathbf{x} \ge \mathrm{V}(U_2;\gamma,\lambda) \\ \dfrac{\lambda^t\mathbf{C}\,\mathbf{x} - \mathrm{V}(L_2;\gamma,\lambda)}{\mathrm{V}(U_2;\gamma,\lambda) - \mathrm{V}(L_2;\gamma,\lambda)}, & \mathrm{V}(L_2;\gamma,\lambda) < \lambda^t\mathbf{C}\,\mathbf{x} < \mathrm{V}(U_2;\gamma,\lambda) \\ 0, & \mathrm{V}(L_2;\gamma,\lambda) \ge \lambda^t\mathbf{C}\,\mathbf{x} \end{cases}$$

$$\mu_{R(3,1)}(\mathbf{x}) = \begin{cases} 1, & \lambda^t\mathbf{C}\,\mathbf{x} \ge \mathrm{V}(U_3;\gamma,\lambda) \\ \dfrac{\lambda^t\mathbf{C}\,\mathbf{x} - \mathrm{V}(L_1;\gamma,\lambda)}{\mathrm{V}(U_3;\gamma,\lambda) - \mathrm{V}(L_1;\gamma,\lambda)}, & \mathrm{V}(L_1;\gamma,\lambda) < \lambda^t\mathbf{C}\,\mathbf{x} < \mathrm{V}(U_3;\gamma,\lambda) \\ 0, & \mathrm{V}(L_1;\gamma,\lambda) \ge \lambda^t\mathbf{C}\,\mathbf{x} \end{cases}$$

$$\mu_{R(4,1)}(\mathbf{x}) = \begin{cases} 1, & \boldsymbol{\lambda}^t \mathbf{C}\,\mathbf{x} \geq \mathrm{V}(U_4;\gamma,\lambda) \\ \dfrac{\boldsymbol{\lambda}^t \mathbf{C}\,\mathbf{x} - \mathrm{V}(L_1;\gamma,\lambda)}{\mathrm{V}(U_4;\gamma,\lambda) - \mathrm{V}(L_1;\gamma,\lambda)}, & \mathrm{V}(L_1;\gamma,\lambda) < \boldsymbol{\lambda}^t \mathbf{C}\,\mathbf{x} < \mathrm{V}(U_4;\gamma,\lambda) \\ 0, & \mathrm{V}(L_1;\gamma,\lambda) \geq \boldsymbol{\lambda}^t \mathbf{C}\,\mathbf{x} \end{cases}$$

and

$$\mu_{R(4,2)}(\mathbf{x}) = \begin{cases} 1, & \boldsymbol{\lambda}^t \mathbf{C}\,\mathbf{x} \geq \mathrm{V}(U_4;\gamma,\lambda) \\ \dfrac{\boldsymbol{\lambda}^t \mathbf{C}\,\mathbf{x} - \mathrm{V}(L_2;\gamma,\lambda)}{\mathrm{V}(U_4;\gamma,\lambda) - \mathrm{V}(L_2;\gamma,\lambda)}, & \mathrm{V}(L_2;\gamma,\lambda) < \boldsymbol{\lambda}^t \mathbf{C}\,\mathbf{x} < \mathrm{V}(U_4;\gamma,\lambda) \\ 0, & \mathrm{V}(L_2;\gamma,\lambda) \geq \boldsymbol{\lambda}^t \mathbf{C}\,\mathbf{x} \end{cases}$$

Table 7.4 Fuzzy Potential solutions for $R(1, 1)$

Bases	$\mathrm{V}(K_i;\lambda,\gamma)$	Bases	$\mathrm{V}(K_i;\lambda,\gamma)$
K_2	$(\lambda_1 \ \ \lambda_2)\begin{pmatrix} 40 & 100 \\ 20 & 50 \end{pmatrix}\begin{pmatrix} \gamma_1 \\ \gamma_2 \end{pmatrix}$	K_{14}	$(\lambda_1 \ \ \lambda_2)\begin{pmatrix} 40 & 20 \\ 10 & 25 \end{pmatrix}\begin{pmatrix} \gamma_1 \\ \gamma_2 \end{pmatrix}$
K_3	$(\lambda_1 \ \ \lambda_2)\begin{pmatrix} 40 & 80 \\ 10 & 40 \end{pmatrix}\begin{pmatrix} \gamma_1 \\ \gamma_2 \end{pmatrix}$	K_{16}	$(\lambda_1 \ \ \lambda_2)\begin{pmatrix} 40 & 80 \\ 30 & 60 \end{pmatrix}\begin{pmatrix} \gamma_1 \\ \gamma_2 \end{pmatrix}$
K_5	$(\lambda_1 \ \ \lambda_2)\begin{pmatrix} 0 & 0 \\ 40 & 20 \end{pmatrix}\begin{pmatrix} \gamma_1 \\ \gamma_2 \end{pmatrix}$	K_{21}	$(\lambda_1 \ \ \lambda_2)\begin{pmatrix} 10 & 70 \\ 51 & 52 \end{pmatrix}\begin{pmatrix} \gamma_1 \\ \gamma_2 \end{pmatrix}$
K_7	$(\lambda_1 \ \ \lambda_2)\begin{pmatrix} 10 & 25 \\ 10 & 25 \end{pmatrix}\begin{pmatrix} \gamma_1 \\ \gamma_2 \end{pmatrix}$	K_{24}	$(\lambda_1 \ \ \lambda_2)\begin{pmatrix} 0 & 80 \\ 40 & 60 \end{pmatrix}\begin{pmatrix} \gamma_1 \\ \gamma_2 \end{pmatrix}$
K_8	$(\lambda_1 \ \ \lambda_2)\begin{pmatrix} -5 & 27.5 \\ 40 & 20 \end{pmatrix}\begin{pmatrix} \gamma_1 \\ \gamma_2 \end{pmatrix}$	K_{25}	$(\lambda_1 \ \ \lambda_2)\begin{pmatrix} 30 & 110 \\ 40 & 30 \end{pmatrix}\begin{pmatrix} \gamma_1 \\ \gamma_2 \end{pmatrix}$
K_9	$(\lambda_1 \ \ \lambda_2)\begin{pmatrix} 0 & 26.7 \\ 40 & 20 \end{pmatrix}\begin{pmatrix} \gamma_1 \\ \gamma_2 \end{pmatrix}$	K_{26}	$(\lambda_1 \ \ \lambda_2)\begin{pmatrix} 0 & 50 \\ 50 & 50 \end{pmatrix}\begin{pmatrix} \gamma_1 \\ \gamma_2 \end{pmatrix}$

Steps 5-6. For each R(i, j) on $\Re$, identify the corresponding fuzzy potential solution with the membership function. This can be readily done by checking $K \in \varsigma\backslash(\Phi \cup \Psi)$ for $\Gamma(U_i) \cap \Gamma(L_j) \subseteq \Gamma(K)$. By Definition 7.2, such a $\mathbf{x}(K, \gamma)$ is a fuzzy potential solution with a membership function $\mu_{R(i,j)}(\mathbf{x})$. Since $\varsigma\backslash(\Phi \cup \Psi) = \{K_2, K_3, K_5, K_6, K_7, K_8, K_9, K_{10}, K_{11}, K_{12}, K_{13}, K_{14}, K_{15}, K_{16}, K_{17}, K_{19}, K_{21}, K_{24}, K_{25}, K_{26}, K_{27}\}$, we obtain all fuzzy potential solution with membership functions for all R(i, j) as shown in Tables 7.4-7.9. Note that a primal potential solution $\mathbf{x}(K, \gamma)$ could become a fuzzy potential solution for several R(i, j). For example, K_3 is a potential solution for all R(i, j) because $\Gamma(K_3) = \{\gamma > 0 \mid 0 \le \gamma_1, \gamma_2 \le 0, \gamma_1 + \gamma_2 = 1\}$ contains γ ranges in all R(i, j). However, the λ ranges of K_3 change with R(i, j). To demonstrate that $\mathbf{x}(K, \gamma)$ is determined by a reformed (γ_1, λ_1) range R(i, j) and the membership function $\mu_{R(i,j)}(\mathbf{x})$, which is a function of (γ_1, λ_1), we see that given $R(3, 1)$ and $\mu_{R(3,1)}(\mathbf{x})$, the corresponding fuzzy potential solutions are $\{\mathbf{x}(K_3, \gamma), \mathbf{x}(K_5, \gamma), \mathbf{x}(K_{16}, \gamma), \mathbf{x}(K_{21}, \gamma), \mathbf{x}(K_{24}, \gamma), \mathbf{x}(K_{26}, \gamma)\}$ in Table 7.7.

Table 7.5 Fuzzy Potential Solutions for R(1, 2)

Bases	$V(K_i; \lambda, \gamma)$	Bases	$V(K_i; \lambda, \gamma)$
K_2	$(\lambda_1 \ \ \lambda_2)\begin{pmatrix} 40 & 100 \\ 20 & 50 \end{pmatrix}\begin{pmatrix} \gamma_1 \\ \gamma_2 \end{pmatrix}$	K_{14}	$(\lambda_1 \ \ \lambda_2)\begin{pmatrix} 40 & 20 \\ 10 & 25 \end{pmatrix}\begin{pmatrix} \gamma_1 \\ \gamma_2 \end{pmatrix}$
K_3	$(\lambda_1 \ \ \lambda_2)\begin{pmatrix} 40 & 80 \\ 10 & 40 \end{pmatrix}\begin{pmatrix} \gamma_1 \\ \gamma_2 \end{pmatrix}$	K_{16}	$(\lambda_1 \ \ \lambda_2)\begin{pmatrix} 40 & 80 \\ 30 & 60 \end{pmatrix}\begin{pmatrix} \gamma_1 \\ \gamma_2 \end{pmatrix}$
K_5	$(\lambda_1 \ \ \lambda_2)\begin{pmatrix} 0 & 0 \\ 40 & 20 \end{pmatrix}\begin{pmatrix} \gamma_1 \\ \gamma_2 \end{pmatrix}$	K_{21}	$(\lambda_1 \ \ \lambda_2)\begin{pmatrix} 10 & 70 \\ 51 & 52 \end{pmatrix}\begin{pmatrix} \gamma_1 \\ \gamma_2 \end{pmatrix}$
K_7	$(\lambda_1 \ \ \lambda_2)\begin{pmatrix} 10 & 25 \\ 10 & 25 \end{pmatrix}\begin{pmatrix} \gamma_1 \\ \gamma_2 \end{pmatrix}$	K_{24}	$(\lambda_1 \ \ \lambda_2)\begin{pmatrix} 0 & 80 \\ 40 & 60 \end{pmatrix}\begin{pmatrix} \gamma_1 \\ \gamma_2 \end{pmatrix}$
K_8	$(\lambda_1 \ \ \lambda_2)\begin{pmatrix} -5 & 27.5 \\ 40 & 20 \end{pmatrix}\begin{pmatrix} \gamma_1 \\ \gamma_2 \end{pmatrix}$	K_{25}	$(\lambda_1 \ \ \lambda_2)\begin{pmatrix} 30 & 110 \\ 40 & 30 \end{pmatrix}\begin{pmatrix} \gamma_1 \\ \gamma_2 \end{pmatrix}$
K_9	$(\lambda_1 \ \ \lambda_2)\begin{pmatrix} 0 & 26.7 \\ 40 & 20 \end{pmatrix}\begin{pmatrix} \gamma_1 \\ \gamma_2 \end{pmatrix}$	K_{26}	$(\lambda_1 \ \ \lambda_2)\begin{pmatrix} 0 & 50 \\ 50 & 50 \end{pmatrix}\begin{pmatrix} \gamma_1 \\ \gamma_2 \end{pmatrix}$

For the illustrative purpose, the membership function $\mu_{R(3,1)}(\mathbf{x})$ of the K_3, a function of (γ_1, λ_1), is shown in Figure 7.4. The graphical representation of the membership functions for other fuzzy potential solutions of R(3, 1) can be similarly shown. Finally, we note that fuzzy potential solutions listed in Tables 7.4-7.9 consist of $\mathfrak{I}$, the set of fuzzy potential solutions for all possible changes of $(\gamma, \lambda) \in \Gamma \times \Lambda$. The bases of Tables 7.4 and 7.5 are the same, while Table 7.6, 7.8, and 7.9 are identical.

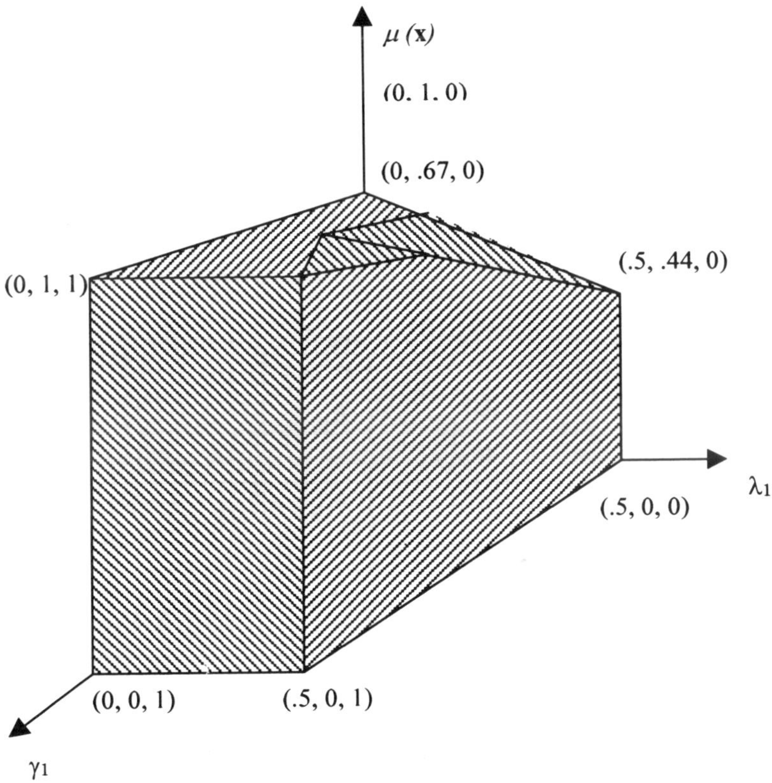

Figure 7. 4 The Membership Function of K_3 for R(3, 1)

Table 7.6 Fuzzy Potential solutions for $R(2, 2)$

Bases	$V(K_i; \lambda, \gamma)$	Bases	$V(K_i; \lambda, \gamma)$
K_3	$\begin{pmatrix}\lambda_1 & \lambda_2\end{pmatrix}\begin{pmatrix}40 & 80\\ 10 & 40\end{pmatrix}\begin{pmatrix}\gamma_1\\ \gamma_2\end{pmatrix}$	K_{21}	$\begin{pmatrix}\lambda_1 & \lambda_2\end{pmatrix}\begin{pmatrix}10 & 70\\ 51 & 52\end{pmatrix}\begin{pmatrix}\gamma_1\\ \gamma_2\end{pmatrix}$
K_5	$\begin{pmatrix}\lambda_1 & \lambda_2\end{pmatrix}\begin{pmatrix}0 & 0\\ 40 & 20\end{pmatrix}\begin{pmatrix}\gamma_1\\ \gamma_2\end{pmatrix}$	K_{24}	$\begin{pmatrix}\lambda_1 & \lambda_2\end{pmatrix}\begin{pmatrix}0 & 80\\ 40 & 60\end{pmatrix}\begin{pmatrix}\gamma_1\\ \gamma_2\end{pmatrix}$
K_{10}	$\begin{pmatrix}\lambda_1 & \lambda_2\end{pmatrix}\begin{pmatrix}40 & 20\\ 0 & 0\end{pmatrix}\begin{pmatrix}\gamma_1\\ \gamma_2\end{pmatrix}$	K_{26}	$\begin{pmatrix}\lambda_1 & \lambda_2\end{pmatrix}\begin{pmatrix}5 & 50\\ 50 & 50\end{pmatrix}\begin{pmatrix}\gamma_1\\ \gamma_2\end{pmatrix}$
K_{16}	$\begin{pmatrix}\lambda_1 & \lambda_2\end{pmatrix}\begin{pmatrix}40 & 80\\ 30 & 60\end{pmatrix}\begin{pmatrix}\gamma_1\\ \gamma_2\end{pmatrix}$	K_{27}	$\begin{pmatrix}\lambda_1 & \lambda_2\end{pmatrix}\begin{pmatrix}60 & 80\\ 30 & 40\end{pmatrix}\begin{pmatrix}\gamma_1\\ \gamma_2\end{pmatrix}$
K_{17}	$\begin{pmatrix}\lambda_1 & \lambda_2\end{pmatrix}\begin{pmatrix}20 & 60\\ -10 & 20\end{pmatrix}\begin{pmatrix}\gamma_1\\ \gamma_2\end{pmatrix}$		

Table 7.7 Fuzzy Potential solutions for $R(3, 1)$

Bases	$V(K_i; \lambda, \gamma)$	Bases	$V(K_i; \lambda, \gamma)$
K_3	$\begin{pmatrix}\lambda_1 & \lambda_2\end{pmatrix}\begin{pmatrix}40 & 80\\ 10 & 40\end{pmatrix}\begin{pmatrix}\gamma_1\\ \gamma_2\end{pmatrix}$	K_{21}	$\begin{pmatrix}\lambda_1 & \lambda_2\end{pmatrix}\begin{pmatrix}10 & 70\\ 51 & 52\end{pmatrix}\begin{pmatrix}\gamma_1\\ \gamma_2\end{pmatrix}$
K_5	$\begin{pmatrix}\lambda_1 & \lambda_2\end{pmatrix}\begin{pmatrix}0 & 0\\ 40 & 20\end{pmatrix}\begin{pmatrix}\gamma_1\\ \gamma_2\end{pmatrix}$	K_{24}	$\begin{pmatrix}\lambda_1 & \lambda_2\end{pmatrix}\begin{pmatrix}0 & 80\\ 40 & 60\end{pmatrix}\begin{pmatrix}\gamma_1\\ \gamma_2\end{pmatrix}$
K_{16}	$\begin{pmatrix}\lambda_1 & \lambda_2\end{pmatrix}\begin{pmatrix}40 & 80\\ 30 & 60\end{pmatrix}\begin{pmatrix}\gamma_1\\ \gamma_2\end{pmatrix}$	K_{26}	$\begin{pmatrix}\lambda_1 & \lambda_2\end{pmatrix}\begin{pmatrix}0 & 50\\ 50 & 50\end{pmatrix}\begin{pmatrix}\gamma_1\\ \gamma_2\end{pmatrix}$

Table 7.8 Fuzzy Potential solutions for $R(4, 1)$

Bases	$V(K_i; \lambda, \gamma)$	Bases	$V(K_i; \lambda, \gamma)$
K_3	$(\lambda_1 \quad \lambda_2)\begin{pmatrix} 40 & 80 \\ 10 & 40 \end{pmatrix}\begin{pmatrix} \gamma_1 \\ \gamma_2 \end{pmatrix}$	K_{21}	$(\lambda_1 \quad \lambda_2)\begin{pmatrix} 10 & 70 \\ 51 & 52 \end{pmatrix}\begin{pmatrix} \gamma_1 \\ \gamma_2 \end{pmatrix}$
K_5	$(\lambda_1 \quad \lambda_2)\begin{pmatrix} 0 & 0 \\ 40 & 20 \end{pmatrix}\begin{pmatrix} \gamma_1 \\ \gamma_2 \end{pmatrix}$	K_{24}	$(\lambda_1 \quad \lambda_2)\begin{pmatrix} 0 & 80 \\ 40 & 60 \end{pmatrix}\begin{pmatrix} \gamma_1 \\ \gamma_2 \end{pmatrix}$
K_{10}	$(\lambda_1 \quad \lambda_2)\begin{pmatrix} 40 & 20 \\ 0 & 0 \end{pmatrix}\begin{pmatrix} \gamma_1 \\ \gamma_2 \end{pmatrix}$	K_{26}	$(\lambda_1 \quad \lambda_2)\begin{pmatrix} 5 & 50 \\ 50 & 50 \end{pmatrix}\begin{pmatrix} \gamma_1 \\ \gamma_2 \end{pmatrix}$
K_{16}	$(\lambda_1 \quad \lambda_2)\begin{pmatrix} 40 & 80 \\ 30 & 60 \end{pmatrix}\begin{pmatrix} \gamma_1 \\ \gamma_2 \end{pmatrix}$	K_{27}	$(\lambda_1 \quad \lambda_2)\begin{pmatrix} 60 & 80 \\ 30 & 40 \end{pmatrix}\begin{pmatrix} \gamma_1 \\ \gamma_2 \end{pmatrix}$
K_{17}	$(\lambda_1 \quad \lambda_2)\begin{pmatrix} 20 & 60 \\ -10 & 20 \end{pmatrix}\begin{pmatrix} \gamma_1 \\ \gamma_2 \end{pmatrix}$		

Table 7.9 Fuzzy Potential solutions for $R(4, 2)$

Bases	$V(K_i; \lambda, \gamma)$	Bases	$V(K_i; \lambda, \gamma)$
K_3	$(\lambda_1 \quad \lambda_2)\begin{pmatrix} 40 & 80 \\ 10 & 40 \end{pmatrix}\begin{pmatrix} \gamma_1 \\ \gamma_2 \end{pmatrix}$	K_{21}	$(\lambda_1 \quad \lambda_2)\begin{pmatrix} 10 & 70 \\ 51 & 52 \end{pmatrix}\begin{pmatrix} \gamma_1 \\ \gamma_2 \end{pmatrix}$
K_5	$(\lambda_1 \quad \lambda_2)\begin{pmatrix} 0 & 0 \\ 40 & 20 \end{pmatrix}\begin{pmatrix} \gamma_1 \\ \gamma_2 \end{pmatrix}$	K_{24}	$(\lambda_1 \quad \lambda_2)\begin{pmatrix} 0 & 80 \\ 40 & 60 \end{pmatrix}\begin{pmatrix} \gamma_1 \\ \gamma_2 \end{pmatrix}$
K_{10}	$(\lambda_1 \quad \lambda_2)\begin{pmatrix} 40 & 20 \\ 0 & 0 \end{pmatrix}\begin{pmatrix} \gamma_1 \\ \gamma_2 \end{pmatrix}$	K_{26}	$(\lambda_1 \quad \lambda_2)\begin{pmatrix} 5 & 50 \\ 50 & 50 \end{pmatrix}\begin{pmatrix} \gamma_1 \\ \gamma_2 \end{pmatrix}$
K_{16}	$(\lambda_1 \quad \lambda_2)\begin{pmatrix} 40 & 80 \\ 30 & 60 \end{pmatrix}\begin{pmatrix} \gamma_1 \\ \gamma_2 \end{pmatrix}$	K_{27}	$(\lambda_1 \quad \lambda_2)\begin{pmatrix} 60 & 80 \\ 30 & 40 \end{pmatrix}\begin{pmatrix} \gamma_1 \\ \gamma_2 \end{pmatrix}$
K_{17}	$(\lambda_1 \quad \lambda_2)\begin{pmatrix} 20 & 60 \\ -10 & 20 \end{pmatrix}\begin{pmatrix} \gamma_1 \\ \gamma_2 \end{pmatrix}$		

7.2 Special Fuzzy Potential Solutions

This section describes a special fuzzy potential solution (Liu and Shi 1994). If we can view the basic variable $\mathbf{x}(J, \gamma) = \mathbf{B}^{-1}\mathbf{D}\gamma$ as a parameter form with respect the parameter γ, then we can set the condition $\gamma_1 + \gamma_2 + \ldots + \gamma_p = 1$ as a constraint in the process of finding the aspiration level for the fuzzy MC^2 objective value. The resulting fuzzy potential solution with a specific γ^0 value is included in one of fuzzy potential solutions discussed in Section 7.1. To discuss the details, we first provide the following definitions and theorems.

Starting from (M4.1), given the $(m \times p)$ constraint level matrix $\mathbf{D}$ with columns $\mathbf{D}_1, \mathbf{D}_2, \ldots, \mathbf{D}_p$, we can denote the convex set $H(\mathbf{D})$ generated by $\mathbf{D}$ (or $\mathbf{D}_1, \mathbf{D}_2, \ldots, \mathbf{D}_p$) or the convex hull of by $\mathbf{D}$ (or $\mathbf{D}_1, \mathbf{D}_2, \ldots, \mathbf{D}_p$), as $H(\mathbf{D}) = \{\mathbf{d} \mid \mathbf{d} = \sum_{k=1}^{p} \gamma_k \mathbf{D}_k,\ \gamma_k \geq 0,\ k = 1, 2, \ldots, p,\ \sum_{k=1}^{p} \gamma_k = 1\}$.
Then, a special MC^2 problem is built as (M7.3):

$$\begin{aligned}
\mathit{Max} \quad & \sum_{i=1}^{q} \lambda_i \mathbf{C}^i \mathbf{x} \\
\mathit{Subject\ to} \quad & \mathbf{A}\mathbf{x} \leq \sum_{k=1}^{p} \gamma_k \mathbf{D}_k \\
& \sum_{k=1}^{p} \gamma_k = 1 \\
& \mathbf{x} \geq 0, \gamma_k \geq 0, k = 1, 2, \ldots, p.
\end{aligned}$$

Note that in (M7.3), $\mathbf{C}^i$ is the ith row of the (qxn) criteria matrix $\mathbf{C} = (\mathbf{C}^1, \mathbf{C}^2, \ldots, \mathbf{C}^q)$ of (M4.1). Model (M7.3) is (M4.1) adding the constraint $\gamma_1 + \gamma_2 + \ldots + \gamma_p = 1$.

Definition 7.4 A nondominated solution of (M7.3) with $\gamma > \mathbf{0}$ is called a potential solution, and a nondominated solution of (M7.3) with $\gamma \geq \mathbf{0}$ is called a weak potential solution.

The above definition can be applied to Definition 4.4 and Theorem 4.1 in forming the weak potential solution for (M4.1).

Theorem 7.4 $\mathbf{x}^* \in X = \{\mathbf{x} \mid \mathbf{A}\mathbf{x} \le \mathbf{d}, \mathbf{x} \ge 0$ for some $\mathbf{d} \in H(\mathbf{D})\}$ is a weak potential solution if and only if there is a $\lambda = (\lambda_1, \lambda_2, \ldots, \lambda_q)^t$ with $0 \le \lambda_i \le 1$, $i \in \{1, 2, \ldots, q\}$, such that $\mathbf{x}^*$solves (M7.3).

Theorem 7.5 For each given $\gamma^0 = (\gamma^0{}_1, \gamma^0{}_2, \ldots, \gamma^0{}_p)^t > 0$, with $\sum_{k=1}^{p} \gamma^0{}_k = 1$, $\mathbf{x}^*$ is a potential solution of (M4.1) if and only if $\mathbf{x}^*$ is a nondominated solution of (M7.4):

$$\begin{array}{ll} \textit{Max} & (\mathbf{C}^1\mathbf{x}, \mathbf{C}^2\mathbf{x}, \ldots, \mathbf{C}^q\mathbf{x}) \\ \textit{Subject to} & \mathbf{A}\mathbf{x} \le \mathbf{D}\gamma^0 \\ & \mathbf{x} \ge 0. \end{array}$$

Theorem 7.5 says that when γ is given, the model (M4.1) becomes a multiple criteria (MC) problem (see Section 3.1).

Corollary 7.1 $\mathbf{x}^* \in X$ is a potential solution of (M7.3) if and only if there is an $\varepsilon > 0$ and $\lambda = (\lambda_1, \lambda_2, \ldots, \lambda_q)^t$ with $0 \le \lambda_i \le 1$, $i \in \{1, 2, \ldots, q\}$, such that $\mathbf{x}^*$solves (M7.5):

$$\begin{array}{ll} \textit{Max} & \sum_{i=1}^{q} \lambda_i \mathbf{C}^i \mathbf{x} \\ \textit{Subject to} & \mathbf{A}\mathbf{x} \le \sum_{k=1}^{p} \gamma_k \mathbf{D}_k \\ & \sum_{k=1}^{p} \gamma_k = 1 \\ & \mathbf{x} \ge 0, \gamma_k \ge \varepsilon > 0, k = 1, 2, \ldots, p. \end{array}$$

Remark 7.4 (i) If a basic solution of (M7.3) is a (weak) potential solution, then its associated basis is a (weak) potential basis. (ii) The optimal objective value of (M7.3) is an upper bound for that of (M7.5). (iii) Similarly to (M7.5), we can build a minimizing problem with $\varepsilon > 0$.

As we pointed out at the beginning of this chapter, fuzzy linear programming is an application of fuzzy set theory in linear decision making problems. This type of problem was first introduced by Zimmermann (1978), which has shown that finding the nondominated solution of MC model (M3.1):

$$\begin{array}{ll} \mathit{Max} & \mathbf{C\,x} \\ \mathit{Subject\ to} & \mathbf{A\,x} \leq \mathbf{d} \\ & \mathbf{x} \geq 0 \end{array}$$

is equivalent to solving the following fuzzy linear program (M7.6):

$$\begin{array}{ll} \mathit{Max} & \alpha \\ \mathit{Subject\ to} & \alpha \leq \dfrac{\mathbf{C}^i\mathbf{x} - l_i^0}{u_i^0 - l_i^0},\ i = 1, 2, \ldots, q \\ & \mathbf{A\,x} \leq \mathbf{b} \\ & \mathbf{x} \geq 0, \alpha \geq 0, \end{array}$$

where u_i^0 is the upper bound and l_i^0 is the lower bound for the ith criterion $\mathbf{C}^i\mathbf{x}$ over the feasible set $\mathbf{Ax} \leq 0$ and $\mathbf{x} \geq 0$.

Recall that the above result is summarized in (iv) of MC solving algorithms in Section 3.1. This leads to the following theorem for a weak potential solution of (M7.3).

Theorem 7.6 Let u_i^0 be the upper bound and l_i^0 be the lower bound for the ith criterion $\mathbf{C}^i\mathbf{x}$ of (M7.3) if $\mathbf{x}^*$ can be obtained from solving the following problem (M7.7):

$$\begin{array}{ll} \textit{Max} & \alpha \\ \textit{Subject to} & \alpha \le \dfrac{\mathbf{C}^i\mathbf{x} - l_i^0}{u_i^0 - l_i^0},\ i = 1, 2, \dots, q \\ & \mathbf{A}\mathbf{x} \le \sum_{k=1}^{p} \gamma_k \mathbf{D}_k \\ & \sum_{k=1}^{p} \gamma_k = 1 \\ & \mathbf{x} \ge 0, \gamma_k \ge 0, \alpha \ge 0, k = 1, 2, \dots, p \end{array}$$

then $\mathbf{x}^* \in \mathbf{X}$ is a weak potential solution of (M7.3).

Proof. If $\mathbf{x}^*$ is obtained from solving (M7.7), then $\mathbf{x}^*$ is a nondominated solution of the following MC linear program (M7.8) (see Definitions 3.2 and 3.3):

$$\begin{array}{ll} \textit{Max} & (\mathbf{C}^1\mathbf{x}, \mathbf{C}^2\mathbf{x}, \dots, \mathbf{C}^q\mathbf{x}) \\ \textit{Subject to} & \mathbf{A}\mathbf{x} \le \mathbf{D}\boldsymbol{\gamma} \\ & \sum_{k=1}^{p} \gamma_k = 1 \\ & \mathbf{x} \ge 0, \boldsymbol{\gamma} \ge 0. \end{array}$$

Since the constraints of (M7.8) form a convex set, there exists a $\lambda > \mathbf{0}$ such that $\mathbf{x}^*$ solve (M7.3). Thus, by Definition 7.4, $\mathbf{x}^*$ is a weak potential solution.

Remark 7.5 $u_i^0\,(l_i^0)$ in Theorem 7.6 can be obtained from solving the linear programming problems (M7.9):

$$\begin{aligned} &Max\,(Min) && \mathbf{C}^i\,\mathbf{x} \\ &Subject\ to && \mathbf{A}\mathbf{x} \le \sum_{k=1}^{p} \gamma_k\,\mathbf{D}_k \\ & && \sum_{k=1}^{p} \gamma_k = 1 \\ & && \mathbf{x} \ge 0,\ \gamma_k \ge 0,\ k = 1, 2, \ldots, p \end{aligned}$$

Corollary 7.2 If $\mathbf{x}^*$ is obtained from solving (M7.7) with $\gamma > \mathbf{0}$ in the solution, then $\mathbf{x}^*$ is a potential solution of (M7.3).

Corollary 7.3 The optimal solution $\mathbf{x}^*$ of the following (M7.10):

$$\begin{aligned} &Max && \alpha \\ &Subject\ to && \alpha \le \frac{\mathbf{C}^i\mathbf{x} - l_i^0}{u_i^0 - l_i^0},\ i = 1, 2, \ldots, q \\ & && \mathbf{A}\mathbf{x} \le \sum_{k=1}^{p} \gamma_k\,\mathbf{D}_k \\ & && \sum_{k=1}^{p} \gamma_k = 1 \\ & && \alpha \ge 0, \\ & && \mathbf{x} \ge 0,\ \gamma_k \ge \varepsilon > 0\ for\ given\ \varepsilon,\ k = 1, 2, \ldots, p \end{aligned}$$

is a potential solution of (M7.3).

Example 7.2 We use the fuzzy problem in Example 7.1 to show the special fuzzy potential solution of (M7.3) as follows.

Instead of using an MC^2 software, we use a well-known standard software package, LINDO (Schrage 1984) to solve the following four (two maximum and two minimum) linear programming problems:

$$Max\,(Min) \quad \begin{pmatrix} 2 & 1 & 1 & 0 & -1 \end{pmatrix} \begin{pmatrix} x_1 \\ x_2 \\ x_3 \\ x_4 \\ x_5 \end{pmatrix}$$

$$Subject\ to \begin{pmatrix} 1 & 0 & 2 & 1 & 0 \\ 0 & 1 & 1 & 2 & 2 \\ 1 & 1 & 0 & 0 & 1 \end{pmatrix} \begin{pmatrix} x_1 \\ x_2 \\ x_3 \\ x_4 \\ x_5 \end{pmatrix} \le \begin{pmatrix} 20 & 50 \\ 40 & 20 \\ 30 & 40 \end{pmatrix} \begin{pmatrix} \gamma_1 \\ \gamma_2 \end{pmatrix}$$

$$\gamma_1 + \gamma_2 = 1, \gamma_1 \ge 0,$$
$$x_i \ge 0, i = 1,\ 2,\ 3,\ 4, 5.$$

$$Max\,(Min) \quad \begin{pmatrix} 1 & 0 & 1 & 2 & 2 \end{pmatrix} \begin{pmatrix} x_1 \\ x_2 \\ x_3 \\ x_4 \\ x_5 \end{pmatrix}$$

$$Subject\ to \begin{pmatrix} 1 & 0 & 2 & 1 & 0 \\ 0 & 1 & 1 & 2 & 2 \\ 1 & 1 & 0 & 0 & 1 \end{pmatrix} \begin{pmatrix} x_1 \\ x_2 \\ x_3 \\ x_4 \\ x_5 \end{pmatrix} \le \begin{pmatrix} 20 & 50 \\ 40 & 20 \\ 30 & 40 \end{pmatrix} \begin{pmatrix} \gamma_1 \\ \gamma_2 \end{pmatrix}$$

$$\gamma_1 + \gamma_2 = 1, \gamma_1 \ge 0,$$
$$x_i \ge 0, i = 1,\ 2,\ 3,\ 4, 5.$$

We obtain $u_1^0 = 85$, $l_1^0 = -20$, $u_2^0 = 60$, $l_2^0 = 0$. Then we solve the following fuzzy programming problem:

$$\begin{aligned}
&Max \quad \alpha \\
&\alpha \le \frac{(2x_1 + x_2 + x_3 - x_5) - (-20)}{85 - (-20)} \\
&\alpha \le \frac{(x_1 + x_3 + 2x_4 + x_5) - (0)}{60 - (0)} \\
&\alpha \ge 0, \\
&Subject\ to \begin{pmatrix} 1 & 0 & 2 & 1 & 0 \\ 0 & 1 & 1 & 2 & 2 \\ 1 & 1 & 0 & 0 & 1 \end{pmatrix} \begin{pmatrix} x_1 \\ x_2 \\ x_3 \\ x_4 \\ x_5 \end{pmatrix} \le \begin{pmatrix} 20 & 50 \\ 40 & 20 \\ 30 & 40 \end{pmatrix} \begin{pmatrix} \gamma_1 \\ \gamma_2 \end{pmatrix} \\
&\gamma_1 + \gamma_2 = 1, \gamma_1 \ge 0, \\
&x_i \ge 0, i = 1,\ 2,\ 3,\ 4, 5.
\end{aligned}$$

The optimum solution to the above (M7.7) is $\alpha = 0.828$, $\gamma_1 = 0$, $\gamma_2 = 1$, $x_1 = 40$, $x_2 = x_5 = 0$, $x_3 = 3.448$ and $x_4 = 3.103$. This is a weak potential solution for the above given (M7.3).

To find a potential solution, we apply Corollary 7.3, choose $\varepsilon = 0.3$, and solve the following linear program:

$$\begin{aligned}
&\mathit{Max} && \alpha \\
& && \alpha \le \frac{(2x_1 + x_2 + x_3 - x_5) - (-20)}{85 - (-20)} \\
& && \alpha \le \frac{(x_1 + x_3 + 2x_4 + x_5) - (0)}{60 - (0)} \\
& && \alpha \ge 0,
\end{aligned}$$

$$\mathit{Subject\ to} \begin{pmatrix} 1 & 0 & 2 & 1 & 0 \\ 0 & 1 & 1 & 2 & 2 \\ 1 & 1 & 0 & 0 & 1 \end{pmatrix} \begin{pmatrix} x_1 \\ x_2 \\ x_3 \\ x_4 \\ x_5 \end{pmatrix} \le \begin{pmatrix} 20 & 50 \\ 40 & 20 \\ 30 & 40 \end{pmatrix} \begin{pmatrix} \gamma_1 \\ \gamma_2 \end{pmatrix}$$

$$\gamma_1 + \gamma_2 = 1, \gamma_1 \ge 0.3, \gamma_2 \ge 0.3,$$

$$x_i \ge 0, i = 1,\ 2,\ 3,\ 4, 5.$$

The solution of this above linear program, by Corollary 7.3, is a potential solution of MC[2] in this example. The values $\alpha = 0.751$, $x_1 = 36.973$, $x_2 = x_3 = 0$, $x_4 = 4.027$ and $x_5 = 0.270$ form an optimal solution of the above linear program, which is a potential solution of the given (M7.3).

Comparing this solution with the fuzzy potential solutions obtained in Algorithm 7.1, one can see that the basis associated with the basic variables x_1, x_4 and x_5 form the potential basis U_3 found by the MC[2] -simplex method (Table 7.2). In the approach of this section $x_1 = 36.973$, $x_2 = x_3 = 0$, $x_4 = 4.027$ and $x_5 = 0.270$ is a potential solution with $\gamma_1 = 0.3$ and $\gamma_2 = 0.7$, while in Example 7.1, it is a potential solution with $0 < \gamma_1, \gamma_2 \le 1$ and $0 < \lambda_1 \le 1/2$, $1/2 < \lambda_2 \le 1$ (see Table 7.2). Thus, the solution of this section is a special fuzzy potential solution for (M7.1).

7.3 Further Remarks

This chapter has extended the MC[2] problems to fuzzy MC[2] problems. The basic assumption is that when the decision makers do not restrict themselves on finding an absolutely optimal solution, the satisfying solution becomes significant. From a given potential solution for the MC^2 problem, this chapter has proposed the methods to construct a fuzzy potential solution as well as special fuzzy potential solution for the fuzzy MC^2 problem. The construction of fuzzy membership function depends on the preferences of the decision makers. All simple membership functions used in this chapter are only for illustrative purposes. The reader should not limit his or her mind on their appearance. However, the basis of the proposed approaches in the chapter is not changed with the form of membership functions. The careful reader can also observe that many fuzzy methods used in linear programming or MC linear programming (Zimmermann 1987) can be applied to the MC^2 problem.

Exercises

7.1 Solve the following Fuzzy MC^2 problem by using Algorithm 7.1:

$$\begin{array}{ll} \textit{Fuzzy-Max} & 10\,x_1 + 12\,x_2 \\ \textit{Fuzzy-Max} & 15\,x_1 + 8\,x_2 \\ \textit{Subject to} & 3\,x_1 + 2\,x_2 \leq (25,\ -35) \\ & 4\,x_1 + 6\,x_2 \leq (24,\ 42) \\ & x_1,\ x_2 \geq 0. \end{array}$$

7.2 Identify the special fuzzy potential solution for the fuzzy MC^2 problem by using the method described in Section 7.2.

7.3 (i) Use the constructive approach to solve the following dual fuzzy MC^2 problem:

$$\begin{array}{ll} \textit{Fuzzy-Min} & 25\,u_1 + 24\,u_2 \\ \textit{Fuzzy-Min} & -35\,u_1 + 42\,u_2 \\ \textit{Subject to} & 3\,u_1 + 4\,u_2 \geq (10,\ 15) \\ & 2\,u_1 + 6\,u_2 \geq (12,\ 8) \\ & u_1,\ u_2 \geq 0. \end{array}$$

(ii) Study the fuzzy weak duality between problem 7.1 and problem 7.3 (i).

7.4 Consider

$$\textit{Fuzzy-Max} \quad 3\,x_1 + 2\,x_2$$

Fuzzy-Max	$6\ x_1 + 8\ x_2$
Subject to	$30\ x_1 + 20\ x_2 \leq (300,\ 250)$
	$5x_1 + 10x_2 \leq (110,\ 160)$
	$x_1, x_2 \geq 0;$

and

Fuzzy-Min	$300\ u_1 + 110\ u_2$
Fuzzy-Min	$250\ u_1 + 160\ u_2$
Subject to	$30\ u_1 + 5\ u_2 \geq (3,\ 6)$
	$20\ u_1 + 10\ u_2 \geq (2,\ 8)$
	$u_1,\ u_2 \geq 0.$

a. Use Algorithm 7.1 to the fuzzy-max problem.
b. Find use the special fuzzy potential solution for the fuzzy-max problem.

Chapter 8
DUALITY OF FUZZY MC² PROBLEMS

Based on Chapter 7, this chapter provides two different methods for exploring the duality of fuzzy MC² problems. The first one called the constructive approach is given in Section 8.1. In this approach, we assume that given decision makers, their primal fuzzy MC² problem is a production process for "better" profit while the corresponding dual fuzzy MC² problem is to sell resource for "better" price. Thus, both primal and dual MC² problems cannot reach the "optimal" status, but only a fuzzy satisficing status. Similar to the primal fuzzy approach in Section 7.1, we can start from a fuzzy dual MC² problem to construct its fuzzy potential solutions. The connections of primal fuzzy MC² problem and dual fuzzy MC² problem are classified as fuzzy week duality and fuzzy complementary slackness. The second duality of fuzzy MC² problem is called the parametric approach in Section 8.2. In this approach, we assume that there is a competitive situation where the decision makers' action is represented by the primal MC² problem and their competitor's action is represented by the dual MC² problem. The win-win strategy to solve this situation is both sides look for the better result out of their worse cases. These cases are the membership functions for the certain fuzzy potential solutions. This section first explores the foundation of the duality for fuzzy MC² problems. Section 8.3 will give a brief remarks about this chapter.

8.1 A Constructive Approach to Fuzzy MC² Duality

In this approach(Liu, Shi and Liu 1995), we assume that given any decision maker, to him or her, the primal fuzzy MC² problem (M7.1) is a production process for "better" profit while the corresponding dual fuzzy MC² problem is to sell resource for "better" price. Thus, both primal and dual MC² problems cannot reach the "optimal" status, but only a fuzzy satisficing status. Similar to the primal fuzzy approach in Section 7.1, we can start from a fuzzy dual MC² problem to construct its fuzzy potential solutions.

In order to illustrate details about the fuzzy duality that we represent here, let us consider (M4.1) as a production design problem.

The q criteria can be viewed as the total sales, profit, customer satisfaction, etc., while the p constraint levels are the p decision makers' preference levels for production resource availability. These decision makers could be production, finance, marketing, and engineering managers. Note that managers' preference levels for production resource availability may be different. For example, the production manager views the resource availability level based on the material, manpower, and equipment under control, while the finance managers views the resource availability level based on the available cash flow. (M4.1) selects a set of potential solutions to produce the selected products by using the available resource $\mathbf{D}\gamma$ intelligently. Note that for each potential solution J for (M4.1), whenever $(\gamma, \lambda) \in \Gamma(J)\times\Lambda(J)$, $\mathbf{x}(J, \gamma) = \mathbf{B}^{-1}\mathbf{D}\gamma$ is the optimal amount of products that we can produce. The fuzzy problem (M7.1) says that instead of finding the potential solutions of (M4.1), we can identify a set of fuzzy potential solutions of (M7.1) for the decision makers. Each fuzzy potential solution is a primal basis K of (M7.1) with the corresponding primal parameter set $\Gamma(K)$ and membership function $\mu_{R(i,j)}(\mathbf{x})$. Whenever $(\gamma, \lambda) \in R(i, j)$, $\mathbf{x}(K, \gamma)$ is the satisficing (fuzzy) amount of products to be produced.

Suppose we want to examine an alternative way to (M7.1). Instead of producing $\mathbf{x}(K, \gamma)$ by using the resource $\mathbf{D}\gamma$ of (M7.1), we may sell all $\mathbf{D}\gamma$ for a total satisficing (fuzzy) "price." This approach can be captured by the following fuzzy program (M8.1):

$$\begin{array}{ll} \textit{Fuzzy-Min} & \mathbf{u}^t\,\mathbf{D}\gamma \\ \textit{Subject to} & \mathbf{u}^t\,\mathbf{A} \geq \lambda^t\,\mathbf{C} \\ & \mathbf{u} \geq 0, \end{array}$$

where $\mathbf{u} = (u_1, u_2, \ldots, u_m)^t$ is a row vector as the "unit price" of the resource $\mathbf{D}\gamma$.

If the total satisficing (fuzzy) "price" is the minimum acceptable "price," then (M8.1) becomes (M4.2), the dual MC^2 problem (see Section 4.5):

$$\begin{array}{ll} \textit{Min} & \mathbf{u}^t\,\mathbf{D}\gamma \\ \textit{Subject to} & \mathbf{u}^t\,\mathbf{A} \geq \lambda^t\,\mathbf{C} \\ & \mathbf{u} \geq 0. \end{array}$$

If the total satisficing (fuzzy) "price" is a maximum "price," the model (M8.1) may be restated as (M8.2):

$$\begin{array}{ll} \textit{Max} & \mathbf{u}'\,\mathbf{D}\gamma \\ \textit{Subject to} & \mathbf{u}'\,\mathbf{A} \geq \lambda'\,\mathbf{C} \\ & \mathbf{u} \geq 0. \end{array}$$

Remark 8.1 In most cases the potential solutions of (M8.2) do not exist (i.e., there are unbounded solutions). For these cases, we can add an upper "price" bound $u_1 + u_2 + \ldots + u_m \leq \lambda'\,\mathbf{b}$ into (M8.2) to guarantee the bounded potential solutions, in which b_k, the component of $\mathbf{b}$, k = 1, . . . , p represents the kth decision maker's desiring "price" bound for $u_1 + u_2 + \ldots + u_m$. We call this condition *the price desirability bound.* From now on, if we mention (M8.2), we assume it has the price desirability bound as a constraint.

In the following, we discuss how to use (M4.2) and (M8.2) to construct fuzzy potential solutions for (M8.1).

By using the MC²-simplex method (Section 4.4), we can identify two sets of all potential solutions over all possible changes of (γ, λ) for (M4.2) and (M8.2), respectively. Let $\xi = \{L^*_1, L^*_2, \ldots, L^*_v\}$ be the set of potential solutions over all possible changes of (γ, λ) for (M4.2) satisfying $(\Gamma(L^*_i)\times\Lambda(L^*_i)) \cap (\Gamma(L^*_g)\times\Lambda(L^*_g)) = \varnothing$ for $i \neq g$; i, g = 1, ..., v; and $\zeta = \{U^*_1, U^*_2, \ldots, U^*_t\}$ be the set of potential solutions over all possible changes of (γ, λ) for (M8.2) satisfying $(\Gamma(U^*_j)\times\Lambda(U^*_j)) \cap (\Gamma(U^*_h)\times\Lambda(U^*_h)) = \varnothing$ for $j \neq h$; j, h = 1, ..., t. Then, given a problem (M8.1) we may view L^*_i, i = 1, . . ., v, as the "lower" potential solution for $\Gamma(L^*_i)\times\Lambda(L^*_i)$ and U^*_j, j = 1, . . ., t, as the "upper" potential solution for $\Gamma(U^*_j)\times\Lambda(U^*_j)$. This means that for each $(\gamma, \lambda) \in \Gamma(U^*_j)\times\Lambda(U^*_j)$, $\mathbf{x}(U^*_j, \gamma)$ maximizes (M7.12), while for $(\gamma, \lambda) \in \Gamma(L^*_i)\times\Lambda(L^*_i)$, $\mathbf{x}(L^*_i, \gamma)$ minimizes (M4.2).

Given $\xi = \{L^*_1, L^*_2, \ldots, L^*_v\}$ and $\zeta = \{U^*_1, U^*_2, \ldots, U^*_t\}$, by Definition 7.1, we see that the space $\Gamma\times\Lambda$ is covered by both ξ and ζ. Let $S(g, h) = (\Gamma(L^*_g)\times\Lambda(L^*_g)) \cap (\Gamma(U^*_h)\times\Lambda(U^*_h)) \neq \varnothing$, g = 1, ..., v; h = 1, ..., t, be the reformed (γ, λ) ranges and $\wp = \cup\,\{S(g, h) \mid \text{for all } (g, h)\}$ be a reformed the (γ, λ) space. Then, similar to Theorem 7.1, we have:

Theorem 8.1

(i) $S(\mathrm{g},\mathrm{h}) \subseteq \Gamma(L^*_{\mathrm{g}})\times\Lambda(L^*_{\mathrm{g}})$ and $S(\mathrm{g},\mathrm{h}) \subseteq \Gamma(U^*_{\mathrm{h}})\times\Lambda(U^*_{\mathrm{h}})$, for all g = 1, ..., v; h = 1, ..., t.

(ii) $S(\mathrm{g},\mathrm{h}) \cap S(\mathrm{i},\mathrm{j}) = \varnothing$ for i ≠ g; j ≠ h; i, g = 1, ..., v; j, h = 1, ..., t.

(iii) $\Gamma\times\Lambda \subseteq \wp$.

Remark 8.2 Since $\Gamma\times\Lambda$ is the space of all (γ, λ), we see that $\Gamma\times\Lambda \supseteq \wp$. By (iii) of Theorem 8.1, $\Gamma\times\Lambda = \wp$.

Theorem 8.2

(i) For a given $S(\mathrm{g},\mathrm{h})$ on $\wp$, $\mathbf{u}(L^*_{\mathrm{g}}, \gamma)$ is its lower potential solution and $\mathbf{u}(U^*_{\mathrm{h}}, \gamma)$ is its upper potential solution.

(ii) For any $(\gamma, \lambda) \in \Gamma\mathrm{x}\Lambda$, there is a corresponding minimum solution resulting from an element of ξ and a corresponding maximum solution resulting from an element of ζ.

The proof of this theorem can be shown by following Theorem 7.2.

Let $\mathbf{B}_{U^*}$ be the basis for $\mathbf{u}(U^*_{\mathrm{h}}, \lambda)$ and the corresponding objective payoff be denoted by $\mathrm{V}(U^*_{\mathrm{h}}; \lambda, \gamma) = \lambda^t \mathbf{C}_{\mathbf{B}}\, \mathbf{B}_{U^*}^{-1}\, \mathbf{D}\gamma$ and let $\mathbf{B}_{L^*}$ be the basis for $\mathbf{u}(L^*_{\mathrm{g}}, \lambda)$ and the corresponding objective payoff be denoted by $\mathrm{V}(L^*_{\mathrm{g}}; \lambda, \gamma) = \lambda^t \mathbf{C}_{\mathbf{B}}\, \mathbf{B}_{L^*}^{-1}\, \mathbf{D}\gamma$. Then, whenever any $(\gamma, \lambda) \in S(\mathrm{g},\mathrm{h})$ are specified, $\mathrm{V}(U^*_{\mathrm{h}}; \lambda, \gamma)$ is the upper aspiration bound with respect U^*_{h}, while $\mathrm{V}(L^*_{\mathrm{g}}; \lambda, \gamma)$ is the lower aspiration bound with respect to L^*_{g}. A linear membership function of any $S(\mathrm{g},\mathrm{h})$ on $\wp$ for (M7.11) is construct as (M8.3):

$$\mu_{S(g,h)}(\mathbf{u}) = \begin{cases} 0, & \mathbf{u}'\mathbf{D}\boldsymbol{\gamma} \geq \mathrm{V}(U_h^*;\boldsymbol{\gamma},\boldsymbol{\lambda}) \\ \dfrac{\mathbf{u}'\mathbf{D}\boldsymbol{\gamma} - \mathrm{V}(L_g^*;\boldsymbol{\gamma},\boldsymbol{\lambda})}{\mathrm{V}(U_h^*;\boldsymbol{\gamma},\boldsymbol{\lambda}) - \mathrm{V}(L_g^*;\boldsymbol{\gamma},\boldsymbol{\lambda})}, & \mathrm{V}(L_g^*;\boldsymbol{\gamma},\boldsymbol{\lambda}) < \mathbf{u}'\mathbf{D}\boldsymbol{\gamma} < \mathrm{V}(U_h^*;\boldsymbol{\gamma},\boldsymbol{\lambda}) \\ 1, & \mathrm{V}(L_g^*;\boldsymbol{\gamma},\boldsymbol{\lambda}) \geq \mathbf{u}'\mathbf{D}\boldsymbol{\gamma} \end{cases}$$

Definition 8.1 Let W be a primal potential basis of (M8.1) with the basic variables $\mathbf{u}(W, \lambda)$ and the corresponding primal parameter set $\Lambda(W)$. $\mathbf{u}(W, \lambda)$ is said to be a dual fuzzy potential solution of S(g, h) for (M7.11) if $\Lambda(L^*_g) \cap \Lambda(U^*_h) \subseteq \Lambda(W)$ and $\mathbf{u}(W, \lambda)$ has a membership function (M8.3). This solution is denoted as $(\mathbf{u}(W, \lambda); \mu_{S(g,h)}(\mathbf{u}))$.

Definition 8.2 Let $(\mathbf{u}(W, \lambda); \mu_{S(g,h)}(\mathbf{u}))$ be a fuzzy potential solution of S(g, h) for (M8.1). The set $\aleph = \{(\mathbf{u}(W, \lambda); \mu_{S(g,h)}(\mathbf{u})) \mid$ for all S(g, h) and $W\}$ is called a set of dual fuzzy potential solutions for all possible changes of $(\gamma, \lambda) \in \Gamma \times \Lambda$.

Theorem 8.3

(i) Given a dual fuzzy potential solution $(\mathbf{u}(W, \lambda); \mu_{S(g,h)}(\mathbf{u}))$, there exists an upper and a lower potential solution for S(g, h).

(ii) Conversely, with an upper and a lower potential solution of S(g, h), a dual fuzzy potential solution $(\mathbf{u}(W, \lambda); \mu_{S(g,h)}(\mathbf{u}))$ can be constructed.

By integrating the above constructive approach to the fuzzy MC^2 duality and the fuzzy approach in Section 7.1, we study the relationships between (M7.1) and (M8.1). In order to do this, we first explore the relationship between the reformed (γ, λ) range R(i, j) for (M7.1) and S(g, h) for (M8.1). Given $\Re = \cup \{R(\text{i, j}) \mid \text{for all (i, j)}\}$ and $\wp = \cup \{S(\text{g, h}) \mid \text{for all (g, h)}\}$, let $\chi = \cup\{R(\text{i, j}) \cap S(\text{g, h}) \mid \text{for all (i, j) and (g, h)}\}$ be a primal-dual (γ, λ) space. Then, we have:

Theorem 8.4

(i) $\chi = \Re \cap \wp$.

(ii) $\chi = \Gamma \times \Lambda$.

Proof. (i) By the Distributive Law of ordinary set theory,

$$\chi = \cup\{R(\text{i, j}) \cap S(\text{g, h}) \mid \text{for all (i, j) and (g, h)}\}$$
$$= \{\cup\{R(\text{i, j}) \mid \text{for all (i, j)}\}\} \cap \{\cup\{S(\text{g, h}) \mid \text{for all (g, h)}\}\}$$

$= \Re \cap \wp$.

(ii) Since $\Re \supseteq \Gamma\times\Lambda$ (Theorem 7.1 (iii)) and $\wp \supseteq \Gamma\times\Lambda$ (Theorem 8.2 (iii)), $\chi \supseteq \Gamma\times\Lambda$. Note that $\chi \subseteq \Gamma\times\Lambda$. Thus, $\chi = \Gamma\times\Lambda$.

Using Theorem 8.4, we can state the ordinary duality relations for (M4.1) and (M4.2) according to the reformed (γ, λ) ranges R(i, j) and S(g, h), which are the special cases of Theorems 4.7 and 4.9:

Theorem 8.5 Let J be a primal potential solution on R(i, j) of (M4.1) and K be a dual potential on S(g, h) of (M4.2), then
(i) (Weak Duality) $V(J; \gamma, \lambda) \leq V(K; \gamma, \lambda)$ for all $(\gamma, \lambda) \in R(i, j) \cap S(g, h)$.
(ii) (Complementary Slackness) J and K are both potential solutions for $(\gamma, \lambda) \in R(i, j) \cap S(g, h)$ if and only if $V(J; \gamma, \lambda) = V(K; \gamma, \lambda)$.

Theorem 8.6 Given a $R(i, j) \cap S(g, h)$, let U_i be a potential solution on R(i, j) of (M4.1) and L^*_g be a potential solution on S(g, h) of (M4.2), then $V(U_i; \gamma, \lambda) = V(L^*_g; \gamma, \lambda)$, for all $(\gamma, \lambda) \in R(i, j) \cap S(g, h)$.

Remark 8.3 Theorem 8.6 can be used to construct the membership functions (M7.3) and (M8.3). It essentially says that the basis L^*_g is the corresponding dual basis of U_i on $R(i, j) \cap S(g, h)$. When we use the MC^2-simplex method to locate U_i, we can find both U_i and L^*_g simultaneously. Thus, when we find the set of all potential solutions for (M4.1), we also obtain the set of all potential solutions for (M4.2). To construct the membership functions (M7.3) and (M8.3), we need to only solve (M4.1), (M7.2), and (M8.2).

Based on the primal fuzzy membership function $\mu_{R(i,j)}(\mathbf{x})$ of (M7.3) and the dual fuzzy membership function $\mu_{S(g,h)}(\mathbf{u})$ of (M8.3), the relationships between (M7.1) and (M8.1) are as follows:

Definition 8.3 Fuzzy weak duality membership functions on a given reformed (γ, λ) range $R(i, j) \cap S(g, h)$ can be defined as

(i) the "min" relation:

$$\mu^1_{R(i,j)\cap S(g,h)}(\mathbf{x},\mathbf{u}) = \min\,(\mu_{R(i,j)}(\mathbf{x})\ ,\mu_{S(g,h)}(\mathbf{u})\ );$$

(ii) the "max" relation:

$$\mu^2_{R(i,j)\cap S(g,h)}(\mathbf{x},\mathbf{u}) = \max\,(\mu_{R(i,j)}(\mathbf{x})\ ,\mu_{S(g,h)}(\mathbf{u})\ );\text{ and}$$

(iii) the "mean" relation:

$$\mu^3_{R(i,j)\cap S(g,h)}(\mathbf{x},\mathbf{u}) = (\mu_{R(i,j)}(\mathbf{x}) + \mu_{S(g,h)}(\mathbf{u}))/2.$$

Remark 8.4 We may use other operators, such as the product operator (Zimmerman 1978) to construct different weak fuzzy duality membership functions. However, fuzzy weak duality membership functions of Definition 8.3 have some explicitly applicative meanings to describe the decision makers' preferences for duality between (M7.1) and (M8.1). The "min" relation could represent the preference of the decision maker who is a risk averter. This kind of people conservatively looks for the minimum value of primal satisficing level ($\mu_{R(i,j)}(\mathbf{x})$) and dual satisficing level ($\mu_{S(g,h)}(\mathbf{u})$). The "max" relation could represent the preference of risk seekers. They positively seek the maximum value of primal satisficing level ($\mu_{R(i,j)}(\mathbf{x})$) and dual satisficing level ($\mu_{S(g,h)}(\mathbf{u})$). Similarly, the "mean" relation could represent the preference of risk-neutral individuals. These people always like the average value of primal satisficing level ($\mu_{R(i,j)}(\mathbf{x})$) and dual satisficing level ($\mu_{S(g,h)}(\mathbf{u})$).

In the fuzzy duality of the MC^{2}problems, ordinary weak duality is replaced by fuzzy weak duality membership functions in Definition 8.3. However, the relationships between ordinary complementary slackness and fuzzy weak duality membership functions can be observed by the following:

Theorem 8.7

(i) Let J be a primal potential solution of (M7.1) and K be a dual potential of (M8.1). If J and K have a complementary slackness for $(\gamma, \lambda) \in R(\mathrm{i}, \mathrm{j}) \cap S(\mathrm{g}, \mathrm{h})$, then $\mu^c_{R(i,j)\cap S(g,h)}(\mathbf{x},\mathbf{u}) = 1$, c = 1, 2, 3.

(ii) Conversely, if $\mu^c_{R(i,j)\cap S(g,h)}(\mathbf{x},\mathbf{u}) = 1$, c = 1, 2, 3, then both the "min" and "mean" relations have a complementary slackness, but the "max" relation may not have a complementary slackness.

Proof. (i) By Theorem 8.5, J and K have a complementary slackness for $(\gamma, \lambda) \in R(\mathrm{i}, \mathrm{j}) \cap S(\mathrm{g}, \mathrm{h})$ means that J and K are both potential solutions for $(\gamma, \lambda) \in R(\mathrm{i}, \mathrm{j}) \cap S(\mathrm{g}, \mathrm{h})$ of (M4.1) and (M4.2), respectively. Thus, $\mathrm{V}(J; \gamma, \lambda) \geq \lambda^t\mathbf{C}\mathbf{x}$ and $\mathrm{V}(K; \gamma, \lambda) \leq \mathbf{u}^t\mathbf{D}\gamma$ for $(\gamma, \lambda) \in R(\mathrm{i}, \mathrm{j}) \cap S(\mathrm{g}, \mathrm{h})$. This implies that J is the upper potential solution of (M7.1), while K is the lower potential solution of (M8.1). By the membership functions (M7.3) and

(M8.3), we have $\mu_{R(i,j)}(\mathbf{x}) = \mu_{S(g,h)}(\mathbf{u}) = 1$. Therefore, $\mu^c{}_{R(i,j)\cap S(g,h)}(\mathbf{x},\mathbf{u}) = 1$, c = 1, 2, 3.

(ii) To show the cases for c = 1 and 3, if $\mu^1{}_{R(i,j)\cap S(g,h)}(\mathbf{x},\mathbf{u}) = 1$ or $\mu^3{}_{R(i,j)\cap S(g,h)}(\mathbf{x},\mathbf{u}) = 1$, by Definition 8.3, we must have $\mu_{R(i,j)}(\mathbf{x}) = \mu_{S(g,h)}(\mathbf{u}) = 1$. This implies that

$$\lambda^t\mathbf{Cx} \geq V(U_i; \gamma, \lambda) \text{ and } V(L^*_g; \gamma, \lambda) \geq \mathbf{u}^t\mathbf{D}\gamma \text{ for all } (\gamma, \lambda) \in R(i, j) \cap S(g, h).$$

By Theorem 8.6, $\lambda^t\mathbf{Cx} \geq V(U_i; \gamma, \lambda) = V(L^*_g; \gamma, \lambda) \geq \mathbf{u}^t\mathbf{D}\gamma$.

Hence, $\lambda^t\mathbf{Cx} \geq \mathbf{u}^t\mathbf{D}\gamma$ for all $(\gamma, \lambda) \in R(i, j) \cap S(g, h)$.

By the ordinary weak duality for MC^{2}linear programming (Theorem 4.7), we know

$$\lambda^t\mathbf{Cx} \leq \mathbf{u}^t\mathbf{D}\gamma \text{ for any } (\gamma, \lambda) \in R(i, j) \cap S(g, h).$$

Therefore, $\lambda^t\mathbf{Cx} = \mathbf{u}^t\mathbf{D}\gamma$ for all $(\gamma, \lambda) \in R(i, j) \cap S(g, h)$, which is a complementary slackness for corresponding potential solutions for (M4.1) and (M4.2), respectively.

To show the case for c =2, we know that there are 3^2 (=9) possible value combinations of $\mu_{R(i,j)}(\mathbf{x})$ and $\mu_{S(g,h)}(\mathbf{u})$ assigning the values of $\mu^2{}_{R(i,j)\cap S(g,h)}(\mathbf{x},\mathbf{u})$.

If $\mu^2{}_{R(i,j)\cap S(g,h)}(\mathbf{x},\mathbf{u}) = 1$, there may be $\mu_{R(i,j)}(\mathbf{x}) =1$and $\mu_{S(g,h)}(\mathbf{u}) = 0$, which implies that there is not a complementary slackness. The complementary slackness happens only when $\mu_{R(i,j)}(\mathbf{x}) = \mu_{S(g,h)}(\mathbf{u}) = 1$.

Both Definition 8.3 and Theorem 8.7 show that the fuzzy MC^2 weak duality membership functions are a kind of extension of the ordinary MC^2 weak duality and complementary slackness.

Example 8.1 Given the fuzzy-max MC^2 problem in Example 7.1, we construct a fuzzy-min MC^2 problem (M8.1) as

$$Fuzzy-Min \quad (\gamma_1 \quad \gamma_2)\begin{pmatrix}20 & 40 & 30\\ 50 & 20 & 40\end{pmatrix}\begin{pmatrix}u_1\\ u_2\\ u_3\end{pmatrix}$$

$$Subject\ to\begin{pmatrix}1 & 0 & 1\\ 0 & 1 & 1\\ 2 & 1 & 0\\ 1 & 2 & 0\\ 0 & 2 & 1\end{pmatrix}\begin{pmatrix}u_1\\ u_2\\ u_3\end{pmatrix}\geq\begin{pmatrix}2 & 1\\ 1 & 0\\ 1 & 1\\ 0 & 2\\ -1 & 2\end{pmatrix}\begin{pmatrix}\lambda_1\\ \lambda_2\end{pmatrix}$$

$$u_i \geq 0, i=1,\ 2,\ 3.$$

We split this problem into the following minimum problem (M4.2) and maximum problem (M8.2). Note that we have added a price desirability bound $u_1 + u_2 + u_3 \leq 35\lambda_1 + 60\lambda_2$ into (M8.2) to avoid the unbounded solutions.

$$Min \quad (\gamma_1 \quad \gamma_2)\begin{pmatrix}20 & 40 & 30\\ 50 & 20 & 40\end{pmatrix}\begin{pmatrix}u_1\\ u_2\\ u_3\end{pmatrix}$$

$$Subject\ to\begin{pmatrix}1 & 0 & 1\\ 0 & 1 & 1\\ 2 & 1 & 0\\ 1 & 2 & 0\\ 0 & 2 & 1\end{pmatrix}\begin{pmatrix}u_1\\ u_2\\ u_3\end{pmatrix}\geq\begin{pmatrix}2 & 1\\ 1 & 0\\ 1 & 1\\ 0 & 2\\ -1 & 2\end{pmatrix}\begin{pmatrix}\lambda_1\\ \lambda_2\end{pmatrix}$$

$$u_i \geq 0, i=1,\ 2,\ 3.$$

$$\textit{Max} \quad (\gamma_1 \quad \gamma_2) \begin{pmatrix} 20 & 40 & 30 \\ 50 & 20 & 40 \end{pmatrix} \begin{pmatrix} u_1 \\ u_2 \\ u_3 \end{pmatrix}$$

$$\textit{Subject to} \begin{pmatrix} 1 & 0 & 1 \\ 0 & 1 & 2 \\ 2 & 1 & 0 \\ 1 & 2 & 0 \\ 0 & 2 & 1 \\ -1 & -1 & -1 \end{pmatrix} \begin{pmatrix} u_1 \\ u_2 \\ u_3 \end{pmatrix} \geq \begin{pmatrix} 2 & 1 \\ 1 & 0 \\ 1 & 1 \\ 0 & 2 \\ -1 & 2 \\ -35 & -60 \end{pmatrix} \begin{pmatrix} \lambda_1 \\ \lambda_2 \end{pmatrix}$$

$$u_i \geq 0, i = 1,\ 2,\ 3.$$

Let $u_i \geq 0$, i = 4, 5, 6, 7, 8, 9, be the possible surplus variables. By Theorem 8.6 and Remark 8.2, we do not need to solve (M4.2). The lower potential solutions of (M8.1) are found from the dual potential solutions of (M4.1) (related to Table 7.2) as shown in Table 8.1. We solve (M8.2) to obtain the upper potential solutions of (M8.1) in Table 8.2.

From ξ and ζ, we form the (γ, λ) ranges $S(g, h) = (\Gamma(L^*_g) \times \Lambda(L^*_g)) \cap (\Gamma(U^*_h) \times \Lambda(U^*_h)) \neq \varnothing$, g = 1, …, 4; h = 1, …, 4, and the (γ, λ) space $\wp = \cup \{S(g, h) \mid \text{for all } (g, h)\}$ as follows:

$S(1, 1) = \{ (\gamma_1, \lambda_1) \mid 1/2 \leq \gamma_1 \leq 2/3,\ 1/2 \leq \lambda_1 \leq 1\}$,
$S(1, 2) = \{ (\gamma_1, \lambda_1) \mid 2/3 \leq \gamma_1 \leq 1,\ 1/2 \leq \lambda_1 \leq 1\}$,
$S(2, 1) = \{ (\gamma_1, \lambda_1) \mid 1/3 \leq \gamma_1 \leq 1/2,\ 3/4 \leq \lambda_1 \leq 1\}$,
$S(2, 4) = \{ (\gamma_1, \lambda_1) \mid 0 < \gamma_1 \leq 1/3,\ 3/4 \leq \lambda_1 \leq 1\}$,
$S(3, 1) = \{ (\gamma_1, \lambda_1) \mid 1/3 \leq \gamma_1 \leq 2/3,\ 0 < \lambda_1 \leq 1/2\}$,
$S(3, 2) = \{ (\gamma_1, \lambda_1) \mid 2/3 \leq \gamma_1 \leq 1,\ 0 < \lambda_1 \leq 1/2\}$,
$S(3, 3) = \{ (\gamma_1, \lambda_1) \mid 0 < \gamma_1 \leq 1/3,\ 0 < \lambda_1 \leq 1/2\}$,
$S(4, 1) = \{ (\gamma_1, \lambda_1) \mid 1/3 \leq \gamma_1 \leq 1/2,\ 1/2 \leq \lambda_1 \leq 3/4\}$; and
$S(4, 4) = \{ (\gamma_1, \lambda_1) \mid 0 < \gamma_1 \leq 1/2,\ 1/2 \leq \lambda_1 \leq 3/4\}$.

Table 8.1 Lower Potential Bases ξ

Bases	$\Gamma(L^*_i)$	$\Lambda(L^*_i)$	$V(L^*_i; \lambda, \gamma)$
L^*_1: $(u_1, u_3, u_6, u_7, u_8)$	$1/2 \le \gamma_1 \le 1$	$1/2 \le \lambda_1 \le 1$	$(\lambda_1 \quad \lambda_2)\begin{pmatrix} 50 & 90 \\ 20 & 50 \end{pmatrix}\begin{pmatrix} \gamma_1 \\ \gamma_2 \end{pmatrix}$
L^*_2: $(u_1, u_3, u_5, u_7, u_8)$	$0 < \gamma_1 \le 1/2$	$3/4 \le \lambda_1 \le 1$	$(\lambda_1 \quad \lambda_2)\begin{pmatrix} 55 & 85 \\ 25 & 45 \end{pmatrix}\begin{pmatrix} \gamma_1 \\ \gamma_2 \end{pmatrix}$
L^*_3: $(u_1, u_2, u_3, u_5, u_6)$	$0 < \gamma_1 \le 1$	$0 < \lambda_1 \le 1/2$	$(\lambda_1 \quad \lambda_2)\begin{pmatrix} 15 & 80 \\ 55 & 60 \end{pmatrix}\begin{pmatrix} \gamma_1 \\ \gamma_2 \end{pmatrix}$
L^*_4: $(u_1, u_3, u_5, u_6, u_8)$	$0 < \gamma_1 \le 1/2$	$1/2 \le \lambda_1 \le 3/4$	$(\lambda_1 \quad \lambda_2)\begin{pmatrix} 60 & 80 \\ 10 & 60 \end{pmatrix}\begin{pmatrix} \gamma_1 \\ \gamma_2 \end{pmatrix}$

The graphical representation of $\wp$ is shown in Figure 8.1. Then, we combine the (γ, λ) space $\Re = \cup\{R(i, j) \mid \text{for all } (i, j)\}$ found in Example 7.1 with $\wp$ to form the primal-dual (γ, λ) space $\chi = \cup\{R(i, j) \cap S(g, h) \mid \text{for all } (i, j) \text{ and } (g, h)\}$ as follows:

$$R(1, 1) \cap S(1, 1) = \{ (\gamma_1, \lambda_1) \mid 1/2 \le \gamma_1 \le 2/3,\ 1/2 \le \lambda_1 \le 2/3\},$$
$$R(1, 1) \cap S(1, 2) = \{ (\gamma_1, \lambda_1) \mid 2/3 \le \gamma_1 \le 1,\ 1/2 \le \lambda_1 \le 2/3\},$$
$$R(1, 2) \cap S(1, 1) = \{ (\gamma_1, \lambda_1) \mid 1/2 \le \gamma_1 \le 2/3,\ 2/3 \le \lambda_1 \le 1\},$$
$$R(1, 2) \cap S(1, 2) = \{ (\gamma_1, \lambda_1) \mid 2/3 \le \gamma_1 \le 1,\ 2/3 \le \lambda_1 \le 1\},$$
$$R(2, 2) \cap S(2, 1) = \{ (\gamma_1, \lambda_1) \mid 1/3 \le \gamma_1 \le 1/2,\ 3/4 \le \lambda_1 \le 1\},$$
$$R(2, 2) \cap S(2, 4) = \{ (\gamma_1, \lambda_1) \mid 0 < \gamma_1 \le 1/3,\ 3/4 \le \lambda_1 \le 1\},$$
$$R(3, 1) \cap S(3, 1) = \{ (\gamma_1, \lambda_1) \mid 1/3 \le \gamma_1 \le 2/3,\ 0 < \lambda_1 \le 1/2\},$$
$$R(3, 1) \cap S(3, 2) = \{ (\gamma_1, \lambda_1) \mid 2/3 \le \gamma_1 \le 1,\ 0 < \lambda_1 \le 1/2\},$$
$$R(3, 1) \cap S(3, 3) = \{ (\gamma_1, \lambda_1) \mid 0 < \gamma_1 \le 1/3,\ 0 < \lambda_1 \le 1/2\},$$
$$R(4, 1) \cap S(4, 4) = \{ (\gamma_1, \lambda_1) \mid 0 < \gamma_1 \le 1/3,\ 1/2 \le \lambda_1 \le 2/3\},$$

Table 8.2 Upper Potential Bases ζ

Bases	$\Gamma(U^*_i)$	$\Lambda(U^*_i)$	$V(U^*_i; \lambda, \gamma)$
U^*_1: (u_2, u_3, u_4, u_5, u_7,u_8)	$1/3 \leq \gamma_1 \leq 2/3$	$0 < \lambda_1 \leq 1$	$(\lambda_1 \quad \lambda_2)\begin{pmatrix} 30 & 40 \\ 1060 & 1380 \end{pmatrix}\begin{pmatrix} \gamma_1 \\ \gamma_2 \end{pmatrix}$
U^*_2: (u_2, u_3, u_5, u_6, u_7,u_8)	$2/3 \leq \gamma_1 \leq 1$	$0 < \lambda_1 \leq 1$	$(\lambda_1 \quad \lambda_2)\begin{pmatrix} 40 & 20 \\ 1380 & 740 \end{pmatrix}\begin{pmatrix} \gamma_1 \\ \gamma_2 \end{pmatrix}$
U^*_3: (u_1, u_3, u_4, u_5, u_6,u_7)	$0 < \gamma_1 \leq 1/3$	$0 < \lambda_1 \leq 1/2$	$(\lambda_1 \quad \lambda_2)\begin{pmatrix} 20 & 50 \\ 690 & 1760 \end{pmatrix}\begin{pmatrix} \gamma_1 \\ \gamma_2 \end{pmatrix}$
U^*_4: (u_1, u_3, u_4, u_6, u_7,u_8)	$0 < \gamma_1 \leq 1/3$	$1/2 \leq \lambda_1 \leq 1$	$(\lambda_1 \quad \lambda_2)\begin{pmatrix} 20 & 50 \\ 710 & 1740 \end{pmatrix}\begin{pmatrix} \gamma_1 \\ \gamma_2 \end{pmatrix}$

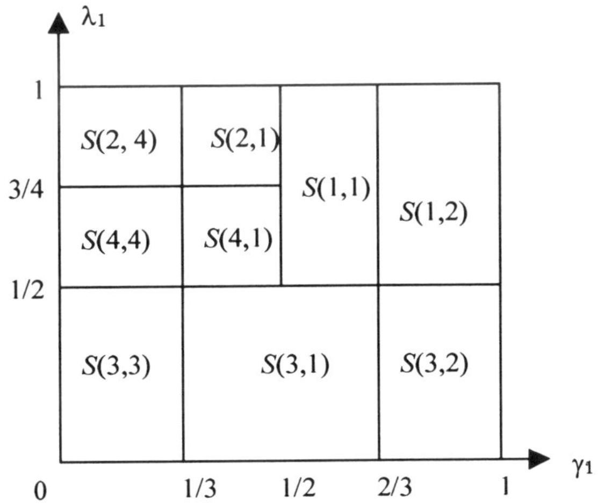

Figure 8.1 All Reformed Ranges S (g, h)

$$R(4,\ 2) \cap S(4,\ 1) = \{ (\gamma_1, \lambda_1) \mid 1/3 \le \gamma_1 \le 1/2,\ 2/3 \le \lambda_1 \le 3/4\},$$

and

$$R(4,\ 2) \cap S(4,\ 4) = \{ (\gamma_1, \lambda_1) \mid 0 < \gamma_1 \le 1/3,\ 2/3 \le \lambda_1 \le 3/4\}.$$

The graphical representation of χ can be similarly shown. By Definitions 7.2 and 8.1, we can construct a fuzzy potential solution of each R(i, j) for (M7.1) and a fuzzy potential solution of each S(g, h) of (M8.1), respectively. For illustrative purposes, let us consider R(3, 1) and S(3, 1). By using (M7.3), the membership function of R(3,1) is built in Example 7.1. The set of primal fuzzy potential solutions with $\mu_{R(3,1)}(\mathbf{x})$ is found in Table 7.7. We use (M8.3), the membership function of S(3, 1) is constructed as

$$\mu_{S(3,1)}(\mathbf{u}) = \begin{cases} 0, & \mathbf{u}'\mathbf{D}\boldsymbol{\gamma} \ge \mathrm{V}(U_1^*;\boldsymbol{\gamma},\boldsymbol{\lambda}) \\ \dfrac{\mathbf{u}'\mathbf{D}\boldsymbol{\gamma} - \mathrm{V}(L_3^*;\boldsymbol{\gamma},\boldsymbol{\lambda})}{\mathrm{V}(U_1^*;\boldsymbol{\gamma},\boldsymbol{\lambda}) - \mathrm{V}(L_3^*;\boldsymbol{\gamma},\boldsymbol{\lambda})}, & \mathrm{V}(L_3^*;\boldsymbol{\gamma},\boldsymbol{\lambda}) < \mathbf{u}'\mathbf{D}\boldsymbol{\gamma} < \mathrm{V}(U_1^*;\boldsymbol{\gamma},\boldsymbol{\lambda}) \\ 1, & \mathrm{V}(L_3^*;\boldsymbol{\gamma},\boldsymbol{\lambda}) \ge \mathbf{u}'\mathbf{D}\boldsymbol{\gamma} \end{cases}$$

Note that in the above form, U^*_1 comes from Table 8.2, while L^*_3 is from Table 8.1.

We identify its corresponding dual fuzzy potential solutions of (M8.1) from the primal bases {W} of (M4.2) as shown in Table 8.3.

Table 8.3 Dual Fuzzy Potential Solutions S(3, 1)

Bases	$\mathrm{V}(W_i;\ \lambda,\ \gamma)$
W_1: $(u_1, u_3, u_4, u_5, u_6)$	$(\lambda_1\ \ \lambda_2)\begin{pmatrix} -30 & -40 \\ 100 & 180 \end{pmatrix}\begin{pmatrix} \gamma_1 \\ \gamma_2 \end{pmatrix}$
W_2: $(u_1, u_2, u_3, u_5, u_8)$	$(\lambda_1\ \ \lambda_2)\begin{pmatrix} 40 & 80 \\ 70 & 60 \end{pmatrix}\begin{pmatrix} \gamma_1 \\ \gamma_2 \end{pmatrix}$

Knowing $\mu_{R(3,1)}(\mathbf{x})$ and $\mu_{S(3,1)}(\mathbf{u})$, we can construct the fuzzy weak duality functions $\mu^c{}_{R(i,j)\cap S(g,h)}(\mathbf{x},\mathbf{u})$. To show this, let us use a pair of fuzzy potential solutions (K_3, W_1) from Tables 7.7 and 8.3, respectively. Assume $\gamma_1 = \gamma_2 = 1/2$ and $\lambda_1 = 1/4$, $\lambda_2 = 3/4$. Then, we can calculate $\mu_{R(3,1)}(\mathbf{x}(K_3)) = 0.610$ and $\mu_{S(3,1)}(\mathbf{u}(W_1)) = 0.048$. We see that

$$\mu^1{}_{R(i,j)\cap S(g,h)}(\mathbf{x}(K_3), \mathbf{u}(W_1)) = 0.048,$$

$$\mu^2{}_{R(i,j)\cap S(g,h)}(\mathbf{x}(K_3), \mathbf{u}(W_1)) = 0.610, \text{ and}$$

$$\mu^3{}_{R(i,j)\cap S(g,h)}(\mathbf{x}(K_3), \mathbf{u}(W_1)) = 0.329.$$

8.2 A Parametric Approach to Fuzzy MC² Duality

The motivation of the parametric approach to fuzzy MC² duality (Zhong and Shi 1999) discussed in this section differs from that of the constructive approach in Section 8.1. The "duality" to the decision makers in the constructive approach can be viewed as a self-alternative "business" strategy (selling the resource, instead of producing the goods), while the "duality" in the parametric approach represents the competitors' behavior. The idea of the parametric approach originally comes from Rödder and Zimmermann (1980) for their fuzzy linear duality (the related studies can be found in Hamacher, Leberling and Zimmermann 1978, Chanas 1989, and Verdegay 1984). If we adopt this idea to the MC² framework, the following story is stated:

Suppose the decision makers (P) plan its production $\mathbf{x}$ when they compete with the rest people in the market (D). On one hand, the (P) side may assume that the (D) behaves in a strictly minimum way for each $\mathbf{x} \geq 0$, rather sub-minimizes probably because of competition. Therefore, (P) has to consider all possible $\mathbf{u} \geq \mathbf{0}$ (the dual variables dependent on $\mathbf{x} \geq 0$) and evaluates them by a membership function $\sigma_{\mathbf{x}}^{P}(\mathbf{u})$, which describes (P)'s own degree of satisfaction for (D)'s decision $\mathbf{u}$ on $\{\mathbf{u} \mid \mathbf{u} \geq 0\}$ for any given $\mathbf{x} \geq 0$. On the other hand, the (P) do not pursue strictly the maximum profits rather satisfied profits with certain aspiration levels. Thus, the decision makers of (P) consider all possible $\mathbf{x} \geq \mathbf{0}$ and evaluate them by their primary profits $\lambda^t\mathbf{Cx}$ with a membership function $\mu^P(\mathbf{x})$, the (P)'s degree of satisfaction for its decision $\mathbf{x}$ on $\{\mathbf{x} \mid \mathbf{x} \geq 0\}$.

Assume that (M7.1) represents the fuzzy problem of (M4.1) for the (P) side and (M8.1) is the fuzzy problem of (M4.2) for the (D) side. For the convenience, let $\{(\mathbf{x}, \mu(\mathbf{x})) \mid \mathbf{x} \in \mathbf{X}\}$ be a fuzzy set on $\mathbf{X} = \{\mathbf{x} \geq \mathbf{0} \mid$

$\mathbf{Ax} \le \mathbf{D}\gamma\}$ and $\{\mathbf{u}, \sigma_{\mathbf{x}}(\mathbf{u}))\,|\,\mathbf{u} \in \mathbf{U}\}$ be a family of fuzzy sets on $\mathbf{U} = \{\mathbf{u} \ge 0 \mid \mathbf{u}^t\mathbf{A} \ge \lambda^t\mathbf{C}\}$. Note that in this notation **x** and **u** can be switched.

Let $\Phi = \{J_1, J_2, \ldots, J_r\}$ be the set of potential solutions over all possible changes of (γ, λ) for (M4.1) satisfying $(\Gamma(J_i) \times \Lambda(J_i)) \cap (\Gamma(J_g) \times \Lambda(J_g)) = \varnothing$ for $i \ne g$; $i, g = 1, \ldots, r$; and let $\xi = \{Q_1, Q_2, \ldots, Q_v\}$ be the set of potential solutions over all possible changes of (γ, λ) for (M4.2) satisfying $(\Gamma(Q_j) \times \Lambda(Q_j)) \cap (\Gamma(Q_h) \times \Lambda(Q_h)) = \varnothing$ for $j \ne h$; $j, h = 1, \ldots, v$. We can form the (γ, λ) range $T(i, j) = (\Gamma(J_i) \times \Lambda(J_i)) \cap (\Gamma(Q_j) \times \Lambda(Q_j)) \ne \varnothing$, $i = 1, \ldots, r$; $j = 1, \ldots, v$, and the (γ, λ) space $\theta = \cup\, \{T(i, j) \mid \text{for all } (i, j)\}$. Then, similar to Theorem 7.1 and Remark 7.1, we have:

Theorem 8.8

(iv) $T(i, j) \subseteq \Gamma(J_i) \times \Lambda(J_i)$ and $T(i, j) \subseteq \Gamma(Q_j) \times \Lambda(Q_j)$, for all $i = 1, \ldots, r$; $j = 1, \ldots, v$.

(v) $T(g, h) \cap T(i, j) = \varnothing$ for $i \ne g$; $j \ne h$; $i, g = 1, \ldots, r$; $j, h = 1, \ldots, v$.

(vi) $\Gamma \times \Lambda = \theta$.

Definition 8.4 Given fuzzy problems (M7.1) of (M4.1) and (M8.1) of (M4.2), let J_i be a potential basis of (M4.1) and Q_j be a potential basis of (M4.2), then

(i) the membership function for (M7.1) is defined as $\underset{\mathbf{x} \in \mathbf{X}}{Max}(\min\{\mu_{J_i}(\mathbf{x}), \sigma^P_{\mathbf{x}}(\mathbf{u})\})$ for J_i; and

(ii) the membership function for (M8.1) is defined as $\underset{\mathbf{u} \in \mathbf{U}}{Max}(\min\{\mu_{Q_j}(\mathbf{u}), \sigma^D_{\mathbf{u}}(\mathbf{x})\})$ for Q_j.

Remark 8.5 (i) Definition 8.4 states a win-win strategy for both decision makers (P) and their competitors (D) in items of looking for the better result out of their worse cases. This could be considered as a special kind of "bounded rationality" of the decision makers when deciding about prices and quantities in competitive situations. (ii) The interpretation of the membership functions in Definition 8.4 can vary according to available information.

A simple form of the membership functions in Definition 8.4 is given as:

For J_i (M8.4):

$$\mu_{J_i}(\mathbf{x}) = \begin{cases} 1, & \textit{if } \boldsymbol{\lambda}^t \mathbf{C}\mathbf{x}^0 \le \boldsymbol{\lambda}^t \mathbf{C}\mathbf{x} \\ 1 - \dfrac{\boldsymbol{\lambda}^t \mathbf{C}\mathbf{x}^0 - \boldsymbol{\lambda}^t \mathbf{C}\mathbf{x}}{\boldsymbol{\lambda}^t \mathbf{C}\mathbf{x}^0 - \min \boldsymbol{\lambda}^t \mathbf{C}\mathbf{x}}, & \textit{otherwise} \end{cases}$$

and (M8.5):

$$\sigma_{\mathbf{x}}^P(\mathbf{u}) = \begin{cases} 0, & \textit{if } \mathbf{u}^t \mathbf{D}\boldsymbol{\gamma} \le \mathbf{u}^t \mathbf{A}\mathbf{x} \\ \dfrac{\mathbf{u}^t \mathbf{D}\boldsymbol{\gamma} - \mathbf{u}^t \mathbf{A}\mathbf{x}}{\boldsymbol{\lambda}^t \mathbf{C}\mathbf{x}^0}, & \textit{otherwise} \end{cases}$$

For Q_j (M8.6):

$$\mu_{Q_i}(\mathbf{u}) = \begin{cases} 1, & \textit{if } \mathbf{u}^t \mathbf{D}\boldsymbol{\gamma} \le (\mathbf{u}^0)^t \mathbf{D}\boldsymbol{\gamma} \\ 1 - \dfrac{\mathbf{u}^t \mathbf{D}\boldsymbol{\gamma} - (\mathbf{u}^0)^t D\gamma}{\max \mathbf{u}^t \mathbf{D}\boldsymbol{\gamma} - (\mathbf{u}^0)^t \mathbf{D}\boldsymbol{\gamma}}, & \textit{otherwise} \end{cases}$$

and (M8.7):

$$\sigma_{\mathbf{u}}^D(\mathbf{x}) = \begin{cases} 0, & \textit{if } \boldsymbol{\lambda}^t \mathbf{C}\mathbf{x} \ge \mathbf{u}^t \mathbf{A}\mathbf{x} \\ -\dfrac{\boldsymbol{\lambda}^t \mathbf{C}\mathbf{x} - \mathbf{u}^t \mathbf{A}\mathbf{x}}{(\mathbf{u}^0)^t \mathbf{D}\boldsymbol{\gamma}}, & \textit{otherwise} \end{cases}$$

where $\mathbf{x}^0$ and $\mathbf{u}^0$, respectively, denote the potential solutions to (M4.1) and (M4.2) on *T*(i, j).

Remark 8.6 From (M8.4), we see that on *T*(i, j), if $\boldsymbol{\lambda}^t \mathbf{C}\mathbf{x} = \boldsymbol{\lambda}^t \mathbf{C}\mathbf{x}^0$, then $\mu_{J_i}(\mathbf{x}) = 1$ and $\mathbf{x} = \mathbf{x}^0(J_i;\gamma)$ is the maximum solution for (M4.1); if $\boldsymbol{\lambda}^t \mathbf{C}\mathbf{x} = \min \boldsymbol{\lambda}^t \mathbf{C}\mathbf{x}$, then $\mu_{J_i}(\mathbf{x}) = 0$; if $\min \boldsymbol{\lambda}^t \mathbf{C}\mathbf{x} \le \boldsymbol{\lambda}^t \mathbf{C}\mathbf{x} \le \boldsymbol{\lambda}^t \mathbf{C}\mathbf{x}^0$, then $\mu_{J_i}(\mathbf{x})$ increases linearly from 0 to 1. Thus we may replace $\min \boldsymbol{\lambda}^t \mathbf{C}\mathbf{x}$ and

$\boldsymbol{\lambda}'\mathbf{Cx}^0$ with the lower and upper aspiration bounds of the objective payoff of (M4.1) for T(i, j). From (M8.5), we see that $\sigma_{\mathbf{x}}^{\mathrm{P}}(\mathbf{u})$ is the evaluation function of the primal decision makers for the dual decision $\mathbf{u}$ of their adversary given a decision $\mathbf{x}$. The primal decision makers are not satisfied if $\mathbf{u}'\ (\mathbf{D}\boldsymbol{\gamma} - \mathbf{Ax}) \le 0$. Their satisfaction degree increases linearly in $\mathbf{u}'\ (\mathbf{D}\boldsymbol{\gamma} - \mathbf{Ax}) / \boldsymbol{\lambda}'\mathbf{Cx}^0$ when $\mathbf{u}'\ (\mathbf{D}\boldsymbol{\gamma} - \mathbf{Ax}) \ge 0$. $\boldsymbol{\lambda}'\mathbf{Cx}^0$ may be replaced by the upper aspiration bounds of the objective payoff of (M4.1) for T(i, j).

Remark 8.7 From (M8.6), we see that on T(i, j), if $\mathbf{u}'\mathbf{D}\boldsymbol{\gamma} = (\mathbf{u}^0)'\mathbf{D}\boldsymbol{\gamma}$, then $\mu_{Q_j}(\mathbf{u})$ =1 and $\mathbf{u} = \mathbf{u}(Q_j, \boldsymbol{\lambda})$ is the minimum solution for (M4.2); if $\mathbf{u}'\mathbf{D}\boldsymbol{\gamma} = \max \mathbf{u}'\mathbf{D}\boldsymbol{\gamma}$, then $\mu_{Q_j}(\mathbf{u})$ =0; if $(\mathbf{u}^0)'\mathbf{D}\boldsymbol{\gamma} \le \mathbf{u}'\mathbf{D}\boldsymbol{\gamma} \le \max \mathbf{u}'\mathbf{D}\boldsymbol{\gamma}$, then $\mu_{Q_j}(\mathbf{u})$ decreases linearly from 1 to 0. Thus we may replace $(\mathbf{u}^0)'\mathbf{D}\gamma$ and $\max \mathbf{u}'\mathbf{D}\boldsymbol{\gamma}$ with the lower and upper aspiration bounds of objective payoff of (M4.2) for T(i, j). From (M8.7), we see that $\sigma_{\mathbf{u}}^{\mathrm{D}}(\mathbf{x})$ is the evaluation function of the competitors for the primal decision $\mathbf{x}$ of their adversary given a decision $\mathbf{u}$. These dual decision makers are not satisfied if $(\boldsymbol{\lambda}'\mathbf{C} - \mathbf{u}'\mathbf{A})\,\mathbf{x} \ge 0$. This satisfaction degree increases linearly in $-(\boldsymbol{\lambda}'\mathbf{C} - \mathbf{u}'\mathbf{A})\mathbf{x}/(\mathbf{u}^0)'\mathbf{D}\boldsymbol{\gamma}$, if $(\boldsymbol{\lambda}'\mathbf{C} - \mathbf{u}'\mathbf{A})\mathbf{x} \le 0$, where $(\mathbf{u}^0)'\mathbf{D}\boldsymbol{\gamma}$ may also be replaced by the lower aspiration bound of the objective payoff of (M4.2) for T(i, j).

Given a T(i, j), the crisp formulation of $\underset{\mathbf{x}\in\mathbf{X}}{Max}(\min\{\mu_{J_i}(\mathbf{x}), \sigma_{\mathbf{x}}^{P}(\mathbf{u})\})$, which finds a fuzzy potential solution for J_i of (M7.1), can be written as (M8.8):

$$\begin{aligned} &Max \quad \delta^P \\ &Suject\ to \quad \delta^P \le 1 - \frac{\boldsymbol{\lambda}^t \mathbf{C}\mathbf{x}^0 - \boldsymbol{\lambda}^t \mathbf{C}\mathbf{x}}{\boldsymbol{\lambda}^t \mathbf{C}\mathbf{x}^0 - \min \boldsymbol{\lambda}^t \mathbf{C}\mathbf{x}}, \textit{ for any given } \mathbf{u} \ge 0 \\ &\qquad \delta^P \le \frac{\mathbf{u}^t \mathbf{D}\boldsymbol{\gamma} - \mathbf{u}^t \mathbf{A}\mathbf{x}}{\boldsymbol{\lambda}^t \mathbf{C}\mathbf{x}^0}, \\ &\qquad \mathbf{x} \ge 0, \delta^P \ge 0. \end{aligned}$$

Similarly, the crisp formulation of $\underset{\mathbf{u}\in\mathbf{U}}{Max}(\min\{\mu_{Q_j}(\mathbf{u}), \sigma_{\mathbf{u}}^D(\mathbf{x})\})$, which finds a fuzzy potential solution for Q_j of (M8.1), can be written as (M8.9):

$$\begin{aligned} &Max \quad \delta^D \\ &Suject\ to \quad \delta^D \ge \frac{\mathbf{u}^t \mathbf{D}\boldsymbol{\gamma} - (\mathbf{u}^0)^t \mathbf{D}\boldsymbol{\gamma}}{\max \mathbf{u}^t \mathbf{D}\boldsymbol{\gamma} - (\mathbf{u}^0)^t \mathbf{D}\boldsymbol{\gamma}} - 1, \textit{ for any given } \mathbf{x} \ge 0 \\ &\qquad \delta^P \ge \frac{\boldsymbol{\lambda}^t \mathbf{C}\mathbf{x} - \mathbf{u}^t \mathbf{A}\mathbf{x}}{(\mathbf{u}^0)^t \mathbf{D}\boldsymbol{\gamma}}, \\ &\qquad \mathbf{u} \ge 0, \delta^D \le 0. \end{aligned}$$

Note that both (M8.8) and (M8.9) are parametric linear programming problems with respect to $(\boldsymbol{\gamma}, \boldsymbol{\lambda})$ and $\mathbf{x}$ or $\mathbf{u}$. This is why we call the fuzzy duality the parametric approach.

Definition 8.5 $(\mathbf{x}^*; \mu_{T(i,j)}(\mathbf{x}^*))$ is a fuzzy potential solution of T(i, j) for (M7.1) if and only if $(\mathbf{x}^*, \delta^P)$ is the optimal solution of (M8.8) and $\mathbf{x}^*$ has a membership function $\mu_{T(i,j)}(\mathbf{x}^*) = \delta^P$.

Definition 8.6 $(\mathbf{u}^*; \mu_{T(i,j)}(\mathbf{u}^*))$ is a dual fuzzy potential solution of T(i, j) for (M8.1) if and only if $(\mathbf{u}^*, \delta^D)$ is the optimal solution of (M8.9) and $\mathbf{u}^*$ has a membership function $\mu_{T(i,j)}(\mathbf{u}^*) = -\delta^D$.

Definition 8.7 Let $(\mathbf{x}^*; \mu_{T(i,j)}(\mathbf{x}^*))$ be fuzzy potential solution of T(i, j) for (M7.1). Then the set $\rho = \{(\mathbf{x}^*; \mu_{T(i,j)}(\mathbf{x}^*)) \mid \text{for all } T(i,j)\}$ is called a set of fuzzy potential solutions for all possible changes of $(\boldsymbol{\gamma}, \boldsymbol{\lambda}) \in \boldsymbol{\Gamma} \times \boldsymbol{\Lambda}$.

Definition 8.8 Let $(\mathbf{u}^*; \mu_{T(i,j)}(\mathbf{u}^*))$ be a dual fuzzy potential solution of T(i, j) for (M8.1). Then the set $\tau = \{(\mathbf{u}^*; \mu_{T(i,j)}(\mathbf{u}^*)) \mid \text{for all } T(i,j)\}$ is called a set of dual fuzzy potential solutions for all possible changes of $(\boldsymbol{\gamma}, \boldsymbol{\lambda}) \in \boldsymbol{\Gamma} \times \boldsymbol{\Lambda}$.

Theorem 8.9

(i) Given a pair of fuzzy potential solutions ($(\mathbf{u}^*; \mu_{T(i,j)}(\mathbf{u}^*))$, $(\mathbf{x}^*; \mu_{T(i,j)}(\mathbf{x}^*))$), there exist a potential solution $\mathbf{x}(J_i)$ of (M4.1) and a potential solution $\mathbf{u}(Q_j)$ of (M4.2) for T(i, j).

(iii) Conversely, with a potential solution $\mathbf{x}(J_i)$ of (M4.1) and a potential solution $\mathbf{u}(Q_j)$ of (M4.2) for T(i, j), a pair of fuzzy potential solutions ($(\mathbf{u}^*; \mu_{T(i,j)}(\mathbf{u}^*))$, ($\mathbf{x}^*; \mu_{T(i,j)}(\mathbf{x}^*)$)) can be identified.

Proof. (i) Given a pair of fuzzy potential solutions ($(\mathbf{u}^*; \mu_{T(i,j)}(\mathbf{u}^*))$, $(\mathbf{x}^*; \mu_{T(i,j)}(\mathbf{x}^*))$), if $\mathbf{x}^*$ and $\mathbf{u}^*$ is a potential solution of (M4.1) and (M4.2), respectively, i.e., $\mathbf{x}(J_i)$ and $\mathbf{u}(Q_j)$, then the proof is complete; otherwise, we solve problem (M4.1) and (M4.2) to obtain the potential solution $\mathbf{x}(J_i)$ and $\mathbf{u}(Q_j)$ for T(i, j).

(ii) This follows from the membership function $\mu_{T(i,j)}(\mathbf{x}^*)$ ($(\mu_{T(i,j)}(\mathbf{u}^*))$ and Definitions 8.5and 8.6.

Theorem 8.9 (ii) allows us to construct a pair of fuzzy potential solutions from given potential solutions of primal-dual problems of (M4.1) and (M4.2) for T(i, j).

Based on the primal fuzzy membership function $\mu_{T(i,j)}(\mathbf{x}^*)$ and the dual fuzzy membership function $\mu_{T(i,j)}(\mathbf{u}^*)$, the relationship between (M7.1) and (M8.1) is as follows:

Theorem 8.10 Fuzzy weak duality membership function on a given reformed (γ, λ) range T(i, j) is

$$\mu_{T(i,j)}(\mathbf{x}^*, \mathbf{u}^*) = \min\{\mu_{T(i,j)}(\mathbf{x}^*), \mu_{T(i,j)}(\mathbf{u}^*)\} = \min\{\delta^P, -\delta^D).$$

Proof. Given a primal fuzzy potential solution $(\mathbf{x}^*; \mu_{T(i,j)}(\mathbf{x}^*))$ and a dual fuzzy potential solution $(\mathbf{u}^*; \mu_{T(i,j)}(\mathbf{u}^*))$ on a given reformed (γ, λ) range T(i, j), by Theorem 8.9 (i), there exist a potential solution $\mathbf{x}(J_i)$ of (M4.1) and a potential solution $\mathbf{u}'(Q_j)$ of (M4.2) for T(i, j) T(i, j). Thus, $\boldsymbol{\lambda}'\mathbf{C}\mathbf{x}^* \leq \boldsymbol{\lambda}'\mathbf{C}\mathbf{x}(J_i)$, $(\mathbf{u}^*)'\mathbf{D}\boldsymbol{\gamma} \geq \mathbf{u}'(Q_j)\mathbf{D}\boldsymbol{\gamma}$. According to the weak duality and complementary slackness of MC^2 problem (Theorem 3.8), $\boldsymbol{\lambda}\ \mathbf{C}\,\mathbf{x}(J_i) = \mathbf{u}(Q_j)\mathbf{D}\boldsymbol{\gamma}$. Therefore, $\boldsymbol{\lambda}'\mathbf{C}\mathbf{x}^* \leq \boldsymbol{\lambda}'\mathbf{C}\,\mathbf{x}(J_i) = \mathbf{u}'(Q_j)\mathbf{D}\boldsymbol{\gamma} \leq (\mathbf{u}^*)'\mathbf{D}\boldsymbol{\gamma}$.

By theory of fuzzy set, we have

$$\mu_{T(i,j)}(\mathbf{x}^*, \mathbf{u}^*) = \min\{\mu_{T(i,j)}(\mathbf{x}^*), \mu_{T(i,j)}(\mathbf{u}^*)\} = \min\{\delta^P, -\delta^D\}.$$

Theorem 8.10 shows that the fuzzy weak duality in the parametric approach is also a kind of extension of the ordinary weak duality.

Example 8.2 Lets use the fuzzy MC^2 problem in Example 7.1 to illustrate the parametric approach. Given the fuzzy (M7.1) in Example 7.1, we solve its max-crisp problem (M4.1) as shown in Table 7.2. After constructing the fuzzy dual problem (M8.1), we also solve its min-crisp problem (M4.2) as shown in Table 7.10 (see Example 7.2).

Based on $\Phi = \{U_1, U_2, U_3, U_4\}$ and $\xi = \{L^*_1, L^*_2, L^*_3, L^*_4\}$, we form the (γ, λ) range $T(\text{i, j}) = (\Gamma(U_i)\times\Lambda(U_i)) \cap (\Gamma(L^*_j)\times\Lambda(L^*_j)) \neq \varnothing$, i = 1, …, 4; j = 1, …, 4, and the (γ, λ) space $\theta = \cup\,\{T(\text{i, j}) \mid \text{for all (i, j)}\}$ as follows:

$T(1, 1) = \{(\gamma_1, \lambda_1) \mid 1/2 \leq \gamma_1 \leq 1,\ 1/2 \leq \lambda_1 \leq 1\}$,
$T(2, 2) = \{(\gamma_1, \lambda_1) \mid 0 < \gamma_1 \leq 1/2,\ 3/4 \leq \lambda_1 \leq 1\}$,
$T(3, 3) = \{(\gamma_1, \lambda_1) \mid 0 < \gamma_1 \leq 1,\ 0 < \lambda_1 \leq 1/2\}$, and

$T(4, 4) = \{ (\gamma_1, \lambda_1) \mid 0 < \gamma_1 \leq 1/2,\ 1/2 \leq \lambda_1 \leq 3/4\}$; and
$\theta = \cup \{T(1, 1), T(2, 2), T(3, 3), T(4, 4)\}$, which is graphically shown in Figure 8.2.

By Theorem 8.15, we can find fuzzy potential solutions of each T(i, j) for (M7.1) and (M8.1), respectively.

For illustration, let us consider $T(3, 3) = \{ (\gamma_1, \lambda_1) \mid 0 < \gamma_1 \leq 1, 0 < \lambda_1 \leq 1/2\}$. According to the above method, $\mathbf{x}^*$, associated with δ^P, of a fuzzy potential solution $(\mathbf{x}^*; \mu_{T(3,3)}(\mathbf{x}^*)) = (\mathbf{x}^*; \delta^1)$ of (M7.1) is the optimal solution of problem (M8.8), i.e.,

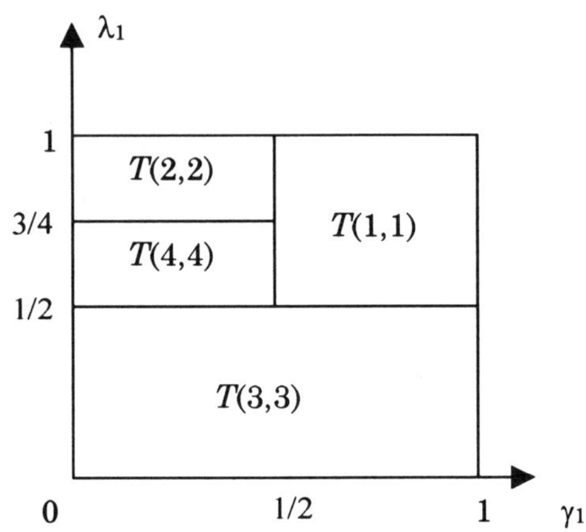

Figure 8.2 All Reformed Ranges $T(i, j)$

$$\begin{aligned}
&Max \quad \delta^P \\
&Suject\ to \quad \delta^P \le 1 - \frac{\boldsymbol{\lambda}^t \mathbf{C}\mathbf{x}^0 - \boldsymbol{\lambda}^t \mathbf{C}\mathbf{x}}{\boldsymbol{\lambda}^t \mathbf{C}\mathbf{x}^0 - \min \boldsymbol{\lambda}^t \mathbf{C}\mathbf{x}},\ for\ any\ given\ \mathbf{u} \ge 0 \\
&\qquad \delta^P \le \frac{\mathbf{u}^t \mathbf{D}\boldsymbol{\gamma} - \mathbf{u}^t \mathbf{A}\mathbf{x}}{\boldsymbol{\lambda}^t \mathbf{C}\mathbf{x}^0}, \\
&\qquad \mathbf{x} \ge 0, \delta^P \ge 0,
\end{aligned}$$

where $\mathbf{A}, \mathbf{C}, \mathbf{D}, \boldsymbol{\lambda}, \boldsymbol{\gamma}, \mathbf{x}$ are the same as in Example 7.1; $\boldsymbol{\lambda}^t\mathbf{C}\mathbf{x}^0 = V(U_3; \lambda, \gamma) = (\lambda_1 \quad \lambda_2) \begin{pmatrix} 15 & 80 \\ 55 & 60 \end{pmatrix} \begin{pmatrix} \gamma_1 \\ \gamma_2 \end{pmatrix}$ (in Table 7.2); and $\min \boldsymbol{\lambda}^t\mathbf{C}\mathbf{x} = 0$ can be obtained by solving problem (M7.2) on $T(3, 3)$. Note that $\min \boldsymbol{\lambda}^t\mathbf{C}\mathbf{x}$ may be replaced by the lower aspiration bound of the decision makers.

Problem (M8.8) can be considered as parametric LP with respect to $(u_1, u_2, u_3) \ge 0$. By MC^2 – simplex method, when $0 < \gamma_1 \le 1$, $0 < \lambda_1 \le 1/2$, $(u_1, u_2, u_3) \ge 0$ subject to:

$7u_1\lambda_1 + 6u_2\lambda_1 + 6u_3\lambda_1 - 2u_1 + 6u_2 - 3u_3 \ge 0$, and

$(25 + 25\lambda_1 + 5u_1 - 15u_2 - 30u_3)\gamma_1 + (45 + 45\lambda_1 - 5u_1 + 25u_2 - 40u_3)\gamma_2 \ge 0$

We can obtain an optimal solution of problem (M8.8) (slack variables are not written) as

$$\delta^P = \frac{\mathbf{u}^t\mathbf{D}\boldsymbol{\gamma}}{\boldsymbol{\lambda}^t\mathbf{C}\mathbf{x}^0}[1 - \frac{(u_1 + u_3)}{2\lambda_1 + \lambda_2 + u_1 + u_3}],$$

$$x_1 = \frac{\mathbf{u}^t\mathbf{D}\boldsymbol{\gamma}}{2\lambda_1 + \lambda_2 + u_1 + u_3}, x_2 = x_3 = x_4 = x_5 = 0.$$

Therefore, we obtain a fuzzy potential solution $(\mathbf{x}^*; \mu_{T(3,3)}(\mathbf{x}^*))$ on $T(3, 3)$ for problem (M7.1):

$$\mathbf{x}^* = (\frac{\mathbf{u}^t\mathbf{D}\boldsymbol{\gamma}}{2\lambda_1 + \lambda_2 + u_1 + u_3}, \quad 0, \quad 0, \quad 0, \quad 0).$$

Its membership function:

$$\mu_{T(3,3)}(\mathbf{x}^*) = \delta^P = \frac{\mathbf{u}'\mathbf{D}\boldsymbol{\gamma}}{\boldsymbol{\lambda}^t\mathbf{C}\mathbf{x}^0}[1 - \frac{(u_1+u_3)}{2\lambda_1+\lambda_2+u_1+u_3}].$$

Especially, let $u_1 = u_2 = u_3 = 1/3$, $\lambda_1 = \gamma_1 = 1/2$, then $\mathbf{x}^* = (200/13,\ 0,\ 0,\ 0,\ 0)$, $\mu_{T(3,3)}(\mathbf{x}^*) = \delta^P = 40/91 \approx 0.44$.

$\mathbf{u}^*$, associated with δ^D, of a fuzzy potential solution $(\mathbf{u}^*;\mu_{T(3,3)}(\mathbf{u}^*)) = (\mathbf{u}^*;-\delta^D)$ of (M8.1) is the optimal solution of problem (M8.9), i.e.,

$$\begin{aligned} &\textit{Max} \quad \delta^D \\ &\textit{Suject to} \quad \delta^D \ge \frac{\mathbf{u}'\mathbf{D}\boldsymbol{\gamma} - (\mathbf{u}^0)'\mathbf{D}\boldsymbol{\gamma}}{\max \mathbf{u}'\mathbf{D}\boldsymbol{\gamma} - (\mathbf{u}^0)'\mathbf{D}\boldsymbol{\gamma}} - 1, \textit{ for any given } \mathbf{x} \ge 0 \\ &\delta^P \ge \frac{\boldsymbol{\lambda}'\mathbf{C}\mathbf{x} - \mathbf{u}'\mathbf{A}\mathbf{x}}{(\mathbf{u}^0)'\mathbf{D}\boldsymbol{\gamma}}, \\ &\mathbf{u} \ge 0, \delta^D \le 0, \end{aligned}$$

where $\mathbf{A}, \mathbf{C}, \mathbf{D}, \boldsymbol{\lambda}, \boldsymbol{\gamma}, \mathbf{u}$ are the same as in Example 7.2; $\mathbf{u}^t\mathbf{D}\boldsymbol{\gamma} = V(L^*_4; \lambda, \gamma) =$
$\begin{pmatrix}\lambda_1 & \lambda_2\end{pmatrix}\begin{pmatrix}15 & 80\\55 & 60\end{pmatrix}\begin{pmatrix}\gamma_1\\\gamma_2\end{pmatrix}$ (in Table 8.1);

and $\max \mathbf{u}'\mathbf{D}\boldsymbol{\gamma} = \begin{pmatrix}\lambda_1 & \lambda_2\end{pmatrix}\begin{pmatrix}20 & 50\\690 & 1760\end{pmatrix}\begin{pmatrix}\gamma_1\\\gamma_2\end{pmatrix}$ (in Table 8.2), which is obtained by solving (M8.2) on $T(3, 3)$.

We also note that $\max \mathbf{u}'\mathbf{D}\boldsymbol{\gamma}$ may be also replaced by the upper aspiration bound of the decision makers.

By MC^2 – simplex method, when $0 < \gamma_1 \le 1$, $0 < \lambda_1 \le 1/2$, $(x_1, x_2, x_3, x_1, x_2) \ge 0$ subject to:

$$\frac{\boldsymbol{\lambda}^t\mathbf{C}\mathbf{x}}{\Delta_1} \ge 1 + \lambda_2/2 - \frac{\max \mathbf{u}'\mathbf{D}\boldsymbol{\gamma}}{\Delta_2}, \quad \text{and}$$

$$(70\gamma_1+50\gamma_2)x_1 - (20\gamma_1+50\gamma_2)x_2 + (60\gamma_1-100\gamma_2)x_3 - (30\gamma_1+200\gamma_2)x_4 - (60\gamma_1+150\gamma_2)x_5 \ge 0,$$

where $\Delta_1 = (\mathbf{u}^0)' \mathbf{D}\boldsymbol{\gamma}, \Delta_2 = \max\ \mathbf{u}' \mathbf{D}\boldsymbol{\gamma} - (\mathbf{u}^0)' \mathbf{D}\boldsymbol{\gamma}$.

We can obtain its an optimal solution (slack variables are not written):

$$\delta^D = [-\max \mathbf{u}' \mathbf{D}\boldsymbol{\gamma} + (2\lambda_2 - \lambda_1)(30\gamma_1 + 40\gamma_2) + (20\gamma_1 + 50\gamma_2)u_1] / \Delta_2 \quad ,$$

and

$$u_1 = \{\boldsymbol{\lambda}\mathbf{C}\mathbf{x} \cdot \Delta_2 + \max \mathbf{u}' \mathbf{D}\boldsymbol{\gamma} \cdot \Delta_1 - (2\lambda_2 - \lambda_1)[(x_1 + x_2 + x_5)\Delta_2 + (30\gamma_1 + 40\gamma_2)\Delta_1]\} /$$

$$[(x_1 + 2x_3 + x_4)\Delta_2 + (20\gamma_1 + 50\gamma_2) \cdot \Delta_1],$$

$u_2 = 0$,

$u_3 = 2\lambda_2 - \lambda_1$.

Therefore, we get a dual fuzzy potential solution $(\mathbf{u}^*; \mu_{R(3,3)}(\mathbf{u}^*))$ on $T(3, 3)$ for problem (M8.1): $\mathbf{u}^* = (u_1, \ u_2, \ u_3)$ and its membership function $\mu_{T(3,3)}(\mathbf{u}^*) = \ \delta^D$.

Especially, let $x_1 = x_3 = 10, x_2 = x_4 = x_5 = 0, \gamma_1 = \lambda_1 = 1/2$, then $\mathbf{u}^* \approx (2.28, \ 0, \ 0.5), \delta^D \approx -0.92$, thus $\mu_{T(3,3)}(\mathbf{u}^*) = 0.92$.

We see that when $\gamma_1 = \lambda_1 = 1/2$, fuzzy weak duality membership function is $\mu_{T(3,3)}(\mathbf{x}^*; \mathbf{u}^*) = \min\{0.44, 0.92\} = 0.92$.

8.3 Further Remarks

The fuzzy duality in this chapter has been studied by two different motivations. One is the duality for the decision makers themselves, and the other is duality from the competitors' point of view. In any case, we see that the key of research is that the constructed fuzzy potential solutions for both primal and dual fuzzy MC^2 problems must be built on the reformed (γ, λ) space which is equivalent to the original (γ, λ) space. The reader should pay their attention on coverage studies for the (γ, λ) ranges.

Exercises

8.1 Given the fuzzy MC^2 problem:

Fuzzy-Max	$10\ x_1 + 12\ x_2$
Fuzzy-Max	$15\ x_1 + 8\ x_2$
Subject to	$3\ x_1 + 2x_2 \leq (25, -35)$
	$4x_1 + 6x_2 \leq (24, 42)$
	$x_1, x_2 \geq 0.$

and its dual fuzzy MC^2 problem:

Fuzzy-Min	$25\ u_1 + 24\ u_2$
Fuzzy-Min	$-35\ u_1 + 42\ u_2$
Subject to	$3\ u_1 + 4u_2 \geq (10, 15)$
	$2u_1 + 6u_2 \geq (12, 8)$
	$u_1, u_2 \geq 0.$

(i) Use the constructive approach to solve them.
(ii) Study the fuzzy weak duality between them.

8.2 (i) Use the parametric approach to solve the primal fuzzy MC^2 problem and the dual fuzzy MC^2 of problem 8.1.
(ii) What is the fuzzy weak duality membership function in this approach?

8.3 Consider

Fuzzy-Max	$3\ x_1 + 2\ x_2$
Fuzzy-Max	$6\ x_1 + 8\ x_2$
Subject to	$30\ x_1 + 20\ x_2 \leq (300, 250)$
	$5\ x_1 + 10\ x_2 \leq (110, 160)$
	$x_1, x_2 \geq 0;$

and

Fuzzy-Min	$300\ u_1 + 110\ u_2$
Fuzzy-Min	$250\ u_1 + 160\ u_2$
Subject to	$30\ u_1 + 5\ u_2 \geq (3, 6)$
	$20\ u_1 + 10\ u_2 \geq (2, 8)$

$$u_1, u_2 \geq 0.$$

(i) Use the constructive approach to solve these problems.
(ii) Use the parametric approach to solve these problems.
(iii) Study the fuzzy weak memberships for both constructive and parametric approaches. Compare the two results.

Part Four: Optimal Linear Designs

Chapter 9
OPTIMAL SYSTEM DESIGNS AND CONTINGENCY PLANS

In the field of system science, traditionally given a system, people try to find an optimal point (solution). This solution only can be used to handle a certain decision situation. In this chapter, we apply the MC^2 framework to design optimal systems for coping with various decision situations. The problem of optimizing a given system originates in "activity analysis" of linear production systems and in optimal engineering design problems. In the activity analysis, given a production system that transforms a set of resources, *inputs*, into a set of products or services, *outputs* (Starr 1966), one can use linear programming to identify the best group of activities as an optimal solution for production (Koopmans 1951 and Churchman 1968). The traditional optimal engineering design problems are usually formulated by mathematical programming which involves a single objective function and a single availability level of resources (Wilde 1978 and Papalambros and Wilde 1988). This approach also finds an optimal point for a given system. Neither the activity analysis nor the engineering design analysis considers two important aspects of formulating and solving optimal system design problems: (i) designing optimal systems and (ii) constructing related contingency plans for each optimal system to cope with various decision situations.

Realizing the importance of designing optimal systems, Zeleny (1975, 1990) used multiple criteria (MC) de novo programming to design an optimal system. His approach, however, does not explicitly address contingency plans for different decision situations. The extension of de novo programming will be discussed in Chapter 14. Y. R. Lee (1986) considered how a set of feasible constraint levels affects a linear system. In an issue of *Management Science*, Lee, Shi and Yu (1990) made the first attempt to propose a basic procedure to design the optimal systems and construct their corresponding contingency plans by using the MC^2-simplex method (see Section 4.4). Since then, a number of research papers have been published along this line.

In Section 9.1, with a set of all possible opportunities, we first use the MC^2 model (M4.1) to formulate the design problems. The significance and complexity of such design problems are discussed. Then we propose a general procedure to identify a *set* of *optimal systems* that contain certain selected opportunities and can optimize the given design problem under certain ranges of decision parameters, such as the contribution of objectives and resource availability levels. Section 9.2 describes a basic procedure of Lee, Shi and Yu (1990) that employs the MC^2-simplex method to identify the set of potential solutions as the *potential good systems* (PGS) for the design problem. However, the system design problem goes beyond finding the potential solutions. For each PGS, we build submodels of the design model to construct optimal contingency plans to overcome changes of the resource availability level parameter. Because the contingency plan for a PGS contains the selected opportunities and some slack resources with purchased external resources (if necessary) and converts infeasible solutions into feasible ones, this kind of contingency plans is called the *primal rigid contingency plans* (PRCP). Finally, using known techniques of decision making under uncertainty, we select the optimal linear system(s) from the set of PGSs and their contingency plans as the final decision. This section also studies the relationships between the contingency plans, shadow price, and resource price in the designing process.

We note that in the process of constructing contingency plans, there are three types of variables can be used: (i) opportunity variables **x**, (ii) slack variables **s**, and (iii) external variables **y**. Whenever a system (basis) is selected, it can contains some **x** and **s**; while its contingency plan may contain **x**, **s**, and **y**. The word "rigid" in PRCP means that there is no flexibility to use the unselected **s** in the designing process. To utilize the unselected (unused resource) **s**, Section 9.3 uses all possible slack variables to construct different contingency plans. To contrast with a PRCP, we call the contingency plan that contains not only the variables in a given PGS, but also other slack variables which are not selected in the PGS a *primal flexible contingency plan* (PFCP). Section 9.4 provides further remarks.

9.1 A General Procedure of Designing Optimal Systems

Given a planning horizon, let N = {1, . . ., n} be n opportunities under consideration. Then, an initial model of designing optimal linear systems can be formulated by applying (M4.1) (M9.1):

$$\begin{array}{ll} \textit{Max} & \boldsymbol{\lambda}^t \mathbf{C}\,\mathbf{x} \\ \textit{Subject to} & \mathbf{A}\,\mathbf{x} \le \mathbf{D}\boldsymbol{\gamma} \\ & \mathbf{x} \ge 0, \end{array}$$

where $\mathbf{C} \in \mathbf{R}^{q\times n}$, $\mathbf{A} \in \mathbf{R}^{m\times n}$, and $\mathbf{D} \in \mathbf{R}^{m\times p}$, are matrices of $q \times n$, mxn, and $m \times p$ dimensions, respectively; $\mathbf{x} \in \mathbf{R}^n$ are decision variables; $\lambda \in \mathbf{R}^q$, $\gamma \in \mathbf{R}^p$ and both (λ, γ) are assumed unknown.

In the above model q situational forces affect the unit contribution, while p situational forces affect resource availability. Both parameters (λ, γ) reflect changes in either internal or external environments of the systems. These environments could be market conditions, economic conditions, social conditions, political conditions, technological conditions, or business philosophy. λ is called the *contribution parameter* and γ the *resource parameter*. Observe that (λ, γ) may be treated as probability distributions over the situational forces. The interpretation should depend on the individual contexts of linear systems.

If parameters (λ, γ) are known ahead of decision time, we can select the best k opportunities from possible n opportunities as the optimal solution by using linear programming techniques (see e.g., Charnes and Cooper 1961, Dantzig 1963). However, if parameters (λ, γ) cannot be known ahead of decision time, the linear programming approach would not be effective, because there are infinitely many possible combinations of (λ, γ). When parameters (λ, γ) are not known before decision time, the change of γ may make the original choice infeasible, while that of λ can render the choice not optimal. Thus, contingency plans need to be constructed to overcome these difficult decision situations. A general procedure of designing the optimal systems is given as (Shi and Yu 1991, Shi *at el.* 1994, Shi and Yu 1995):

Algorithm 9.1

Step 1. Locate a number of good systems (subset or combination of opportunities) associated with feasible (λ, γ) as candidates for the optimal system;
Step 2. For each candidate, construct the corresponding optimal contingency plans that offer the optimal contribution under various changes of (λ, γ) situations;

Step 3. Use a well-known technique of decision making under uncertainty to select the optimal system(s) from the set of candidates and the corresponding contingency plans as the final decision.

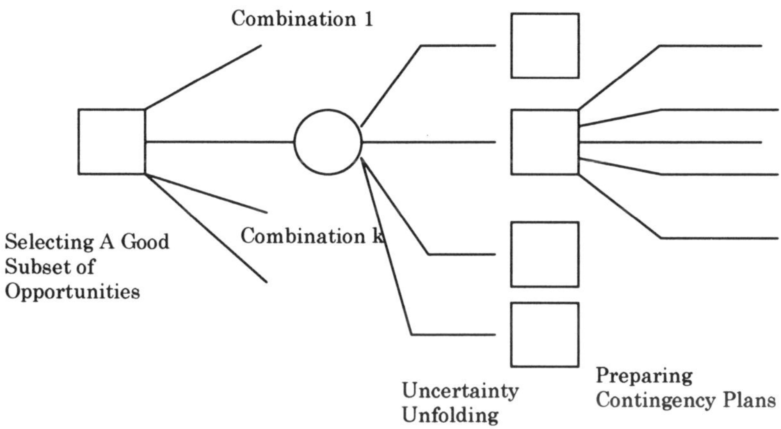

Figure 9.1 A General Procedure of Selecting Optimal Systems

Because of the complex nature of model (M9.1), there could be a number of methods to perform the general procedure. When n (the number of possible opportunities), q (number of criteria), and p (the number of resource availability levels) are large, massive computation may be involved in selecting candidates and constructing their contingency plans. Thus, a computer-aided support system is needed to design the optimal systems (see Chapter 12).

Model (M9.1) can abstractly be viewed as a "linear program" with parameters (λ, γ). It is well known that if there is an optimal solution, then there is a basic optimal solution that has m basic variables. This suggests the following heuristic assumptions:

Assumption 9. 1

(i) The number of selected opportunities, k, in a good system for model (M9.1) should not exceed the number of resources under consideration, m.

(ii) The selected k opportunities should be able to "optimize" Model (M9.1) under some possible ranges of (λ, γ).

With assumption 9.1, we can use the MC^2-simplex method to locate a set of potential solutions $\vartheta = \{J_1, J_2, \ldots, J_r\}$ as PGSs. Once a PGS J is determined, all other opportunities j $\notin J$ are rejected or not undertaken. The opportunity set of interest reduces to the opportunities in J from N. From statement 9.1, we see that given a PGS J with basic variables $\mathbf{x}(J)$, when $\gamma \notin \Gamma(J)$, J is not feasible; and when $\lambda \notin \Lambda(J)$, J is not optimal (recall Section 4.2). Thus, we need to prepare (construct) the corresponding contingency plans for each J to cope with the difficulty (i.e., to make the system feasible and optimal in some sense) by building related submodels of model (M9.l). This leads to research on exploring *various* methods to construct contingency plans for PGSs.

9.2 Primal Rigid Designs and Contingency Plans

9.2.1 Finding Primal Rigid Contingency Plans

Given basic variables $\{x_{j1}, \ldots, x_{jm}\}$ for (M1), we denote the index set of the basic variables by $J = \{j_1, \ldots, j_m\}$. Then , we can write $\mathbf{x}(J) = \{x_{j1}, \ldots, x_{jm}\}$. Note that $\mathbf{x}(J)$ may contain slack variables. Without confusion, J is called a *basis* for (M9.1). Let **I** be an $(m \times m)$-dimensional identity matrix and **0** be a $(q \times m)$-dimensional zero matrix. Given a basis J with its basic variables $\mathbf{x}(J)$, we define the *associated basis matrix* $\mathbf{B}_J$ as the submatrix of [**A**, **I**] with column index of J (i.e., column j of [**A**, **I**] is in $\mathbf{B}_J$ if and only if $j \in J$), and the *associated objective function coefficient* $\mathbf{C}_J$ as the submatrix of [**C**, **0**] with column index of J. We note that the notations used here are similar to those in Section 4.2. To avoid confusion with parameter sets of the future models, we use the subscript one for both primal parameter set $\Gamma_1(J)$ and dual parameter set $\Lambda_1(J)$ of model (M9.1) in Definition 4.4.

Given a PGS J for model (M9.1), by Definition 4.5, the basic solution $\mathbf{x}(J) = \mathbf{B}_J^{-1}\mathbf{D}\gamma$ is feasible if and only if $\gamma \in \Gamma_1(J)$. This means that $\mathbf{B}_J\,\mathbf{x}(J) = \mathbf{D}\gamma$. If some $\gamma \notin \Gamma_1(J)$, we need to purchase external resources $\mathbf{y}$ so as $\mathbf{B}_J\mathbf{x}(J) = \mathbf{D}\gamma + \mathbf{y}$ to make the designing process feasible (Lee, Shi and Yu 1990). This leads to build a submodel for constructing *primal rigid contingency plans* (PRCPs) for a given PGS J. A PRCP for a PGS contains only selected opportunities, slack resources and purchased external resources, if necessary. It converts infeasible solutions to feasible ones and optimizes the submodel related to the PGS under certain ranges of (λ, γ). Note that the PRCP has no *flexibility* to use other possible slack variables (resources) that are not in a PGS.

To construct the PRCPs for a given PGS J, we solve the following submodel (M9.2):

$$\begin{aligned} &\textit{Max} && \boldsymbol{\lambda}^t \mathbf{C}_J\, \mathbf{x}(J) - \boldsymbol{\alpha}^t\, \mathbf{y} \\ &\textit{Subject to} && \mathbf{B}_J\, \mathbf{x}(J) = \mathbf{D}\boldsymbol{\gamma} + \mathbf{y} \\ & && \mathbf{x}(J), \mathbf{y} \geq 0, \end{aligned}$$

where $\alpha = (\alpha_1, \ldots, \alpha_m) \in \mathbf{R}^m$ is a row vector representing the given unit price of purchasing external resources $\mathbf{y} = (y_1, \ldots, y_m) \in \mathbf{R}^m$ (a column vector), and (λ, γ) are presumed.

The tableau presentation of model (M9.2) is Table 9.1.

Table 9.1 The Initial Simplex Tableau of (M9.2)

$\mathbf{x}(J)$	$\mathbf{y}$	RHS	
$\mathbf{B}_J$	$-\mathbf{I}$	$\mathbf{D}\gamma$	(9.1)
$-\lambda^t\mathbf{C}_J$	α	$\mathbf{0}$	(9.2)

By using the MC^2-simplex method, we use a row linear transformation on Table 9.1:
(9.3) = $\mathbf{B}_J^{-1}$ (9.1) (i.e., premultiply (9.1) by $\mathbf{B}_J^{-1}$), and
(9.4) = $\lambda^t\mathbf{C}_J$ (9.3) + (9.2). Then, Equation (9.1)-(9.2) can be rewritten as

Table 9.2 The Simplex Tableau of (M9.2)

$\mathbf{x}(J)$	$\mathbf{y}$	RHS	
$\mathbf{I}$	$-\mathbf{B}_J^{-1}$	$\mathbf{B}_J^{-1}\mathbf{D}\gamma$	(9.3)
$\mathbf{0}$	$-\lambda^t\mathbf{C}_J\mathbf{B}_J^{-1} + \lambda\, \mathbf{e_q}$ α	$\lambda^t\mathbf{C}_J\mathbf{B}_J^{-1}\mathbf{D}\gamma$	(9.4)

where $\mathbf{e_q} = (1, 1, \ldots, 1) \in \mathbf{R}^q$ is a column vector.

Definition 9.1 In identifying PRCP for PGS J, we call

(i) $\Gamma_2(J) = \{\gamma > 0 \mid \mathbf{B}_J^{-1}\,\mathbf{D}\gamma \geq 0\}$ the primal parameter set of model (M9.2); and

(ii) $\Lambda_2(J, \alpha) = \{\lambda > 0 \mid -\lambda^t(\mathbf{C}_J\mathbf{B}_J^{-1} - \mathbf{e}_q\alpha) \geq 0\}$ the dual parameter set of model (M9.2).

Definition 9.2 Given a PGS J, let $\{K_1(J), \ldots, K_h(J)\}$ be the set of potential solutions obtained from (M8.2) by using the MC²-simplex method. When $(\gamma, \lambda) \in \Gamma_2(K_i(J)) \mathrm{x} \Lambda_2(K_i(J))$, i = 1, . . .,h, we call $\{K_1(J), \ldots, K_h(J)\}$ the *primal rigid contingency plans* (PRCPs) for J.

Remark 9.1 From the above discussion, we see that the linear system design problem in this chapter is an advanced study of MC² linear programming. If finding a PGS is the "first order" problem on MC² linear programming (i.e., solving (M9.1) or (M4.1)), then identifying its contingency plans (i.e., solving (M9.2)) is the "second order" problem.

However, using model (M9.2) may not be effective to construct all PRCPs for each PGS of ϑ since there may be r related models which need to be built to complete the task! In order to effectively facilitate the computation of identifying all PRCPs for each PGS, we develop the following *primal augmented model* (M9.3):

$$\begin{aligned} &\mathit{Max} && \boldsymbol{\lambda}'\mathbf{C}\,\mathbf{x} - \boldsymbol{\alpha}'\,\mathbf{y} \\ &\mathit{Subject\ to} && \mathbf{A}\,\mathbf{x} \leq \mathbf{D}\gamma + \mathbf{y} \\ & && \mathbf{x}, \mathbf{y} \geq 0. \end{aligned}$$

The initial tableau of model (M9.3) is Table 9.3 and the simplex tableau is Table 9.4.

Table 9.3 The Initial Simplex Tableau of (M9.3)

$\mathbf{x}(J)$	$\mathbf{x}(J')$	$\mathbf{y}$	RHS	
$\mathbf{B}_J$	$\mathbf{R}$	$-\mathbf{I}$	$\mathbf{D}\gamma$	(9.5)
$-\lambda^t\mathbf{C}_J$	$-\lambda^t\mathbf{C}_{J'}$	α	$\mathbf{0}$	(9.6)

Table 9.4 The Simplex Tableau of (M9.3)

$\mathbf{x}(J)$	$\mathbf{x}(J')$	$\mathbf{y}$	RHS	
$\mathbf{I}$	$\mathbf{B}_J^{-1}\mathbf{R}$	$-\mathbf{B}_J^{-1}$	$\mathbf{B}_J^{-1}\mathbf{D}\gamma$	(9.7)
$\mathbf{0}$	$\lambda^t\mathbf{C}_J\mathbf{B}_J^{-1}\mathbf{R} - \lambda^t\mathbf{C}_{J'}$	$-\lambda^t\mathbf{C}_J\mathbf{B}_J^{-1} + \lambda^t\mathbf{e_q}\alpha$	$\lambda^t\mathbf{C}_J\mathbf{B}_J^{-1}\mathbf{D}\gamma$	(9.8)

In Tables 9.3-9.4, $\mathbf{x}(J')$ are the nonbasic variables, $\mathbf{R}$ is the nonbasis matrix with no column index of $[\mathbf{A}, \mathbf{I}]$ in J, and $-\lambda^t\mathbf{C}_{J'}$ is the subvector of $[-\lambda^t\mathbf{C}, \mathbf{0}]$ with no column index in J.

Definition 9.3 Given a PGS J, we call

(i) $\Gamma_3(J) = \{\gamma > 0 \mid \mathbf{B}_J^{-1}\mathbf{D}\gamma \geq 0\}$ the primal parameter set of model (M9.3); and

(ii) $\Lambda_3(J, \alpha) = \{\lambda > 0 \mid \lambda^t [(\mathbf{C}_J\mathbf{B}_J^{-1}\mathbf{R} - \mathbf{C}_{J'}), -(\mathbf{C}_J\mathbf{B}_J^{-1} - \mathbf{e_q}\alpha)] \geq 0\}$ the dual parameter set of model (M9.3).

Remark 9.2 We observe that both (M9.1) and (M9.2) are submodels of model (M9.3) because model (M9.1) can be obtained by dropping columns of $\mathbf{y}$ from model (M9.3) and model (M9.2) can be obtained by dropping columns $\mathbf{x}(J')$ associated with opportunities that are not selected in PGS J from N (i.e., N\\J).

To be precise, let us look at the relationship between model (9.2) and model (9.3). For a given basis K, without confusion, we decompose its basic variables into $(\mathbf{x}(K), \mathbf{y}(K))$ with $\mathbf{x}(K)$ being selected variables x_j and $\mathbf{y}(K)$ being those y_j. Then, we have:

Theorem 9.1 If K is a basis for model (M9.2), then K is also a basis for model (M9.3). Conversely, if K is a basis for model (M9.3) and contains none of $\mathbf{x}(J')$ (the nonbasic variables with respect to a basis J) as basic variables, then K is also a basis for model (M9.2). Furthermore, the basic feasible solution for a given γ, $(\mathbf{x}(K, \gamma), \mathbf{y}(K, \gamma))$, is identical for the both (M9.2) and (M9.3) in either case.

Corollary 9.1

(i) $\Gamma_2(K(J)) = \Gamma_3(K(J))$; and

(ii) $\Lambda_2(K(J), \alpha) \supseteq \Lambda_3(K(J), \alpha)$, $\alpha \in \mathbf{R}^m$.

Because (λ, γ) are unknown at the design time, we have a decision problem under uncertainty with a finite number of choices; that is, PGSs. The problem of selecting the final optimal systems involves the assessment of the likelihood of (λ, γ) to occur at the various points of its range. With proper assumptions, the uncertainty may be represented by random variables with some known probability distributions. A number of known criteria, such as "maximizing expected payoff", "minimizing the variance of the payoff", "maximin payoff", and "maximizing the probability of achieving a targeted payoff" (Keeney and Raiffa 1976, Ziemba and Vickson 1975, Yu 1985 and Shi 1991), can be used to identify the nondominated solutions and/or construct the utility functions as to find the final optimal system.

To illustrate the determination of the final optimal system, given a PGS, we assume that a setup cost is needed to build the facility for producing the selected opportunities. Since the setup cost for a PGS can be broadly defined as the fixed cost associated with patents, technologies, production, marketing, etc., this cost could include initial costs for purchasing the production facility, equipment, and assembly lines. Suppose that we use maximizing expected payoff as the criterion for designing the final optimal system in a given planning horizon. Let **EV**(J) be the expected payoff for a given PGS J. Let the set of all contingency plans for J be $\{K_1, K_2, \ldots, K_h\}$, then

$$\mathbf{EV}(J) = \sum_{i=1}^{h} \int_{S_i} \mathbf{V}(K_i(J))\, dF(\gamma, \lambda),$$

where $\mathbf{V}(K_i \mid J)$ is the payoff (a function of (λ, γ)) when using the contingency plan K_i, I = 1, . . . , h; $F(\lambda, \gamma)$ is the joint cumulative probability distribution function which is assumed to be known by some estimation method; and $S_i = \{(\lambda, \gamma) > 0 \mid \gamma \in \Gamma_2(K(J)_i),\ \lambda \in \Lambda_2(K(J)_i)\}$ is the optimal situation set of (λ, γ) for K_i.

Definition 9.4 Let $f(J)$ be the initial setup cost for using J and δ be the target rate of return of the investment over the given planning horizon. Then, the opportunity cost related to the setup cost may be estimated as $\delta f(J)$ and the net expected payoff for using J is given by:

$$\mathbf{NEV}(J) = \sum_{i=1}^{h} \int_{S_i} \mathbf{V}(K_i(J))\, dF(\gamma, \lambda) - \delta f(J).$$

Based on the above, all PGSs and their PRCPs can be systematically and effectively located by using the MC^2-simplex method. This is captured as the following basic algorithm (Lee, Shi and Yu 1990):

Algorithm 9.2

Step 1. Use the MC^2-simplex method to find a set of PGSs of model (M8.1) as candidates for the optimal system.
Step 2. Apply Theorem 9.1 and Corollary 9.1 to solve model (M9.3) and identify the PRCPs for each PGS of step 1.
Step 3. Use a well-known technique of decision making under uncertainty to select the "best" system from the set of PGSs and their PRCPs as the final decision.

The following example is used to illustrate this basic procedure for designing the optimal system:

Example 9.1 Consider a design problem with two criteria and two constraint levels for five possible opportunities:

$$Max \;\; (\lambda_1 \;\; \lambda_2) \begin{pmatrix} 3 & 2 & 1 & 1 & 0 \\ 0 & 1 & 2 & 3 & 3 \end{pmatrix} \begin{pmatrix} x_1 \\ x_2 \\ x_3 \\ x_4 \\ x_5 \end{pmatrix}$$

$$Subject\ to \begin{pmatrix} 1 & 0 & 2 & 1 & 0 \\ 0 & 1 & 1 & 2 & 2 \\ 1 & 1 & 0 & 0 & 1 \end{pmatrix} \begin{pmatrix} x_1 \\ x_2 \\ x_3 \\ x_4 \\ x_5 \end{pmatrix} \leq \begin{pmatrix} -10 & 20 \\ 40 & 20 \\ 30 & 40 \end{pmatrix} \begin{pmatrix} \gamma_1 \\ \gamma_2 \end{pmatrix}$$

$$x_j \geq 0, \quad j=1, 2, 3, 4, 5.$$

Let s_i be slack variables for constraints 1, 2, 3 respectively. Then Algorithm 9.2 is conducted as follows:

Step 1. Using the computer software of the MC^2-simplex method (Section 4.4), we find the set of all PGSs, denoted by $\vartheta = \{J_1, J_2, J_3, J_4, J_5\}$ as shown in Table 9.5. Here, J_1 has (x_1, x_2, s_3) as the basic variables. If J_1 is chosen, then the opportunities {1, 2} will be used while s_3 is the slack resource. Assuming $\gamma_1 + \gamma_2 = 1$ and $\lambda_1 + \lambda_2 = 1$, we see that J_1 is optimal whenever $0 < \gamma_1 \leq 2/3$ and $1/5 \leq \lambda_1 \leq 1$. However, if $2/3 < \gamma_1 \leq 1$, J_1 is becomes infeasible and if $0 < \lambda_1 < 1/5$, J_1 is not optimal. Thus, we need to construct the corresponding contingency plans for PGS J_1 (see Figure 9.2). The meanings of J_2, J_3, J_4, and J_5 can be similarly explained.

Table 9.5 Potential Good Systems

PGS	Bases	$\Gamma_1(J_i)$	$\Lambda_1(J_i)$
J_1	(x_1, x_2, s_3)	$0 < \gamma_1 \leq 2/3$	$1/5 \leq \lambda_1 \leq 1$
J_2	(x_1, x_2, x_4)	$0 < \gamma_1 \leq 2/3$	$1/4 \leq \lambda_1 \leq 1$
J_3	(x_1, x_2, x_5)	$0 < \gamma_1 \leq 2/3$	$1/5 \leq \lambda_1 \leq 1/4$
J_4	(x_1, x_5, s_3)	$0 < \gamma_1 \leq 2/3$	$1/11 \leq \lambda_1 \leq 1/5$
J_5	(x_3, x_5, s_3)	$0 < \gamma_1 \leq 2/3$	$0 < \lambda_1 \leq 1/11$

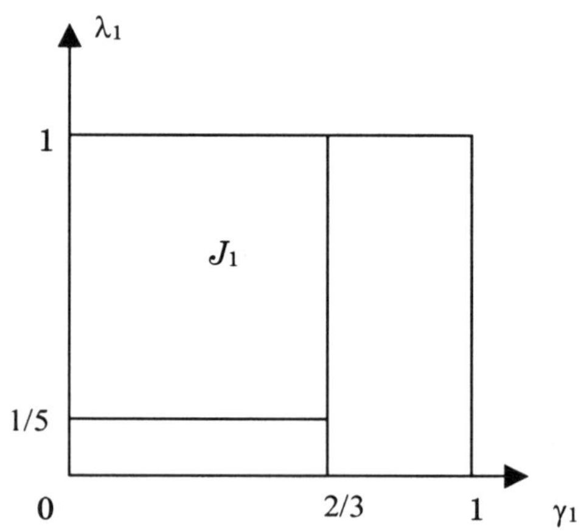

Figure 9.2 Not All Contingency Plans for J_1

Step 2. In order to effectively construct PRCPs for J_i, i = 1, 2, 3, 4, 5, we solve the corresponding augmented problem:

$$Max \quad (\lambda_1 \quad \lambda_2) \begin{pmatrix} 3 & 2 & 1 & 1 & 0 \\ 0 & 1 & 2 & 3 & 3 \end{pmatrix} \begin{pmatrix} x_1 \\ x_2 \\ x_3 \\ x_4 \\ x_5 \end{pmatrix} - (\alpha_1 \quad \alpha_2 \quad \alpha_3) \begin{pmatrix} y_1 \\ y_2 \\ y_3 \end{pmatrix}$$

$$Subject\ to \begin{pmatrix} 1 & 0 & 2 & 1 & 0 \\ 0 & 1 & 1 & 2 & 2 \\ 1 & 1 & 0 & 0 & 1 \end{pmatrix} \begin{pmatrix} x_1 \\ x_2 \\ x_3 \\ x_4 \\ x_5 \end{pmatrix} \le \begin{pmatrix} -10 & 20 \\ 40 & 20 \\ 30 & 40 \end{pmatrix} \begin{pmatrix} \gamma_1 \\ \gamma_2 \end{pmatrix} + \begin{pmatrix} 1 & 0 & 0 \\ 0 & 1 & 0 \\ 0 & 0 & 1 \end{pmatrix} \begin{pmatrix} y_1 \\ y_2 \\ y_3 \end{pmatrix}$$

$$x_j \ge 0, \quad j = 1, 2, 3, 4, 5; \text{ and } y_i \ge 0, \quad i = 1, 2, 3.$$

According to Theorem 2.1, we first obtain the set of all primal bases, denoted by $\upsilon = \{K_1, \ldots, K_{44}\}$ (Table 9.6). Then, for each J_i, i = 1, 2, 3, 4, 5, we identify the set of all potential bases (PRCPs) for Model (M9.2) by examining those K of υ such that $\mathbf{x}(J) \supseteq \mathbf{x}(K)$. Since the parameter set $\Lambda_2(K(J), \alpha)$ is a function of α, for illustration, let $\alpha = (2, 2, 2)$ in the process. Table 9.7 shows the set of all PRCPs for J_1, where basis $K_{40} = J_1$. We see that in terms of the ranges of (γ_1, λ_1), PGS J_1 has four alternative contingency plan sets: $\{K_{40}, K_5, K_{22}\}$, $\{K_7, K_{22}\}$, $\{K_5, K_7, K_{22}\}$, $\{K_{40}, K_7, K_{22}\}$. Similarly, we can find the PRCPs for J_2, J_3, J_4, and J_5 (see Shi and Yu 1995).

Step 3. Suppose we use the "maximizing expected payoff" as the criterion to selecting the final optimal system from ϑ and the PRCPs. For simplicity, let the initial setup cost for using J, $f(J) = 0$. We assume that γ_1 is independent on λ_1, and (γ_1, λ_1) have the bi-uniform probability distribution $F(\gamma_1, \lambda_1)$.

Table 9.6 Primal Potential Bases for Augmented Model

Bases	$\mathbf{x}(K_i, \gamma)$	Bases	$\mathbf{x}(K_i, \gamma)$
K_1	(x_1, s_2, s_3)	K_{23}	(x_2, s_1, s_3)
K_2	(x_1, x_3, y_1)	K_{24}	(x_3, s_2, s_3)
K_3	(x_1, x_4, s_3)	K_{25}	(x_3, y_1, s_3)
K_4	(x_1, x_4, y_1)	K_{26}	(x_3, x_5, y_2)
K_5	(x_1, x_2, y_1)	K_{27}	(x_3, x_4, s_3)
K_6	(x_1, x_2, y_2)	K_{28}	(x_4, x_5, s_3)
K_7	(x_1, x_2, y_3)	K_{29}	(y_1, s_2, s_3)
K_8	(x_1, x_2, s_1)	K_{30}	(x_4, y_1, s_3)
K_9	(x_1, x_4, s_2)	K_{31}	(x_4, x_5, y_2)
K_{10}	(x_1, x_2, x_3)	K_{32}	(x_4, y_2, s_3)
K_{11}	(x_2, x_3, y_1)	K_{33}	(x_4, s_1, s_3)
K_{12}	(x_2, x_3, y_2)	K_{34}	(x_4, s_2, s_3)
K_{13}	(x_2, x_3, s_3)	K_{35}	(x_5, s_1, s_3)
K_{14}	(x_1, x_5, y_2)	K_{36}	(x_5, y_2, s_1)
K_{15}	(x_1, x_5, y_1)	K_{37}	(x_5, y_1, y_2)
K_{16}	(x_1, y_1, s_2)	K_{38}	(x_5, y_1, s_3)
K_{17}	(x_2, y_2, s_1)	K_{39}	(x_2, x_4, y_1)
K_{18}	(x_2, x_5, y_1)	$K_{40} = J_1$	(x_1, x_2, s_3)
K_{19}	(x_2, x_4, y_2)	$K_{41} = J_2$	(x_1, x_2, x_4)
K_{20}	(x_2, x_4, s_3)	$K_{42} = J_3$	(x_1, x_2, x_5)
K_{21}	(x_2, y_1, s_2)	$K_{43} = J_4$	(x_1, x_5, s_3)
K_{22}	(x_2, y_1, y_3)	$K_{44} = J_5$	(x_3, x_5, s_3)

Then, the expected payoff $\mathbf{EV}(J_1)$ with respect to the contingency plan set $\{K_{40}, K_5, K_{22}\}$ (Figure 8.3) can be expressed as

$$\mathbf{EV}(J_1) = \int_{S_1} \mathbf{V}(K_{40}(J_1))\,dF(\gamma_1,\lambda_1) + \int_{S_2} \mathbf{V}(K_5(J_1))\,dF(\gamma_1,\lambda_1)$$
$$\int_{S_3} \mathbf{V}(K_{22}(J_1))\,dF(\gamma_1,\lambda_1),$$

where $\mathbf{V}(K_{40}(J_1))$, $\mathbf{V}(K_5(J_1))$, and $\mathbf{V}(K_{22}(J_1))$; and the (γ_1, λ_1) ranges $S_j = \Gamma_4(K_j(J_1)) \mathrm{x} \Lambda_4(K_j(J_1))$, j = 1, 2, 3, are shown in Table 9.3.

Table 9.7 Primal Rigid Contingency Plans for J_1 with $\alpha = (2, 2, 2)$

Bases	$\Gamma_2(K_j(J_1))$	$\Lambda_2(K_j(J_1))$	$\mathrm{V}(K_j(J_1))$
$K_{40}(J_1)$	$0 < \gamma_1 \leq 2/3$	$0 < \lambda_1 \leq 2/3$	$(\lambda_1 \quad \lambda_2)\begin{pmatrix} 50 & 100 \\ 40 & 20 \end{pmatrix}\begin{pmatrix} \gamma_1 \\ \gamma_2 \end{pmatrix}$
$K_5(J_1)$	$0 < \gamma_1 \leq 2/3$	$2/3 \leq \lambda_1 \leq 1$	$(\lambda_1 \quad \lambda_2)\begin{pmatrix} 50 & 100 \\ 40 & 20 \end{pmatrix}\begin{pmatrix} \gamma_1 \\ \gamma_2 \end{pmatrix}$
$K_7(J_1)$	$0 < \gamma_1 \leq 2/3$	$0 < \lambda_1 \leq 1$	$(\lambda_1 \quad \lambda_2)\begin{pmatrix} 50 & 100 \\ 40 & 20 \end{pmatrix}\begin{pmatrix} \gamma_1 \\ \gamma_2 \end{pmatrix}$
$K_{22}(J_1)$	$2/3 < \gamma_1 \leq 1$	$0 < \lambda_1 \leq 1$	$(\lambda_1 \quad \lambda_2)\begin{pmatrix} 50 & 120 \\ 0 & 100 \end{pmatrix}\begin{pmatrix} \gamma_1 \\ \gamma_2 \end{pmatrix}$

The computing result is $\mathbf{EV}(J_1) = 48.45$. If we use other contingency plan sets for J_1, the result is the same. Similarly, we can get $\mathbf{EV}(J_2) = 72.21$, $\mathbf{EV}(J_3) = 49.83$, $\mathbf{EV}(J_4) = 32.88$, and $\mathbf{EV}(J_5) = 21.66$. Thus, in terms of the "maximizing expected payoff," J_2 is the "best" (optimal) system we should choose. Note that no matter what decision situation will change in the future, we always have an optimal contingency to deal with. That is, for every (γ_1, λ_1), there is a PRCP of J_2 to optimally undertake the opportunities {1, 2, 4}.

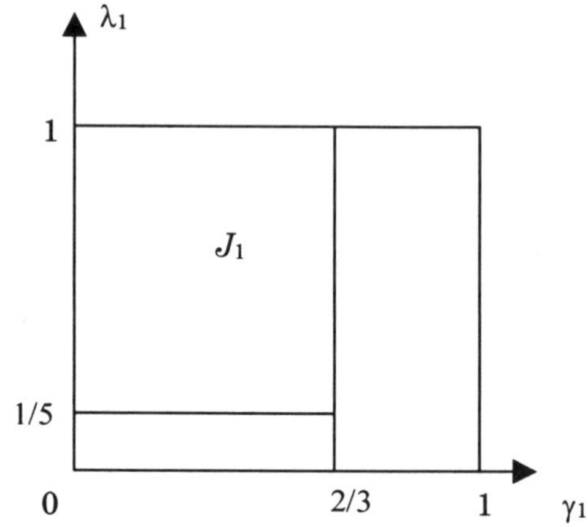

Figure 9.3 Primal Rigid Contingency Plan for J_1

9.2.2 Analysis of External Resource Price

An interesting economic problem in constructing contingency plans for the optimal systems is the analysis of external resource price of model (M9.2) (Shi and Yu 1991). Recall that given a basis K for the contingency model (M9.2) corresponding to PGS J, the dual parameter set $\Lambda_2(K(J), \alpha)$ is a function of α, which is the price of $\mathbf{y}$. Suppose that $\mathbf{y}(K, \alpha) > \mathbf{0}$ and that α is very large. Then the decision makers may not purchase $\mathbf{y}$ for preparing contingency plans and K may not be a contingency plan for J. Thus, the change of α value can affect the determination of contingency plans.

Let the primal and dual parameter sets of PGS J for (M9.2) be $\Gamma_2(J)$ and $\Lambda_2(J, \alpha)$, respectively. Then we have the following results (see Shi and Yu 1991):

Theorem 9.2 Given a PGS for model (M9.1), the comparison of model (M9.1) to model (M9.2) indicates that

(i) $\Gamma_1(J) = \Gamma_2(J)$; and

(ii) if $(\alpha')^t \geq \lambda^t \mathbf{C}_J \mathbf{B}_J^{-1} \geq \mathbf{0}$ for all $\lambda \in \Lambda_1(J)$, $\alpha' \in \mathbf{R}^m$, then $\Lambda_2(J, \alpha) \supseteq \Lambda_1(J)$.

The condition of $(\alpha')^t \geq \lambda^t \mathbf{C}_J \mathbf{B}_J^{-1} \geq \mathbf{0}$ has an important economic meaning here. By using the terminology of linear programming Dantzig (1963), we can view $\lambda^t \mathbf{C}_J \mathbf{B}_J^{-1} \geq \mathbf{0}$ as the *shadow prices* of PGS J for model (M9.1). Then, Theorem 9.2 says that if the price of external resource is larger than or equal to the shadow price of PGS J, the dual parameter set of J for model (M9.2) contains the dual parameter set of J for model (M9.1). Because there exist infinitely many values of α satisfying $\alpha' \geq \lambda^t \mathbf{C}_J \mathbf{B}_J^{-1} \geq \mathbf{0}$, the theorem also intuitively indicates that if α is large enough, there is no incentive to buy any external resource $\mathbf{y}$ and/or change the basis for new optimality.

To discuss more details, let Γ be the γ parameter space and Λ be the λ parameter space. Then, the Cartesian space of (γ, λ) is denoted by $\Gamma \times \Lambda$.

Definition 9.5 Given a PGS J of model (M9.1),

(i) the space Γ is said to be covered by J if $\Gamma_1(J) \supseteq \Gamma$;

(ii) the space Λ is said to be covered by J if $\Lambda_1(J) \supseteq \Lambda$; and

(iii) the space $\Gamma \times \Lambda$is said to be covered by J if $\Gamma_1(J) \times \Lambda_1(J)$ $\supseteq \Gamma \times \Lambda$.

From Section 9.2.1, we can know that if Γ is covered by PGS J, then J is the contingency plan of (M9.2) for PGS J because after deleting all $j \notin J$ from N, J covers both Λ and $\Gamma \times \Lambda$with respect to (M9.2) (see Shi 1991). Here we are interested in the case when Γ is *not* covered by the contingency plan J selected by model (M9.2) for PGS J. What levels of α value can make the contingency plan J cover Γ with respect to model (M9.2) so that the purchase of the external resource $\mathbf{y}$ is needed only in preparing some PRCPs to cover $(\Gamma \backslash \Gamma_1(J)) \times \Lambda$?

Theorem 9.3 Given a PGS J, suppose that its contingency plan J selected by model (M9.2) covers Λ, but not Γ. Then there exists a $\alpha^* \in \mathbf{R}^m$ such that whenever $\alpha \geq \alpha^*$, Γ is covered by the J with respect to (M9.3). Here, no purchase of $\mathbf{y}$ is needed in constructing the contingency plan J for PGS J.

Remark 9.3 For a given PGS J, let $A_1 = \{\alpha \geq 0 \mid \mathbf{e_q}\alpha^t \geq \mathbf{C}_J\mathbf{B}_J^{-1}, \alpha \in \mathbf{R}^m\}$. Then, the "greatest lower bound" of $A_1$is $\alpha^* = (\alpha_1^*, \ldots, \alpha_m^*)^t$, where $\alpha_i^* = \max\{(\mathbf{C}_J\mathbf{B}_J^{-1})_{ki} \mid 1 \leq k \leq q\}$, $i = 1, \ldots, m$. To verify theorem 9.3, we can choose $\alpha^* = \alpha$.

Let A be the space of parameter α and $A_2 = \{\alpha \geq 0 \mid \alpha^t \geq \lambda^t\mathbf{C}_J\mathbf{B}_J^{-1}, \alpha \in \mathbf{R}^m, \lambda \in \Lambda_1(J)\}$. We define $A_2' = \{\alpha \geq 0 \mid \alpha^t < \lambda^t\mathbf{C}_J\mathbf{B}_J^{-1}, \alpha \in \mathbf{R}^m, \lambda \in \Lambda_1(J)\}$ as the complement set of A_2. Based on Theorem 9.3, we have:

Theorem 9.4 Given a PGS J, the possible cases to purchase the external resource y for preparing contingency plans selected by model (M9.2) are:

(i) if $\alpha \in A_2 \backslash A_1$ and Γ is covered by J, then $\mathbf{y}$ is needed to cover $\Gamma \times (\Lambda \backslash \Lambda_2(J, \alpha))$.

(ii) if $\alpha \in A_1$ and Γ is *not* covered by J, then $\mathbf{y}$ is needed to cover $(\Gamma \backslash \Gamma_2(J)) \times \Lambda$.

(iii) if $\alpha \in A_2 \backslash A_1$and Γ is *not* covered by J, then $\mathbf{y}$ is needed to cover $(\Gamma \times \Lambda) \backslash (\Gamma_2(J) \times \Lambda_2(J, \alpha))$.

(iv) if $\alpha \in A_2'$, then J is not a contingency plan and $\mathbf{y}$ is needed to cover $\Gamma \times \Lambda$.

Example 9.2

We use Example 9.1 to illustrate Theorem 9.3 and Theorem 9.4 (ii) as follows.

Consider the contingency plan J_4 of model (M9.2) which does not cover Γ for PGS J_4. From the simplex tableau of basis J_4 (i.e., Table 9.8), we see that

$$\mathbf{C}_{J_4}\mathbf{B}_{J_4}^{-1} = \begin{pmatrix} 3 & 0 & 0 \\ 0 & 1.5 & 0 \end{pmatrix}.$$

Table 9.8 Simplex Tableau of J_4

x_1	x_2	x_3	x_4	x_5	s_1	s_2	s_3	RHS	
1	0	2	1	0	1	0	0	-10	20
0	0.5	-2.5	-2	0	-1	-0.5	1	20	10
0	0.5	0.5	1	1	0	0.5	0	20	10
0	-2	5	2	0	3	0	0	-30	60
0	0.5	-0.5	0	0	0	1.5	0	60	30

Then, $\alpha^* = (3, 1.5, 0)^t \in A_1$. Let $\alpha^* = \alpha$, we have

$$\mathbf{e}_q\,(\boldsymbol{\alpha}^*)^t - \mathbf{C}_{J_4}\mathbf{B}_{J_4}{}^{-1} = \begin{pmatrix} 0 & 1.5 & 0 \\ 3 & 0 & 0 \end{pmatrix}.$$

Thus, $\Lambda_2(J_4, \alpha^*) = \{\lambda_1 > 0 \mid 0 < \lambda_1 \le 1,\ \lambda_1 + \lambda_2 = 1\} \supseteq \Lambda$. This means that Λ is covered by the contingency plan J_4 selected by model (M9.2) if we choose $\alpha^* = \alpha$. In this case, no purchase of **y** is needed to cover $\Gamma_2(J_4) \times \Lambda$. This illustrates Theorem 9.3. However, we still need to purchase **y** to prepare other PRCPs from model (M9.2) to cover $(\Gamma \backslash \Gamma_2(J_4)) \times \Lambda$. This illustrates Theorem 9.4 (ii). Other cases in Theorem 9.4 can be similarly demonstrated.

9.3 Primal Flexible Designs and Contingency Plans

As described in Sections 9.2, when we construct a contingency plan for PGS, three types of variables can be possibly selected. They are opportunity variables **x**, slack variables **s**, and external variables **y**. In selecting a basic basis for PGS, the ways of utilizing slack variables (unused resources) lead to different approaches to constructing different contingency plans. To contrast with a PRCP, we call the contingency plan that contains not only the variables in a given PGS, but also other slack variables which are not selected in the PGS a *primal flexible contingency plan* (PFCP) (Shi and Yu 1992). The PFCP provides flexibility for the decision makers to use all possible slack resources in designing the optimal system.

To construct the PFCPs for a given PGS J, let the corresponding non-basic variables of J be $\mathbf{x}(J')$. We decompose $\mathbf{x}(J')$ into $\mathbf{x}^1(J')$ and $\mathbf{x}^2(J')$, where $\mathbf{x}^1(J')$ is the set of non-basic variables consisting of those slack variables that are not selected in J, and $\mathbf{x}^2(J')$ is the set of non-basic variables consisting of those opportunity variables of N that are not selected in J. Then, in addition to considering $\mathbf{x}(J)$ in the designing process, we also use the slack resource $\mathbf{x}^1(J')$. The design constraints become $\mathbf{B}_J\,\mathbf{x}(J) + \mathbf{R}^1\mathbf{x}^1(J') = \mathbf{D}\gamma + \mathbf{y}$, where $\mathbf{R}^1$ is the submatrix of $[\mathbf{A}, \mathbf{I}]$ associated with $\mathbf{x}^1(J')$. We can construct the PFCPs for the J by solving the following submodel (M9.4):

$$
\begin{aligned}
&\textit{Max} && \lambda' \mathbf{C}_J\, \mathbf{x}(J) - \alpha'\, \mathbf{y} \\
&\textit{Subject to} && \mathbf{B}_J\, \mathbf{x}(J) + \mathbf{R}^1\, \mathbf{x}^1(J') = \mathbf{D}\gamma + \mathbf{y} \\
& && \mathbf{x}(J), \mathbf{x}^1(J'), \mathbf{y} \geq 0.
\end{aligned}
$$

The simplex tableau analysis is shown as Tables 9.9 and 9.10.

Table 9.9 The Initial Simplex Tableau of (M9.4)

$\mathbf{x}(J)$	$\mathbf{x}^1(J')$	$\mathbf{y}$	RHS	
$\mathbf{B}_J$	$\mathbf{R}^1$	$-\mathbf{I}$	$\mathbf{D}\gamma$	(9.9)
$-\lambda^t\mathbf{C}_J$	$\mathbf{0}$	α	$\mathbf{0}$	(9.10)

Table 9.10 The Simplex Tableau of (M9.4)

$\mathbf{x}(J)$	$\mathbf{x}^1(J')$	$\mathbf{y}$	RHS	
$\mathbf{I}$	$\mathbf{B}_J^{-1}\,\mathbf{R}^1$	$-\mathbf{B}_J^{-1}$	$\mathbf{B}_J^{-1}\,\mathbf{D}\gamma$	(9.11)
$\mathbf{0}$	$\lambda^t\mathbf{C}_J\mathbf{B}_J^{-1}\mathbf{R}^1$	$-\lambda^t\mathbf{C}_J\mathbf{B}_J^{-1} + \lambda\,\mathbf{e_q}\,\alpha$	$\lambda^t\mathbf{C}_J\mathbf{B}_J^{-1}\,\mathbf{D}\gamma$	(9.12)

Definition 9.6 Given a PGS J, we call

(i) $\Gamma_4(J) = \{\gamma > 0 \mid \mathbf{B}_J^{-1}\mathbf{D}\gamma \geq 0\}$ the primal parameter set of model (M9.4); and

(ii) $\Lambda_4(J, \alpha) = \{\lambda > 0 \mid \lambda^t[(\mathbf{C}_J\mathbf{B}_J^{-1}\mathbf{R}^1), -(\mathbf{C}_J\mathbf{B}_J^{-1} - \mathbf{e}_q\alpha^t)] \geq 0\}$ the dual parameter set of model (M9.4).

Remark 9.4 Model (M9.4) is a submodel of model (M9.3) since model (M9.4) can be obtained by dropping columns $\mathbf{x}^2(J')$, the non-basic variables.

For a given basis K, we decompose its basic variables into $(\mathbf{x}(K), \mathbf{x}^1(K(J')), \mathbf{y}(K))$ with $\mathbf{x}(K)$ being selected variables x_j, $\mathbf{x}^1(K(J'))$ being those s_j, and $\mathbf{y}(K)$ being those y_j. Then, we have:

Theorem 9.5 If K is a basis for model (M9.4), then K is also a basis for model (M9.3). Conversely, if K is a basis for model (M9.3) and contains none of $\mathbf{x}^2(J')$ as basic variables, then K is also a basis for model (M9.4). Furthermore, the basic feasible solution for a given γ, $(\mathbf{x}(K, \gamma), \mathbf{x}^1(K(J', \gamma)), \mathbf{y}(K, \gamma))$, is identical for the both (M9.4) and (M9.3) in either case.

Corollary 9.2

(i) $\Gamma_4(K(J)) = \Gamma_3(K(J))$; and

(ii) $\Lambda_4(K(J), \alpha) \supseteq \Lambda_3(K(J), \alpha), \alpha \in \mathbf{R}^m$.

We recall that from Remark 9.2, the relationship between model (M9.1) and model (M9.3) can be stated as:

Theorem 9.6

(i) $\Gamma_1(K(J)) = \Gamma_3(K(J))$; and

(ii) $\Lambda_1(K(J) \supseteq \Lambda_3(K(J), \alpha), \alpha \in \mathbf{R}^m$.

In Algorithm 9.2, to avoid the confusion, we only use model (M9.3) to identify PRCPs for a PGS. To effectively utilize model (M9.3) in the design process, we can first find all primal potential solutions of (M9.3), then use them to identify all PGSs and their PFCPs. This idea is employed to construct the following new algorithm:

Algorithm 9.3

Step 1. Find all primal potential solutions for model (M9.3). Denote the set of such solutions by $\upsilon = \{K_1, \ldots, K_v\}$. Each K is generically defined as a basis and its corresponding basic variables in **x** and **y** are $\mathbf{x}(K)$ and $\mathbf{y}(K)$, respectively.
Step 2. Identify the set of PGS, $\vartheta = \{J_1, J_2, \ldots, J_r\}$, from $\upsilon = \{K_1, \ldots, K_v\}$. This can be done by dropping those K which contain $\mathbf{y}(K)$ in the simplex tableau. Then, use Definition 4.4 and Theorem 4.1 to find the PGSs.
Step 3. For each PGS J, identify those $K \in \upsilon$ for which $\mathbf{x}(K) \subseteq (\mathbf{x}(J), \mathbf{x}^1(J'))$. Denote the resulting subset as $\upsilon\,(J)$. Given a value of α^*, find $\Gamma_4(K(J))$ and $\Lambda_4(K(J), \alpha^*)$ so that $K(J)$ is a PFCP.
Step 4. Use a well-known technique of decision making under uncertainty to select the "best" system from the set of PGSs and their PFCPs as the final decision.

Example 9.3

We continue use Example 9.1 to demonstrate Algorithm 9.3:

Step 1. We first find all primal potential solutions from the corresponding augmented problem (M9.3) in Example 9.1 as $\upsilon = \{K_1, \ldots, K_{44}\}$ (see Table 9.6).

Step 2. After dropping those K which contain $\mathbf{y}(K)$ in the simplex tableau associated with Table 8.6, we use Definition 4.4 and Theorem 4.1 to find the PGSs, denoted by $\vartheta = \{J_1, J_2, J_3, J_4, J_5\}$ (see Table 9.5).

Step 3. For each PGS J, identify those $K \in \upsilon$ for which $\mathbf{x}(K) \subseteq (\mathbf{x}(J), \mathbf{x}^1(J'))$. Denote the resulting subset as $\upsilon^F(J)$. From Table 9.6, they are:

$\upsilon^F(J_1) = \{J_1, K_1, K_5, K_6, K_7, K_8, K_9, K_{16}, K_{17}, K_{21}, K_{22}, K_{23}, K_{29}\}$;

$\upsilon^F(J_2) = \{J_2, K_1, K_3, K_4, K_5, K_6, K_7, K_8, K_9, K_{16}, K_{17}, K_{19}, K_{20}, K_{21}, K_{22}, K_{23}, K_{29}, K_{30}, K_{32}, K_{33}, K_{34}, K_{39}\}$;

$\upsilon^F(J_3) = \{J_1, K_1, K_5, K_6, K_7, K_8, K_9, K_{14}, K_{15}, K_{16}, K_{17}, K_{18}, K_{21}, K_{22}, K_{23}, K_{29}, K_{35}, K_{36}, K_{37}, K_{38}\}$;

$\upsilon^F(J_4) = \{J_4, K_1, K_{14}, K_{15}, K_{16}, K_{29}, K_{35}, K_{36}, K_{37}, K_{38}\}$; and

$\upsilon^F(J_5) = \{J_5, K_{24}, K_{25}, K_{26}, K_{29}, K_{35}, K_{36}, K_{37}, K_{38}\}$.

For illustration, let $\alpha^* = (4, 4, 4)$. By Definition 8.6, we find $\Gamma_4(K(J))$ and $\Lambda_4(K(J), \alpha^*)$ so that $K(J)$ is a PFCP. For example, the PFCPs of J_1 is found as in Table 9.11 (Figure 9.4). Note that the set of all PRCPs for J_1 in Table 9.7 differs from the set of all PFCPs for J_1 in Table 9.11. Similarly, we can find the PFCPs for J_2, J_3, J_4, and J_5. For the sake of space, we will not list them here (see Shi and Yu 1992 for the details).

Step 4. Let us also use the "maximizing expected payoff" as the criterion to selecting the final optimal system from ϑ and the PFCPs. Let the initial setup cost for using J, $f(J) = 0$. If γ_1 is independent on λ_1, and (γ_1, λ_1) have the bi-uniform probability distribution $F(\gamma_1, \lambda_1)$, the expected payoff $\mathbf{EV}(J_1)$ with respect to the PFCP set $\{J_1, K_{21}\}$ (obtained from Figure 8.4) can be expressed as

$$\mathbf{EV}(J_1) = \int_{S_1} \mathbf{V}(J_1)\, dF(\gamma_1, \lambda_1) + \int_{S_2} \mathbf{V}(K_{21}(J_1))\, dF(\gamma_1, \lambda_1),$$

where $\mathbf{V}(J_1)$ and $\mathbf{V}(K_{21}(J_1))$; and the (γ_1, λ_1) ranges $S_j = \Gamma_4(K_j(J_1)) \times \Lambda_4(K_j(J_1))$, $j = 1, 2$, are shown in Table 9.11.

We have the $\mathbf{EV}(J_1) = 59.45$, which is higher than the $\mathbf{EV}(J_1) = 48.45$ of PRCPs in Example 8.1. If we use other PFCP sets for J_1, the result is the same (the matrix of $\mathbf{V}(J_1)$ is the same as those of $\mathbf{V}(K_8(J_1))$ and $\mathbf{V}(K_9(J_1))$ in Table 8.11). Similarly, we can get $\mathbf{EV}(J_2) = 59.83$, $\mathbf{EV}(J_3) = 55.44$, $\mathbf{EV}(J_4) = 48.05$, and $\mathbf{EV}(J_5) = 21.66$. Thus, when the "maximizing expected payoff" is used as the decision criterion, J_2 is also the "best" (optimal) system. Note that this happened by chance. In many cases, the final choice of PRCPs is different from that of PFCPs.

Table 9.11 Primal Flexible Contingency Plans for J_1 with $\alpha^* = (4, 4, 4)$

Bases	$\Gamma_4(K_j(J_1))$	$\Lambda_4(K_j(J_1))$	$V(K_j(J_1))$
J_1	$0 < \gamma_1 \le 2/3$	$0 < \lambda_1 \le 1$	$(\lambda_1 \quad \lambda_2)\begin{pmatrix} 50 & 100 \\ 40 & 20 \end{pmatrix}\begin{pmatrix} \gamma_1 \\ \gamma_2 \end{pmatrix}$
$K_8(J_1)$	$0 < \gamma_1 \le 2/3$	$0 < \lambda_1 \le 1/2$	$(\lambda_1 \quad \lambda_2)\begin{pmatrix} 50 & 100 \\ 40 & 20 \end{pmatrix}\begin{pmatrix} \gamma_1 \\ \gamma_2 \end{pmatrix}$
$K_9(J_1)$	$0 < \gamma_1 \le 2/3$	$1/2 \le \lambda_1 \le 1$	$(\lambda_1 \quad \lambda_2)\begin{pmatrix} 50 & 100 \\ 40 & 20 \end{pmatrix}\begin{pmatrix} \gamma_1 \\ \gamma_2 \end{pmatrix}$
$K_{21}(J_1)$	$2/3 < \gamma_1 \le 1$	$0 < \lambda_1 \le 1$	$(\lambda_1 \quad \lambda_2)\begin{pmatrix} 40 & 120 \\ 10 & 100 \end{pmatrix}\begin{pmatrix} \gamma_1 \\ \gamma_2 \end{pmatrix}$

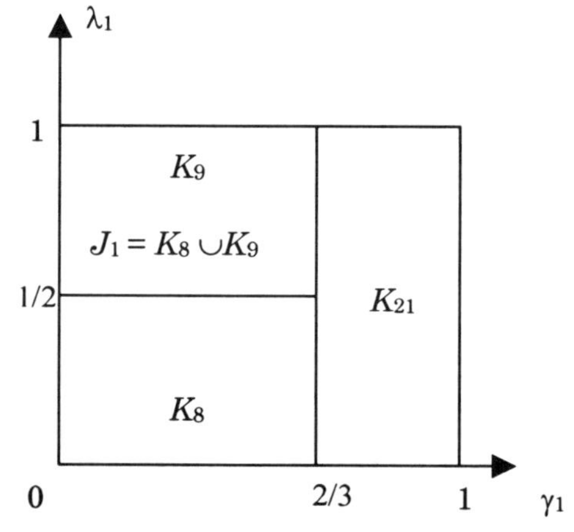

Figure 9.4 Primal Flexible Contingency Plans for J_1

9.4 Further Remarks

If finding a set of potential solutions is called the "first order" of MC^2 problems, then designing optimal systems can be viewed as a "second order" of the MC^2 problems. Given a set of *potential good systems* (potential solutions), this chapter goes beyond these solutions and investigates how to utilize the slack resources. When we only use the slack resource within a PGS, we the *primal rigid contingency plans* (PRCP). When we want to use all possible slack resources, the *primal flexible contingency plan* (PFCP) is an alternative. We hope the reader can get this insight from the chapter.

Exercises

9.1 Use Algorithm 9.2 to construct PRCPs for each PGS J of the following optimal design problem:

Max	$10\, x_1 + 12\, x_2$
Max	$15\, x_1 + 8\, x_2$
Subject to	$3\, x_1 + 2\, x_2 \leq (24, -35)$
	$4\, x_1 + 6\, x_2 \leq (15, 42)$
	$x_1, x_2 \geq 0.$

9.2 Conduct an analysis of external resource price for problem 9.1 (refer to Section 9.2.2).

9.3 Use Algorithm 9.3 to construct PFCPs for each PGS J of problem 9.1.

Chapter 10
GENRALIZED GOOD SYSTEMS AND CONTINGENCY PLANS

This chapter explores a procedure of selecting the optimal systems by using *generalized good systems* (GGSs). These systems contain a large set of selected opportunities and may offer a higher "payoff" than any PGS under certain distributions of decision parameters. A mathematical algorithm to generate the GGSs using union operations is discussed for finding the contingency plans in Section 10.1. Then, Section 10.2 is finding PRCPs for all GGSs, while Section 10.3 identifies PFCPs for a given GGS. Finally, Section 10.4 concludes the chapter with further remarks.

10.1 Generating New Designs Using Union Operators

Given a system design problem (M9.1), we can obtain a set of all PGSs, $\vartheta = \{J_1, J_2, \ldots, J_r\}$, by using the MC^2-simplex method. Under the restriction of assumption 9.1 (i), the number of selected opportunities, k, in a PGS J cannot exceed the number of available resources, m. For some design problems with a large number of possible opportunities, where n farther exceeds k, this assumption may rule many opportunities out of consideration for the optimal system. To *remove* or *relax* the assumption, we can take the *unions* of subsets of ϑ. It is possible to generate some new good systems that can contain more opportunities than any PGS and also have a higher "payoff" than any PGS under certain distribution of (λ, γ). We call such a new system the *generalized good system* (GGS), because it is generated by some PGSs.

A natural approach to find all distinct unions, which can be considered as GGSs, of subsets of ϑ is to check all 2^r possible unions of subsets of ϑ. For example, when $r = 10$, we have to check all 1024 ($=2^r$) unions in order to find all distinct unions! When r is large, this may be a prohibitive task. To overcome these difficulties, Shi, Yu and Zhang (1994) proposed an efficient algorithm of finding all distinct unions from ϑ.

Abstractly, we can state the problem of generating GGS from a given basic set of PGSs as follows:

Let X = $\{x_1, x_2, \ldots, x_z\}$ be a set of variables to be chosen for the design problem, and $\vartheta = \{J_1, J_2, \ldots, J_r\}$ be a set of PGSs, which are subsets of X, selected by some systematic method. That is, $J_i = \{x_{i1}, x_{i2}, \ldots, x_{in}\}$, $0 < n \leq z$, $i = 1, \ldots, r$, $J_i \subseteq X$. We want to find all new designs or *distinct* unions of the PGSs in ϑ.

Note that to develop an efficient algorithm for generating all such distinct unions of given PGSs and discarding the redundant unions is *not* a trivial task, since we take unions over all subsets of ϑ (i.e., the power set of ϑ), in which some subsets contain the same variables.

Given a set of PGSs, $\vartheta = \{J_1, J_2, \ldots, J_r\}$, for $1 \leq i_1 < \ldots < i_k \leq g$, we denote $n(i_1, \ldots, i_k) = \{J_t \mid t = i_1, \ldots, i_k\}$. Let $\Theta = P(\vartheta)$ be the power set of ϑ. Then $\Theta = \{n(i_1, \cdots, i_k) \mid 1 \leq i_1 < \ldots < i_k \leq r, 0 \leq k \leq r\}$. Denote $\Delta = P(\{J_1 \cup \ldots \cup J_r\})$, the power set of $\{J_1 \cup \ldots \cup J_r\}$.

Definition 10.1 The mapping u: $\Theta \rightarrow \Delta$ such that $u(n(i_1, \ldots, i_k)) = \cup\{J_t \mid t = i_1, \ldots, i_k\} \in \Delta$, for all $n(i_1, \ldots, i_k) \in \Theta$ is called the *union operation* over the PGSs.

Denote the range of the union operation by $u(\Theta)$. Then, finding all distinct unions of subsets of ϑ is equivalent to searching for all distinct elements of $u(\Theta)$. In the following, we build a tree structure to represent all elements of Θ and order these elements by traversing the corresponding nodes in the tree. This will help us develop an algorithm to search for all distinct elements of $u(\Theta)$, based on the ordering of elements of Θ.

We briefly sketch some basic concepts about "tree structure" (see, for example, Aho, Hopcroft and Ullman 1987).

A tree is a collection of elements called *nodes,* one of which is distinguished as a root, along with a relation called *parenthood* that places a hierarchical structure on the nodes. Formally, a tree can be defined recursively in the following manner:

(i) A *null* tree is a tree with no nodes. It is represented by $\emptyset$.

(ii) A *single node* by itself is a tree. This node is also the root of the tree.

(iii) Suppose n_0 is a node and $T_1, T_2, \ldots, T_k$ are trees with roots $n_1, n_2 \ldots, n_k$, respectively. One can construct a new tree by making n_0 be the parent of nodes $n_1, n_2 \ldots, n_k$. In this tree, n_0 is the root and $T_1, T_2, \ldots, T_k$ are *subtrees* of the root. Nodes $n_1, n_2 \ldots, n_k$ are called the *children* of node n_0. Nodes $n_1, n_2 \ldots, n_k$ are also called the *siblings* to each other because they have the same parent node n_0. A node n_r (or n_l) *is* called the right (or left) sibling node of a given node n if n_r (or n_l) has the same parent node np as n and lies to the right (or left) of n in the tree ordering of children of n_p. Given two nodes n_{i0} and n_{i1}, we say that n_{i0} is *a grandparent* of n_{i1}, and n_{i1} is a *grandchild* of n_{i0} if there are nodes n_{j1}, n_{j2}, ..., and n_{jk} ($k \geq 1$), such that n_{i0} is the parent of n_{j1}, n_{jp} is the parent of $n_{jp} + 1$ ($p = 1, 2, \ldots, k - 1$), and n_{jk}, is the parent of n_{i1}.

Assume that a tree contains at least two nodes. Then, we see that a tree consists of a number of subtrees which themselves consist of further subtrees. To order all nodes of a tree, one may use the order determined by the well-known preorder *listing* in data structure, which is defined recursive*ly* as follows:

Let T be a tree with root n_0 and subtrees $T_1, T_2, \ldots, T_k$. which have roots n_1, $n_2 \ldots, n_k$, respectively, such that $T_i \cap T_k = \emptyset$, $i \neq k$. Note that the root n_k of T_k is a child of n_0. Then, the order of the nodes of T is determined by the following rule:

(i) n_0 is the least (beginning) node; that is; $n_0 \succ n_i$, for any $i = 1, \ldots, k$;

(ii) If $n_i \in T_i$ and $n_k \in T_k$, then $n_i \succ n_k$. (i.e., n_i is ordered prior to n_k) if and only if $i < k$; and

(iii) Recursively, given any subtree T_j with root at n_j, suppose that T_j consists of further subtrees T_{j1}, $T_{j2}, \ldots, T_{jr}$. Then n_j is prior to any n_{jk}, $k = 1, \ldots, r$, and for any two nodes $n_{ji} \in T_{ji}$, $n_{jk} \in T_{jk}$, we have $n_{ji} \succ n_{jk}$, if and only if $i < k$.

Recall that Θ is the power set of ϑ. We now define a special tree *structure* on Θ as follows.

Definition 10.2 Given Θ, we define

(i) The root of the tree by $n(0) = \emptyset$;

(ii) The children of $n(0)$ by $n(i)$, $i = 1, \ldots, r$, and $n(j)$ is the right sibling of (i) if $j \in \{i+1, \ldots, r\}$;

(iii) The children of $n(i_1)$ by $n(i_1, i_2)$, $i_1 < i_2$, $i_1, i_2 \in \{1, \ldots, r\}$, and $n(i_1, i_r)$ is the right sibling of $n(i_1, i_2)$, if $i_2 < i_r$, for $i_r \in \{i_2+1, \ldots, r\}$; and

(iv) In general, the children of $n(i_1, \ldots, i_{k-1})$ by $n(i_1, \ldots, i_{k-1}, i_k)$, $i_1 < \ldots < i_k$, $i_1, \ldots, i_k \in \{1, \ldots, g\}$, and $n(i_1, \ldots, i_{k-1}, i_p)$ is the right sibling of $n(i_1, \ldots, i_{k-1}, i_k)$ if $i_k < i_p$ for $i_p \in \{i_{k+1}, \ldots, r\}$.

Let us use the following example to illustrate the tree structure of Θ.

Example 10.1

Given a set of PGSs, $\vartheta = \{J_1, J_2, J_3, J_4, J_5, J_6\}$, where $J_1 = \{x_1, x_2, x_4\}$, $J_2 = \{x_3, x_5, x_6\}$, $J_3 = \{x_3, x_4, x_6\}$, $J_4 = \{x_1, x_2, x_6\}$, $J_5 = \{x_1, x_2, x_5\}$, and $J_6 = \{x_1, x_5, x_6\}$. By Definition 10.2, we have the tree structure with 2^6 (= 64) possible nodes of Θ, as shown in Figure 10.1.

We can use the preorder listing to list all nodes in the tree structure of Figure 10.1. For illustration, we list the nodes in the second subtree of Figure 10.1 as follows:

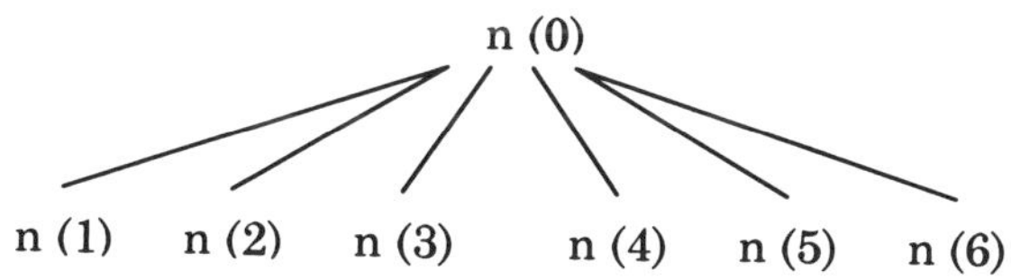

The Main Tree

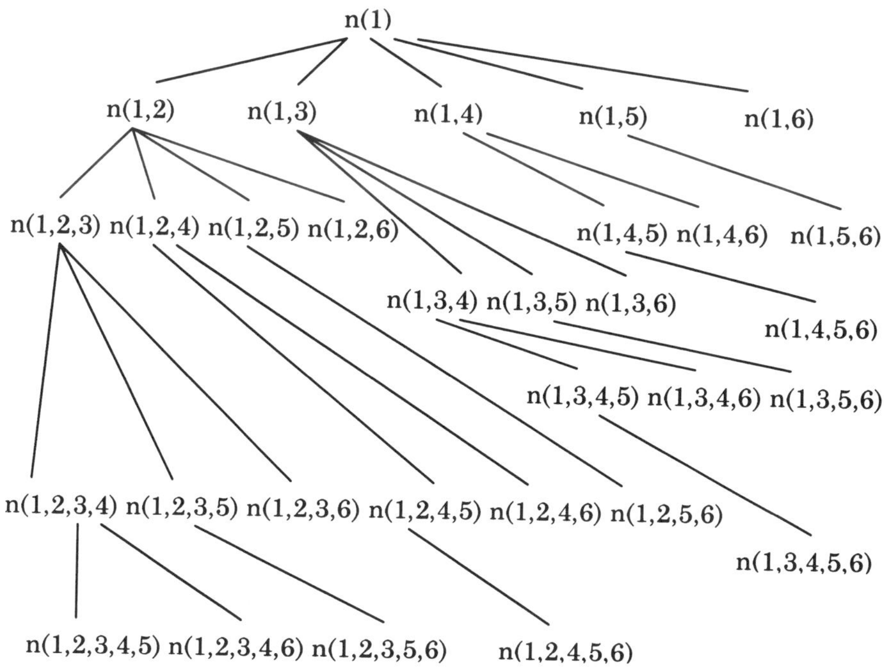

Subtree 1

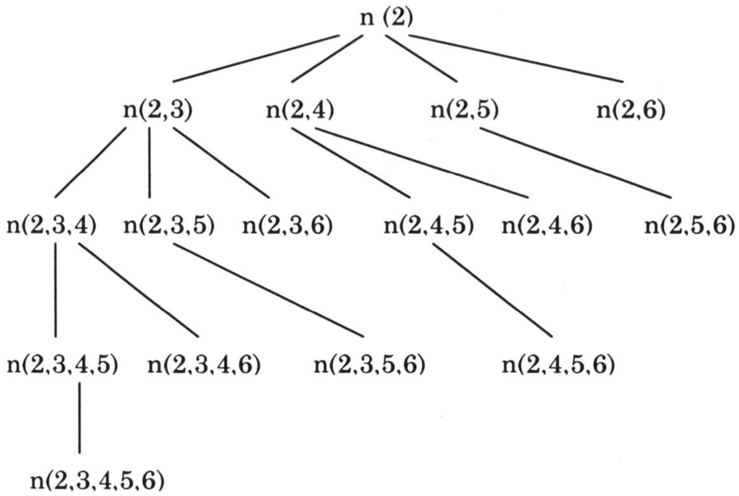

Subtree 2

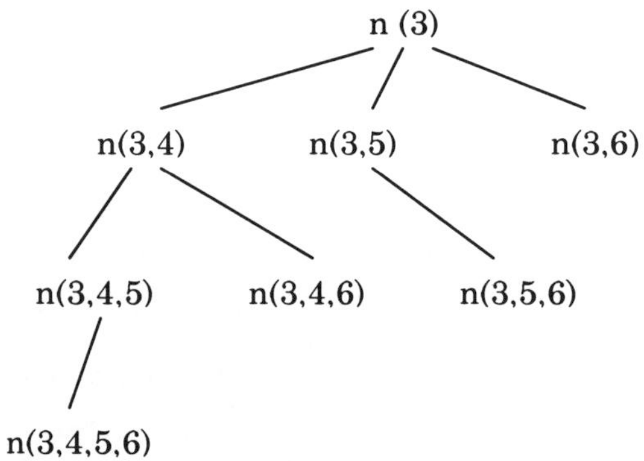

Subtree 3

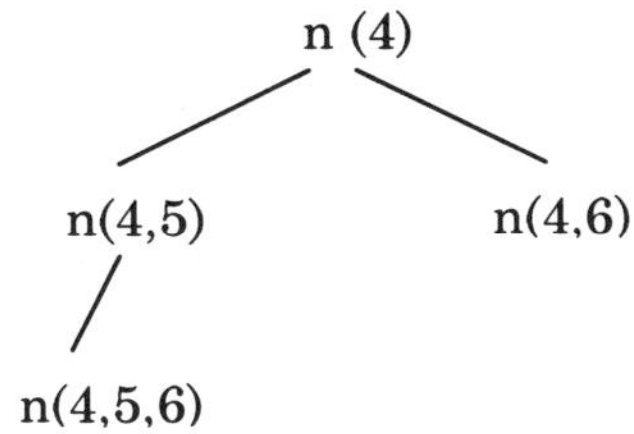

Subtree 4

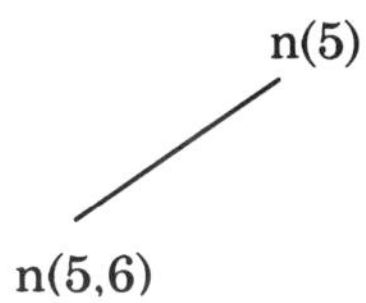

Subtree 5

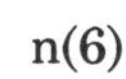

Subtree 6

Figure 10.1

$$\begin{aligned} n(2) &\succ n(2,3) \succ n(2,3,4) \succ n(2,3,4,5) \succ n(2,3,4,5,6) \succ n(2,3,4,6) \\ &\succ n(2,3,5) \succ n(2,3,5,6) \succ n(2,3,6) \succ n(2,4) \succ n(2,4,5) \\ &\succ n(2,4,5,6) \succ n(2,4,6) \succ n(2,5) \succ n(2,5,6) \succ n(2,6). \end{aligned}$$

Remark 10.1 (i) Given $n(i_1, \ldots, i_k)$, $n(j_1, \ldots, j_s) \in \Theta$, like the ordering of words in a dictionary, we say that $n(i_1, \ldots, i_k)$ is *lexicographically prior to* $n(j_1, \ldots, j_s)$, denoted by that $n(i_1, \ldots, i_k) \succ_L n(j_1, \ldots, j_s)$ if and only if

(1) $i_1 < j_1$, or
(2) there is some p, so that $i_q = j_q$ for $q = 1,\ldots, p\text{-}1$ and $i_p < j_p$, or
(3) $i_v = j_v$ for $v = 1,\ldots,k$ and $s > k$.

Note that we can verify that the ordering $\{\succ_L\}$ on Θ is a strict ordering (that is, $\{\succ_L\}$ is asymmetric, transitive, and complete) and the pair $(\Theta, \{\succ_L\})$ is an ordered set. From the preorder listing in Example 10.1, we observe that all nodes of Θ in the listing enjoy the ordering $\{\succ_L\}$. Thus, given the tree structure of Θ, the preorder listing $\{\succ\}$ is equivalent to the lexicographic ordering $\{\succ_L\}$.

(ii) Given $n(i_1, \ldots, i_k) \in \Theta$, its index set, $\{i_1, \ldots, i_k\}$, enjoys a monotonicity which means that for any $i_q, i_r \in \{i_1, \ldots, i_k\}$ $i_q < i_r$ if $q < r$. Both the lexicographic ordering and the monotonicity are important in the following discussion.

Without confusion, in the following, we shall use Θ to represent the ordered set $(\Theta, \{\succ_L\})$. Since $\{\succ_L\}$ is a strict ordering on Θ, we can rename elements of Θ as $n_1, n_2, \ldots,$ according to this ordering. For instance, in Example 10.1, n_1 is $n(0)$, n_2 is $n(1)$, n_3 is $n(1, 2)$, and so on. In this way, any $n(i_1, \ldots, i_k) \in \Theta$ is denoted by an n_i, where $1 \le i \le 2^r$. Note that for all $n_i, n_j \in \Theta$, $n_I \succ n_j$ if and only if $i < j$ and $1 \le i \le 2^r$.

Now we develop a streamlined algorithm for finding all distinct elements of $u(\Theta)$.

Intuitively, this algorithm starts from the most left branch of the tree from the root and we verify whether the generated design is new or redundant (i.e., it is not new or generated before). If it is new, we move to the next node to generate other new designs. Otherwise, when it is redundant, the entire subtree from the redundant node can be removed from further consideration. When the tree is large, the algorithm will allow us to remove a large portion of the tree for generating new designs and save a lot of computation time. In the following, we will precisely and systematically describe the procedure.

Recall that for any $n_i, n_j \in \Theta$, it is possible that $n_i \neq n_j$, but $u(n_i) = u(n_j)$. For instance, in Example 10.1, $n(1,2) \neq n(1,2,3)$, but $u(n(1, 2)) = u(n(1,2,3)) = \{x_1, x_2, x_3, x_4, x_5, x_6\}$. We want to find an efficient procedure to locate $u(\Theta)$ without checking each element of Θ.

Denote θ as a subset of Θ, such that $u(n_i) \neq u(n_j)$ for all $n_i, n_j \in \theta$ $(i \neq j)$ and $u(\Theta) = u(\theta)$. Note that the subset θ, so defined, may not be unique. Our goal is to find one such θ without checking all 2^r elements of Θ. To do so, let us check all the nodes according to the preorder listing of the tree structure of Θ.

Denote $n^k \in \Theta$ as the node that needs to be checked at the kth step of the checking process (see Algorithm 10.1). Let θ_k be the subset of Θ before the kth step of the checking, such that for any n_i, $n_j \in \theta_k$, $i \neq j$, $u(n_i) \neq u(n_j)$. Let N_k be the subset of Θ containing all remaining nodes that may need to be checked before the kth step of the checking. Note that $n^k \in N_k$, but $n^k \notin \theta_k$. Denote C_k as the set of all nodes but n^k in the subtree T_k with root n^k. Then C_k contains all children and grandchildren of n^k.

Algorithm 10.1

Step 0. Set $k = 0$. Let $\theta_0 = \varnothing$. Denote $\Theta\backslash\varnothing$ by N_0. Let $n^0 = n(1) \in N_0$.

Step 1. Find θ_k and $n^k \in N_k$, where n^k is the first node in N_k (according to the preorder) that needs to be checked. If $u(n^k) \in u(\theta_k)$, let $\theta_{k+1} = \theta_k$ and $N_{k+1} = N_k\backslash(\{n^k\}\cup C_k)$, and go to Step 2. If $u(n^k) \notin u(\theta_k)$, let $\theta_{k+1} = \theta_k \cup \{n^k\}$ and $N_{k+1} = N_k\backslash\{n^k\}$, and go to Step 2.

Step 2. If $N_{k+1} = \varnothing$, stop. Let $\theta_{k+1} = \theta^*$. Otherwise, set $k = k + 1$ and go to Step 1.

Note that as the tree is finite, the algorithm will terminate in a finite number of steps.

We need to show that θ^* identified by Algorithm 8.4 is what we want. In order to do this, we first need to know whether $u(\theta^*) = u(\Theta)$. Then, we need to know whether all elements of θ^* are distinct (i.e., θ^* satisfies that for any n_i, $n_j \in \theta^*$, $u(n_i) \neq u(n_j)$, $i \neq j$).

Theorem 10.1 For any $n \in \Theta$, there is $n' \in \theta^*$ such that $u(n) = u(n')$.

Theorem 10.1 is the main theoretical support for Algorithm 10.1. Its proof looks obvious. But, to prove it precisely demands a lot of work because of its combinatorial complexity. To avoid a distraction for readers, we skip it. For those who are really interested, please check Shi, Zhang and Yu (1994).

Remark 10.2 (i) From Theorem 10.1, we see that $u(\Theta) \subseteq u(\theta^*)$. Since $u(\theta^*) \subseteq u(\Theta)$, we have $u(\theta^*) = u(\Theta)$. (ii) From Algorithm 10.1, all elements of θ^* are distinct. Results of (i) and (ii) imply that θ^* is what we want.

In order to facilitate the computation of searching for the entire θ^* by using Algorithm 10.1, we use an incidence matrix to represent the relationship between index x_i and elements of Θ (Yu 1985). Given ϑ, let $\mathbf{x}(\vartheta) = J_1 \cup \ldots \cup J_r = \{x_1,\ldots,x_s\}$, reordering if necessary, be the set of all distinct indices involved in ϑ. Note that $\mathbf{x}(\vartheta) \subseteq X$.

Definition 10.3

(i) For any $\aleph = \{x_{c1},\ldots,x_{cq}\} \subseteq \mathbf{x}(\vartheta)$, we define the incidence vector of $\aleph$ by $W(\aleph) = (w_1,\ldots,w_s)$, *where* $w_i = 1$, if $x_i \in \aleph$; $w_i = 0$, if $x_i \notin \aleph$, $i = 1,\ldots s$.

(ii) An incidence matrix between $\mathbf{x}(\vartheta)$ and $S = \{n_{e1},\ldots,n_{er}\} \subseteq \Theta$ is defined by $W = \{w_{ij}\}_{sxr}$, whose columns, denoted by W_j, $j = 1, \ldots$, r, are the incidence vectors of $u(n_{ej})$, $n_{ej} \in S$, where u is the union operation as in Definition 10.1.

Note that rows and columns of W correspond to elements of $\mathbf{x}(\vartheta)$ and elements of S, respectively. Because of $\vartheta \subseteq \theta^* \subseteq \Theta$ and Definition 10.3, the incidence matrix between $\mathbf{x}(\vartheta)$ and θ^* will be denoted by W^* and the incidence matrix between $\mathbf{x}(\vartheta)$ and ϑ will be denoted by W^0. Note W^0 is a submatrix of W^*. Since ϑ is given, we shall call columns of W^0 the original columns.

The following discussion essentially says that searching for the entire θ^* is "equivalent" to generating W^*.

Definition 10.4 For any two incidence column vectors $W_j = (w_{1j\ldots}, w_{sj})$ and $W_p = (w_{1p\ldots}, w_{sp})$, we define $W_j \vee W_p = (\max(w_{1j}, w_{1p}),\ldots, \max(w_{sj}, w_{sp}))$.

Note that the operation $\vee$ in Definition 10.4 is often used in formal logic. It is known that the operations in logic can be transformed into the operations in set theory and vice versa (Stoll 1963 and Rueff and Jeger 1970). Based on the structure of W in Definition 10.3, we shall show that the operation $\cup$ used in Definition 10.1 is equivalent to the operation $\vee$ in Definition 10.4.

Lemma 10.1 If W_j and W_p are the incidence vectors of $u(n_j)$ and $u(n_p)$, respectively, then the incidence vector of $u(n_j) \cup u(n_p)$ is $W_j \vee W_p$.

Proof. Given $u(n_j)$ and $u(n_p)$, $j \neq p$, let $u(n_j) = \{x_{a1},\ldots,x_{ak}\}$ and $u(n_p) = \{x_{b1},\ldots,x_{bq}\}$. Let $W_j = (w_{1j\ldots}, w_{sj})$ and $W_p = (w_{1p\ldots}, w_{sp})$ be the columns corresponding to n_j and n_p, respectively. Let $W_j \vee W_p = (w_{1jp\ldots}, w_{sjp})$. If $w_{ijp} = 1$, $i = 1, \ldots, s$, then by Definition 10.4, we have that $w_{ij} = 1$ or w_{ip}

= 1. This implies that $x_i \in u(n_j)$ or $x_i \in u(n_p)$, where $x_i \in \mathbf{x}(\vartheta)$, $i = 1, \ldots, s$. Thus, $x_i \in u(n_j) \cup u(n_p)$. If $w_{ijp} = 0$, then by Definition 10.4, we see that $w_{ij} = 0$ and $w_{ip} = 0$. This implies that $x_i \notin u(n_j)$ and $x_i \notin u(n_p)$, where $x_i \in \mathbf{x}(\vartheta)$, $i = 1, \ldots, s$. Thus, $x_i \notin u(n_j) \cup u(n_p)$. By Definition 10.3, the proof is complete.

Based on Lemma 10.1, we can generate any columns of W by using the known original columns.

Lemma 10.2 Given $u(n_j)$ and $u(n_p)$, $j \neq p$, we have that
(i) $W_j \leq W_p$, if and only if $u(n_j) \subseteq u(n_p)$;
(iii) $W_j = W_p$, if and only $u(n_j) = u(n_p)$.

Proof. Given $u(n_j)$ and $u(n_p)$, $j \neq p$, let $u(n_j) = \{x_{a1},\ldots,x_{ak}\}$ and $u(n_p) = \{x_{b1},\ldots,x_{bq}\}$. Let $W_j = (w_{1j\ldots,}\ w_{sj})$ and $W_p = (w_{1p\ldots,}\ w_{sp})$ be the columns corresponding to n_j and n_p, respectively.

We first prove (i). Suppose $W_j \leq W_p$. If $x_i \in u(n_j)$, which implies $w_{ij} = 1$, then $w_{ip} = 1$, which implies that $x_i \in u(n_p)$, $x_i \in \mathbf{x}(\vartheta)$, $i = 1, \ldots, s$. Thus, $u(n_j) \subseteq u(n_p)$. Suppose that $u(n_j) \subseteq u(n_p)$. If $w_{ij} = 1$, which implies $x_i \in u(n_j)$, then $x_i \in u(n_p)$, where $x_i \in \mathbf{x}(\vartheta)$, $i = 1, \ldots, s$. This implies that $w_{ip} = 1$. Thus, $W_j \leq W_p$.

The result of (ii) follows from that of (i).

Let W^{**} be the incidence matrix between $\mathbf{x}(\vartheta)$ and Θ. Let W^k be the submatrix of W^{**} before the kth step of the checking such that for any two columns W_p, $W_j \in W_{k.}$, $p \neq j$, $W_p \neq W_j$. From Definitions 10.3 and 10.4 and Lemmas 10.1 and 10.2, Algorithm 10.1 can be rewritten as the following equivalent form.

Algorithm 10.2

Step 0. Set $k = 0$. Let $W^0 = \emptyset$. Denote $\Theta/\emptyset$ by N_0. Let $n^0 = n(1) \in N_0$.

Step 1. Find W^k and $n^k \in N_k$, where n^k is the first node in N_k (according to the preorder) that needs to be checked. Let W_k be the column of W^{**} corresponding to n^k. If there is a column W_j of W^k such that $W_k = W_j$, then we denote $W^{k+1} = W^k$ and $N_{k+1} = N_k \backslash (\{n^k\} \cup C_k)$, and go to Step 2. If for any column W_j of W^k, $W_k \neq W_j$, then we record W_k and denote $W^{k+1} = W^k \cup \{W_k\}$ (which means that the columns of W^{k+1} consist of all columns of W^k and W_k) and $N_{k+1} = N_k \backslash \{n^k\}$, and go to Step 2.

Step 2. If $N_{k+1} = \varnothing$, stop. Let $W^{k+1} = W^*$. Otherwise, set $k = k + 1$ and go to Step 1.

Remark 10.3 The information from the incidence matrix W makes it easier to search for the entire θ^*. To check the relation between $u(n_j)$ and $n(n_p)$, $j \neq p$, we just simply check two columns W_p and W_j by using Lemma 10.2. Using Algorithm 10.2, if the generated W_j corresponding to n_j is recorded as a column of W^*, then $n_j \in \theta^*$ is identified. By generating the columns of W^*, we can easily keep the track of searching for the entire θ^*. Once all columns of W^* are recorded, the entire θ^* is found.

In the following example, we use the incidence matrix to illustrate Algorithm 10.2 for searching for the entire θ^*.

Example 10.2 (Based On Example 10.1)

Given $\vartheta = \{J_1, J_2, J_3, J_4, J_5, J_6\}$, let $\mathbf{x}(\vartheta) = \{ x_1, x_2, x_3, x_4, x_5, x_6\}$. Note that Θ is given in Figure 10.1. By using Algorithm 10.2, we have:
(0) Let $W^0 = \varnothing$. Denote $\Theta\backslash\varnothing$ by N_0. Let $n^0 = n(1) = J_1 = \{ x_1, x_2, x_4\}$. Then $u(n(1)) = \{ x_1, x_2, x_4\}$. Let W^0 be the column corresponding to $n(1)$. Since $W^0 = \varnothing$, W_0 is recorded as the first column of Table 10.1. Note that W_0 is an original column.
(1) Check $n(1, 2)$: Let W_1 be the column corresponding to $n(1, 2)$. Since $W_0 \neq W_1$, we record W_1 as the second column of Table 10.1.
(2) Check $n(1,2,3)$: Let W_2 be the column corresponding to $n(1,2,3)$. Since $W_2 = W_1$, we do not record W_2 and eliminate the subtree with root $n(1, 2, 3)$ from the checking process.
(3) Check $n(1,2,4)$: Let W_3 be the column corresponding to $n(1,2,4)$. Since $W_3 = W_1$, we do not record W_3 and eliminate the subtree with root $n(1, 2, 4)$ from the checking process.
(4) Check $n(1, 2, 5)$: Let W_4 be the column corresponding to $n(1, 2, 5)$. Since $W_4 = W_1$, we do not record W_4 and eliminate the subtree with root $n(1, 2, 5)$ from the checking process.
(5) Check $n(1, 2, 6)$: Let W_5 be the column corresponding to $n(1, 2, 6)$. Since $W_5 = W_1$, we do not record W_5.
(6) Check $n(1,3)$: Let W_6 be the column corresponding to $n(1,3)$. Since $W_6 \neq W_0, W_1$, we record W_6 as the third column of Table 10.1.
(7) Check $n(1, 3, 4)$: Let W_7 be the column corresponding to $n(1, 3, 4)$. Since $W_7 = W_6$, we do not record W_7 and eliminate the subtree with root $n(1, 3, 4)$ from the checking process.

(8) Check $n(1, 3, 5)$: Let W_8 be the column corresponding to $n(1, 3, 5)$. Since $W_8 = W_1$, we do not record W_8 and eliminate the subtree with root $n(1, 3, 5)$ from the checking process.
(9) Check $n(1, 3, 6)$: Let W_9 be the column corresponding to $n(1, 3, 6)$. Since $W_9 = W_1$, we do not record W_9.
(10) Check $n(1,4)$: Let W_{10} be the column corresponding to $n(1,4)$. Since $W_{10} \neq W_0, W_1, W_6$, we record W_{10} as the fourth column of Table 10.1.
(11) Check $n(1,4,5)$: Let W_{11} be the column corresponding to $n(1,4,5)$. Since $W_{11} \neq W_0, W_1, W_6, W_{10}$, we record W_{11} as the fifth column of Table 10.1.
(12) Check $n(1, 4, 5, 6)$: Let W_{12} be the column corresponding to $n(1, 4, 5, 6)$. Since $W_{12} = W_{11}$, we do not record W_{12}.
(13) Check $n(1,4,6)$: Let W_{13} be the column corresponding to $n(1,4,6)$. Since $W_{13} = W_{11}$, we do not record W_{13}.
(14) Check $n(1,5)$: Let W_{14} be the column corresponding to $n(1, 5)$. Since $W_{14} \neq W_0, W_1, W_6, W_{10}, W_{11}$, we record W_{14} as the sixth column of Table 10.1.
(15) Check $n(1,5, 6)$: Let W_{15} be the column corresponding to $n(1,5,6)$. Since $W_{15} = W_{11}$, we do not record W_{15}.
(16) Check $n(1, 6)$: Let W_{16} be the column corresponding to $n(1, 6)$. Since $W_{16} = W_{11}$, we do not record W_{16}.

We have checked the left-most subtree with root $n(1)$. Continuing in this manner, the checking process terminates when $n(6)$, the last node in the preorder listing of Θ, is checked. All 16 distinct columns are shown in Table 10.1. Note that among them, the columns corresponding to $n(1)$, $n(2)$, $n(3)$, $n(4)$, $n(5)$, and $n(6)$ are the original columns. In the checking process they are automatically recorded as distinct columns. Thus, the number of distinct generated columns is 10. Rearranging the order, we can write

$$\theta^* = \{ n(1), n(2), n(3), n(4), n(5), n(6), n(1,2), n(1,3), n(1,4), n(1,4,5), n(1, 5), n(2, 3), n(2, 3, 6), n(2, 4), n(2, 6), n(4, 5)\}.$$

Thus,

$$u(\theta^*) = \{ J_1, J_2, J_3, J_4, J_5, J_6, u(n(1,2)), u(n(1, 3)), u(n(1,4)), u(n(1,4,5)), u(n(1, 5)), u(n(2, 3)), u(n(2, 3, 6)), u(n(2, 4)), u(n(2, 6)), u(n(4, 5))\},$$

where $J_1 = \{x_1, x_2, x_4\}$, $J_2 = \{x_3, x_5, x_6\}$, $J_3 = \{x_3, x_4, x_6\}$, $J_4 = \{x_1, x_2, x_6\}$, $J_5 = \{x_1, x_2, x_5\}$, $J_6 = \{x_1, x_5, x_6\}$, $u(n(1,2)) = \{x_1, x_2, x_3, x_4, x_5, x_6\}$, $u(n(1, 3)) = \{x_1, x_2, x_3, x_4, x_6\}$, $u(n(1,4)) = \{x_1, x_2, x_4, x_6\}$, $u(n(1,4,5)) = \{x_1, x_2, x_4, x_5, x_6\}$, $u(n(1, 5)) = \{x_1, x_2, x_4, x_5\}$,

$u(n(2, 3)) = \{x_3, x_4, x_5, x_6\}$, $u(n(2, 3, 6)) = \{x_1, x_3, x_4, x_5, x_6\}$,
$u(n(2, 4)) = \{x_1, x_2, x_3, x_5, x_6\}$, $u(n(2, 6)) = \{x_1, x_3, x_5, x_6\}$,
$u(n(4, 5)) = \{x_1, x_2, x_5, x_6\}$.

Table 10.1 Search results for Subset θ*

	W_0 $n(1)$	W_1 $n(1,2)$	W_6 $n(1,3)$	W_{10} $n(1,4)$	W_{11} $n(1,4,5)$	W_{14} $n(1,5)$	W_{17} $n(2)$	W_{18} $n(2,3)$
x_1	1	1	1	1	1	1	0	0
x_2	1	1	1	1	1	1	0	0
x_3	0	1	1	0	0	0	1	1
x_4	1	1	1	1	1	1	0	1
x_5	0	1	0	0	1	1	1	1
x_6	0	1	1	1	1	0	1	1
	W_{21} $n(2,3,6)$	W_{22} $n(2,4)$	W_{26} $n(2,6)$	W_{27} $n(3)$	W_{31} $n(4)$	W_{32} $n(4,5)$	W_{35} $n(5)$	W_{37} $n(6)$
x_1	1	1	1	0	1	1	1	1
x_2	0	1	0	0	1	1	1	0
x_3	1	1	1	1	0	0	0	0
x_4	1	0	0	1	0	0	0	0
x_5	1	1	1	0	0	1	1	1
x_6	1	1	1	1	1	1	0	1

In the above example using Algorithm 10.2, the number of nodes that we have checked is 31, compared with a total number of nodes of 63 ($n(0) = \varnothing$ is excluded). We did not check 32 nodes. Thus, the method in Algorithm 10.2 (or Algorithm 10.1) is more efficient than the natural method, which needs 63 checks for finding all distinct elements of $u(\Theta)$.

One may expect that as the number of primitives increases, the efficiency of the algorithm will become even more obviously powerful. However, this needs to be further studied and verified.

10.2 Finding Primal Rigid Contingency Plans for Generalized Good Systems

Let $\Omega = u(n(i_1,\dots, i_k))$ be a GGS (generalized good system) found by Algorithm 10.2. Since Ω is generated from some of PGSs, a PGS J is a special Ω. We denote the set of all PGSs and GGSs by $\wp$. Then $\wp$

represents all possible candidates for the optimal system. For any given index subset H of opportunity variables and/or slack variables that is not selected as an element of $\wp$, we want to know whether there is an element of $\wp$, say Ω_0, such that Ω_0 *is preferred to* the H with respect to optimality (see Definition 10.7) for model (M9.1). If so, this allows us to ignore the all non-selected subsets of variables from further consideration in designing the optimal system (Shi, Yu and Zhang 1996).

Let $N^* = \{1, . . . , n+m\}$ be the index set of all possible opportunity variables x_j and slack variables s_i. Note that $N^* \supseteq N$. Suppose that for the given model (M9.1), we have obtained $\wp$ by taking unions of subsets of ϑ. We have:

Definition 10.5 Given a PGS J of model (M9.1), we define its *optimal situation set* by $S(J) = \{(\lambda, \gamma) > 0 \mid \gamma \in \Gamma_1(J), \lambda \in \Lambda_1(J)\}$.

Note that we have used the concept of the optimal situation set in Definition 8.4 to compute the expected payoff of PGS J.

Definition 10.6 For any subset H, $H \subseteq N^*$, its *optimal situation set* is defined by $S(H) = \cup\{ S(J_k) \mid J_k \subseteq H\}$.

Definition 10.7 For any subsets H_1 and H_2, $H_1, H_2 \subseteq N^*$, *an optimality preference* with respect to model (M9.1) is defined by

(i) $H_1 \succ H_2$, if and only if $S(H_1) \supset S(H_2)$ and

(ii) $H_1 \sim H_2$, if and only if $S(H_1) = S(H_2)$.

By Definition 10.7, the induced preference $\{\succsim\}$ is defined by $H_1 \succsim H_2$ if and only if $S(H_1) \supseteq S(H_2)$ for all $H_1, H_2 \subseteq N^*$.

Theorem 10.1 Given a subset H as possible candidate for optimal system, if $H \notin \wp$, there is a Ω_k such that $\Omega_k \succsim H$.

Proof. For a given $H \notin \wp$, there are two possible cases about $S(H)$:

(i) Suppose $S(H) = \varnothing$. Then for any $\Omega_k \in \wp$, $S(\Omega_k) \neq \varnothing$, we have $S(\Omega_k) \supset S(H)$. Thus, $\Omega_k \succ H$.

(ii) Suppose $S(H) = \cup\{ S(J_k) \mid J_k \subseteq H\} \neq \varnothing$. Without loss of generality, let $S(H) = S(J_{i1}) \cup\ldots\cup S(J_{ie})$, where {i1, ..., ie} is the subset of {1, ..., r} and $J_{i1}, \ldots J_{ie} \in \vartheta$. Let $\Omega_k \supseteq J_{i1}\cup\ldots\cup J_{ie} \in \wp$. Thus, $S(\Omega_k) \supseteq S(H) = S(J_{i1}) \cup\ldots\cup S(J_{ie})$ and $\Omega_k \succsim H$.

Theorem 10.3 For any subset H that contains some opportunity j which is not in any $J_i \in \vartheta$, there is a $\Omega_k \in \wp$ such that $\Omega_k \succsim H$.

Proof. For any H which contains some opportunity j, $j \notin J_i$ for any $J_i \in \vartheta$, we must have $H \notin \wp$. The conclusion then follows immediately from Theorem 10.2.

Theorems 10.2 and 10.3 guarantee that $\wp$ is *the set of all possible candidates* for the optimal system. Therefore, we can concentrate on constructing the corresponding contingency plans for each element of $\wp$ without paying attention to any non-selected subsets.

As we have discussed how to construct PRCPs for all PGSs in Section 9.2, we focus on preparing PRCPs for all GGSs as follows.

Since the number of selected opportunities in a GGS Ω may exceed m, the number of resources under consideration, Ω may *not* satisfy Assumption 9.1 (i) and (ii). However, by using the MC^2-simplex method, some bases for model (M9.1) contained in Ω can optimize (M9.1) under certain ranges of (λ, γ). To construct PRCPs for a GGS Ω, we are restricted in using the opportunity variables x_j and slack variables s_i involved in Ω and purchase external resources y, if necessary. Given a GGS Ω, let $\mathbf{x}_\Omega$ be the corresponding variables involved in Ω. Then, for a given $\mathbf{x}_\Omega$, we denote $\mathbf{A}_\Omega$ as the submatrix of [**A**, **I**] associated with Ω and $\mathbf{C}_\Omega$ as the submatrix of [C, 0] associated with Ω.

In constructing the PRCPs for a given GGS Ω, we solve the following submodel (M10.1):

$$
\begin{aligned}
&\textit{Max} && \boldsymbol{\lambda}'\mathbf{C}_\Omega \mathbf{x}_\Omega - \boldsymbol{\alpha}'\mathbf{y} \\
&\textit{Subject to} && \mathbf{A}_\Omega \mathbf{x}_\Omega = \mathbf{D}\boldsymbol{\gamma} + \mathbf{y} \\
& && \mathbf{x}_\Omega, \mathbf{y} \geq 0.
\end{aligned}
$$

The initial tableau of model (M10.1) is Table 10.2.

Table 10.2 The Initial Simplex Tableau of (M10.1)

$\mathbf{x}_\Omega$	$\mathbf{y}$	RHS	
$\mathbf{A}_\Omega$	$-\mathbf{I}$	$\mathbf{D}\gamma$	(10.1)
$-\lambda^t\mathbf{C}_\Omega$	α^t	$\mathbf{0}$	(10.2)

Given a GGS Ω, let G be a basis of model (M9.1) with corresponding basis variables denoted by $\mathbf{x}_\Omega(G)$. We decompose $\mathbf{A}_\Omega$ into $[\mathbf{B}_G(\Omega), \mathbf{R}_G(\Omega)]$, where $\mathbf{B}_G(\Omega)$ is the basis matrix corresponding to $\mathbf{x}_\Omega(G)$ and $\mathbf{R}_G(\Omega)$is a submatrix of $\mathbf{A}_\Omega$ corresponding to the nonbasic variables $\mathbf{x}'_\Omega(G)$. Then, Table 10.2 can be rewritten as Table 10.3, where $\mathbf{C}_G(\Omega)$ and $\mathbf{C}'_G(\Omega)$ are the submatrices of $\mathbf{C}_\Omega$ for $\mathbf{x}_\Omega(G)$ and $\mathbf{x}'_\Omega(G)$, respectively.

Table 10.3 The Revised Tableau of (M10.1)

$\mathbf{x}_\Omega(G)$	$\mathbf{x}'_\Omega(G)$	$\mathbf{y}$	RHS	
$\mathbf{B}_G(\Omega)$	$\mathbf{R}_G(\Omega)$	$-\mathbf{I}$	$\mathbf{D}\gamma$	(10.3)
$-\lambda^t\mathbf{C}_G(\Omega)$	$-\lambda^t\mathbf{C}'_G(\Omega)$	α^t	$\mathbf{0}$	(10.4)

Table 10.3 can be transformed as Table 10.4, where (10.5) = $\mathbf{B}_G(\Omega)^{-1}$ (10.3), and (10.6) = $\lambda^t\mathbf{C}_G(\Omega)$(10.5) + (10.4).

Definition 10.8 In identifying PRCP G for GGS Ω, we call

(i) $\Gamma_5(G(\Omega)) = \{\gamma > 0 \mid \mathbf{B}_G(\Omega)^{-1}\mathbf{D}\gamma \geq 0\}$ the primal parameter set of model (M10.1); and

(ii) $\Lambda_5(G(\Omega), \alpha) = \{\lambda > 0 \mid \lambda^t[\mathbf{C}_G(\Omega)\mathbf{B}_G(\Omega)^{-1}\mathbf{R}_G(\Omega)-\mathbf{C}'_G(\Omega), \mathbf{e}_q\alpha^t-\mathbf{C}_G(\Omega)\mathbf{B}_G(\Omega)^{-1}] \geq 0\}$ the dual parameter set of model (M10.1).

Table 10.4 The Simplex Tableau of (M10.1)

$\mathbf{x}_\Omega(G)$	$\mathbf{x}'_\Omega(G)$	$\mathbf{y}$	RHS	
$\mathbf{I}$	$\mathbf{B}_G(\Omega)^{-1}\mathbf{R}_G(\Omega)$	$-\mathbf{B}_G(\Omega)^{-1}$	$\mathbf{B}_G(\Omega)^{-1}\mathbf{D}\gamma$	(10.5)
$\mathbf{0}$	$\lambda^t[\mathbf{C}_G(\Omega)\mathbf{B}_G(\Omega)^{-1}\mathbf{R}_G(\Omega)-\mathbf{C}'_G(\Omega)]$	$\lambda^t[\mathbf{e}_q\alpha^t-\mathbf{C}_G(\Omega)\mathbf{B}_G(\Omega)^{-1}]$	$\lambda^t\mathbf{C}_G(\Omega)\mathbf{B}_G(\Omega)^{-1}\mathbf{D}\gamma$	(10.6)

Definition 10.9 Given a GGS Ω, let $\{G_1(\Omega), \ldots, G_h(\Omega)\}$ be the set of potential solutions obtained from (M10.1) by using the MC²-simplex method. When $(\gamma, \lambda) \in \Gamma_5(G_i(\Omega))\mathrm{x}\Lambda_5(G_i(\Omega))$, $i = 1, \ldots, h$, we call $\{G_1(\Omega), \ldots, G_h(\Omega)\}$ the *primal rigid contingency plans* (PRCPs) for Ω.

Remark 10.3 Model (M10.1) is a submodel of model (M9.3) since model (M10.1) can be obtained by dropping columns of $\mathbf{x}_{\Omega'}$ that represent the variables that are not selected in Ω from model (M9.3).

Algorithm 10.3

Step 1. Find all primal potential solutions for model (M9.3). Denote the set of such solutions by $\upsilon = \{K_1, \ldots, K_v\}$. Each K is generically defined as a basis and its corresponding basic variables in $\mathbf{x}$ and $\mathbf{y}$ are $\mathbf{x}(K)$ and $\mathbf{y}(K)$, respectively.

Step 2. Use Algorithm 10.1 (or 10.2) to find the set of generalized good systems, $\wp$.

Step 3. For each GGS Ω of $\wp$, identify those $G \in \upsilon$ for which $\mathbf{x}(G) \subseteq \mathbf{x}_\Omega(G)$. Denote the resulting subset as $\upsilon(\Omega)$. Given a value of α^*, find $\Gamma_5(G(\Omega))$ and $\Lambda_5(G(\Omega), \alpha^*)$ so that $G(\Omega)$ is a PRCP for Ω.

Step 4. Use a well-known technique of decision making under uncertainty to select the "best" system from the set of GGSs and their PRCPs as the final decision.

Example 10.3

We reconsider a design problem with two criteria and two constraint levels for five possible opportunities in Example 9.1. Instead of finding all PRCPs for the set of PGSs in Table 9.5, we use Algorithm 10.3 to identify all PRCPs for all possible GGSs as follows:

Step 1. We solve the augmented problem of Example 9.1 and obtain all primal solutions as shown in Table 9.6.

Step 2. Applying Algorithm 10.1 (or 10.2), we find the set of generalized good systems, $\wp$ in Table 10.5 based on PGSs J_i, i = 1, 2, 3, 4, 5, in Table 9.5.

Table 10.5 A Set of Generalized Good Systems $\wp$

GGS	$\mathbf{x}(\Omega_i, \gamma)$	GGS	$\mathbf{x}(\Omega_i, \gamma)$
Ω_1	(x_1, x_2, x_4, s_3)	Ω_7	(x_1, x_3, x_5, s_3)
Ω_2	$(x_1, x_2, x_4, x_5, s_3)$	J_1	(x_1, x_2, s_3)
Ω_3	$(x_1, x_2, x_3, x_4, x_5, s_3)$	J_2	(x_1, x_2, x_4)
Ω_4	(x_1, x_3, x_5, s_3)	J_3	(x_1, x_2, x_5)
Ω_5	$(x_1, x_2, x_3, x_5, s_3)$	J_4	(x_1, x_5, s_3)
Ω_6	(x_1, x_2, x_4, x_5)	J_5	(x_3, x_5, s_3)

Step 3. For each GGS Ω of $\wp$, identify those $K \in \upsilon$ for which $\mathbf{x}(K) \subseteq \mathbf{x}_\Omega(K)$. Denote the resulting subset as $\upsilon^R(\Omega_i)$, i = 1, 2, 3, 4, 5, 6, 7. We have:

$\upsilon^R(\Omega_1) = \{J_1, K_3, K_4, K_5, K_6, K_7, K_{19}, K_{20}, K_{22}, K_{30}, K_{32}, K_{39}\}$;

$\upsilon^R(\Omega_2) = \{J_1, J_2, J_3, J_4, K_3, K_4, K_5, K_6, K_7, K_{14}, K_{15}, K_{18}, K_{19}, K_{20}, K_{22}, K_{28}, K_{30}, K_{31}, K_{32}, K_{37}, K_{38}, K_{39}\}$;

$\upsilon^R(\Omega_3) = \{J_1, J_2, J_3, J_4, J_5, K_3, K_4, K_5, K_6, K_7, K_{10}, K_{11}, K_{12}, K_{13}, K_{14}, K_{15}, K_{18}, K_{19}, K_{20}, K_{22}, K_{25}, K_{26}, K_{27}, K_{28}, K_{30}, K_{31}, K_{37}, K_{38}, K_{39}\}$;

$\upsilon^R(\Omega_4) = \{J_1, J_3, J_4, K_5, K_6, K_7, K_{14}, K_{15}, K_{18}, K_{22}, K_{30}, K_{37}, K_{38}\}$;

$\upsilon^R(\Omega_5) = \{J_1, J_3, J_4, J_5, K_2, K_5, K_6, K_7, K_{10}, K_{11}, K_{12}, K_{13}, K_{14}, K_{15}, K_{18},$

$K_{22}, K_{25}, K_{26}, K_{27}, K_{37}, K_{38}\}$;

$\upsilon^R(\Omega_6) = \{J_2, J_3, J_4, K_5, K_6, K_7, K_{14}, K_{15}, K_{18}, K_{19}, K_{22}, K_{31}, K_{37}, K_{39}\}$;

and

$\upsilon^R(\Omega_7) = \{J_4, J_5, K_2, K_{15}, K_{25}, K_{26}, K_{27}, K_{37}, K_{38}\}$.

Since PGS is a special GGS, we can find $\upsilon^R(J_i)$ for PGS J_i, i = 1, 2, 3, 4, 5 (see Exercise 10.2). Let $\alpha^* = (2, 2, 2)^t$, we also find $\Gamma_5(K(\Omega))$ and $\Lambda_5(K(\Omega), \alpha^*)$ so that $K(\Omega)$ is a PRCP for each Ω. For example, the PRCPs of Ω_1 is identified as Table 10.6.

Step 4. Let the initial setup cost for using Ω, $f(\Omega) = 0$. We assume that γ_1 is independent on λ_1, and (γ_1, λ_1) have the bi-uniform probability distribution $F(\gamma_1, \lambda_1)$. Then, the expected payoff $\mathbf{EV}(\Omega_1)$ with respect to the contingency plan set $\{J_1, K_3, K_5, K_{20}, K_{21}, K_{39}\}$ (Figure 10.2) can be expressed as

$$\begin{aligned}\mathbf{EV}(\Omega_1) = & \int_{S_1} \mathbf{V}(J_1)\, dF(\gamma_1, \lambda_1) + \int_{S_2} \mathbf{V}(K_3(J_1))\, dF(\gamma_1, \lambda_1) \\ & \int_{S_3} \mathbf{V}(K_5(J_1))\, dF(\gamma_1, \lambda_1) + \int_{S_4} \mathbf{V}(K_{20}(J_1))\, dF(\gamma_1, \lambda_1) + \\ & \int_{S_5} \mathbf{V}(K_{22}(J_1))\, dF(\gamma_1, \lambda_1) + \int_{S_6} \mathbf{V}(K_{39}(J_1))\, dF(\gamma_1, \lambda_1),\end{aligned}$$

where $\mathbf{V}(J_1(\Omega_1))$, $\mathbf{V}(K_3(\Omega_1))$, $\mathbf{V}(K_5(\Omega_1))$, $\mathbf{V}(K_{20}(\Omega_1))$, $\mathbf{V}(K_{22}(\Omega_1))$, and $\mathbf{V}(K_{39}(\Omega_1))$; and the (γ_1, λ_1) ranges $S_j = \Gamma_5(K_j(\Omega_1)) \times \Lambda_5(K_j(\Omega_1))$, j = 1, 2, 3, 4, 5, 6 are shown in Table 10.6. The result is $\mathbf{EV}(\Omega_1) = 78.04$.

Table 10.6 Primal Rigid Contingency Plans for Ω_1 with $\alpha^* = (2, 2, 2)^t$

Bases	$\Gamma_5(K_j(\Omega_1))$	$\Lambda_5(K_j(\Omega_1))$	$V(K_j(\Omega_1))$
$J_1(\Omega_1)$	$0 < \gamma_1 \le 2/3$	$1/7 \le \lambda_1 \le 2/3$	$(\lambda_1 \quad \lambda_2)\begin{pmatrix} 50 & 100 \\ 40 & 20 \end{pmatrix}\begin{pmatrix} \gamma_1 \\ \gamma_2 \end{pmatrix}$
$J_2(\Omega_1)$	$0 < \gamma_1 \le 2/3$	$1/7 \le \lambda_1 \le 1$	$(\lambda_1 \quad \lambda_2)\begin{pmatrix} 50 & 100 \\ 40 & 20 \end{pmatrix}\begin{pmatrix} \gamma_1 \\ \gamma_2 \end{pmatrix}$
$K_5(\Omega_1)$	$0 < \gamma_1 \le 2/3$	$2/3 \le \lambda_1 \le 1$	$(\lambda_1 \quad \lambda_2)\begin{pmatrix} 50 & 100 \\ 40 & 20 \end{pmatrix}\begin{pmatrix} \gamma_1 \\ \gamma_2 \end{pmatrix}$
$K_3(\Omega_1)$	$0 < \gamma_1 \le 1/4$	$0 < \lambda_1 \le 1/7$	$(\lambda_1 \quad \lambda_2)\begin{pmatrix} -70 & 40 \\ 60 & 30 \end{pmatrix}\begin{pmatrix} \gamma_1 \\ \gamma_2 \end{pmatrix}$
$K_{20}(\Omega_1)$	$1/4<\gamma_1\le 2/3$	$0 < \lambda_1 \le 1/7$	$(\lambda_1 \quad \lambda_2)\begin{pmatrix} 110 & -20 \\ 30 & 40 \end{pmatrix}\begin{pmatrix} \gamma_1 \\ \gamma_2 \end{pmatrix}$
$K_{22}(\Omega_1)$	$2/3 < \gamma_1 \le 1$	$3/4 \le \lambda_1 \le 1$	$(\lambda_1 \quad \lambda_2)\begin{pmatrix} 40 & 120 \\ 0 & 100 \end{pmatrix}\begin{pmatrix} \gamma_1 \\ \gamma_2 \end{pmatrix}$
$K_{39}(\Omega_1)$	$2/3 < \gamma_1 \le 1$	$0 < \lambda_1 \le 3/4$	$(\lambda_1 \quad \lambda_2)\begin{pmatrix} 35 & 130 \\ 15 & 70 \end{pmatrix}\begin{pmatrix} \gamma_1 \\ \gamma_2 \end{pmatrix}$

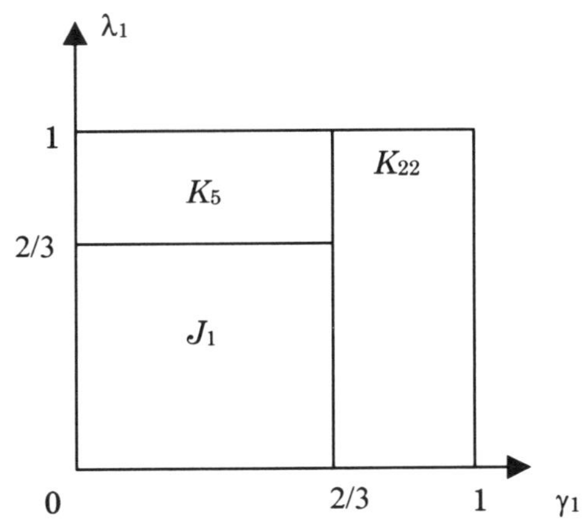

Figure 10.2 Primal Rigid Contingency Plans for Ω_1

Similarly, we have $\mathbf{EV}(\Omega_2)$ = 55.67, $\mathbf{EV}(\Omega_3)$ = 58.21, $\mathbf{EV}(\Omega_4)$ = 55.64, $\mathbf{EV}(\Omega_5)$ = 58.21, $\mathbf{EV}(\Omega_6)$ = 55.64, and $\mathbf{EV}(\Omega_7)$ = 35.74. Together with $\mathbf{EV}(J_1)$ = 59.45, $\mathbf{EV}(J_2)$ = 59.83, $\mathbf{EV}(J_3)$ = 55.44, $\mathbf{EV}(J_4)$ = 36.05, and $\mathbf{EV}(J_5)$ = 21.66 under PRCPs from Example 9.1, we conclude that Ω_1 is the optimal system we should take.

10.3 Finding Primal Flexible Contingency Plans for Generalized Good Systems

Parallel to Section 9.3, this section discusses how to construct a *primal flexible contingency plan* (PFCP) for GGS (Shi and Zhang 1993).

To construct the PFCPs for a given GGS Ω, let the corresponding non-selected variables of Ω be $\mathbf{x}_{\Omega'}$. We decompose $\mathbf{x}_{\Omega'}$ into $\mathbf{x}_{\Omega'}$ and $\mathbf{x}_{\Omega''}$, where $\mathbf{x}_{\Omega'}$ is the set of non-selected slack variables and $\mathbf{x}_{\Omega''}$ is the set of non-basic variables consisting of those opportunity variables of N that are not selected in Ω. Then, the PFCPs for the Ω can be identified by solving the following submodel (M10.2):

$$
\begin{array}{ll}
\textit{Max} & \boldsymbol{\lambda}^t \mathbf{C}_{\Omega}\, \mathbf{x}_{\Omega} - \boldsymbol{\alpha}^t \mathbf{y} \\
\textit{Subject to} & \mathbf{A}_{\Omega}\, \mathbf{x}_{\Omega} + \mathbf{A}_{\Omega'}\, \mathbf{x}_{\Omega'} = \mathbf{D}\boldsymbol{\gamma} + \mathbf{y} \\
 & \mathbf{x}_{\Omega}, \mathbf{x}_{\Omega'}, \mathbf{y} \geq 0,
\end{array}
$$

where $\mathbf{A}_{\Omega'}$ is the submatrix of $[\mathbf{A}, \mathbf{I}_m]$ corresponding to $\mathbf{x}_{\Omega'}$.

Given GGS Ω, we call the set of potential solutions obtained from model (M10.2) the PFCPs for Ω. Note that by increasing y if necessary, model (M10.2) always has feasible solutions; that is, the PFCPs for Ω can always be feasible for all possible changes of (λ, γ). Furthermore, these contingency plans can ensure both feasibility and optimality if a proper value of price α^* is chosen. For the sake of space, we skip the detailed discussion on the simplex tableau analysis of model (M10.2) and leave it to the reader as an exercise (see Exercise 10.3).

Remark 10.4 Model (M10.2) is a submodel of model (M9.3) since model (M10.2) can be obtained by dropping columns of $\mathbf{x}_{\Omega''}$ from model (M9.3).

Applying Remark 10.4, the careful reader can develop an algorithm (similar to Algorithm 9.3) to effectively and systematically identify the PFCPs for all GGSs (see Exercise 10.4 or Shi and Zhang 1993). The analysis of external resource price for GGSs (parallel to Section 9.2.2) can be also carried out (see Exercise 10.5 or Shi, Yu and Zhang 1996).

10.4 Further Remarks

The main idea of the *generalized good systems* (GGSs) is to break the restriction of finding a basis in linear programming: the number of selected opportunities, k, in a basis cannot exceed the number of constraints, m. This break allows us to consider a large number of possible opportunities, where n farther exceeds k, for various possible potential good systems. The task of designing optimal systems with GGSs expands the domain of linear system designs to a new challenging level.

Exercises

10.1 Apply Algorithm 10.2 to identify all GGSs for the optimal design problem in Example 9.1.

10.2 Find the subsets $\upsilon^R(J_i)$ for each PGS J_i, i = 1, 2, 3, 4, 5 in Example 10.3.

10.3 Perform the simplex tableau analysis of model (M10.2) in constructing PFCPs for GGS Ω.

10.4 Develop an algorithm (similar to Algorithm 9.3) to effectively and systematically identify the PFCPs for all GGSs of the following optimal design problem:

$$\begin{array}{ll} \textit{Max} & 6x_1 + 5x_2 + x_3 + 4x_4 + 3x_5 \\ \textit{Max} & 7x_1 + 4x_2 + 3x_3 + 2x_4 + 7x_5 \\ \textit{Subject to} & 4x_1 + 2x_2 + 3x_3 + x_4 + x_5 \le (60, -40) \\ & 2x_1 + x_2 + 2x_3 + 4x_4 + 3x_5 \le (80, 120) \\ & x_1, x_2 \ge 0. \end{array}$$

10.5 Conduct an analysis of external resource price for GGSs of problem 10.4.

Chapter 11
OPTIMAL SYSTEM DESIGNS AND DUAL CONTINGENCY PLANS

From Chapters 9 and 10, we observe that both PRCPs and PFCPs for a given PGS or GGS are constructed by adding external resources **y** for the purpose of converting infeasible solutions into feasible and optimal ones. However, for some change of the unit contribution of selected opportunities, these opportunities in a given PGS or GGS may not satisfy the optimality condition under consideration. For instance, in Table 9.5 of Example 9.1, if $0 < \lambda_1 < 1/5$, J_1 is not optimal. In this case, we need to construct the corresponding contingency plans for PGS J_1 to ensure the optimality. In the real-world applications, correcting the non-optimal situations for designing the optimal system is a significant problem. For example, using market promotion or persuasion as a tool, we can avoid the decline of business contributions such as profit and market share from situation to situation. In this chapter, we will first discuss how to correct these non-optimal situations for designing the optimal system by exploring dual contingency plans. Section 11.1 is to find the contingency plans that converts possible non-optimal solutions into optimal ones for the PGS is called *dual rigid contingency plans* (DRCPs), because they only use the selected slack resources, but not all possible slack resources in the process of constructing contingency plans. Section 11.2 deals with the method of identifying *dual flexible contingency plans* (DFCPs) for a PGS. Section 11.3 will provides some further remarks.

11.1 Optimal System Design and Dual Rigid Contingency Plans

This section is going to show a method of adjusting the unit contribution of selected opportunities in a given PGS to ensure the optimality condition over decision situations (Shi 1998). The resulting contingency plans that converts possible non-optimal solutions into optimal ones for the PGS is called *dual rigid contingency plans* (DRCPs), because they only use the selected slack resources, but not all possible slack resources in the process of constructing contingency plans.

Let us consider PGS J as a possible optimal system for implementation. The submodel of using the PGS J is (M11.1):

$$\begin{array}{ll} \mathit{Max} & \boldsymbol{\lambda}'\mathbf{C}_J\ \mathbf{x}(J) \\ \mathit{Subject\ to} & \mathbf{B}_J\ \mathbf{x}(J) = \mathbf{D}\boldsymbol{\gamma} \\ & \mathbf{x}(J) \geq 0. \end{array}$$

The dual model of this submodel is (M11.2):

$$\begin{array}{ll} \mathit{Min} & \mathbf{u}'\ \mathbf{D}\boldsymbol{\gamma} \\ \mathit{Subject\ to} & \mathbf{u}'\mathbf{B}_J \geq \boldsymbol{\lambda}'\mathbf{C}_J \\ & \mathbf{u}\ \text{unrestricted.} \end{array}$$

Suppose that for some λ, $\mathbf{x}(J)$ has no optimal solution. In order to construct dual contingency plans to recover the optimality, we can increase β_i units of the selected opportunity i in J. Let $\beta(J) = (\beta_1, \ldots, \beta_m)^t$ be the column vector of increments of the m selected opportunities in J (Generally, $\beta(J) \geq 0$. If $\beta(J) < 0$, we say $\beta(J)$ is the decrement of opportunity i). Let the unit contribution of these newly acquired $\beta(J)$ be $\mathbf{w}(J) = (w_1, \ldots, w_m)^t$ (Note that if $\mathbf{x}(J)$ contain a slack variable s_j as a component, then we may say that the corresponding β_j is the additional amount of the jth resource saved, which is worth w_j per unit). Then, the unit contribution $\lambda^t\mathbf{C}_J$ becomes $\lambda^t\mathbf{C}_J + \mathbf{w}(J)$. After adding $\mathbf{w}(J)$ into the right-hand side of the above dual model (M11.2), the total extra contribution is $\mathbf{w}(J)^t\beta(J)$. Thus, the total net implicit value is $\mathbf{u}^t\mathbf{D}\gamma$ - $\mathbf{w}(J)^t\beta(J)$. Therefore, the submodel of finding the DRCPs is (M11.3):

$$\begin{array}{ll} \mathit{Min} & \mathbf{u}'\mathbf{D}\boldsymbol{\gamma} - \mathbf{w}(J)'\ \boldsymbol{\beta}(J) \\ \mathit{Subject\ to} & \mathbf{u}'\mathbf{B}_J \geq \boldsymbol{\lambda}'\mathbf{C}_J + \mathbf{w}'(J) \\ & \mathbf{u}\ \text{unrestricted}, \mathbf{w}(J) \geq 0. \end{array}$$

We can use the MC^2-simplex method to solve model (M11.3). By adding the dual surplus variable $\mathbf{u}(J)$ complementary to $\mathbf{x}(J)$ into (M11.3), its initial simplex tableau is:

Table 11.1 The Initial Simplex Tableau of (M11.3)

$\mathbf{u}$	$\mathbf{u}(J)$	$\mathbf{w}(J)$	RHS	
$(\mathbf{B}_J)^t$	$-\mathbf{I}_m$	$-\mathbf{I}_m$	$(\mathbf{C}_J)^t \lambda$	(11.1)
$\gamma^t\mathbf{D}^t$	$\mathbf{0}$	$-\beta^t(J)$	$\mathbf{0}$	(11.2)

Using $\mathbf{B}_J^{-1}$ for linear transformation, we rewrite Table 11.1 as:

Table 11.2 The Simplex Tableau of (M11.3)

$\mathbf{u}$	$\mathbf{u}(J)$	$\mathbf{w}(J)$	RHS	
$-\mathbf{I}_m$	$-(\mathbf{B}_J^{-1})^t$	$-(\mathbf{B}_J^{-1})^t$	$(\mathbf{B}_J^{-1})^t(\mathbf{C}_J)^t \lambda$	(11.3)
$\mathbf{0}$	$\gamma^t\mathbf{D}^t(\mathbf{B}_J^{-1})^t$	$\gamma^t\mathbf{D}^t(\mathbf{B}_J^{-1})^t - \gamma^t \mathbf{e_p} \beta^t(J)$	$-\gamma^t\mathbf{D}^t(\mathbf{B}_J^{-1})^t(\mathbf{C}_J)^t \lambda$	(11.4)

where $\mathbf{e_p} = (1, 1, \ldots, 1)^t \in \mathbf{R}^p$ is a column vector.

Definition 11.1 Given a basis $U(J)$ for model (M11.3), define its corresponding

(i) *primal parameter set* by

$\Lambda_6(U(J)) = \{\lambda > 0 \mid \lambda^t \mathbf{C}_J \mathbf{B}_J^{-1} \geq 0\}$; and

(ii) *dual parameter set* by

$\Gamma_6(U(J), \beta(J)) = \{\gamma > 0 \mid [\mathbf{B}_J^{-1}\mathbf{D}, -\beta(J)(\mathbf{e_p})^t + \mathbf{B}_J^{-1}\mathbf{D}]\,\gamma \geq 0\}$.

Definition 11.2. For a given PGS J, the corresponding dual basis $U(J)$ is said to be a dual rigid contingency plan (DRCP) of J if $\Lambda_6(U(J)) \times \Gamma_6(U(J), \beta(J)) \neq \varnothing$ for some $\beta(J)$.

According to Definition 11.2, given a PGS J, the set of all potential bases $\{U_1(J), \ldots, U_h(J)\}$ obtained from model (M11.3) by using the MC^2-simplex method is called the set of DRCPs for PGS J. Similar to Section 9.2.1, given a set of PGSs $\vartheta = \{J_1, J_2, ..., J_r\}$, to find all DRCPs for each of J in ϑ, we have to solve r related models (M11.3). In order to avoid this ineffective approach, we develop an augmented dual model to locate all DRCPs for each of J in ϑ.

In the following discussion we use the primal basis matrix to represent the dual basis matrix of an augmented dual model. To do so, we write the (D) model of model (9.1) as (M11.4):

$$\begin{array}{ll} \textit{Min} & \mathbf{u}'\,\mathbf{D}\boldsymbol{\gamma} \\ \textit{Subject to} & \mathbf{u}'\,[\mathbf{A},\mathbf{I}_m] \geq \boldsymbol{\lambda}'\,[\mathbf{C},\mathbf{0}] \\ & \mathbf{u}\ \text{unrestricted.} \end{array}$$

In an augmented dual model for the RDCPs, we try to identify potential bases from variables **u** and **w** (not just **u**) as follows (M11.5):

$$\begin{array}{ll} \textit{Min} & \mathbf{u}'\,\mathbf{D}\boldsymbol{\gamma} - \mathbf{w}'\boldsymbol{\beta} \\ \textit{Subject to} & \mathbf{u}'\,[\mathbf{A},\mathbf{I}_m] \geq \boldsymbol{\lambda}'\,[\mathbf{C},\mathbf{0}] + \mathbf{w}' \\ & \mathbf{u}\ \text{unrestricted and}\ \mathbf{w} \geq 0, \end{array}$$

where $\mathbf{w} \in \mathbf{R}^{n+m}$ is a (n+m)-dimensional vector and $\beta \in \mathbf{R}^{n+m}$ is a (n+m)-dimensional parameter.

The term $\mathbf{u}^t\mathbf{D}\gamma - \mathbf{w}^t\beta$ in model (M11.5) is the net total implicit value of the resources $\mathbf{D}\gamma$ after deducting the extra contribution of **w** from the total implicit value of the resources, while $\mathbf{u}^t[\mathbf{A}, \mathbf{I}] \geq \lambda^t[\mathbf{C}, \mathbf{0}] + \mathbf{w}^t$ is constraints.

By putting surplus variables $\mathbf{u}_S$ into model (M11.5), it becomes (M11.6):

$$\begin{array}{ll} \textit{Min} & \mathbf{u}'\mathbf{D}\boldsymbol{\gamma} - \mathbf{w}'\boldsymbol{\beta} \\ \textit{Subject to} & \mathbf{u}'\,[\mathbf{A},\mathbf{I}_m] - \mathbf{u}'_s\mathbf{I}_{n+m} - \mathbf{w}'\mathbf{I}_{n+m} = \boldsymbol{\lambda}'\,[\mathbf{C},\mathbf{0}] + \mathbf{w}' \\ & \mathbf{u}\ \text{unrestricted and}\ \mathbf{w},\ \mathbf{u}_s \geq 0, \end{array}$$

where $\mathbf{u}_S \in \mathbf{R}^{n+m}$ is a (n+m)-dimensional vector.

The initial tableau of (M11.6) can be written as Table 11.3:

Table 11.3 The Simplex Tableau of (M11.6)

$\mathbf{u}$	$\mathbf{u}^1_s$	$\mathbf{u}^2_s$	$\mathbf{w}^1$	$\mathbf{w}^2$	RHS	
$\mathbf{A}^t$	$-\mathbf{I}_n$	$\mathbf{0}$	$-\mathbf{I}_n$	$\mathbf{0}$	$\mathbf{C}^t\lambda$	(11.5)
$\mathbf{I}_m$	$\mathbf{0}$	$-\mathbf{I}_m$	$\mathbf{0}$	$-\mathbf{I}_m$	$\mathbf{0}$	(11.5)
$\gamma^t\mathbf{D}^t$	$\mathbf{0}$	$\mathbf{0}$	$-(\beta^1)^t$	$-(\beta^2)^t$	$\mathbf{0}$	(11.6)

where $\mathbf{u}_s$, $\mathbf{w}$, and β are decomposed as $(\mathbf{u}^1_s, \mathbf{u}^2_s)$, $(\mathbf{w}^1, \mathbf{w}^2)$,and (β^1, β^2), respectively.

Given a PGS J with its basis $\mathbf{B}_J$ and the primal basic variables $\mathbf{x}(J)$, we partition $(\mathbf{u}, \mathbf{u}^1_s, \mathbf{u}^2_s)$ into $(\mathbf{u}, \mathbf{u}(J), \mathbf{u}(J'))$, where $\mathbf{u}$ are called *the dual basic variables corresponding to the primal constraints*, $\mathbf{u}(J)$ are called *the dual surplus variables complementary to* $\mathbf{x}(J)$, and $\mathbf{u}(J')$ are called the *dual surplus variables complementary to* $\mathbf{x}(J')$; and $(\mathbf{w}^1, \mathbf{w}^2)$ (= $\mathbf{w}$) into $(\mathbf{w}(J), \mathbf{w}(J'))$, where $\mathbf{w}(J)$ have m components of $\mathbf{w}$, which are the unit contribution of newly acquired units $\beta(J)$ of $\mathbf{x}(J)$ and $\mathbf{w}(J')$ are the n other components of $\mathbf{w}$ after excluding $\mathbf{w}(J)$ from $\mathbf{w}$. (β^1, β^2) are also partitioned into $(\beta(J), \beta(J'))$ associated with $(\mathbf{w}(J), \mathbf{w}(J'))$, respectively. Thus, Table 11.3 is rewritten as:

Table 11.4 The Simplex Tableau of (M11.6)

$\mathbf{u}$	$\mathbf{u}(J')$	$\mathbf{u}(J)$	$\mathbf{w}(J)$	$\mathbf{w}(J')$	RHS	
$(\mathbf{B}_J)^t$	$\mathbf{0}$	$-\mathbf{I}_m$	$-\mathbf{I}_m$	$\mathbf{0}$	$(\mathbf{C}_J)^t\lambda$	(11.5)
$\mathbf{R}^t$	$-\mathbf{I}_n$	$\mathbf{0}$	$\mathbf{0}$	$-\mathbf{I}_n$	$(\mathbf{C}_{J'})^t\lambda$	(11.5)
$\gamma^t\mathbf{D}^t$	$\mathbf{0}$	$\mathbf{0}$	$-\gamma^t\mathbf{e_p}\,\beta^t(J)$	$-\gamma^t\mathbf{e_p}\,\beta^t(J')$	$\mathbf{0}$	(11.6)

where $\mathbf{C}_J$ and $\mathbf{C}_{J'}$ are the coefficients of $[\mathbf{C}, \mathbf{0}]$ for $\mathbf{x}(J)$ and $\mathbf{x}(J')$, respectively.

It is well-known that in this case (see Bazaraa, Jarvis and Sherali 1990), the dual basis of model (M11.6) corresponding to the primal basis $\mathbf{B}_J$ is

$$\mathbf{B}_J^* = \begin{pmatrix} (\mathbf{B}_J)^t & \mathbf{0} \\ \mathbf{R}^t & -\mathbf{I}_n \end{pmatrix}$$

and the inverse matrix is

$$\left(\mathbf{B}_J^*\right)^{-1} = \begin{pmatrix} (\mathbf{B}_J^{-1})^t & \mathbf{0} \\ \mathbf{R}^t(\mathbf{B}_J^{-1})^t & -\mathbf{I}_n \end{pmatrix}.$$

Using this inverse matrix, Table 11.4 becomes Table 11.5.

Table 11.5 The Simplex Tableau of (M11.6)

u	u(J′)	u(J)	w(J)	w(J′)	RHS	
$-\mathbf{I}_m$	$\mathbf{0}$	$-(\mathbf{B}_J{}^{-1})^t$	$-(\mathbf{B}_J{}^{-1})^t$	$\mathbf{0}$	$(\mathbf{B}_J{}^{-1})^t\,(\mathbf{C}_J)^t\lambda$	(11.7)
$\mathbf{0}$	$-\mathbf{I}_n$	$-\mathbf{R}^t\,(\mathbf{B}_J{}^{-1})^t$	$-\mathbf{R}^t\,(\mathbf{B}_J{}^{-1})^t$	$-\mathbf{I}_n$	$[\mathbf{R}^t(\mathbf{B}_J{}^{-1})^t(\mathbf{C}_J)^t - (\mathbf{C}_{J'})^t]\,\lambda$	(11.7)
$\mathbf{0}$	$\mathbf{0}$	$\gamma^t\mathbf{D}^t\,(\mathbf{B}_J{}^{-1})^t$	$\gamma^t\mathbf{D}^t(\mathbf{B}_J{}^{-1})^t - \gamma^t\mathbf{e_p}\,\beta^t\,(J)$	$-\gamma^t\mathbf{e_p}\,\beta^t\,(J')$	$-\gamma^t\mathbf{D}^t\,(\mathbf{B}_J{}^{-1})^t\,\mathbf{C}^t_J\,\lambda$	(11.8)

where (11.7) = $((\mathbf{B}_{J^*})^{-1})^t\cdot$ (11.5), and (11.8) $=[-\gamma^t\mathbf{D}^t,\ \mathbf{0}]\cdot$ (11.7) + (11.6).

From the first row block of (11.7), we can read out $\mathbf{u}(J, \lambda) = (\mathbf{B}_J{}^{-1})^t(\mathbf{C}_J)^t\,\lambda$ (or $\lambda\mathbf{C}_J\,\mathbf{B}_J{}^{-1}$) which is a (D)-feasible solution for all $\lambda \in \Lambda(J)$ (see Remark 4.5). The optimal payoff value of the given design model (M9.1) is the negativity of $-\gamma^t\mathbf{D}^t(\mathbf{B}_J{}^{-1})^t(\mathbf{C}_J)^t\lambda$ (or $-\lambda\mathbf{C}_J\mathbf{B}_J{}^{-1}\,\mathbf{D}\gamma$) in (11.8).

Comparing Table (11.5) with Table (11.2), we see that Table (11.2) can be easily obtained from Table (11.5) by dropping the columns of $\mathbf{u}(J')$ and $\mathbf{w}(J')$; and the second row block of (11.7). Furthermore, by using the same pivoting sequence each basic feasible solution (which may contain some of u_j of $\mathbf{u}(J)$ and some w_i of $\mathbf{w}(J)$ as the basic variables) that can be derived from Table (11.2) can also be derived from Table (11.5).

Let U be a basis with $\mathbf{u}(U)$ and $\mathbf{w}(U)$ as the corresponding basic variables in $(\mathbf{u}, \mathbf{u}(J))$ and $\mathbf{w}(J)$ respectively. Let Q be another basis which also contains $\mathbf{u}(U)$ and $\mathbf{w}(U)$ as the basic variables. If U is a basis for model (M11.3), then both $(\mathbf{u}, \mathbf{u}(J)) \supseteq \mathbf{u}(U)$ and $\mathbf{w}(J) \supseteq \mathbf{w}(U)$ hold. This observation offers the following result:

Theorem 11.1. If U is a basis for model (M11.3), then U must be contained in a basis Q for model (M11.6). Conversely, if Q is a basis for model (M11.6) and Q contains a basis U for model (M11.3), then Q can be reduced to U by deleting the variables of $\mathbf{u}(J')$ and/or $\mathbf{w}(J')$ from Q. Furthermore, the basic feasible solution for model (M11.3) for a given λ, $(\mathbf{u}(U, \lambda), \mathbf{w}(U, \lambda))$, can be obtained by dropping the second row block of (11.7) from the MC²-simplex tableau of model (M11.6).

Let $\Lambda_7(U(J))$ be the primal parameter set for model (M11.6) and let $\Gamma_7(U(J, \beta))$ be the dual parameter set for model (M11.6). Recall that Table (11.5) has no less rows and columns than that of Table (11.2) in defining the primal and dual parameter sets. Thus, using Definition 11.1 and Theorem 11.1, we have:

Theorem 11.2 For any basis U for model (M11.3) and any basis Q that contains U for model (M11.6),

(i) $\Lambda_6(U(J)) \supseteq \Lambda_7(Q(J))$; and

(ii) $\Gamma_6(U(J, \beta)) \supseteq \Gamma_7(Q(J, \beta))$, where $\beta \in \mathbf{R}^{\mathbf{m}}$.

Remark 11.1 (i) Theorem 11.1 essentially says that model (M11.3) can be viewed as a submodel of model (M11.6). From Theorem 11.2, we see that if $\lambda \in \Lambda_7(Q(J))$, then basis U is feasible for model (M11.3) and Q is feasible for model (M11.6). When $\lambda \in \Lambda_6(U(J)) \backslash \Lambda_7(Q(J))$, U is feasible for model (M11.3) but Q is not feasible for model (M11.6). Similarly, if $\gamma \in \Gamma_7(Q(J, \beta))$, then basis U satisfies the optimality condition for model (M11.3) while Q is optimal for model (M11.6). If $\gamma \in \Gamma_6(U(J, \beta)) \backslash \Gamma_7(Q(J, \beta))$, then U satisfies the optimality condition for model (M11.3) but Q is not optimal for model (M11.6). Nevertheless, since Table (11.2) can be easily derived from Tableau (11.5), we can easily obtain $\Lambda_6(U(J))$ and $\Gamma_6(U(J, \beta))$ from model (M8.12) by dropping the second row block of (11.7) and the columns of $\mathbf{u}(J')$ and $\mathbf{w}(J')$.

(ii) The dual model (M11.2) is also a submodel of model (M11.6) because the simplex tableau of model (M11.2) can be obtained by dropping the columns of both $\mathbf{w}(J)$ and $\mathbf{w}(J')$ from Table (11.4).

Based on Theorems 11.1-11.2, and Remark 11.1, a method of effectively and systematically locating all DRCPs for all PGSs and selecting the final optimal linear system is proposed in the following procedure:

Algorithm 11.1

Step 1. Find all PGSs (potential solutions) for the original model (M9.1). Denote these solutions by $\vartheta = \{J_1, J_2, ..., J_r\}$. Without confusion, J will designate such a basis, the corresponding basic variables are denoted by $\mathbf{x}(J)$. Note that $\mathbf{x}(J)$ can contain slack variables $\mathbf{s}$.

Step 2. Find all possible bases for the augmented dual model (M11.6). Denote these bases by $\rho = \{Q_1, Q_2, ..., Q_t\}$. Without confusion, Q will designate such a basis, and the corresponding basic variables in $(\mathbf{u}, \mathbf{u}(J))$ and $\mathbf{w}(J)$ will be denoted by $\mathbf{u}(Q)$ and $\mathbf{w}(Q)$ respectively.

Step 3. For each PGS J of Step 1, identify its corresponding dual basis from ρ. Using (ii) of Remark 11.1, this can be done by dropping the columns of both $\mathbf{w}(J)$ and $\mathbf{w}(J')$ from Table (11.5) associated with PGS J. Denote such dual bases by $\{Q(J_1), Q(J_2), ..., Q(J_r)\}$. For each $Q(J)$, identify those $Q \in \rho$ of Step 2 which has $(\mathbf{u}, \mathbf{u}(J)) \supseteq \mathrm{u}(Q)$ (i.e., the set of basic variables $\mathbf{u}(Q)$ is contained in $(\mathbf{u}, \mathbf{u}(J))$ and $\mathbf{w}(J) \supseteq \mathbf{w}(Q)$. Denote such subsets of ρ by $\rho(J)$.

Step 4. For each PGS J, identify the primal bases of model (M11.3) from $\rho(J)$ of Step 3. This can be easily done by dropping the second row block of (11.7) from the simplex tableaus of $\rho(J)$. Denote such reduced bases by $\hat{\boldsymbol{u}}(J) = \{U_1(J), \ldots, U_k(J)\}$. Whenever $\lambda \in \Lambda_6(U_j(J))$, $(\mathbf{u}(U_j, \lambda), \mathbf{w}(U_j, \lambda))$ will offer a primal feasible solution for model (M11.3) with respect to system J.

Step 5. For each PGS J, with a given β^0 value, find the dual parameter set $\Gamma_6(U_j(J, \beta^0))$ for each $U_j(J)$ of $\hat{\boldsymbol{u}}(J)$ for model (M11.3). Recall that this can be easily done by dropping the columns of $\mathbf{u}(J')$ and $\mathbf{w}(J')$ from Table (11.4).

Step 6. Rearrange PGS J, $U_j(J) \in \hat{\boldsymbol{u}}(J)$, $\Lambda_6(U_j(J))$ and $\Gamma_6(U_j(J, \beta^0))$ in good order. Then when $\lambda \in \Lambda_6(U_j(J))$ and $\gamma \in \Gamma_6(U_j(J, \beta^0))$, $(\mathbf{u}(U_j, \lambda), \mathbf{w}(U_j, \lambda))$ is the DRCP for J.

Step 7. Evaluate each PGS J and its corresponding DRCPs in terms of a *criterion* preferred by decision makers. The final optimal system is selected based on the evaluation.

Example 11.1

Step 1. Given the following design problem:

$$Max \quad (\lambda_1 \quad \lambda_2)\begin{pmatrix} 6 & 2 \\ 1 & 1 \end{pmatrix}\begin{pmatrix} x_1 \\ x_2 \end{pmatrix}$$

$$Subject\ to \begin{pmatrix} 1 & 3 \\ 2 & 1 \end{pmatrix}\begin{pmatrix} x_1 \\ x_2 \end{pmatrix} \le \begin{pmatrix} 40 & 30 \\ 10 & 15 \end{pmatrix}\begin{pmatrix} \gamma_1 \\ \gamma_2 \end{pmatrix}$$

$$x_j \ge 0, \quad j=1,\ 2.$$

Let s_1 and s_2 be slack variables for constraints 1 and 2, respectively. Then using the computer software of the MC^2-simplex method (Hao and Shi 1996), we find the set of all PGSs, denoted by $\vartheta = \{J_1, J_2, J_3\}$ as shown in Table 11.6.

Table 11.6 Potential Good Systems

PGS	Bases	$\Gamma_1(J_i)$	$\Lambda_1(J_i)$
J_1	(x_1, x_2)	$0 < \gamma_1 \le 3/5$	$0 < \lambda_1 \le 1/3$
J_2	(x_1, s_1)	$0 < \gamma_1 \le 1$	$1/3 \le \lambda_1 \le 1$
J_3	(x_2, s_1)	$3/5 < \gamma_1 \le 1$	$0 < \lambda_1 \le 1/3$

Step 2. Let β_i be the increments of the ith product and its unit contribution be w_i, i = 1, 2, 3, 4. Then the corresponding augmented dual model is as follows: we solve the corresponding augmented problem:

$$Min \quad \begin{pmatrix}\gamma_1 & \gamma_2\end{pmatrix} \begin{pmatrix}40 & 10\\ 30 & 15\end{pmatrix} \begin{pmatrix}u_1\\ u_2\end{pmatrix} - \begin{pmatrix}\beta_1 & \beta_2 & \beta_3 & \beta_4\end{pmatrix} \begin{pmatrix}w_1\\ w_2\\ w_3\\ w_4\end{pmatrix}$$

$$Subject\ to \begin{pmatrix}1 & 2\\ 3 & 1\\ 1 & 0\\ 0 & 1\end{pmatrix} \begin{pmatrix}u_1\\ u_2\end{pmatrix} \geq \begin{pmatrix}6 & 1\\ 2 & 1\\ 0 & 0\\ 0 & 0\end{pmatrix} \begin{pmatrix}\lambda_1\\ \lambda_2\end{pmatrix} + \begin{pmatrix}1 & 0 & 0 & 0\\ 0 & 1 & 0 & 0\\ 0 & 0 & 1 & 0\\ 0 & 0 & 0 & 1\end{pmatrix} \begin{pmatrix}w_1\\ w_2\\ w_3\\ w_4\end{pmatrix}$$

u_j, $j = 1, 2$ are unrestricted; and $w_i \geq 0$, $i = 1, 2, 3, 4$.

Let $u_j \geq 0$, j = 3, 4, 5, 6, be surplus variables. Then by using the computer program of Hao and Shi (1996), we solve the above problem and obtain 85 bases, denoted by $\rho = \{Q_1, \ldots, Q_{85}\}$.

Step 3. For each J_i of ϑ, i = 1, 2, 3, we identify their corresponding dual bases from ρ by dropping the columns of both $\mathbf{w}(J_i)$ and $\mathbf{w}(J'_i)$ from Table (11.4) associated with PGS J_i. We denote the resulting dual bases by $\{Q(J_1), Q(J_2), Q(J_3)\}$, where $Q(J_1) = Q_8$, $Q(J_2) = Q_{77}$, and $Q(J_3) = Q_{46}$.

We also find that
for $Q(J_1)$, $(\mathbf{u}, \mathbf{u}(J_1)) = \{u_1, u_2, u_5, u_6\}$ and $\mathbf{w}(J_1) = \{w_1, w_2\}$;
for $Q(J_2)$, $(\mathbf{u}, \mathbf{u}(J_2)) = \{u_1, u_2, u_4, u_6\}$ and $\mathbf{w}(J_2) = \{w_1, w_3\}$; and
for $Q(J_3)$, $(\mathbf{u}, \mathbf{u}(J_3)) = \{u_1, u_2, u_3, u_6\}$ and $\mathbf{w}(J_3) = \{w_2, w_3\}$.

Then for each $Q(J_i)$, i = 1, 2, 3, we identify those $Q \in \rho$ of Step 2 for which has $(\mathbf{u}, \mathbf{u}(J_i)) \supseteq \mathbf{u}(Q)$ and $\mathbf{w}(J_i) \supseteq \mathbf{w}(Q)$ as in Tables 11.7-11.9, respectively. Note that we denote such subsets of ρ by $\rho(J_i)$.

Step 4. For each PGS J_i, i = 1, 2, 3, we identify the primal bases of Model (M11.6) from $\rho(J_i)$ of Step 3 by dropping the second row block of (11.7) from the simplex tableaus of $\rho(J_i)$. Then we have:

$\hat{u}(J_1) = \{U_8(J_1), U_{32}(J_1), U_{33}(J_1), U_{34}(J_1)\}$;

$\hat{u}(J_2) = \{U_{77}(J_2), U_{38}(J_2), U_{39}(J_2), U_{70}(J_2)\}$; and

$\hat{u}(J_3) = \{U_{46}(J_3), U_{32}(J_3), U_{37}(J_3), U_{38}(J_3)\}$. Note that the basis $U_8(J_1)$ with basic variables (u_5, u_6) is obtained by deleting (u_1, u_2), which corresponds the first row block of (11.7) in the MC^2-simplex tableau, from U_8 in Table 11.7. The other primal bases of Model (M11.3) can be similarly explained.

Table 11.7 The Subset $\rho(J_1)$

Bases	Variables	Bases	Variables
Q_8	(u_1, u_2, u_5, u_6)	Q_{33}	(u_1, u_2, w_1, u_6)
Q_{18}	(u_1, w_2, u_5, u_6)	Q_{34}	(u_1, u_2, w_2, u_6)
Q_{19}	(u_1, w_1, u_5, u_6)	Q_{63}	(u_2, w_1, u_5, u_6)
Q_{25}	(u_1, w_1, w_2, u_6)	Q_{64}	(u_2, w_2, u_5, u_6)
Q_{27}	(u_1, u_2, w_1, w_2)	Q_{66}	(u_2, w_1, w_2, u_5)
Q_{31}	(u_1, u_2, w_1, u_5)	Q_{75}	(w_1, w_2, u_5, u_6)
Q_{32}	(u_1, u_2, w_2, u_6)		

Table 11.8 The Subset $\rho(J_2)$

Bases	Variables	Bases	Variables
Q_{77}	(u_1, w_3, u_4, u_6)	Q_{33}	(u_1, u_2, w_1, u_6)
Q_9	(u_1, u_2, u_4, u_6)	Q_{38}	(u_1, u_2, w_3, u_6)
Q_{10}	(u_1, w_3, u_4, u_6)	Q_{39}	(u_1, u_2, w_3, u_4)
Q_{20}	(u_1, w_1, u_4, u_6)	Q_{69}	(u_2, w_1, w_3, u_4)
Q_{24}	(u_1, w_1, w_3, u_6)	Q_{70}	$u_2, w_1, w_3, u_6)$
Q_{28}	(u_1, u_2, w_1, w_3)	Q_{73}	(w_1, w_3, u_4, u_6)
Q_{30}	(u_1, u_2, w_1, u_4)		

Table 11.9 The Subset $\rho(J_3)$

Bases	Variables	Bases	Variables
Q_{46}	(u_1, u_2, u_3, u_6)	Q_{37}	(u_1, u_2, w_2, w_3)
Q_{11}	(u_1, w_2, u_3, u_6)	Q_{38}	(u_1, u_2, w_3, u_6)
Q_{12}	(u_1, u_2, w_5, u_3)	Q_{47}	(u_2, w_3, u_3, u_6)
Q_{16}	(u_1, w_2, w_3, u_3)	Q_{51}	(u_2, w_2, w_3, u_3)
Q_{17}	(u_1, w_2, u_3, u_6)	Q_{52}	(u_2, w_2, w_3, u_6)
Q_{32}	(u_1, u_2, w_2, u_6)	Q_{82}	(w_2, w_3, u_3, u_6)
Q_{35}	(u_1, u_2, w_2, u_3)		

Steps 5-6. For each PGS J_i, i = 1, 2, 3, we find $\Lambda_6(U_j(J))$ and $\Gamma_6(U_j(J, \beta^0))$, which is a function of β^0, for each $U(J_i)$ of $\hat{\boldsymbol{u}}(J_i)$ for Model (M11.3). This can be easily done by dropping the columns of $\mathbf{u}(J')$ and $\mathbf{w}(J')$ from Table (11.4) in the related MC^2-simplex tableaus of Model (M11.6). For illustrative purposes, let $\beta^0_i = 3$, i = 1, 2, 3. We obtain all DRCPs, denoted by $\boldsymbol{u}^*(J_1) = \{U_{32}(J_1), U_{33}(J_1)\}$ for J_1 as in Table 11.10; $\boldsymbol{u}^*(J_2) = \{U_{77}(J_2), U_{38}(J_2), U_{70}(J_2)\}$ for J_2 as in Table 11.11; and $\boldsymbol{u}^*(J_3) = \{U_{46}(J_3), U_{32}(J_3), U_{38}(J_3)\}$ for J_3 as in Table 11.12. The graphical representation of Tables 11.10 is Figure 11.1.

Table 11.10, when compared with Table 11.6, shows that if J_1 is chosen for commitment, values of (γ_1, λ_1) vary in two possible ranges:

(i) when $0 < \lambda_1 \leq 1/3$ and $0 < \gamma_1 \leq 1$, we use basis $U_{32}(J_1)$ to produce x_1 and x_2, in which x_1 has increased three units (i.e., $\beta^0_i = 3$) and the corresponding increment of the unit contribution is $w_1 = -2\lambda_1 + \lambda_2$ from the related MC^2-simplex of $U_{32}(J_1)$ in Model (M11.3), while x_2 and its unit contribution have no adjustment; and

(ii) when $1/3 < \lambda_1 \leq 1$ and $0 < \gamma_1 \leq 1$, we use basis $U_{33}(J_1)$ to produce x_1 and x_2, in which x_2 has increased three units (i.e., $\beta^0_i = 3$) and the corresponding increment of the unit contribution is $w_2 = \lambda_1 - .5\lambda_2$ from the MC^2-simplex of $U_{33}(J_1)$ in Model (M11.3), while x_1 and its unit contribution have no adjustment.

The meaning of Tables 11.12-11.13 can be similarly explained.

Table 11.10 Dual Rigid Contingency Plans for J_1 with $\beta^0 = (3, 3, 3)^t$

Bases	$\Gamma_6(U_j(J_1))$	$\Lambda_6(U_j(J_1))$	$\mathbf{V}(U_j(J_1))$
$U_{32}(J_1)$ $= (w_1, u_6)$	$0 < \gamma_1 \leq 1$	$0 < \lambda_1 \leq 1/3$	$(\lambda_1 \quad \lambda_2)\begin{pmatrix}26 & 36\\ 7 & 12\end{pmatrix}\begin{pmatrix}\gamma_1\\ \gamma_2\end{pmatrix}$
$U_{33}(J_1)$ $= (w_2, u_6)$	$0 < \gamma_1 \leq 1$	$1/3 < \lambda_1 \leq 1$	$(\lambda_1 \quad \lambda_2)\begin{pmatrix}27 & 42\\ 6.5 & 9\end{pmatrix}\begin{pmatrix}\gamma_1\\ \gamma_2\end{pmatrix}$

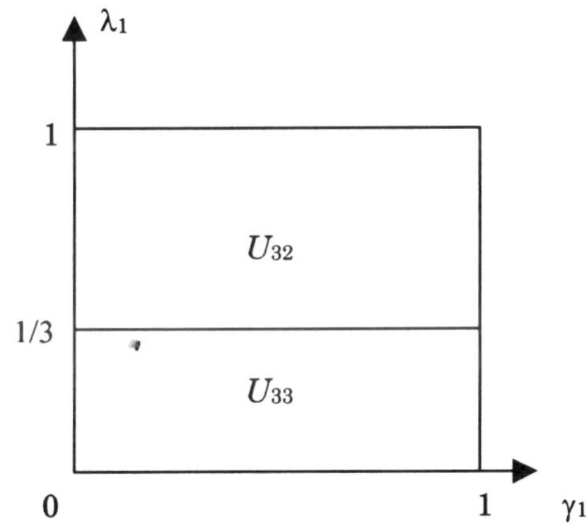

Figure 11.1 Dual Rigid Contingency Plans for J_1

Table 11.11 Dual Rigid Contingency Plans for J_2 with $\beta^0 = (3, 3, 3)^t$

Bases	$\Gamma_6(U_j(J_2))$	$\Lambda_6(U_j(J_2))$	$\mathbf{V}(U_j(J_2))$
$U_{77}(J_2) =$ (u_1, u_6)	$0 < \gamma_1 \leq 1$	$1/3 < \lambda_1 \leq 1/3$	$(\lambda_1 \quad \lambda_2)\begin{pmatrix} 30 & 45 \\ 5 & 7.5 \end{pmatrix}\begin{pmatrix} \gamma_1 \\ \gamma_2 \end{pmatrix}$
$U_{38}(J_1) =$ (u_1, u_6)	$0<\gamma_1 \leq 3/25$	$0 < \lambda_1 \leq 1/3$	$(\lambda_1 \quad \lambda_2)\begin{pmatrix} 17.2 & 37.2 \\ 11.4 & 11.4 \end{pmatrix}\begin{pmatrix} \gamma_1 \\ \gamma_2 \end{pmatrix}$
$U_{70}(J_2) =$ (w_1, u_6)	$3/25<\gamma_1 \leq 1$	$0 < \lambda_1 \leq 1/3$	$(\lambda_1 \quad \lambda_2)\begin{pmatrix} 26 & 36 \\ 7 & 12 \end{pmatrix}\begin{pmatrix} \gamma_1 \\ \gamma_2 \end{pmatrix}$

Table 11.12 Dual Rigid Contingency Plans for J_3 with $\beta^0 = (3, 3, 3)^t$

Bases	$\Gamma_6(U_j(J_3))$	$\Lambda_6(U_j(J_3))$	$\mathbf{V}(U_j(J_3))$
$U_{46}(J_3) =$ (u_3, u_6)	$18/25<\gamma_1\leq 1$	$0 < \lambda_1 \leq 1$	$(\lambda_1 \quad \lambda_2)\begin{pmatrix} 20 & 30 \\ 10 & 15 \end{pmatrix}\begin{pmatrix} \gamma_1 \\ \gamma_2 \end{pmatrix}$
$U_{32}(J_3) =$ (w_2, u_6)	$0<\gamma_1 \leq 1$	$1/3 < \lambda_1 \leq 1$	$(\lambda_1 \quad \lambda_2)\begin{pmatrix} 27 & 42 \\ 6.5 & 9 \end{pmatrix}\begin{pmatrix} \gamma_1 \\ \gamma_2 \end{pmatrix}$
$U_{36}(J_3) =$ (w_3, u_6)	$0<\gamma_1\leq 18/25$	$0 < \lambda_1 \leq 1/3$	$(\lambda_1 \quad \lambda_2)\begin{pmatrix} 17.2 & 37.2 \\ 11.4 & 11.4 \end{pmatrix}\begin{pmatrix} \gamma_1 \\ \gamma_2 \end{pmatrix}$

Step 7. For illustrative purposes, let us use the *maximizing expected payoff* as the criterion to select the final optimal designs from $\boldsymbol{u}^*(J_1)$, $\boldsymbol{u}^*(J_2)$ and $\boldsymbol{u}^*(J_3)$.

Let the initial setup cost for using J, $f(J) = 0$. We assume that γ_1 is independent on λ_1, and (γ_1, λ_1) have the bi-uniform probability distribution $F(\gamma_1, \lambda_1)$. Given $\boldsymbol{u}^*(J_1)$ for PGS J_1, to find the expected payoff $\mathbf{EV}(J_1)$, we compute:

$$\mathbf{EV}(J_1) = \int_{S_1} \mathbf{V}(U_{32}(J_1))\, dF(\gamma_1, \lambda_1) + \int_{S_2} \mathbf{V}(U_{33}(J_1))\, dF(\gamma_1, \lambda_1)$$

where $\mathbf{V}(U_{32}(J_1))$ and $\mathbf{V}(U_{33}(J_1))$; and the (γ_1, λ_1) ranges $S_j = \Gamma_6(U_j(J_1)) \times \Lambda_6(U_j(J_1))$, $j = 1, 2$, are shown in Table 11.10.

Then, we get $\mathbf{EV}(J_1) = 39.77$. Similarly, we can get:
$\mathbf{EV}(J_2) = 38.50$ and $\mathbf{EV}(J_3) = 61.46$.

Thus in view of *maximizing expected payoff*, J_3 is the best candidate and can be recommended to the decision maker as the final optimal system, where the set of all DRCPs for J_3 is $\boldsymbol{u}^*(J_3) = \{U_{46}(J_3), U_{32}(J_3), U_{38}(J_3)\}$ in Table 11.12.

11.2 Optimal System Design and Dual Flexible Contingency Plans

In this section, we explore how to construct *dual flexible contingency plans* (DFCP) for a given PGS, in which the number of selected opportunities, the contribution of the opportunities, and the units of all possible slack resources are flexibly adjusted to meet the optimality (Shi 1995).

To guarantee the DFCP is meaningful in the design process, we need the following definition:

Definition 11.3 A PGS J is said to be a *non-trivial* PGS if J contains at least one opportunity i.

From Definition 11.3, when a PGS J does not contain any opportunities (that is, J is an initial potential basis) J is trivial. To construct DFCPs for the optimal systems, we should consider the set of all non-trivial PGSs.

Recall Section 9.3, given PGS J, let the corresponding non-basic variables of J be $\mathbf{x}(J')$. We decompose $\mathbf{x}(J')$ into $\mathbf{x}^1(J')$ and $\mathbf{x}^2(J')$, where $\mathbf{x}^1(J')$ is the set of non-basic variables consisting of those slack variables that are not selected in J, and $\mathbf{x}^2(J')$ is the set of non-basic variables consisting of those opportunity variables of N that are not selected in J. Then, in addition to considering $\mathbf{x}(J)$ in the designing process, we also use the slack resource $\mathbf{x}^1(J')$. A flexible design process can be $\mathbf{B}_J\,\mathbf{x}(J)$ +

$\mathbf{R}^1\mathbf{x}^1(J')$ = $\mathbf{D}\gamma$, where $\mathbf{R}^1$ is the submatrix of [**A**, **I**] associated with $\mathbf{x}^1(J')$. A primal model to reflect this process is (M11.7):

$$
\begin{array}{ll}
\textit{Max} & \lambda^t \mathbf{C}_J\ \mathbf{x}(J) \\
\textit{Subject to} & \mathbf{B}_J\ \mathbf{x}(J) + \mathbf{R}^1\ \mathbf{x}^1(J') = \mathbf{D}\gamma \\
 & \mathbf{x}(J),\ \mathbf{x}^1(J') \geq 0.
\end{array}
$$

The dual model of model (M11.7) is (M11.8):

$$
\begin{array}{ll}
\textit{Min} & \mathbf{u}^t\ \mathbf{D}\gamma \\
\textit{Subject to} & \mathbf{u}^t\ [\mathbf{B}_J, \mathbf{R}^1] \geq \lambda^t\ [\mathbf{C}_J, \mathbf{0}] \\
 & \mathbf{u} \text{ unrestricted.}
\end{array}
$$

Let d be the number of nonbasic slack variables in $\mathbf{x}^1(J')$. Then n - d *is* the number of nonbasic slack variables in $\mathbf{x}^2(J')$, since $\mathbf{x}(J')$ has n non-basic variables. Suppose that model (M11.8) has no optimal solution for some λ. To overcome this situation, let $\beta(J) = (\beta_1, \ldots, \beta_m)^t$ be the column vector of increments of the m selected opportunities $\mathbf{x}(J)$ and $\beta^1(J') = (\beta_{m-1}, \ldots, \beta_{m-d})^t$ be the column vector of increments of $\mathbf{x}^1(J')$. Let the unit contribution of these newly acquired $\beta(J)$ be $\mathbf{w}(J) = (w_1, \ldots, w_m)^t$, while $\beta^1(J')$ is worth $\mathbf{w}^1(J') = (w_{m-1}, \ldots, w_{m-d})^t$. Then we add these iterms into model (M11.8). Note that $\mathbf{u}^1(J')$ is the *dual surplus variables complementary to* $\mathbf{x}^1(J')$ and $\mathbf{u}(J)$ is the *dual surplus variables complementary to* $\mathbf{x}(J)$. Considering $\mathbf{u}^1(J')$ and $\mathbf{u}(J)$, model (M11.8) becomes (M11.9):

$$
\begin{array}{ll}
\textit{Min} & \mathbf{u}^t\ \mathbf{D}\gamma - (\boldsymbol{\beta}^1(J'))^t\ \mathbf{w}^1(J') - (\boldsymbol{\beta}(J))^t\ \mathbf{w}(J) \\
\textit{Subject to} & \mathbf{u}^t\ [\mathbf{B}_J, \mathbf{R}^1] - (\mathbf{u}^1(J'))^t [\mathbf{0}, \mathbf{I}_d] - \mathbf{u}(J)[\mathbf{I}_m, \mathbf{0}] \\
 & \quad = \lambda^t\ [\mathbf{C}_J, \mathbf{0}] + (\mathbf{w}^1(J'))^t - (\mathbf{w}(J))^t \\
 & \mathbf{u} \text{ unrestricted}; \mathbf{u}^1(J'), \mathbf{u}(J), \mathbf{w}^1(J'), \mathbf{w}(J) \geq \mathbf{0}.
\end{array}
$$

The initial tableau of the above model is

Table 11.13 The Simplex Tableau of (M11.9)

$\mathbf{u}$	$\mathbf{u}^1(J')$	$\mathbf{u}(J)$	$\mathbf{w}(J)$	$\mathbf{w}^1(J')$	RHS	
$(\mathbf{B}_J)^t$	$\mathbf{0}$	$-\mathbf{I}_m$	$-\mathbf{I}_m$	$\mathbf{0}$	$(\mathbf{C}_J)^t\lambda$	(11.9)
$(\mathbf{R}^1)^t$	$-\mathbf{I}_d$	$\mathbf{0}$	$\mathbf{0}$	$-\mathbf{I}_d$	$(\mathbf{C}_{J'})^t\lambda$	(11.9)
$\gamma^t\mathbf{D}^t$	$\mathbf{0}$	$\mathbf{0}$	$-\gamma^t\mathbf{e}_{p}(\beta(J))^t$	$-\gamma^t\mathbf{e}_{p}(\beta(J'))^t$	$\mathbf{0}$	(11.10)

The dual basis of model (M11.9) corresponding to the primal basis $\mathbf{B}_J$ is

$$\mathbf{B}_J^{**} = \begin{pmatrix} (\mathbf{B}_J)^t & \mathbf{0} \\ (\mathbf{R}^1)^t & -\mathbf{I}_d \end{pmatrix}$$

and the inverse matrix is

$$\left(\mathbf{B}_J^{**}\right)^{-1} = \begin{pmatrix} (\mathbf{B}_J^{-1})^t & \mathbf{0} \\ (\mathbf{R}^1)^t(\mathbf{B}_J^{-1})^t & -\mathbf{I}_d \end{pmatrix}.$$

Applying the above inverse matrix, Table 11.13 is written as Table 11.14.

Table 11.14 The Simplex Tableau of (M11.9)

u	$u^1(J')$	$u(J)$	$w(J)$
$-I_m$	0	$-(B_J^{-1})^t$	$-(B_J^{-1})^t$
0	$-I_d$	$-(R^1)^t(B_J^{-1})^t$	$-(R^1)^t(B_J^{-1})^t$
0	0	$\gamma^t D^t(B_J^{-1})^t$	$\gamma^t D^t(B_J^{-1})^t - \gamma^t e_p(\beta(J))^t$

$w^1(J')$	RHS	
0	$(B_J^{-1})^t(C_J)^t\lambda$	(11.11)
$-I_d$	$[R^t(B_J^{-1})^t(C_J)^t-(C_{J'})^t]\lambda$	(11.11)
$-\gamma^t e_p(\beta^1(J'))^t$	$-\gamma^t D^t(B_J^{-1})^t(C_J)^t\lambda$	(11.12)

where (11.11) = $(B_J^{**})^{-1}$ · (8.27), and (8.30) =$[-\gamma^t D^t, \mathbf{0}]$ · (11.11) + (11.10).

Definition 11.4 Given a basis $U(J)$ for model (M11.8), define its corresponding

(i) *primal parameter set* by

$\Lambda_8(U(J)) = \{\lambda > 0 \mid \lambda^t C_J B_J^{-1} \geq 0\}$; and

(ii) *dual parameter set* by

$\Gamma_8(U(J), \beta(J), \beta^1(J))$

$$= \{\gamma > 0 \mid [B_J^{-1} D, -\beta(J)(e_p)^t + B_J^{-1} D, -\beta^1(J)(e_p)^t]\gamma \geq 0\}.$$

Definition 11.5. For a given PGS J, the corresponding dual basis $U(J)$ is said to be a dual flexible contingency plan (DFCP) of J if $\Lambda_8(U(J)) \times \Gamma_8(U(J), \beta(J), \beta^1(J)) \neq \emptyset$ for some $\beta(J)$ and $\beta^1(J)$.

From Definition 11.5, given a non-trivial PGS J, the set of all potential bases $\{U_1(J), \ldots, U_s(J)\}$ obtained from model (M11.8) by using the MC²-simplex method is called the set of DFCPs for PGS J. We also can use the augmented dual model (M11.5) to locate all DFCPs for each of J in ϑ.

After putting surplus varaibles $\mathbf{u}_s$ into model (M11.5), we have model (M11.6) that produces Table 11.3. To decompose Table 11.3 for the analysis, we do the following variable partitions:

Given a non-trivial PGD J with its basis $\mathbf{B}_J$, and the primal basic variables $\mathbf{x}(J)$, we partition $(\mathbf{u}, \mathbf{u}^1_s, \mathbf{u}^2_s)$ into $(\mathbf{u}, \mathbf{u}(J), \mathbf{u}^1(J'), \mathbf{u}^2(J'))$, where $\mathbf{u}$ are the *dual variables corresponding to the primal constraints;* $\mathbf{u}(J)$ are the *dual surplus variables complementary to* $\mathbf{x}(J)$; $\mathbf{u}^1(J')$ are the *dual surplus variables complementary to* $\mathbf{x}^1(J')$; and $\mathbf{u}^2(J')$ are the *dual surplus variables complementary to* $\mathbf{x}^2(J')$. Correspondingly, we also partition $(\mathbf{w}^1, \mathbf{w}^2)$ $(= \mathbf{w})$ into $(\mathbf{w}(J), \mathbf{w}^1(J'), \mathbf{w}^2(J'))$, and (β^1, β^2) into $(\beta(J), \beta^1(J'), \beta^2(J'))$, where $\mathbf{w}(J)$ are the unit contributions of newly acquired units $\beta(J)$ of $\mathbf{x}(J)$; $\mathbf{w}^1(J')$ are the unit price of newly saved units $\beta^1(J')$ of $\mathbf{x}^1(J')$; and $\mathbf{w}^2(J')$ are the remaining n - d components of $\mathbf{w}$ associated with $\beta^2(J')$, after excluding $\mathbf{w}(J)$ and $\mathbf{w}^1(J')$ from $\mathbf{w}$. Then, Table 11.3 can be rewritten as Table 11.15.

Table 11.15 The Simplex Tableau of (M11.6)

$\mathbf{u}$	$\mathbf{u}^1(J')$	$\mathbf{u}^2(J')$	$\mathbf{u}(J)$	$\mathbf{w}(J)$	$\mathbf{w}^1(J')$	$\mathbf{w}^2(J')$	RHS	
$(\mathbf{B}_J)^t$	$\mathbf{0}$	$\mathbf{0}$	$-\mathbf{I}_m$	$-\mathbf{I}_m$	$\mathbf{0}$	$\mathbf{0}$	$(\mathbf{C}_J)^t\lambda$	(11.13)
$(\mathbf{R}^1)^t$	$-\mathbf{I}_d$	$\mathbf{0}$	$\mathbf{0}$	$\mathbf{0}$	$-\mathbf{I}_d$	$\mathbf{0}$	$(\mathbf{C}^1{}_{J'})^t\lambda$	(11.13)
$(\mathbf{R}^2)^t$	$\mathbf{0}$	$-\mathbf{I}_{n-d}$	$\mathbf{0}$	$\mathbf{0}$	$\mathbf{0}$	$-\mathbf{I}_{n-d}$	$(\mathbf{C}^2{}_{J'})^t\lambda$	(11.13)
$\gamma^t\mathbf{D}^t$	$\mathbf{0}$	$\mathbf{0}$	$\mathbf{0}$	$-\gamma^t\mathbf{e_p}\,\beta(J)^t$	$-\gamma^t\mathbf{e_p}(\beta^1(J'))^t$	$-\gamma^t\mathbf{e_p}(\beta^2(J'))^t$	$\mathbf{0}$	(11.14)

where $\mathbf{C}_J$, $\mathbf{C}^1{}_{J'}$ and $\mathbf{C}^2{}_{J'}$ are the coefficients of $[\mathbf{C}, \mathbf{0}]$ for $\mathbf{x}(J)$, $\mathbf{x}^1(J')$ and $\mathbf{x}^2(J')$, respectively.

The dual basis of the above format of model (M11.6) corresponding to the primal basis $\mathbf{B}_J$ is

$$\mathbf{B}_J^{***} = \begin{pmatrix} (\mathbf{B}_J)^t & \mathbf{0} & \mathbf{0} \\ (\mathbf{R}^1)^t & -\mathbf{I}_d & \mathbf{0} \\ (\mathbf{R}^2)^t & \mathbf{0} & -\mathbf{I}_{n-d} \end{pmatrix}$$

and the inverse matrix is

$$\left(\mathbf{B}_J^{***}\right)^{-1} = \begin{pmatrix} (\mathbf{B}_J^{-1})^t & \mathbf{0} & \mathbf{0} \\ (\mathbf{R}^1)^t (\mathbf{B}_J^{-1})^t & -\mathbf{I}_d & \mathbf{0} \\ (\mathbf{R}^2)^t (\mathbf{B}_J^{-1})^t & \mathbf{0} & -\mathbf{I}_{n-d} \end{pmatrix}.$$

Using this inverse matrix, Table 11.15 is written as Table 11.16.

Table 11.16 The Simplex Tableau of (M11.6)

$\mathbf{u}$	$\mathbf{u}^1(J')$	$\mathbf{u}^2(J')$	$\mathbf{u}(J)$	$\mathbf{w}(J)$
$-\mathbf{I}_m$	$\mathbf{0}$	$\mathbf{0}$	$-(\mathbf{B}_J^{-1})^t$	$-(\mathbf{B}_J^{-1})^t$
$\mathbf{0}$	$-\mathbf{I}_d$	$\mathbf{0}$	$-(\mathbf{R}^1)^t(\mathbf{B}_J^{-1})^t$	$-(\mathbf{R}^1)^t(\mathbf{B}_J^{-1})^t$
$\mathbf{0}$	$\mathbf{0}$	$-\mathbf{I}_{n-d}$	$-(\mathbf{R}^2)^t(\mathbf{B}_J^{-1})^t$	$-(\mathbf{R}^2)^t(\mathbf{B}_J^{-1})^t$
$\mathbf{0}$	$\mathbf{0}$	$\mathbf{0}$	$\gamma^t\mathbf{D}^t(\mathbf{B}_J^{-1})^t$	$\gamma^t\mathbf{D}^t(\mathbf{B}_J^{-1})^t - \gamma^t\mathbf{e_p}(\beta(J))^t$

$\mathbf{w}^1(J')$	$\mathbf{w}^2(J')$	RHS	
$\mathbf{0}$	$\mathbf{0}$	$(\mathbf{B}_J^{-1})^t(\mathbf{C}_J)^t\lambda$	(11.15)
$-\mathbf{I}_d$	$\mathbf{0}$	$[(\mathbf{R}^1)^t(\mathbf{B}_J^{-1})^t(\mathbf{C}_J)^t - (\mathbf{C}^1_{J'})^t]\lambda$	(11.15)
$\mathbf{0}$	$-\mathbf{I}_{n-d}$	$[(\mathbf{R}^2)^t(\mathbf{B}_J^{-1})^t(\mathbf{C}_J)^t - (\mathbf{C}^2_{J'})^t]\lambda$	(11.15)
$-\gamma^t\mathbf{e_p}(\beta^1(J'))^t$	$-\gamma^t\mathbf{e_p}(\beta^2(J'))^t$	$-\gamma^t\mathbf{D}^t(\mathbf{B}_J^{-1})^t(\mathbf{C}_J)^t\lambda$	(11.16)

where (11.15) = $(\mathbf{B}_J^{***})^{-1}$ · (11.13), and (11.16) $=[-\gamma^t\mathbf{D}^t, \mathbf{0}, \mathbf{0}]$ · (11.15) + (11.14).

Comparing Table (11.14) with Table (11.16), we see that Table (11.14) can be easily obtained from Table (11.16) by dropping the columns of $\mathbf{u}^2(J')$ and $\mathbf{w}^2(J')$; and the third row block of (11.15). Furthermore, by using the same pivoting sequence each basic feasible solution (which may contain some of u_j of $\mathbf{u}(J)$ and $\mathbf{u}^1(J')$, and some w_i of $\mathbf{w}(J)$ and $\mathbf{w}^1(J')$ as the basic variables) that can be derived from Table (11.14) can also be derived from Table (11.16).

Let U be a basis with $\mathbf{u}(U)$ and $\mathbf{w}(U)$ as the corresponding basic variables in $(\mathbf{u}, \mathbf{u}(J), \mathbf{u}^1(J'))$ and $(\mathbf{w}(J), \mathbf{w}^1(J'))$ respectively. Let Q be another basis which also contains $\mathbf{u}(U)$ and $\mathbf{w}(U)$ as the basic variables. If U is a basis for model (M11.9), then both $(\mathbf{u}, \mathbf{u}(J), \mathbf{u}^1(J')) \supseteq \mathbf{u}(U)$ and $(\mathbf{w}(J), \mathbf{w}^1(J')) \supseteq \mathbf{w}(U)$ hold. This leads to the following result:

Theorem 11.3. If U is a basis for model (M11.9), then U must be contained in a basis Q for model (M11.6). Conversely, if Q is a basis for model (M11.6) and Q contains a basis U for model (M11.9), then Q can be reduced to U by deleting the variables of $\mathbf{u}^2(J')$ and/or $\mathbf{w}^2(J')$ from Q. Furthermore, the basic feasible solution for model (M11.9) for a given λ, $(\mathbf{u}(U, \lambda), \mathbf{w}(U, \lambda))$, can be obtained by dropping the third row block of (11.15) from the MC^2-simplex tableau of model (M11.6).

Let $\Lambda_9(U(J))$ be the primal parameter set for model (M11.6) in Table 11.16 and let $\Gamma_9(U(J, \beta, \beta^1))$ be the dual parameter set for model (M11.6) in Table 11.16. Using Definition 11.4 and Theorem 11.3, we have:

Theorem 11.4 For any basis U for model (M11.9) and any basis Q that contains U for model (M11.6),

(i) $\Lambda_8(U(J)) \supseteq \Lambda_9(Q(J))$; and

(ii) $\Gamma_8(U(J, \beta, \beta^1)) \supseteq \Gamma_9(Q(J, \beta, \beta^1))$.

Theorems 11.3-11.4 provide a method of effectively and systematically locating all DFCPs for all PGSs:

Algorithm 11.2

Step 1. Find all non-trivial PGSs (potential solutions) for the original model (M9.1). Denote these solutions by $\vartheta = \{J_1, J_2, ..., J_r\}$. Without confusion, J will designate such a basis, the corresponding basic variables are denoted by $\mathbf{x}(J)$. Note that $\mathbf{x}(J)$ can contain slack variables $\mathbf{s}$.

Step 2. Find all possible bases for the augmented dual model (M11.6). Denote these bases by $\rho = \{Q_1, Q_2,...., Q_t\}$. Without confusion, Q will designate such a basis, and the corresponding basic variables in $(\mathbf{u}, \mathbf{u}(J), \mathbf{u}^1(J'))$ and $(\mathbf{w}(J), \mathbf{w}^1(J'))$ will be denoted by $\mathbf{u}(Q)$ and $\mathbf{w}(Q)$ respectively.

Step 3. For each PGS J of Step 1, identify its corresponding dual basis from ρ. This can be done by dropping the columns of $\mathbf{w}(J)$, $\mathbf{w}^1(J')$ and $\mathbf{w}^2(J')$ from Table (11.16) associated with PGS J. Denote such dual bases by $\{Q(J_1), Q(J_2), ..., Q(J_r)\}$. For each $Q(J)$, identify those $Q \in \rho$ of Step 2 which has $(\mathbf{u}, \mathbf{u}(J), \mathbf{u}^1(J')) \supseteq \mathrm{u}(Q)$ and $(\mathbf{w}(J), \mathbf{w}^1(J')) \supseteq \mathbf{w}(Q)$. Denote such subsets of ρ by $\rho^*(J)$.

Step 4. For each PGS J, identify the primal bases of model (M11.9) from $\rho^*(J)$ of Step 3. This can be easily done by dropping the third row block of (11.15) from the simplex tableaus of $\rho^*(J)$. Denote such reduced bases by $\ddot{u}(J) = \{U_1(J), \ldots, U_d(J)\}$. Whenever $\lambda \in \Lambda_8(U_j(J))$, $(\mathbf{u}(U_j, \lambda), \mathbf{w}(U_j, \lambda))$ will offer a primal feasible solution for model (M11.9) with respect to system J.

Step 5. For each PGS J, with a given β^0 and β^{10} value, find the dual parameter set $\Gamma_8(U_j(J, \beta^0, \beta^{10}))$ for each $U_j(J)$ of $\ddot{u}(J)$ for model (M11.9). Recall that this can be easily done by dropping the columns of $\mathbf{u}^2(J')$ and $\mathbf{w}^2(J')$ from Table (11.16).

Step 6. Rearrange PGS J, $U_j(J) \in \ddot{u}(J)$, $\Lambda_8(U_j(J))$ and $\Gamma_8(U_j(J, \beta^0, \beta^{10}))$ in good order. Then when $\lambda \in \Lambda_8\ (U_j(J))$ and $\gamma \in \Gamma_8(U_j(J, \beta^0, \beta^{10}))$, $(\mathbf{u}(U_j, \lambda), \mathbf{w}(U_j, \lambda))$ is the DFCP for J.

Step 7. Evaluate each PGS J and its corresponding DFCPs in terms of a *criterion* preferred by decision makers. The final optimal system is selected based on the evaluation.

Example 11.2 (Algorithm 11.2)

Step 1. Given the following design problem:

$$\text{Max} \quad (\lambda_1 \quad \lambda_2) \begin{pmatrix} -4 & -2 \\ 1 & 1 \end{pmatrix} \begin{pmatrix} x_1 \\ x_2 \end{pmatrix}$$

$$\text{Subject to} \begin{pmatrix} 1 & 3 \\ 2 & 1 \end{pmatrix} \begin{pmatrix} x_1 \\ x_2 \end{pmatrix} \leq \begin{pmatrix} 40 & 30 \\ 10 & 15 \end{pmatrix} \begin{pmatrix} \gamma_1 \\ \gamma_2 \end{pmatrix}$$

$$x_j \geq 0, \quad j = 1, 2.$$

Let s_1 and s_2 be slack variables for constraints 1 and 2, respectively. Then using the computer software of the MC^2-simplex method (Hao and Shi 1996), we find the set of all PGSs, denoted by $\vartheta = \{J_1, J_2, J_3\}$ as shown in Table 11.17.

Table 11.17 Potential Good Systems

PGS	Bases	$\Gamma_1(J_i)$	$\Lambda_1(J_i)$
J_1	(x_2, s_1)	$3/5 < \gamma_1 \leq 1$	$0 < \lambda_1 \leq 1/3$
J_2	(x_1, x_2)	$0 < \gamma_1 \leq 3/5$	$0 < \lambda_1 \leq 1/6$
J_3	(x_2, s_2)	$0 < \gamma_1 \leq 3/5$	$1/6 < \lambda_1 \leq 1/3$

Step 2. Let β_i be the increments of the ith product and its unit contribution be w_i, i = 1, 2, 3, 4. Then the corresponding augmented dual model is as follows: we solve the corresponding augmented problem:

$$Min \quad \begin{pmatrix}\gamma_1 & \gamma_2\end{pmatrix} \begin{pmatrix}40 & 10\\30 & 15\end{pmatrix} \begin{pmatrix}u_1\\u_2\end{pmatrix} - \begin{pmatrix}\beta_1 & \beta_2 & \beta_3 & \beta_4\end{pmatrix} \begin{pmatrix}w_1\\w_2\\w_3\\w_4\end{pmatrix}$$

$$Subject\ to \begin{pmatrix}1 & 2\\3 & 1\\1 & 0\\0 & 1\end{pmatrix} \begin{pmatrix}u_1\\u_2\end{pmatrix} \geq \begin{pmatrix}-4 & 1\\-2 & 1\\0 & 0\\0 & 0\end{pmatrix} \begin{pmatrix}\lambda_1\\\lambda_2\end{pmatrix} + \begin{pmatrix}1 & 0 & 0 & 0\\0 & 1 & 0 & 0\\0 & 0 & 1 & 0\\0 & 0 & 0 & 1\end{pmatrix} \begin{pmatrix}w_1\\w_2\\w_3\\w_4\end{pmatrix}$$

$$u_j, \quad j=1,\ 2 \text{ are unrestricted; and } w_i \geq 0, \quad i=1,\ 2,\ 3, 4.$$

Let $u_j \geq 0$, j = 3, 4, 5, 6, be surplus variables. Then by using the computer program of Hao and Shi (1996), we solve the above problem and obtain 89 bases, denoted by $\rho = \{Q_1, \ldots, Q_{89}\}$ (see Shi 1995).

Step 3. For each J_i of ϑ, i = 1, 2, 3, we identify their corresponding dual bases from ρ by dropping the columns of both $\mathbf{w}(J_i)$, $\mathbf{w}^1(J'_i)$ and $\mathbf{w}^2(J'_i)$ from Table (11.16) associated with PGS J_i. As an example, for PGS J_1, we have $\mathbf{x}(J_1) = (x_2, s_1)$, $\mathbf{x}^1(J'_1) = s_2$, and $\mathbf{x}^2(J'_1) = x_1$ The partition of corresponding complementary dual variables are $\mathbf{u}(J_1) = (u_4, u_5)$, $\mathbf{u}^1(J'_1) = u_6$ and $\mathbf{u}^2(J_1) = u_3$, respectively. The dual variable $\mathbf{u} = (u_1, u_2)$ are complementary to the two primal constraints. Denote such dual bases by $\{Q(J_1), Q(J_2), \ldots, Q(J_r)\}$. For each $Q(J)$, identify those $Q \in \rho$ of Step 2 which has $(\mathbf{u}, \mathbf{u}(J), \mathbf{u}^1(J')) \supseteq u(Q)$ and $(\mathbf{w}(J), \mathbf{w}^1(J')) \supseteq \mathbf{w}(Q)$. Denote such subsets of ρ by $\rho^*(J_i)$.

Step 4. For each PGS J_i, i = 1, 2, 3, we identify the primal bases of Model (M11.6) from $\rho^*(J_i)$ of Step 3 by dropping the third row block of (11.15) from the simplex tableaus of $\rho^*(J_i)$. Then we have $\ddot{u}(J_1)$, $\ddot{u}(J_2)$, and $\ddot{u}(J_3)$.

Steps 5-6 For each PGS J_i, i = 1, 2, 3, we find $\Lambda_8(U_j(J))$ and $\Gamma_8(U_j(J, \beta^0, \beta^{10}))$, which is a function of β^0 and β^{10} for each $U(J_i)$ of $\hat{u}(J_i)$ for Model (M11.9). For illustrative purposes, let $\beta^0_i = \beta^{10}_i = 2$, i = 1, 2, 3. We obtain all DFCPs in Table 11.18-11.19. The graphical representation of Tables 11.18 is Figure 11.2.

Table 11.18 Dual Flexible Contingency Plans for J_1 with $\beta^0_i = \beta^{10}_i = 2$

Bases	$\Gamma_8(U_j(J_1))$	$\Lambda_8(U_j(J_1))$	$\mathbf{V}(U_j(J_1))$
$U_{13}(J_1)=(u_1,u_6,w_2)$	$0 < \gamma_1 \leq 1$	$1/3 < \lambda_1 \leq 1$	$(\lambda_1 \quad \lambda_2)\begin{pmatrix}-4 & -4\\ 2 & 2\end{pmatrix}\begin{pmatrix}\gamma_1\\ \gamma_2\end{pmatrix}$
$U_{40}(J_1) = (u_1,w_2,w_4)$	$0 < \gamma_1 \leq 1$	$1/3 < \lambda_1 \leq 1$	$(\lambda_1 \quad \lambda_2)\begin{pmatrix}-4 & -4\\ 2 & 2\end{pmatrix}\begin{pmatrix}\gamma_1\\ \gamma_2\end{pmatrix}$
$U_{83}(J_1) = (w_2,w_3,w_4)$	$0 < \gamma_1 \leq 1$	$1/3 < \lambda_1 \leq 1$	$(\lambda_1 \quad \lambda_2)\begin{pmatrix}-4 & -4\\ 2 & 2\end{pmatrix}\begin{pmatrix}\gamma_1\\ \gamma_2\end{pmatrix}$
$U_{17}(J_1) = (u_1,u_2,w_3)$	$0<\gamma_1\leq 11/25$	$0 < \lambda_1 \leq 1/3$	$(\lambda_1 \quad \lambda_2)\begin{pmatrix}-25.3 & -18.7\\ 12.7 & 9.3\end{pmatrix}\begin{pmatrix}\gamma_1\\ \gamma_2\end{pmatrix}$
$U_{46}(J_1) = (u_1,w_3,w_4)$	$0<\gamma_1\leq 11/25$	$0 < \lambda_1 \leq 1/3$	$(\lambda_1 \quad \lambda_2)\begin{pmatrix}-25.3 & -18.7\\ 12.7 & 9.3\end{pmatrix}\begin{pmatrix}\gamma_1\\ \gamma_2\end{pmatrix}$
$U_{23}(J_1) = (u_1,u_2,w_4)$	$11/25<\gamma_1\leq 1$	$0 < \lambda_1 \leq 1/3$	$(\lambda_1 \quad \lambda_2)\begin{pmatrix}-16 & -26\\ 8 & 13\end{pmatrix}\begin{pmatrix}\gamma_1\\ \gamma_2\end{pmatrix}$
$U_{72}(J_1) = (u_2,w_3,w_4)$	$11/25<\gamma_1\leq 1$	$1/3 < \lambda_1 \leq 1$	$(\lambda_1 \quad \lambda_2)\begin{pmatrix}-16 & -26\\ 8 & 13\end{pmatrix}\begin{pmatrix}\gamma_1\\ \gamma_2\end{pmatrix}$

If J_1 is chosen for commitment, then Table 11.18 can be explained as follows:

(i) When $1/3 < \lambda_1 \leq 1$ and $0 < \gamma_1 \leq 1$ occur, we use a basis U_{13}, U_{40}, or U_{83} to produce product {2}. When U_{13} is used, x_2 has increased two units (i.e., $\beta_2 = 2$) and the corresponding increment of the unit contribution is $w_2 = 2\lambda_1 - \lambda_2$ from the related MC^2-simplex tableau in model (M11.9). When U_{40} is used, x_2 has increased two units (i.e., $\beta_2 = 2$) and the corresponding increment of the unit contribution is $w_2 = 2\lambda_1 - \lambda_2$. The slack resource s_2 can be saved two more units (i.e., $\beta_4 = 2$), which are worth nothing because $w_4 = 0$ from the related MC²-simplex tableau of U_{40}. When U_{83} is used, x_2 has increased two units (i.e., $\beta_2 = 2$) and the corresponding increment of the unit contribution is $w_2 = 2\lambda_1 - \lambda_2$; and both slack resources s_1 and s_2 can save two more units (i.e., $\beta_3 = 2$, $\beta_4 = 2$), respectively. However, the prices of β_3 and β_4 are zero because $w_2 = w_4 = 0$ from the related MC²-simplex tableau of U_{83}.

(ii) When $0 < \lambda_1 \leq 1/3$ and $0 < \gamma_1 \leq 11/25$ occur, we use either U_{17} or U_{46} to produce product {2}. When U_{17} is used, x_2 has no increment; but the slack resource s_1 can be saved two more units (i.e., $\beta_3 = 2$), which are worth $w_3 = -0.67\ \lambda_1 + 0.33\lambda_2$ from the related MC²-simplex tableau of U_{17}. When U_{46} is used, x_2 has no increment; the slack resource s_1 can be saved two more units (i.e., $\beta_3 = 2$), which are worth $w_3 = -0.67\ \lambda_1 + 0.33\ \lambda_2$; and the slack resource s_2 can be saved two more units (i.e., $\beta_4 = 2$), which are worth nothing because $w_4 = 0$ from the related MC²-simplex tableau of U_{46}.

(iii) When $0 < \lambda_1 \leq 1/3$ and $11/25 < \gamma_1 \leq 1$ occur, we use either U_{23} or U_{72} to produce product {2}. When U_{23} is used, x_2 has no increment, but the slack resource s_2 can be saved two more units (i.e., $\beta_4 = 2$) which are worth $w_4 = -2\lambda_1 + \lambda_2$ from the related MC²-simplex tableau of U_{23}. When U_{72} is used, x_2 has no increment. The slack resource s_2 can be saved two more units (i.e., $\beta_4 = 2$), which are worth $w_4 = -2\lambda_1 + \lambda_2$, and the slack resource s_1 can be saved two more units (i.e., $\beta_3 = 2$), which are worth nothing because $w_3 = 0$ from the related MC²-simplex tableau of U_{72}.

The set of *all* FDCPs for the PGD J_1 is $\{U_{13}, U_{40}, U_{83}, U_{17}, U_{46}, U_{23}, U_{72}\}$. For the space (λ_1, γ_1), this set produces twelve alternative FDCP sets for J_1: $\{U_{13}, U_{17}, U_{23}\}$, $\{U_{13}, U_{17}, U_{72}\}$, $\{U_{13}, U_{46}, U_{23}\}$, $\{U_{13}, U_{46}, U_{72}\}$, $\{U_{40}, U_{17}, U_{23}\}$, $\{U_{40}, U_{17}, U_{72}\}$, $\{U_{40}, U_{46}, U_{23}\}$, $\{U_{40}, U_{46}, U_{72}\}$, $\{U_{13}, U_{17}, U_{23}\}$, $\{U_{83}, U_{17}, U_{72}\}$, $\{U_{83}, U_{46}, U_{23}\}$, and $\{U_{83}, U_{46}, U_{72}\}$. Similarly, Tables 11.19-11.20 can be explained.

Step 7. For illustrative purposes, let us also use the *maximizing expected payoff* as the criterion to select the final optimal designs.

Let the initial setup cost for using J, $f(J) = 0$. We assume that γ_1 is independent on λ_1, and (γ_1, λ_1) have the bi-uniform probability distribution $F(\gamma_1, \lambda_1)$. Given the contingency set $\{U_{13}, U_{17}, U_{23}\}$ for PGS J_1, to find the expected payoff $\mathbf{EV}(J_1)$, we compute:

Table 11.19 Dual Flexible Contingency Plans for J_2 with $\beta^0_i = \beta^{10}_i = 2$

Bases	$\Gamma_8(U_j(J_2))$	$\Lambda_8(U_j(J_2))$	$\mathbf{V}(U_j(J_2))$
$U_5(J_2)=$ (u_1,u_2,w_1,w_2)	$0<\gamma_1\leq 1$	$1/3<\lambda_1\leq 1$	$(\lambda_1 \ \ \lambda_2)\begin{pmatrix}-12 & -12\\ 4 & 4\end{pmatrix}\begin{pmatrix}\gamma_1\\ \gamma_2\end{pmatrix}$
$U_{31}(J_2)=$ (u_1,u_2,w_2,w_4)	$0<\gamma_1\leq 1$	$1/3<\lambda_1\leq 1$	$(\lambda_1 \ \ \lambda_2)\begin{pmatrix}-12 & -12\\ 4 & 4\end{pmatrix}\begin{pmatrix}\gamma_1\\ \gamma_2\end{pmatrix}$
$U_{64}(J_2)=$ (u_2,w_1,w_2,w_4)	$0<\gamma_1\leq 1$	$1/3<\lambda_1\leq 1$	$(\lambda_1 \ \ \lambda_2)\begin{pmatrix}-12 & -12\\ 4 & 4\end{pmatrix}\begin{pmatrix}\gamma_1\\ \gamma_2\end{pmatrix}$
$U_{65}(J_2)=$ (u_1,w_1,w_2,w_5)	$0<\gamma_1\leq 1$	$0<\lambda_1\leq 1/3$	$(\lambda_1 \ \ \lambda_2)\begin{pmatrix}-12 & -12\\ 4 & 4\end{pmatrix}\begin{pmatrix}\gamma_1\\ \gamma_2\end{pmatrix}$
$U_6(J_2)=$ (u_1,u_2,w_1,w_3)	$0<\gamma_1\leq 1/25$	$1/6<\lambda_1\leq 1/3$	$(\lambda_1 \ \ \lambda_2)\begin{pmatrix}-32 & -25.3\\ 14 & 10.7\end{pmatrix}\begin{pmatrix}\gamma_1\\ \gamma_2\end{pmatrix}$
$U_{32}(J_2)=$ (u_1,w_1,w_3,w_4)	$0<\gamma_1\leq 1/25$	$1/6<\lambda_1\leq 1/3$	$(\lambda_1 \ \ \lambda_2)\begin{pmatrix}-32 & -25.3\\ 14 & 10.7\end{pmatrix}\begin{pmatrix}\gamma_1\\ \gamma_2\end{pmatrix}$
$U_{16}(J_2)=$ (u_1,u_2,w_3,w_4)	$0<\gamma_1\leq 1/25$	$0<\lambda_1\leq 1/6$	$(\lambda_1 \ \ \lambda_2)\begin{pmatrix}-16 & -26\\ 10.8 & 10.8\end{pmatrix}\begin{pmatrix}\gamma_1\\ \gamma_2\end{pmatrix}$
$U_7(J_2)=$ (u_1,u_2,w_1,w_4)	$1/25<\gamma_1\leq 1$	$0<\lambda_1\leq 1/3$	$(\lambda_1 \ \ \lambda_2)\begin{pmatrix}-16 & -26\\ 6 & 11\end{pmatrix}\begin{pmatrix}\gamma_1\\ \gamma_2\end{pmatrix}$
$U_{63}(J_2)=$ (u_2,w_1,w_3,w_4)	$1/25<\gamma_1\leq 1$	$0<\lambda_1\leq 1/3$	$(\lambda_1 \ \ \lambda_2)\begin{pmatrix}-16 & -26\\ 6 & 11\end{pmatrix}\begin{pmatrix}\gamma_1\\ \gamma_2\end{pmatrix}$

$$\mathbf{EV}(J_1) = \int_{S_1} \mathbf{V}(U_{13}(J_1))\, dF(\gamma_1, \lambda_1) + \int_{S_2} \mathbf{V}(U_{17}(J_1))\, dF(\gamma_1, \lambda_1) + \int_{S_3} \mathbf{V}(U_{23}(J_1))\, dF(\gamma_1, \lambda)$$

where $\mathbf{V}(U_{13}(J_1))$, $\mathbf{V}(U_{17}(J_1))$ and $\mathbf{V}(U_{23}(J_1))$; and the (γ_1, λ_1) ranges $S_j = \Gamma_8(U_j(J_1))\times\Lambda_8(U_j(J_1))$, j = 1, 2, 3 are shown in Table 11.18.

Table 11.20 Dual Flexible Contingency Plans for J_3 with $\beta^0_i = \beta^{10}_i = 2$

Bases	$\Gamma_8(U_j(J_1))$	$\Lambda_8(U_j(J_1))$	$\mathbf{V}(U_j(J_1))$
$U_{13}(J_1)$= (u_1, u_6, w_2)	$0 < \gamma_1 \le 1$	$1/3 < \lambda_1 \le 1$	$(\lambda_1 \quad \lambda_2)\begin{pmatrix} -4 & -4 \\ 2 & 2 \end{pmatrix}\begin{pmatrix} \gamma_1 \\ \gamma_2 \end{pmatrix}$
$U_{40}(J_1)$ = (u_1, w_2, w_4)	$0 < \gamma_1 \le 1$	$1/3 < \lambda_1 \le 1$	$(\lambda_1 \quad \lambda_2)\begin{pmatrix} -4 & -4 \\ 2 & 2 \end{pmatrix}\begin{pmatrix} \gamma_1 \\ \gamma_2 \end{pmatrix}$
$U_{83}(J_1)$ = (w_2, w_3, w_4)	$0 < \gamma_1 \le 1$	$1/3 < \lambda_1 \le 1$	$(\lambda_1 \quad \lambda_2)\begin{pmatrix} -4 & -4 \\ 2 & 2 \end{pmatrix}\begin{pmatrix} \gamma_1 \\ \gamma_2 \end{pmatrix}$
$U_{17}(J_1)$ = (u_1, u_2, w_3)	$0<\gamma_1\le 11/25$	$0 < \lambda_1 \le 1/3$	$(\lambda_1 \quad \lambda_2)\begin{pmatrix} -25.3 & -18.7 \\ 12.7 & 9.3 \end{pmatrix}\begin{pmatrix} \gamma_1 \\ \gamma_2 \end{pmatrix}$
$U_{46}(J_1)$ = (u_1, w_3, w_4)	$0<\gamma_1\le 11/25$	$0 < \lambda_1 \le 1/3$	$(\lambda_1 \quad \lambda_2)\begin{pmatrix} -25.3 & -18.7 \\ 12.7 & 9.3 \end{pmatrix}\begin{pmatrix} \gamma_1 \\ \gamma_2 \end{pmatrix}$
$U_{23}(J_1)$ = (u_1, u_2, w_4)	$11/25<\gamma_1\le 1$	$0 < \lambda_1 \le 1/3$	$(\lambda_1 \quad \lambda_2)\begin{pmatrix} -16 & -26 \\ 8 & 13 \end{pmatrix}\begin{pmatrix} \gamma_1 \\ \gamma_2 \end{pmatrix}$
$U_{72}(J_1)$ = (u_2, w_3, w_4)	$11/25<\gamma_1\le 1$	$1/3 < \lambda_1 \le 1$	$(\lambda_1 \quad \lambda_2)\begin{pmatrix} -16 & -26 \\ 8 & 13 \end{pmatrix}\begin{pmatrix} \gamma_1 \\ \gamma_2 \end{pmatrix}$

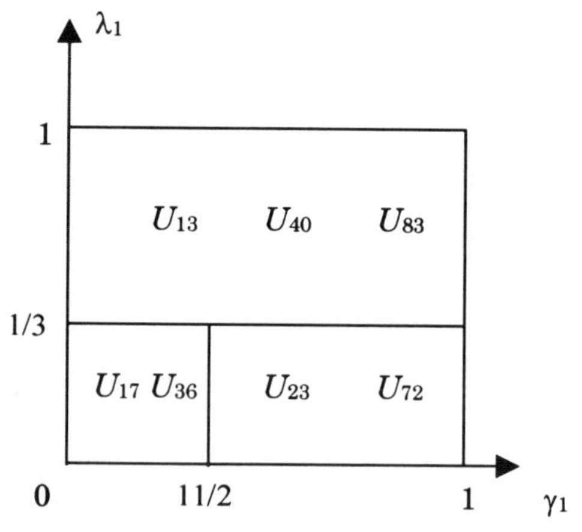

Figure 11.2 Dual Flexible Contingency Plans for J_1

Then, we get **EV**(J_1) = -12.09. Similarly, we can get:

EV(J_2) = 37.06 and **EV**(J_3) = -12.09.

Therefore, in view of *maximizing expected payoff*, J_2 should be the final optimal system preferred by the decision makers, where the set of all DFCPs are shown in Table 11.19.

11.3 Further Remarks

The reason to both *dual rigid contingency plans* (DRCPs) and *dual flexible contingency plans* (DFCPs) for a PGS is to avoid the non-optimal designs caused by some value λ. The meanings of "rigid " and "flexible" similar to that of Chapter 9 depend on how to use the slack resources. If we use the slack resource outside a PGS, we have DRCPs. Otherwise, we find DFCPs. Although the process of identifying these contingency plans is tedious, the reader can understand its importance in linear design context.

Exercises

11.1 Use Algorithm 11.1 to construct DRCPs for each PGS J of the following optimal design problem:

$$\begin{aligned}
&\textit{Max} && -4\,x_1 - 2\,x_2 \\
&\textit{Max} && x_1 + x_2 \\
&\textit{Subject to} && x_1 + 3\,x_2 \le (40,\ 30) \\
& && 2x_1 + x_2 \le (10,\ 15) \\
& && x_1,\ x_2 \ge 0.
\end{aligned}$$

11.2 Use Algorithm 11.2 to construct DFCPs for each PGS J of the following optimal design problem:

$$\begin{aligned}
&\textit{Max} && 6\,x_1 + 2\,x_2 \\
&\textit{Max} && x_1 + x_2 \\
&\textit{Subject to} && x_1 + 3\,x_2 \le (40,\ 30) \\
& && 2x_1 + x_2 \le (10,\ 15) \\
& && x_1,\ x_2 \ge 0.
\end{aligned}$$

Chapter 12
GENERALIZED GOOD SYSTEMS AND DUAL CONTINGENCY PLANS WITH COMPUTER AIDED SYSTEM

The approach of this chapter is complimentary to Section 10.2 and 10.3. Instead of finding the primal contingency plans for a generalized optimal system (GGS) Ω, we extend the techniques of finding DRCP (Section 11.1) and DFCP (section 11.2) for PGS J to construct DRCP and DFCP for Ω, respectively. For the sake of space, we will only outline the main ideas of the approach, not elaborate the details. Section 12.1 comes from Shi and He (1994), while Section 12.2 is based on Shi (1998b). Section 12.3 integrates all of different contingency plan approaches into an augmented model. A computer-aided system for optimal system designs is developed to effectively handle the design complexity. Section 12.4 concludes the chapter with further remarks.

12.1 Finding Dual Rigid Contingency Plans for Generalized Good Systems

We recall that in Section 10.2, we denote $\wp$ as *the set of all possible* GGSs, which including the set of PGSs, ϑ. Given a GGS Ω, let $\mathbf{x}_\Omega$ be the corresponding variables involved in Ω. Then, for a given $\mathbf{x}_\Omega$, we denote $\mathbf{A}_\Omega$ as the submatrix of $[\mathbf{A}, \mathbf{I}_m]$ associated with Ω and $\mathbf{C}_\Omega$ as the submatrix of $[\mathbf{C}, \mathbf{0}]$ associated with Ω. The submodel of model (M9.1) for GGS Ω is (M12.1):

$$\begin{aligned} &\mathit{Max} && \boldsymbol{\lambda}'\mathbf{C}_\Omega \mathbf{x}_\Omega \\ &\mathit{Subject\ to} && \mathbf{A}_\Omega \mathbf{x}_\Omega = \mathbf{D}\boldsymbol{\gamma} \\ & && \mathbf{x}_\Omega \geq 0. \end{aligned}$$

To construct the DRCPs for the given Ω, we solve the dual model of the model (12.1) which is defined by (M12.2):

$$\begin{array}{ll} \textit{Min} & \mathbf{u}^t\,\mathbf{D}\gamma \\ \textit{Subject to} & \mathbf{u}^t\mathbf{A}_{\Omega} \geq \lambda^t\,\mathbf{C}_{\Omega} \\ & \mathbf{u} \text{ unrestricted.} \end{array}$$

Note that the model (M12.2) is a submodel of the model (M11.4). We can get the model (M12.2) from the model (M11.4) by deleting columns of $\mathbf{A}_{\Omega'}$ and $\mathbf{C}_{\Omega'}$ from the model (M11.4), where $N^* = \Omega \cup \Omega'$ (see Section 10.2).

Let c be the number of variables in $\mathbf{x}_{\Omega}$ (note that $c \geq m$). Then $n + m - c$ is the number of variables in $\mathbf{x}_{\Omega'}$, which are not selected in GGS Ω. Let $\mathbf{u}_{\Omega}$ be the dual surplus variables complementary to $\mathbf{x}_{\Omega}$. Since $\mathbf{x}_{\Omega}$ is not a basis (which may have more than m variables), for a basis $G(\Omega)$, we define $\mathbf{x}_G(\Omega)$ and $\mathbf{x}_{G'}(\Omega)$ be the basic variables and non-basic variables, respectively in Ω. Then, we decompose $\mathbf{u}_{\Omega}$ into $(\mathbf{u}^1(\Omega), \mathbf{u}^2(\Omega))$, where $\mathbf{u}^1(\Omega)$ is the dual surplus variables complementary to $\mathbf{x}_G(\Omega)$ and $\mathbf{u}^2(\Omega)$ is the dual surplus variables complementary to $\mathbf{x}_{G'}(\Omega)$.

To avoid non-optimal situations, we let $\beta^1(\Omega) = (\beta_1, \ldots, \beta_m)^t$ be the vector of increments of the primal basic variables $\mathbf{x}_G(\Omega)$, and $\beta^2(\Omega) = (\beta_{m+1}, \ldots, \beta_c)^t$ be the vector of increments of the primal non-basic variables $\mathbf{x}_{G'}(\Omega)$. Let the unit contribution of these newly acquired, $\beta^1(\Omega)$ be $\mathbf{w}^1(\Omega) = (w_1, \ldots, w_m)^t$, and that of $\beta^2(\Omega)$ be $\mathbf{w}^2(\Omega) = (w_{m+1}, \ldots, w_c)^t$. (Note that for a slack variable s_j, we say that, β_j *is* the additional amount of the jth resource saved, which is worth w_j per unit). Then we add $\mathbf{u}^1(\Omega)$, $\mathbf{u}^2(\Omega)$, $\mathbf{w}^1(\Omega)$ and $\mathbf{w}^2(\Omega)$ into the model (M12.2), and the submodel of finding the DRCPs for a GGS Ω becomes (M12.3):

$$\begin{array}{ll} \textit{Min} & \mathbf{u}^t\,\mathbf{D}\gamma - \mathbf{w}^1(\Omega)^t\,\boldsymbol{\beta}^1(\Omega) - \mathbf{w}^2(\Omega)^t\,\boldsymbol{\beta}^2(\Omega) \\ \textit{Subject to} & \mathbf{u}^t\,\mathbf{A}_{\Omega} - (\mathbf{u}^1)^t(\Omega)[\mathbf{I}_m, \mathbf{0}] - (\mathbf{u}^2)^t(\Omega)[0, \mathbf{I}_{c-m}] \\ & \qquad = \boldsymbol{\lambda}^t\,\mathbf{C}_{\Omega} + \mathbf{w}^1(\Omega)^t + \mathbf{w}^2(\Omega)^t \\ & \mathbf{u} \text{ unrestricted}, \mathbf{u}^1(\Omega), \mathbf{u}^2(\Omega), \mathbf{w}^1(\Omega), \mathbf{w}^2(\Omega) \geq 0. \end{array}$$

Given an Ω of $\wp$, let $\Lambda_{10}(G(\Omega))$ be the primal parameter set for model (M12.3) and let $\Gamma_{10}(G(\Omega, \beta^1, \beta^2))$ be the dual parameter set for model (M12.3). We call the set of all non-trivial potential solutions obtained from the model (M12.3) by using the MC²-simplex method the DRCPs selected by the model (M12.3). By adjusting $\mathbf{w}^1(\Omega)$ and $\mathbf{w}^2(\Omega)$, we can always have feasible solutions. Thus, by using the model (M12.3), we can always have the optimal DRCPs for all possible situations of (λ, γ).

Using the augmented dual model (M11.4), we can discuss an effective and systematic method of identifying all DRCPs for the set of $\wp$. Let us start with (M12.3):

$$\begin{aligned} &Min && \mathbf{u}^t\,\mathbf{D}\gamma - \mathbf{w}^t\boldsymbol{\beta} \\ &Subject\ to && \mathbf{u}^t\,[\mathbf{A}, \mathbf{I}_m] - (\mathbf{u}_s)^t\,\mathbf{I}_{n+m} - \mathbf{w}^t\,\mathbf{I}_{n+m} = \boldsymbol{\lambda}^t\,[\mathbf{C}, \mathbf{0}] + \mathbf{w}^t \\ & && \mathbf{u} \text{ unrestricted and } \mathbf{w},\ \mathbf{u}_s \geq 0, \end{aligned}$$

where $\mathbf{u}_s \in \mathbf{R}^{n+m}$ is a (n+m)-dimensional vector.

Given a basis $G(\Omega)$ for Ω of $\wp$, we decompose $[\mathbf{A}, \mathbf{I}_m]$ into $[\mathbf{A}^1{}_\Omega, \mathbf{A}^2{}_\Omega, \mathbf{A}_{\Omega'}]$, and $[\mathbf{C}, \mathbf{0}]$ into $[\mathbf{C}^1{}_\Omega, \mathbf{C}^2{}_\Omega, \mathbf{C}_{\Omega'}]$. For the basis $G(\Omega)$, we partition $\mathbf{u}_s$ into $(\mathbf{u}^1(\Omega), \mathbf{u}^2(\Omega), \mathbf{u}(\Omega'))$, where $\mathbf{u}^1(\Omega)$ is the dual surplus variables complementary to $\mathbf{x}_G(\Omega)$, $\mathbf{u}^2(\Omega)$ is the dual surplus variables complementary to $\mathbf{x}_{G'}(\Omega)$, and $\mathbf{u}(\Omega')$ is the dual surplus variables complementary to $\mathbf{x}(\Omega')$. Correspondingly, we also partition $\mathbf{w}$ into $(\mathbf{w}^1(\Omega), \mathbf{w}^2(\Omega), \mathbf{w}(\Omega'))$,, and β into $(\beta^1(\Omega), \beta^2(\Omega), \beta(\Omega'))$, where $\mathbf{w}(\Omega')$are the rest n+m-c components of w associated with $\beta(\Omega')$ after excluding $\mathbf{w}^1(\Omega)$ and $\mathbf{w}^2(\Omega)$ from w. Then, the model (M11.4) becomes (M12.4):

$$\begin{aligned} &Min && \mathbf{u}^t\,\mathbf{D}\gamma - \mathbf{w}^1(\Omega)^t\,\beta^1(\Omega) - \mathbf{w}^2(\Omega)^t\,\beta^2(\Omega) - \mathbf{w}(\Omega')^t\,\beta(\Omega') \\ &Subject\ to && \mathbf{u}^t\,[\mathbf{A}^1{}_\Omega, \mathbf{A}^2{}_\Omega, \mathbf{A}_{\Omega'}] - \mathbf{u}^1(\Omega)^t\,[\mathbf{I}_m, \mathbf{0}, \mathbf{0}] - \mathbf{u}^2(\Omega)^t\,[\mathbf{0}, \mathbf{I}_{c-m}, \mathbf{0}] \\ & && -\mathbf{u}(\Omega')^t\,[\mathbf{0}, \mathbf{0}, \mathbf{I}_{n+m-c}] = \boldsymbol{\lambda}^t\,[\mathbf{C}^1{}_\Omega, \mathbf{C}^2{}_\Omega, \mathbf{C}_{\Omega'}] + \mathbf{w}^1(\Omega)^t \\ & && + \mathbf{w}^2(\Omega)^t + \mathbf{w}(\Omega')^t \\ & && \mathbf{u} \text{ unrestricted}, \mathbf{u}^1(\Omega), \mathbf{u}^2(\Omega), \mathbf{u}(\Omega'), \mathbf{w}^1(\Omega), \mathbf{w}^2(\Omega), \mathbf{w}(\Omega') \geq 0. \end{aligned}$$

Remark 12.1 The dual model (M12.3) is a submodel of model (M12.4) because the simplex tableau of model (M12.4) can be obtained by dropping the columns of $\mathbf{w}(\Omega')$ and the rows of $\mathbf{A}_{\Omega'}$ from (M12.4). Note that $\mathbf{A}_\Omega = [\mathbf{A}^1{}_\Omega, \mathbf{A}^2{}_\Omega]$.

Applying Remark 12.1, the reader can develop a method of effectively and systematically locating all DRCPs for all GGSs and selecting the final optimal linear system (see Exercise 12.1).

12.2 Finding Dual Flexible Contingency Plans for Generalized Good Systems

Shi (1998b) showed a method of finding DFCPs for a GGS Ω. Given a GGS Ω of $\wp$, the variables $(\mathbf{x}, \mathbf{s})$ are decomposed into $\mathbf{x}_\Omega$, $\mathbf{x}_{\Omega'}$, and $\mathbf{x}_{\Omega''}$, where $\mathbf{x}_\Omega$ are the selected variables corresponding to Ω; $\mathbf{x}_{\Omega'}$ are non-selected slack variables; and $\mathbf{x}_{\Omega''}$ are non-selected opportunity variables (Note that $N^* = \Omega \cup \Omega' \cup \Omega''$). Then, $[\mathbf{A}, \mathbf{I}_m]$ is decomposed into $[\mathbf{A}_\Omega, \mathbf{A}_{\Omega'}, \mathbf{A}_{\Omega''}]$, where $\mathbf{A}_\Omega$ is the submatrix of $[\mathbf{A}, \mathbf{I}_m]$ associated with Ω, $\mathbf{A}_{\Omega'}$ is the submatrix of $[\mathbf{A}, \mathbf{I}_m]$ associated with Ω', and $\mathbf{A}_{\Omega''}$ is the submatrix of $[\mathbf{A}, \mathbf{I}_m]$ associated with Ω''. Similarly, $[\mathbf{C}, \mathbf{0}]$ is decomposed into $[\mathbf{C}_\Omega, \mathbf{C}_{\Omega'}, \mathbf{C}_{\Omega''}]$ where $\mathbf{C}_{\Omega'}$ is a zero matrix with proper dimensions. If Ω is chosen, we use $\Omega^* = \Omega \cup \Omega'$ to build the following primal model (M12.5):

$$
\begin{aligned}
&\textit{Max} && \boldsymbol{\lambda}^t \mathbf{C}_\Omega \mathbf{x}_\Omega - \boldsymbol{\alpha}^t \mathbf{y} \\
&\textit{Subject to} && \mathbf{A}_\Omega \mathbf{x}_\Omega + \mathbf{A}_{\Omega'} \mathbf{x}_{\Omega'} = \mathbf{D}\boldsymbol{\gamma} \\
& && \mathbf{x}_\Omega, \mathbf{x}_{\Omega'} \ge 0.
\end{aligned}
$$

To construct the DFCP for Ω^*, we solve the dual model of (M12.5) as (M12.6):

$$\begin{array}{ll} \textit{Min} & \mathbf{u}'\,\mathbf{D}\gamma \\ \textit{Subject to} & \mathbf{u}[\mathbf{A}_{\Omega},\mathbf{A}_{\Omega'}] \geq \lambda[\mathbf{C}_{\Omega},\mathbf{0}] \\ & \mathbf{u}\text{ unrestricted.} \end{array}$$

Let c be the number of variables in $\mathbf{x}_{\Omega}$ and p be the number of variables in $\mathbf{x}_{\Omega'}$ (note that c+p ≥m). Then n+m-c-p is the number of variables in $\mathbf{x}_{\Omega''}$. Let $\mathbf{u}_{\Omega}$ be the dual surplus variables complementary $\mathbf{x}_{\Omega}$ and $\mathbf{u}_{\Omega'}$ be the dual surplus variables complementary to $\mathbf{x}_{\Omega'}$.

Given a basis $L(\Omega^*)$ in model (M12.5), for the basic variables $\mathbf{x}_L(\Omega^*)$, we denote the corresponding non-basic variables by $\mathbf{x}_{L'}(\Omega^*)$. Note that $\mathbf{x}_L(\Omega^*) \in \mathbf{R}^m$ $\mathbf{x}_{L'}(\Omega^*) \in \mathbf{R}^{c+p-m}$. Then, $(\mathbf{u}_{\Omega}, \mathbf{u}_{L'})$ is further decomposed into $(\mathbf{u}^1(\Omega^*), \mathbf{u}^2(\Omega^*))$, where $\mathbf{u}^1(\Omega^*)$ are the dual surplus variables complementary to $\mathbf{x}_L(\Omega^*)$, and $\mathbf{u}^2(\Omega^*)$ are the dual surplus variables complementary to $\mathbf{x}_{L'}(\Omega^*)$. Let $\beta^1(\Omega^*) = (\beta_1, \ldots, \beta_m)^t$ be the vector of increments of the primal basic variables $\mathbf{x}_L(\Omega^*)$,and $\beta^2(\Omega^*) = (\beta_{m+1}, \ldots, \beta_{c+p})$ be the vector of increments of the primal non-basic variables $\mathbf{x}_{L'}(\Omega^*)$. Let the unit contribution of $\beta^1(\Omega^*)$ be $\mathbf{w}^1(\Omega^*) = (w_1, \ldots, w_m)$, and that of $\beta^2(\Omega^*)$ be $\mathbf{w}^2(\Omega^*) = (w_{m+1}, \ldots, w_{c+p})^t$. Adding $\mathbf{u}^1(\Omega^*)$, $\mathbf{u}^2(\Omega^*)$, $\mathbf{w}^1(\Omega^*)$ and $\mathbf{w}^2(\Omega^*)$ into the model (M12.6), we build the submodel of finding the DFCPs for a GGS Ω as (M12.7):

$$\begin{array}{ll} \textit{Min} & \mathbf{u}'\,\mathbf{D}\gamma - \mathbf{w}^1(\Omega^*)'\,\boldsymbol{\beta}^1(\Omega^*) - \mathbf{w}^2(\Omega^*)'\,\boldsymbol{\beta}^2(\Omega^*) \\ \textit{Subject to} & \mathbf{u}'\,[\mathbf{A}_{\Omega},\mathbf{A}_{\Omega'}] - \mathbf{u}^1(\Omega^*)'\,[\mathbf{I}_m,\mathbf{0}] - \mathbf{u}^2(\Omega^*)'\,[0,\mathbf{I}_{c+p-m}] \\ & \quad = \boldsymbol{\lambda}'\,[\mathbf{C}_{\Omega},\mathbf{0}] + \mathbf{w}^1(\Omega^*)' + \mathbf{w}^2(\Omega^*)' \\ & \mathbf{u}\text{ unrestricted}, \mathbf{u}^1(\Omega^*), \mathbf{u}^2(\Omega^*), \mathbf{w}^1(\Omega^*), \mathbf{w}^2(\Omega^*) \geq 0. \end{array}$$

Let $\Lambda_{11}(L(\Omega^*))$ be the *primal parameter set* of a basis $L(\Omega^*)$ for model (M12.7), and $\Gamma_{11}(L(\Omega^*, \beta^1, \beta^2))$ be the *dual parameter set* of a basis $L(\Omega^*)$ for model (M12.7). Then the set of all potential solutions obtained from (M12.7) is the set of DFCPs selected from model (M12.7). By adding $\mathbf{w}^1(\Omega^*)$ and $\mathbf{w}^2(\Omega^*)$, (M12.7) can always result in feasible solutions. Therefore, using model (M12.7), there is always a DFCP for a GGS Ω under all possible situations of (λ, γ).

When the augmented dual model (M11.4) (or (M11.6)) is used to identify all DRCPs for the set of $\wp$, we need the following variables and/or matrices partitions:

Given a basis $L(\Omega^*)$ for Ω of $\wp$, we decompose $[\mathbf{A}, \mathbf{I}_m]$ into $[\mathbf{A}_\Omega, \mathbf{A}_{\Omega'}, \mathbf{A}_{\Omega''}]$, and $[\mathbf{C}, \mathbf{0}]$ into $[\mathbf{C}_\Omega, \mathbf{C}_{\Omega'}, \mathbf{C}_{\Omega''}]$. For the basis $L(\Omega^*)$, we partition the surplus variables $\mathbf{u}_s$ into $(\mathbf{u}^1(\Omega^*), \mathbf{u}^2(\Omega^*), \mathbf{u}(\Omega''))$, where $\mathbf{u}^1(\Omega^*)$ is the dual surplus variables complementary to $\mathbf{x}_L(\Omega^*)$, $\mathbf{u}^2(\Omega^*)$ is the dual surplus variables complementary to $\mathbf{x}_{L'}(\Omega^*)$, and $\mathbf{u}(\Omega'')$ is the dual surplus variables complementary to $\mathbf{x}(\Omega'')$. We also partition $\mathbf{w}$ into $(\mathbf{w}^1(\Omega^*), \mathbf{w}^2(\Omega^*), \mathbf{w}(\Omega''))$,, and β into $(\beta^1(\Omega^*), \beta^2(\Omega^*), \beta(\Omega''))$, where $\mathbf{w}(\Omega'')$ has the n+m-c-p components of w associated with $\beta(\Omega'')$. Then, the dual model (M11.6) is written as (M12.8):

$$
\begin{aligned}
&\textit{Min} && \mathbf{u}'\,\mathbf{D}\gamma - \mathbf{w}^1(\Omega^*)'\,\beta^1(\Omega^*) - \mathbf{w}^2(\Omega^*)'\,\beta^2(\Omega^*) - \mathbf{w}(\Omega'')'\,\beta(\Omega'') \\
&\textit{Subject to} && \mathbf{u}'\,[\mathbf{A}_\Omega, \mathbf{A}_{\Omega'}, \mathbf{A}_{\Omega''}] - \mathbf{u}^1(\Omega^*)'\,[\mathbf{I}_m, \mathbf{0}, \mathbf{0}] - \mathbf{u}^2(\Omega^*)'\,[\mathbf{0}, \mathbf{I}_{c+p-m}, \mathbf{0}] \\
& && - \mathbf{u}(\Omega'')'\,[\mathbf{0}, \mathbf{0}, \mathbf{I}_{n+m-c-p}] \\
& && \quad = \lambda^t\,[\mathbf{C}_\Omega, \mathbf{C}_{\Omega'}, \mathbf{C}_{\Omega''}] + \mathbf{w}^1(\Omega^*)' + \mathbf{w}^2(\Omega^*)' + \mathbf{w}(\Omega'')' \\
& && \mathbf{u}\ \text{unrestricted}, \mathbf{u}^1(\Omega^*), \mathbf{u}^2(\Omega^*), \mathbf{u}(\Omega''), \mathbf{w}^1(\Omega^*), \mathbf{w}^2(\Omega^*), \mathbf{w}(\Omega'') \geq 0.
\end{aligned}
$$

Remark 12.2 The dual model (M12.7) is also a submodel of model (M12.8) since the simplex tableau of model (M12.7) can be obtained by dropping the columns of $\mathbf{w}(\Omega'')$ and the rows of $\mathbf{A}_{\Omega''}$ from (M12.8).

The reader can readily apply Remark 12.2 to construct an algorithm for effectively and systematically locating all DFCPs for all GGSs (see Exercise 12.2).

12.3 A Computer-Aided System for Optimal System Designs

In this section, by integrating Chapter 9-11 and Section 12.1-12.2, we first summarize the relationships between the primal augmented model (M9.3) and its submodels for primal contingency plans and the dual augmented model (M11.4) and its submodels for dual contingency plans, respectively (Section 12.3.1). Then, based on Algorithm 9.1, we propose a detailed procedure to solve a problem of designing optimal systems (Section 12.3.2). For the design of a Computer Aided System (CAS), we shall not discuss how to write and implement the program of the CAS by using some particular computer language, but focus on exploring the related flowcharts of the algorithms for the CAS (section 12.3.3), a top-down design of the CAS and the pseudo-language programs (Section 12.3.4).

12.3.1 Modeling Relationships

In constructing primal (rigid and flexible) contingency plans, models (M9.1), (M9.2), (M9.4), (M10.1) and (M10.2) are the various submodels of model (M9.3).

First let us add slack variables s to model (M9.3), then we observe:

(i) We can obtain model (M9.1) to identify the set of PGSs by deleting the columns of **y** from model (M9.3). Thus, model (M9.1) is a submodel of model (M9.3).

(ii) From model (M9.3), we can obtain model (M9.2) to identify PRCPs for PGS J by deleting the columns of $\mathbf{x}(J')$. Thus, model (M9.2) is a submodel of model (M9.3).

(iii) From model (M9.3), we can obtain model (M9.4) to identify PFCPs for PGS J by deleting the columns of $\mathbf{x}^2(J')$. Thus, model (M9.4) is a submodel of model (M9.3).

(iv) From model (M9.3), we can obtain model (M10.1) to identify PRCPs for GGS Ω by deleting the columns of $\mathbf{x}_{\Omega'}$. Thus, model (M10.1) is a submodel of model (M9.3).

(v) From model (M9.3), we can obtain model (M10.2) to identify PFCPs for GGS Ω by deleting the columns of $\mathbf{x}^2{}_{\Omega'}$. Thus, model (M10.2) is also a submodel of model (M9.3).

In constructing dual (rigid and flexible) contingency plans, models (M11.3), (M11.4), (M11.9), (M12.3) and (M12.7) are the various submodels of dual augmented model (M11.5).

The observation is as follows:

(i) From model (M11.5), we can obtain model (M11.3) to identify DRCPs for PGS J by deleting the columns of both u(J') and w(J') and the second row block of (11.7). Thus, model (M11.3) is a submodel of model (M11.5).

(ii) We can obtain model (M11.4) to identify the dual bases for the set of PGSs by deleting the columns of **w** from model (M11.5). Thus, model (M11.4) is a submodel of model (M11.5).

(iii) From model (M11.5), we can obtain model (M11.9) to identify DFCPs for PGS J by deleting the columns of $\mathbf{u}^2(J')$ and $\mathbf{w}^2(J')$ as well as the third row block of (11.16). Thus, model (M11.9) is a submodel of model (M11.5).

(iv) From model (M11.5) (or (M12.4)), we can obtain model (M12.3) to identify DRCPs for GGS Ω by deleting the columns of $\mathbf{w}(\Omega')$ and the rows of $\mathbf{A}_{\Omega'}$. Thus, model (M12.3) is a submodel of model (M11.5).

(v) From model (M11.5) (or (M12.8)), we can obtain model (M12.7) to identify PFCPs for GGS Ω by deleting the columns of $\mathbf{w}(\Omega'')$ and the rows of $\mathbf{A}_{\Omega''}$. Thus, model (M12.7) is also a submodel of model (M11.5).

12.3.2 A Solution Procedure for CAS

From our discussion in previous sections, a detailed procedure for solving problems of designing optimal systems can be described as follows:

Algorithm 12.1

Step 1. Given a problem of optimal system designs (M9.1), find all *potential solutions or* PGS's. Denote the set of such solutions by $\vartheta = \{J_1, J_2, \ldots, J_r\}$.

Step 2. Use Algorithm 10.1 (or 10.2) to find the set of generalized good systems, $\wp$, which includes ϑ. If we need to construct the primal contingency plans for some γ, go to Step 3. If we need to construct the primal contingency plans for some λ, go to Step 6.

Step 3. Find all primal potential solutions for the primal augmented model (M9.3). Denote the set of such solutions by $\upsilon = \{K_1, \ldots, K_v\}$. Each K is generically defined as a basis and its corresponding basic variables in **x** and **y** are $\mathbf{x}(K)$ and $\mathbf{y}(K)$, respectively. If we want to PRCP's for each GGS Ω of $\wp$, go to Step 4. If we want to PFCP's for each *GGS* Ω of $\wp$, *go* to Step 5.

Step 4. For each GGS Ω of $\wp$, identify those $G \in \upsilon$ for which $\mathbf{x}(G) \subseteq \mathbf{x}_\Omega(G)$. Denote the resulting subset as $\upsilon(\Omega)$. Given a value of α^*, find $\Gamma_5(G(\Omega))$ (or $\Gamma_3(G(J))$and $\Lambda_5(G(\Omega), \alpha^*)$ (or $\Lambda_3(G(J), \alpha^*)$), so that $G(\Omega)$ (or $G(J)$) is a PRCP for Ω.

Step 5. For each GGS Ω of $\wp$, identify those $G \in \upsilon$ for which $\mathbf{x}(G) \subseteq (\mathbf{x}_\Omega(G), \mathbf{x}^1_{\Omega'}(G))$. Denote the resulting subset as $\upsilon^*(\Omega)$. Given a value of α^*, find the primal parameter set for $G(\Omega)$ (or $\Gamma_4(G(J))$and the dual parameter set for $G(\Omega)$ at α^* (or $\Lambda_4(G(J), \alpha^*)$), so that $G(\Omega)$ (or $G(J)$) is a PFCP for Ω.

Step 6. Find all possible bases for the augmented dual model (M11.5). Denote these bases by $\rho = \{Q_1, Q_2, \ldots, Q_t\}$. Without confusion, Q will designate such a basis, and the corresponding basic variables in $(\mathbf{u}, \mathbf{u}(J))$ and $\mathbf{w}(J)$ will be denoted by $\mathbf{u}(Q)$ and $\mathbf{w}(Q)$ respectively. If we want to DRCP's for each GGS Ω of $\wp$, go to Step 7. If we want to DFCP's for each GGS Ω of $\wp$, go to Step 9.

Step 7. For each GGSΩ of $\wp$, identify its corresponding dual basis from ρ. This can be done by dropping the columns of $\mathbf{w}$ from model (M11.5). Denote such dual bases by $\{Q(\Omega_1), Q(\Omega_2), \ldots, Q(\Omega_r)\}$. For each $Q(\Omega)$, identify those $Q \in \rho$ of Step 6 which has $(\mathbf{u}, \mathbf{u}_\Omega) \supseteq u(Q)$ and $\mathbf{w}(\Omega) \supseteq \mathbf{w}(Q)$. Denote such subsets of ρ by $\rho(\Omega)$.

Step 8. For each GGS Ω of $\wp$, identify the primal bases of model (M12.3) (or (M11.3)) from $\rho(\Omega)$ of Step 7. This can be easily done by dropping the rows of $\mathbf{A}_{\Omega'}$ from (M12.3) (or the second row block of (11.7)) from the simplex tableaus of $\rho(\Omega)$. Denote such reduced bases by $\hat{\mathbf{u}}(\Omega) = \{U_1(\Omega), \ldots, U_k(\Omega)\}$. With a given β^0 value, find the dual parameter set $\Gamma_{10}(U_j(\Omega, \beta^0))$ (or $\Gamma_6(U_j(J, \beta^0))$ and the primal parameter set $\Lambda_{10}(U_j(\Omega))$ (or $\Lambda_6(U_j(J)))$ so that $U_j(\Omega)$ is a DRCP for Ω.

Step 9. For each GGS Ω of $\wp$, identify its corresponding dual basis from ρ. This can be done by dropping the columns of $\mathbf{w}$ from model (M11.5). Denote such dual bases by $\{Q(\Omega_1), Q(\Omega_2), \ldots, Q(\Omega_r)\}$. For each $Q(\Omega)$, identify those $Q \in \rho$ of Step 2 which has $(\mathbf{u}, \mathbf{u}^1(\Omega^*), \mathbf{u}^2(\Omega^*)) \supseteq u(Q)$ and $\mathbf{w}(\Omega) \supseteq \mathbf{w}(Q)$. Denote such subsets of ρ by $\rho^*(\Omega)$.

Step 10. For each GGS Ω, identify the primal bases of model (M12.7) (or (M11.9)) from $\rho^*(\Omega)$ of Step 9. This can be easily done by dropping the rows of $\mathbf{A}_{\Omega''}$ from (M12.8) from the simplex tableaus of $\rho^*(\Omega)$. Denote such reduced bases by $\ddot{\mathbf{u}}(\Omega) = \{U_1(\Omega), \ldots, U_d(\Omega)\}$. With a given β^0 and β^{10} value, find the dual parameter set $\Gamma_{11}(U_j(\Omega, \beta^0, \beta^{10}))$ (or $\Gamma_8(U_j(J, \beta^0, \beta^{10})))$ and the primal parameter set $\Lambda_{11}(U_j(\Omega))$ (or $\Lambda_8(U_j(J)))$ so that $U_j(\Omega))$ is a DFCP for Ω.

Step 11. Evaluate each PGS J and GGS Ω and their corresponding contingency plans from Step 4, Step 5, Step 8, and Step 10 in terms of a criterion preferred by the decision makers (Definition 9.4). The final optimal systems are selected based on these results of evaluation.

The above algorithm is not only a detailed explanation for Algorithm 8.1, but also a summary of Algorithms 9.2-11.2.

12.3.3 Flowcharts of the Algorithm

Based on Algorithm 12.1, *a heuristic* algorithm of the CAS for selecting optimal production systems is presented in Figures 12.1-12.3. Figure 12.1 is the main flowchart of the algorithm. Figure 12.2 is a sub-flowchart that explains box (4) of Figure 12.1 in detail. Furthermore, the details of box (7) in Figure 12.1 are shown by Figure 12.3.

Although the CAS is self-explanatory, in the following we provide a description of the algorithm to emphasize connections between the CAS and Algorithm 12.1.

In Figure 12.1, given a system problem, box (1) deals with all possible input data including the number of possible product variables, the number of criteria, the number of resource levels, the number of constraints, matrices **C**, **A**, **D**, the price of additional resources α, the increments of the opportunities β, and the probability distributions to (γ, λ), etc.

According to the information of box (1), we first locate the set of all PGS's. Then, we decide if we need to prepare the primal contingency plans for some γ (box (2)), or to prepare the dual contingency plans for some λ (box (3)). In the approaches to the primal contingency plans, we can locate PRCP's for all PGS's and GGS's (box (4)) or PFCP's for all PGS's and GGS's (box (5)).

Similarly, in the approaches to the dual contingency plans, we can locate DRCP's for all PGS's and GGS's (box (6)) or DFCP's for all PGS's and GGS's (box (7)). For the sake of space, we only provide the detailed discussion for box (4) in Figure 12.2 and box (7) in Figure 12.3. Figure 12.2 is related to Step 1-4 and Step 11 in Algorithm 12.1, while Figure 12.3 reflects Step 1-2, Step 6, and Step 9-11 in Algorithm 12.1. Note that these boxes correspond some models we discussed in this chapter. The details of box (5) and box (6) can be similarly explained.

After optimal systems have been selected from boxes (4)-(7), the decision makers will judge them in box (8). If the decision makers are not satisfied with the result, box (9) revises the data for box (1). The process is not terminated at box (10) until the decision makers are satisfied with the selected optimal systems.

12.3.4 A Top-Down Design

The *top-down* design is a well-known technique to develop a large and complex computer program. It divides all tasks of solving a target problem by levels in terms of a "tree structure." The target problem is put on the top level. Then, each subproblem is put on a lower level. The tasks on a lower level are more specific. Once the tasks on the lower level are accomplished or the subproblems are solved, we can solve the sub-problems on the higher level and eventually solve the target problem. Thus, the date structures of a computer program for solving the target problem can be developed from the lower levels to the higher levels (see Dale and Orshalick 1983).

A top-down design of the CAS for selecting the optimal production systems is proposed in Figure 12.4. The first level of Figure 12.4 is the Target System Problem (box 1). The second level has three tasks: Input (box 2), Computation (box 3), and Output (box 4). On the third level, Read Data (box 5) and Revise Data (box 6) are subtasks of Input; Primal Design (box 7) and Dual Design (box 8) are subtasks of Computation; and Tableau (box 9) and Expected Payoff (box 10) are subtasks of Output. Continuing in this manner, we get a picture containing every detailed subtask needed to develop the CAS. On the forth level, box 11 of Figure 12.4 is for the primal contingency plans for PGS, which involves models (M9.2) and (M9.3); while box 12 contains models (M10.1) and (M10.2) for the primal contingency plans for GGS. The subtasks of box 8 are two parallel tasks: models (M11.3) and (M11.9) for the dual contingency plans for PGS; and models (M12.3) and (M12.7) for the dual contingency plans for GGS. The subtasks of box 9 are four items: Basic Variable; Parameter Set Γ(.); Parameter set Λ(.); and Payoff **V**(.).

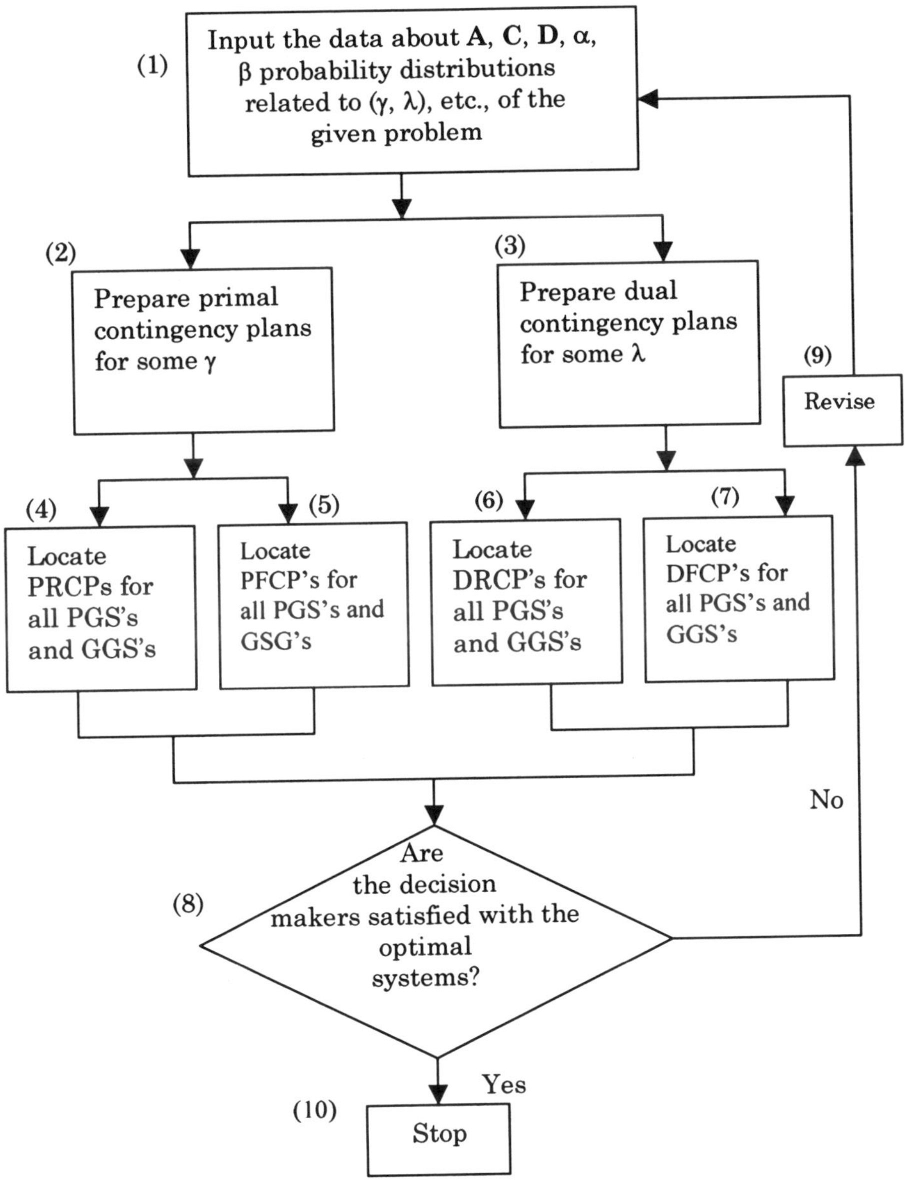

Figure 12. 1 Main Flowchart of CAS for Selecting Optimal Systems

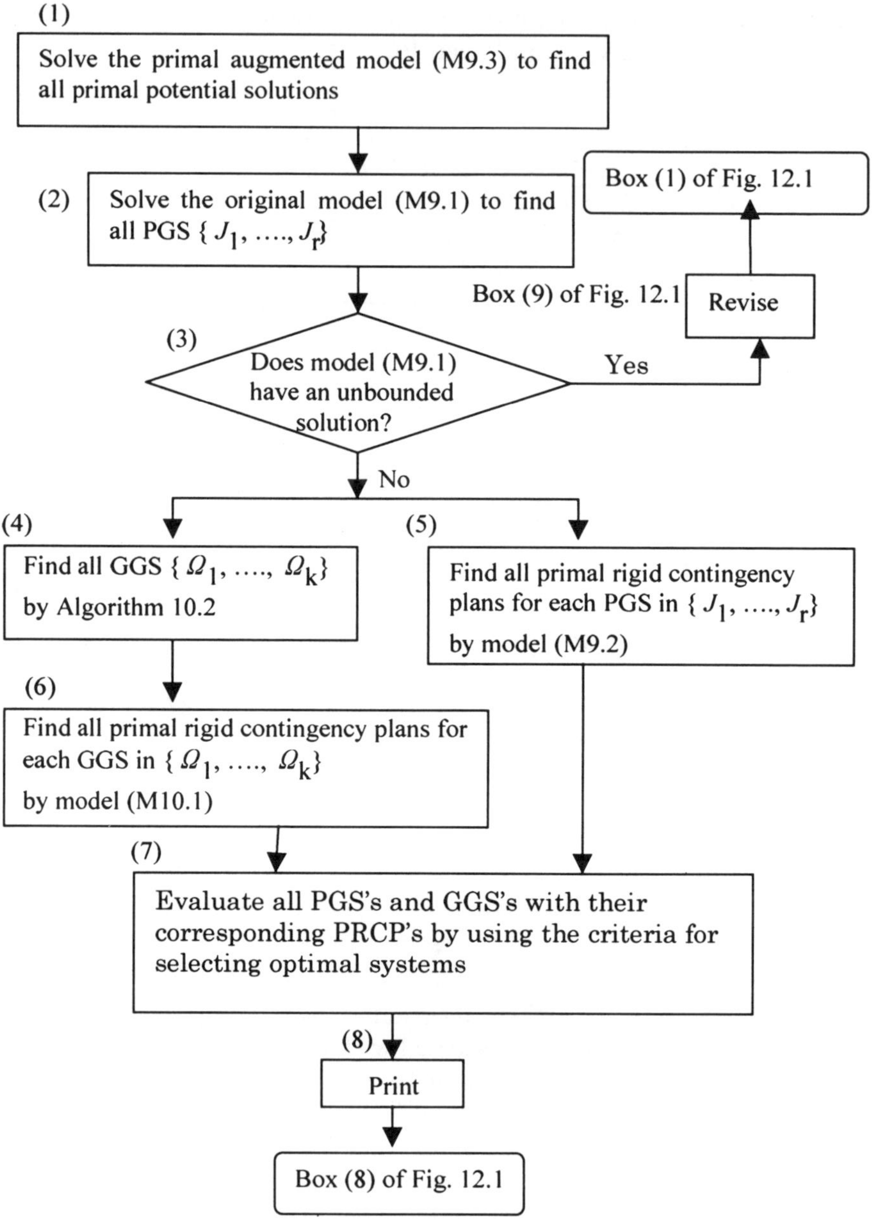

Figure 12. 2 Flowchart Corresponding to Box (4) of Figure 12.1

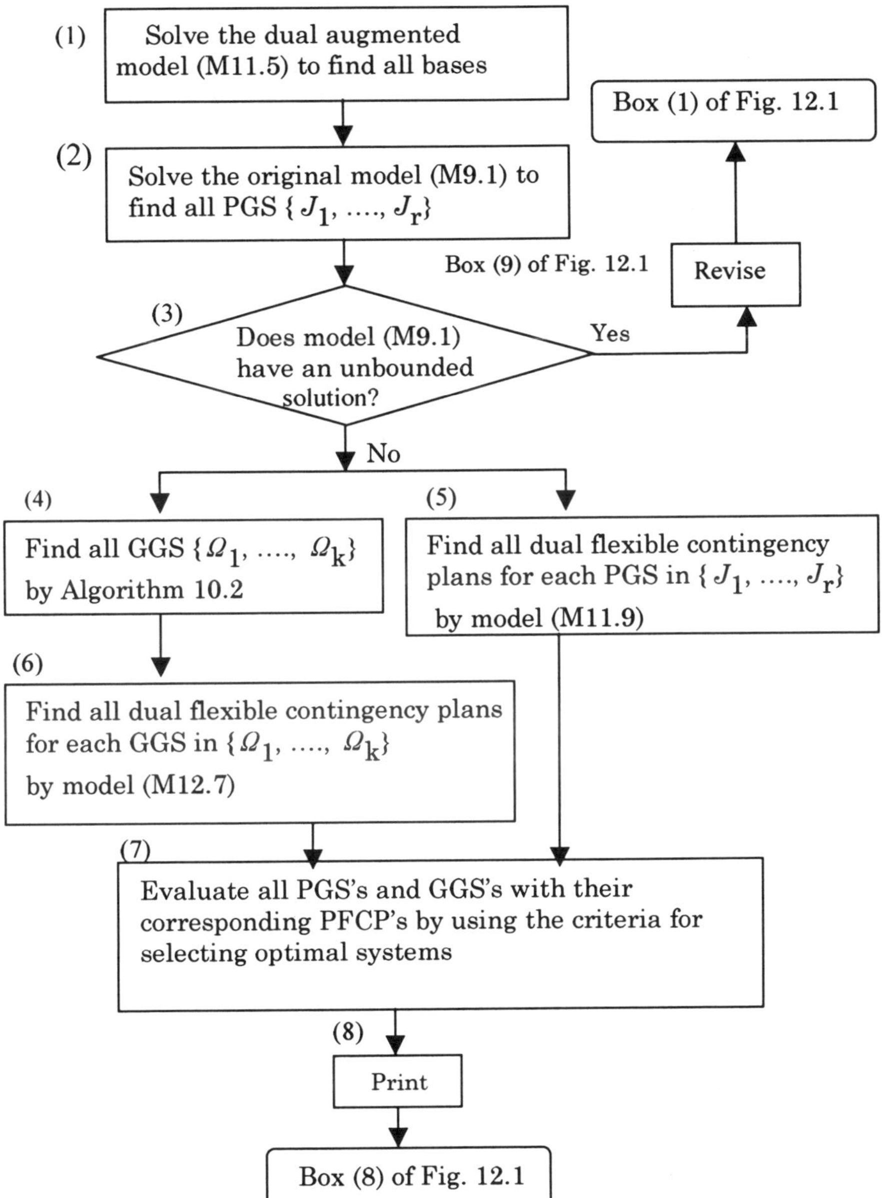

Figure 12.3 Flowchart Corresponding to Box (7) of Figure 12. 1

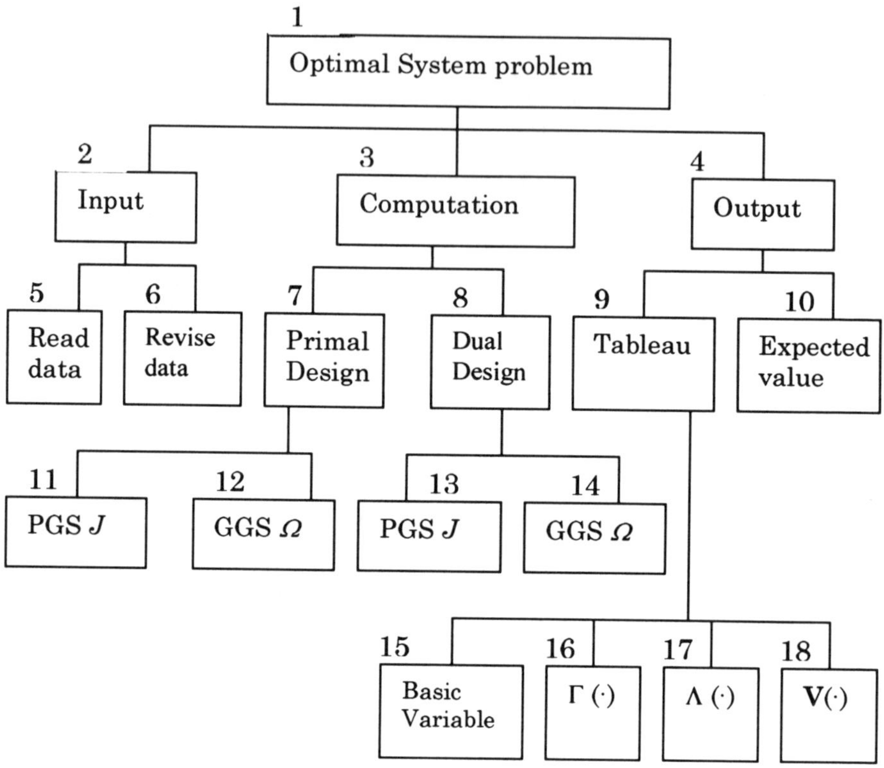

Figure 12.4 Top-Down Design of CAS

According to Figure 12.4, we can develop a series of computer procedures for developing the entire CAS for selecting the optimal production systems. The following are some of the computer procedures needed in the CAS.

(a) Procedures that have an error-check capability to read the given data of a target system problem (box 5) and to revise or update the data (box 6).

(b) Procedure to identify potential solutions for model (M9.1) (box 3).

(c) Procedure to search for a set of all GGS's (box 3). (Note that Algorithm 10.2 can be applied here).

(d) Procedure to identify all bases for model (M9.3) (box 3).

(e) Procedure to identify all bases for model (M11.5) (box 3).

(f) Procedures to identify potential solutions for models (M9.2), (M9.4), (M10.1), and (M10.2) (box 7).

(g) Procedures to identify potential solutions for models (M11.3), (M11.9), (M12.3), and (M12.7) (box 8).

(h) Procedure to evaluate the set of all PGS's and GGS's with their corresponding contingency plans for choosing optimal systems as the final decision (box 4).

(i) Procedures to print out the basic variables, the primal and dual parameter sets, the payoff matrix, and the expected payoff value for each of optimal systems (boxes 10, 15, 16, 17, and 18).

12.3.5 A Pseudo-Language Program

Based on the discussion in Sections 12.3.1-12.3.4, the whole program of the CAS for selecting the optimal systems can be carried out by using a pseudo-language. The pseudo-language program can be converted into any particular computer language for implementation through a series of stepwise refinements (Aho *et al.* 1987). For illustrative purposes, in the following, we write a sub pseudo-language program for selecting optimal systems with PRCPs (refer to boxes (1), (2), (4), (8), and (10) of Figure 12.1, Figure 12.2, and Figure 12.4).

```
Program of Selecting Optimal Systems with Primal Rigid Contingency
Plans;
Begin
   Date_Change: = True;
   While Date Change do
   Begin
      Read Data;
      Print Simplex Tableau; Augment: = True;
      While Augment do
         Begin
            If Primal Pivoting needed then
               Repeat
                  Primal Pivoting;
                  Print Result
                Until done;
```

```
        If Primal Potential Testing needed then
          Repeat
            Primal Potential Testing;
            Print Result
          Until done;
        If Finding Primal Effective and Null Constraint needed
        then
          Repeat
            Primal Effective and Null Constraint Testing;
            Print Result
          Until done;
        If all Primal Solutions of Model (M9.3) found then
          Augment: = False
      end;
    Repeat
      Identify Simplex Tableau of Model (M9.1);
      Print Tableau
    Until all Tableaus of Model (M9.1) found;
    If Model (M9.1) has unbounded solutions then
    go to Read Data;
    For each PGS J do
      Begin
      Repeat
        Identify Simplex Tableau of Model (M9.2);
        Print Tableau
      Until all Tableaus of Model (M9.2) found
      If Model (M9.2) has unbounded solutions then
      go to Read Data;
      end;
    Evaluate the contingency plans for all PGS's;
    Print Expected Payoff
    Find all GGS's;
    For each GGS Ω do
      Begin
      Repeat
        Identify Simplex Tableau of Model (M10.1);
        Print Tableau
      Until all Tableaus of Model (M10.1) found;
      If Model (M10.1) has unbounded solutions then
      go to Read Data;
      end;
```

```
        Evaluate the contingency plans for all GGS's;
        Print Expected Payoff;
        end;
    If Data does not need to be revised then
    Data _ Change: = False
    end:
```

Note that if we want to select optimal systems with PFCPs, the corresponding sub pseudo-language program can be obtained by changing Model (M9.2) to Model (M9.4), and Model (M10.1) to Model (M10.2), respectively in the above pseudo language program. Similarly, we can develop a pseudo-language program for preparing DRCPs and DFCPs. Also, from the main flowchart in Figure 12.1, we can write the entire pseudo-language program of the CAS for selecting optimal systems by merging these pseudo-language programs.

In summary, based on flowcharts of the algorithm (Section 12.3.3), the top-down design (Section 12.3.4), and the pseudo-language program (Section 12.3.5), professional programmers should be able to implement the CAS.

12.4 Further Remarks

The author and his colleagues have extensively studied linear optimal system designs in the MC^2 framework in the last one and half decade. The research project, on one hand, has a significant breakthrough from the traditional optimal design scenario for engineers; on the other hand, is mathematically tedious and boring-some for practitioners. It is why the length of this chapter exceeds other chapters. However, for the reader who is only interested in the general ideas of this chapter, the concise paper of Shi and Yu (1995) will be a good choice. Other related topics, such as De Novo Programming approaches and optimal system designs will be discussed in Chapter 14.

Exercises

12.1 Develop a method of effectively and systematically locating DRCPs for all GGSs of the following design problem (see Shi and He (1994)):

Max	$-4\, x_1 - 2\, x_2$
Max	$x_1 + x_2$
Subject to	$x_1 + 3\, x_2 \leq (40, 30)$
	$2\, x_1 + x_2 \leq (10, 15)$
	$x_1, x_2 \geq 0.$

12.2 Develop a method of effectively and systematically locating DFCPs for all GGSs of problem 12.1 (see Shi (1998a)).

12.3 Write three pseudo-language programs for selecting optimal systems with PFCPs, DRCPs, and DFCPs, respectively.

Part Five: Related Approaches

Chapter 13
SATISFICING AND COMPROMISE MODELS

It has been recognized that each human being has a set of ideal goals to achieve and maintain (Yu 1981, Yu 1985 and Shi and Yu 1989). Such goals include: (i) survival and security of the self: physiological health, safety, and freedom from danger; (ii) perpetuation of the species: sex, giving birth to the next generation, family love and welfare; (iii) feelings of self-importance: self-respect, self-esteem, accumulation of wealth and power, dominance, recognition, prestige, achievement, etc.; (iv) social approval: esteem and respect from others, affiliation, conformity with a desired group, giving and accepting sympathy and protectiveness; (v) sensual-gratification: sexual, visual, auditory, smell, taste, tactile; (vi) cognitive consistency and curiosity: consistency in thought and opinion, exploring and acquiring knowledge, truth, beauty and religion; (vii) self-actualization: the ability to accept and depend on the self, to rely on one's own standard, to cease identifying with others.

Note that each goal in abstract is a function and the ideal (ideal value or the ideal goal) is a particular value of the (goal) function at which we want to achieve and maintain. The goal functions are under our continual monitoring, consciously or subconsciously.

Observe that some goals, such as accumulation of wealth, reputation and achievement, are self-suggested, adjustable and can be characterized by "more is better". Other goal settings, such as body temperature and other biochemical states of the body are fairly fixed without much room for self-adjustment or change. Up too much or down too much from the ideal setting is not desirable. The deviation can be disastrous. Once a perceived state (or value) has deviated from its target or ideal state (or value), a charge (tension or pressure) will be produced to prompt an action to reduce or eliminate the deviations. This behavior of taking actions, including adjustment of the ideal values, to move the perceived states to the targeted ideal states will be called *goal-seeking behavior*.

This goal-seeking behavior that has been well documented in psychological literature (for instance, see Lindsay and Norman 1972), has an important and pervasive impact on human decision making. In this chapter, we shall focus on two concepts that define human goal-seeking behavior (Yu 1973, Yu 1985, and Shi and Yu 1989). The first one is related to satisficing models, which are treated in Section 13.1, the other is related to compromise models including goal programming which are included in Section 13.2. Finally, Section 13.3 offers further remarks. All of the theoretical foundation of this chapter heavily depends on Chapter 4 of Yu (1985).

13.1 Satisficing Models

To facilitate our discussion let us consider a simple multiple criteria decision making (MCDM) problem:

Example 13.1

Mr. Davis is planning to make extra money on the coming Saturday. He has 14 hours available for work and leisure. And he would like to work at least five hours and spend no more than eight hours for leisure activities. Assume that he can earn \$10 per hour and he can work as long as he wants at the same rate. In addition to making money, Mr. Davis has a life enjoyment goal that specifies one and five units of satisfaction respectively for each hour spent in work and leisure. Suppose Mr. Davis wants to make at least \$145 and to reach the total enjoyment of life of at least 50 units whenever possible.

Can Mr. Davis achieve his Goals? And if not, what would you suggest him to do?

According to Mr. Davis' story, let x_1 and x_2 be his time (measured in hours) allocated for work and leisure respectively. The Davis' problem can be formulated as model (M13.1):

His two objective functions or goals can be specified by:

$$y_1 = f_1(\mathbf{x}) = 10\, x_1$$
$$y_2 = f_2(\mathbf{x}) = x_1 + 5\, x_2$$

the decision space x as a subset of $\mathbf{R}^2$ is defined by

$$g_1(\mathbf{x}) = x_1 + x_2 \leq 14$$
$$g_2(\mathbf{x}) = x_1 \geq 5$$
$$g_3(\mathbf{x}) = x_2 \leq 8$$
$$\text{and } x_1,\, x_2 \geq 0.$$

His ideal goals are specified by

$$y_1 = f_1(\mathbf{x}) = 10\, x_1 \geq 145, \text{ and}$$
$$y_2 = f_2(\mathbf{x}) = x_1 + 5\, x_2 \geq 50$$

From the example, we see that Mr. Davis' goal function are "more is better" and the goal values are adjustable. We shall continue on this example after we give a formal description of our models.

13.1.1 Goal Setting

Definition 13.1 Goal setting for satisficing models is defined as the procedure of identifying a satisficing set **S** such that, whenever the decision outcome is an element of set **S**, the decision maker will be happy and satisfied and is assumed to have reached the optimal solution.

Let our MC problem be specified by the criteria $\boldsymbol{f} = (f_1, \ldots, f_q)$ and the feasible set (decision space in Definition 3.1) is $\mathbf{X} = \{ \mathbf{x} \in \mathbf{R}^n \mid \boldsymbol{g}(\mathbf{x}) \leq 0\}$, where $\boldsymbol{g} = (g_1, \ldots, g_m)$. There are two kinds of goal functions. In the first kind, goal values are adjustable and "more is better" (or "less is better"). For this kind of goal function, say $f_i(\mathbf{x})$, one may specify two critical values a_i and b_i (a_i for above and b_i for below), $b_i \leq a_i$, such that whenever $f_i(\mathbf{x}^0) < b_1$ (i.e. $\mathbf{x}^0$ yields a value below the lower critical value for f_i), $\mathbf{x}^0$ is unacceptable; and whenever $f_i(\mathbf{x}^0) \geq a_i$ (i.e. $\mathbf{x}^0$ yields a value above the upper critical value for f_i), $\mathbf{x}^0$ is unacceptable with respect to f_i (although some other criteria may turn $\mathbf{x}^0$ down). Note that since f_i is an adjustable criterion, its corresponding critical values $\{b_i, a_i\}$ may be subject to change as time and events change. For criteria of this kind, we may add

$$f_i(\mathbf{x}) \geq b_i \tag{13.1}$$

to the constraints, so defining a new feasible set **X**; and

$$f_i(\mathbf{x}) \geq a_i \tag{13.2}$$

will be used to define the satisficing set **S**.

Observe that b_i and a_i may be interpreted as "the minimum limit" and the "aspiration level" for f_i, respectively. Also observe that whenever (13.2) is satisfied, (13.1) is satisfied. In satisficing solutions, (13.2) is important for modeling; in compromising solutions, (13.1) is also important in the formulation. Note that when $b_i = a_i$, the two inequalities (13.1) and (13.2) become the same.

If the monotonicity is reversed, f_i exhibits "less is better". Then (13.1) and (13.2) can be replaced by (recall that $b_i \leq a_i$)

$$f_i(\mathbf{x}) \leq b_i\,, \tag{13.3}$$

$$f_i(\mathbf{x}) \leq a_i\,, \tag{13.4}$$

where (13.3) defines the satisficing set **S** and (13.4) becomes a new constraint for **X**.

Note that in this case b_i becomes the aspiration level, and a_i becomes the maximum limit for f_i. Also note that by using $-f_i$ (instead of f_i), and $-b_i$, and $-a_i$ (instead of b_i, a_i), (13.3) and (13.4) reduce to (13.1) and (13.2).

The second kind of goal functions are those of which the target goal values are fairly fixed and not subject to much change; too much or too little is not acceptable. For this kind of goal function, say $f_j(\mathbf{x})$, one may again identify two values $\{b_j, a_j\}$ with $a_j \geq b_j$ such that x is acceptable if and only if the following inequalities hold:

$$b_j \leq f_j(\mathbf{x}) \leq a_j \,. \tag{13.5}$$

Note that whenever $a_j = b_j$, (13.5) reduces to an inequality. Observe that a_j and b_j are two critical values of f_j so that whenever $f_j(\mathbf{x})$ is *below* b_j or *above* a_j, x is unacceptable.

In contrast to (13.5), there may be a criterion, say f_k for which one may specify two parameters a_k and b_k, $a_k \geq b_k$, such that x is acceptable if and only if

$$f_k(\mathbf{x}) \leq b_k, \tag{13.6}$$

or

$$f_k(\mathbf{x}) \geq a_k, \tag{13.7}$$

For instance, $f_k(\mathbf{x})$ could be a position measurement from a potentially disastrous location, say a nuclear power plant, for which one may specify that any acceptable site of a factory must satisfy

$$|f_k(\mathbf{x})| \geq a_k,$$

where a_k is the minimum distance from the undesirable place. Using $b_k = a_k$, the above inequality reduces to (13.6) and (13.7).

For future reference, let us summarize the above discussion in the following remark.

Remark 13.1 If the decision maker considers each criterion individually, one at a time, he can reach a satisficing set **S** which satisfies one or all of the inequalities of the types (13.1)-(13.7). Note that **S** so defined is a "rectangle" or the union of a number of "rectangles" (see (13.6) and (13.7)). Also note that the satisficing set so defined does not reflect any consideration of tradeoffs among the criteria. When such consideration is important, one can define **S** by the union of

$$\mathbf{S}_k = \{f \mid G_k(f) \geq 0\}, \; k=1,\ldots, r, \tag{13.8}$$

where $\boldsymbol{f}$ is a point in S_k, and G_k is a vector function which reflects the trade off over $\boldsymbol{f}$. Note that when the union contains a single set (i.e., r=1), $S = S_1$ and S is defined by a system of inequalities.

Example 13.2 Let us continue on Mr. Davis' problem. From example 13.1 we see that $S = \{\boldsymbol{f} \mid f_1(\mathbf{x}) - 145 \geq 0,\ f_2(\mathbf{x}) - 50 \geq 0\}$, specifies Mr. Davis' satisficing set.

Example 13.3 Let us considering the problem of selecting an athlete for a basketball team. Assume quickness, f_1, and accuracy of shooting, f_2, of the player essential. Let both f_1 and f_2 be indexed from 0 to 10, where the higher the index, the better the player. Assume that a player, $\mathbf{x}$, will be selected with satisfaction if $f_1(\mathbf{x}) \geq 9$ or $f_2(\mathbf{x}) \geq 9$, or $f_1(\mathbf{x}) + f_2(\mathbf{x}) \geq 15$. Now set $G_1(\boldsymbol{f}) = f_1(\mathbf{x}) - 9$, $G_2(\boldsymbol{f}) = f_2(\mathbf{x}) - 9$, and $G_3(\boldsymbol{f}) = f_1(\mathbf{x}) + f_2(\mathbf{x}) - 15$. We see that each $S_i = \{\boldsymbol{f} \mid G_i(\boldsymbol{f}) \geq 0\}$, i = 1, 2, 3, specifies a satisficing set and $S = S_1 \cup S_2 \cup S_3$ is not convex.

13.1.2 Satisficing Model and its Interactive Methods

Once S, the satisficing set has been determined, finding a satisficing solution $\mathbf{x}^0$, such that $\boldsymbol{f}(\mathbf{x}^0) \in S$, is a mathematical programming problem. Recall from Section 13.1.1.1 that when each criterion is considered individually the satisficing set S will be defined by inequalities of the types of (13.1)-(13.7); and when the criteria are jointly considered, then S may be defined by a form such as (13.8). As (13.8) is the most general form, we shall assume that S is defined by the union of S_k, k=1,..., r, with each S_k defined as in (13.8). Let the criteria space be $Y = \{\mathbf{y} = f(\mathbf{x}) \mid \mathbf{x} \in X\}$ (See Definition 3.1), then with $S = \bigcup_{k=1}^{r} S_k$, $Y \cap S \neq \varnothing$ if and only if there is an S_k such that $Y \cap S_k \neq \varnothing$. As we need only one point of $Y \cap S$ when it is nonempty, we could verify individually if $Y \cap S_k \neq \varnothing$, k = 1,..., r, and find one point from the nonempty intersections.

Note that to verify whether or not $Y \cap S_k$ is empty and to find a point in $Y \cap S_k$, if it is nonempty, we can use the following mathematical program.

Rewrite each set in (13.8) as

$$S_k = \{\boldsymbol{f}(\mathbf{x}) \mid G_{kj}(\boldsymbol{f}(\mathbf{x})) \geq 0,\ j=1,\dots, h_k\}, \qquad (13.9)$$

where G_{kj}, j = 1,..., h_k , are components of G_k and h_k is the number of components of G_k. Recall that all of the goal constraints (13.1)-(13.7), after properly rearranging the signs and terms, can be written as (13.9).

The first LP problem is model (M13.2):

$$\mathbf{V}_k = Min \sum_{j=1}^{h_k} d_j, \tag{13.10}$$

$$Subject\ to\ G_{kj}(f(\mathbf{x})) + d_j \geq 0,\ j=1,\ldots,h_k \tag{13.11}$$

$$\mathbf{x} \in \mathbf{X},\ d_j \geq 0 \tag{13.12}$$

$$j=1,\ldots,h_k.$$

It is readily verified that $\mathbf{Y} \cap \mathbf{S}_k \neq \emptyset$ if and only if $\mathbf{V}_k = \mathbf{0}$ in (13.10). We thus have the following theorem:

Theorem 13.1 Given $\mathbf{S} = \bigcup_{k=1}^{r} \mathbf{S}_k$, a satisficing solution exists if and only if there is at least one k ∈ {1,..., r} such that model (M13.2) yields $\mathbf{V}_k = \mathbf{0}$.

Remark 13.2 The objective function in (13.10) can be replaced by any strictly increasing function of d_j, j= 1..., h_k. In particular, one may use $\sum_{j=1}^{h_k} d_j^p, p \geq 1$. Then the resulting $\mathbf{V}_k$ measures the minimum l_p–distance (see Section 13.2 for further discussion) between $\mathbf{Y}$ and $\mathbf{S}_k$. From a computational point of view the form (13.10), in which p=1, has the advantage of simplicity. Especially when G_{kj}, f, and g are all linear, model (M13.2) can be reduced to a linear program (i.e., a linear goal program which will be discussed in Section 13.1.2).

Remark 13.3 In a degenerate case, the satisficing levels may be reduced to a point and (13.9) becomes

$$\mathbf{S}_k = \{f(\mathbf{x}) \mid G_{kj}(f(\mathbf{x})) = 0,\ j=1,\ldots,h_k\}. \tag{13.13}$$

In this case, model (M13.2) can be replaced by the following model (M13.3):

$$\mathbf{V}_2 = Min \sum_{j=1}^{h_k} \left| G_{kj}(f(\mathbf{x})) \right|$$

Subject to $\mathbf{x} \in \mathbf{X}$.

Then the goal, **0**, is obtained if and only if $\mathbf{V}_2 = \mathbf{0}$. We thus have the following theorem.

Theorem 13.2 If $\mathbf{S}_k$ is defined as in (13.13), then $\mathbf{Y} \cap \mathbf{S}_k \neq \varnothing$ if and only if model (M13.3) yields $\mathbf{V}_2 = \mathbf{0}$.

Remark 13.4 Comparing model (M13.2) with (M13.3), we note that (M13.3) has a simpler form of the objective function and may in general be easier to solve computationally than (M13.2). However, model (M13.3) is not valid for a general satisficing set defined as in (13.9).

Remark 13.5 Observe that when one satisficing solution is found, the decision problem is solved, at least temporarily. Otherwise the decision maker can activate either *a positive problem solving* or *negative problem avoidance* to re-structure the problem. The former will enable careful restudy and restructuring of the problem so as to find a solution, perhaps a new one, which lies in the satisficing set. The latter will try to reduce the aspiration levels or play down the importance of making a good decision, thus "lowering" the satisficing set to have a nonempty intersection with **Y**. While the psychological attitude in problem solving and in problem avoidance may be different, the final consequence is the same. That is, eventually, the newly structured problem enables the decision maker to have $\mathbf{Y} \cap \mathbf{S} \neq \varnothing$.

We shall not discuss the restructuring of a problem to obtain a satisficing solution. Instead we will examine helpful information to aid the decision maker in restructuring his satisficing set **S** and feasible set Y. The following provides some helpful information:

Depending on the formation of **S**, if model (M13.1) is used to identify a satisficing solution, one can produce $\mathbf{x}^k$, k=1, ..., r, which solves (M13.2) with $\mathbf{S}_k$ as the satisficing set. One can then compute $f(\mathbf{x}^k)$, $G_k(f(\mathbf{x}^k))$, the distance from $f(\mathbf{x}^k)$ to $\mathbf{S}_k$, and tradeoff among the $G_k(f(\mathbf{x}^k))$, j=1,... , h_k, at $\mathbf{x}^k$. All of these can be helpful for the decision maker to reframe $\mathbf{S}_k$. If model (M13.3) is used, **0** is the target for $G_k(f(\mathbf{x}))$

to attain. The solution $\mathbf{x}^k$ will thus give a minimum distance from $G_k(\boldsymbol{f}(\mathbf{x}))$ to 0. Again, information on $\boldsymbol{f}(\mathbf{x}^k)$, $G_k(\boldsymbol{f}(\mathbf{x}^k))$, and the minimum distance from $G_k(\boldsymbol{f}(\mathbf{x}^k))$ to 0, k=1, ..., r, and the tradeoff among $G_k(\boldsymbol{f}(\mathbf{x}^k))$, j=1,... , h_k , at $\mathbf{x}^k$ can prove valuable. The decision maker can use this information to realize what he can attain and how far away those attainable points are from his satisficing set.

Suppose that it has been revealed or established that "more is better " for each criterion. Then the revealed preference $\{\succ\}$ contains at least the Pareto preference (see Chapter 3). One can then generate the value $f^*_k = \max\{ f_k(\mathbf{x}) \mid \mathbf{x} \in \mathbf{X}\}$, k=1, ..., q, which gives the best value of $f_k(\mathbf{x})$ over $\mathbf{x}$. This information can help the decision maker to re-structure his **S**. Relaxation of the target goal with reference to the ideal point $(f^*_1, \ldots, f^*_q)$ sequentially as well as iteratively with the decision maker, until a satisficing solution is obtained, is certainly an art. It is especially useful when x and f are fairly fixed and not subject to change.

If possible, (for instance, in bi-criteria cases) displaying **Y** and **S** even if only partially, can help the decision maker conceptualize where **Y** and **S** are, to make it easier to re-structure the problem.

We summarize the concept of satisficing solutions and the procedure of using it to obtain a final satisficing solution in Figure 13.1.

Observe that Box (1) is clear. Box (2) is described in Sections 13.1.1 and 13.1.2. Boxes (5), (7) and (9) are clear for a conclusion of decision analysis. Note that related information includes those points in $\mathbf{Y} \cap \mathbf{S}$ and the corresponding points in **X**. For Boxes (6) and (8), one can refer to Remark 13.5.

One most keep in mind that with **X**, $\boldsymbol{f}(\mathbf{x})$ and **S**, one can generate as much information as one wishes, just as one can generate as many statistics from sample values as he wants. However, relevant information (e.g., useful statistics) that can positively affect the problem solving may not be too much. Irrelevant information can become a burden for decision analysis and decision making.

Remark 13.6 Everyone is subject to suggestions, including self-suggestion, to change his perception of decision outcomes and satisficing sets. As each individual is unique, there is no universal way that is most effective for every individual for changing the satisficing set and obtaining a final solution. Changing or relaxing the aspiration levels of the criteria one at a time, two at a time or all simultaneously, will depend on the individual. A good decision aid should be able to supply relevant information to facilitate proper change of the satisficing set in order to get a good final solution.

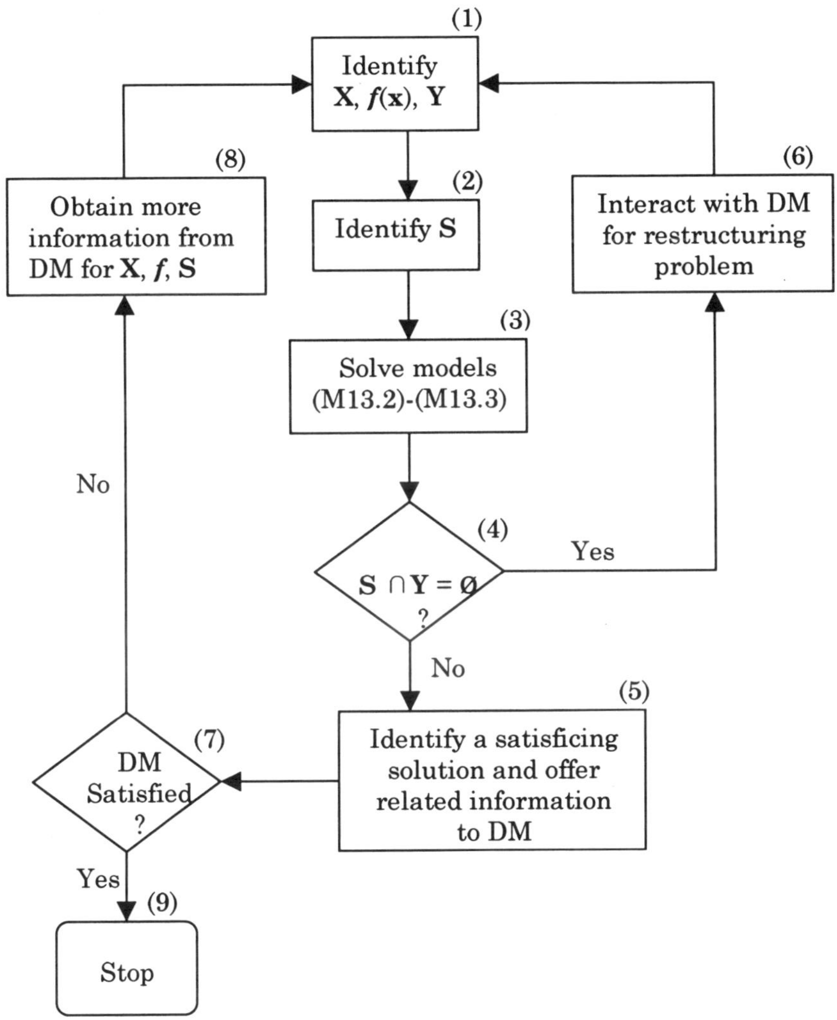

Figure 13.1 Interactive Process

Now we are ready to handle Mr. Davis' problem.

Example 13.2

For illustration purposes, we draw the decision space **X** and the outcome space **Y** of Mr. Davis' problem in Figure 13.2 and Figure 13.3 respectively.

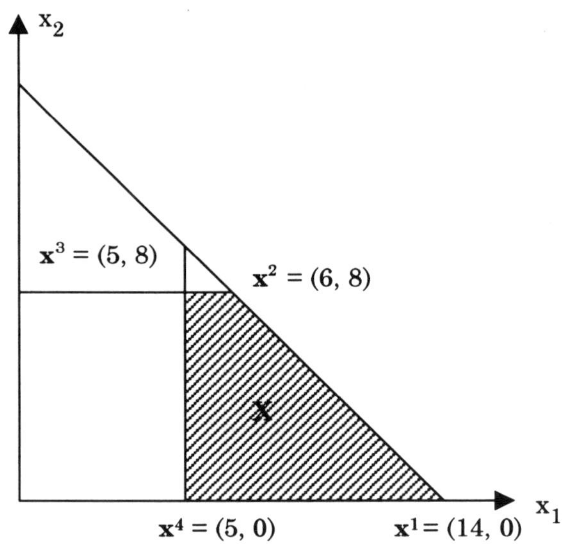

Figure 13.2 Decision Space **X**

In Example 13.2, we have known that Mr. Davis' initial satisficing set is $\mathbf{S} = \{ f_1(\mathbf{x}) \mid 10x_1 - 145 \geq 0,\ x_1 + 5x_2 - 50 \geq 0\}$. Then using (M13.2), we solve the following program:

$$
\begin{aligned}
\mathbf{V}_1 = &\ Min\ d_1 + d_2 \\
\text{Subject to}\ &\ 10x_1 + d_1 \geq 145 \\
&\ x_1 + 5x_2 + d_2 \geq 50 \\
&\ \mathbf{x} \in \mathbf{X} \\
&\ d_1, d_2 \geq 0,
\end{aligned}
$$

where **X** defined in the constraints of model (M13.1).

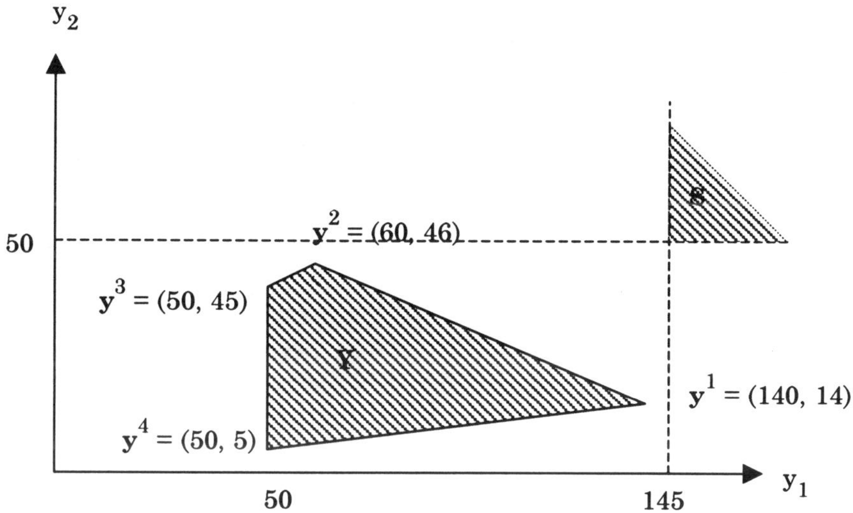

Figure 13.3 Outcome Space **Y**

Observe that from Figure 13.3, $\mathbf{Y} \cap \mathbf{S} = \varnothing$ as the solution to the above program $\mathbf{V}_1 = 41 > 0$ indicated. Thus Mr. Davis cannot reach his goals. We may suggest that he restructure, $\boldsymbol{f}$ and/or $\mathbf{S}$ to find a new satisficing solution. Now suppose that Mr. Davis restructures his satisficing by lowering down the aspiration levels, said $\mathbf{S}' = \{ f(\mathbf{x}) \mid 10x_1 - 135 \geq 0,\ x_1 + 5x_2 - 16 \geq 0\}$. We see that the program becomes

$$\mathbf{V}_1' = Min\ d_1' + d_2'$$

$$\text{Subject to} \quad 10x_1 + d_1' \geq 135$$

$$x_1 + 5x_2 + d_2' \geq 16$$

$$\mathbf{x} \in \mathbf{X}$$

$$d_1', d_2' \geq 0,$$

which has the optimal solution $\mathbf{V}_1' = 0$. That means $\mathbf{Y} \cap \mathbf{S}' \neq \varnothing$ and there is a satisficing solution $\mathbf{y} = (135, 16)$.

13.2 Compromise Models

In this section, mathematical formulation of goal seeking behavior in terms of a distance function will be described. Because it is simple to understand and compute, the concept has much general appeal to many scholars. We shall start with the simplest form and then gradually extend the concept to more complex cases. In Section 13.2.1, the basic concept and preference structure of compromise models are introduced (Freimer and Yu 1976 and Yu 1985). Section 13.2.2 discusses the issues of compromise solutions. While Section 13.2.2.1 is devoted to computational aspects including goal programming and *maximin* programming and Section 13.2.2.2 to a discussion on dynamic adjustment of ideal points for a final optimal solution. Finally, Section 13.2.3 provides an interactive method for compromise models.

13.2.1 Regret Functions and Compromise Solutions

Let the satisficing set **S** of Section 2.2 be a set of only one point $\mathbf{y}^*$. That is, $\mathbf{y}^*$ is a unique target. Typically, when each criterion f_i, i = 1, …, q, is characterized by "more is better", one can set $\mathbf{y}^* = (y^*_1, \ldots, y^*_q)$ where $y^*_i = \sup \{f_i(\mathbf{x}) \mid \mathbf{x} \in \mathbf{X}\}$. In this case $\mathbf{Y}^*$ is called an *ideal* (or *utopia*) point because $\mathbf{y}^*$ is usually not attainable. Note that individually the y^*_i may be attainable. But to find a point $\mathbf{y}^*$ which can simultaneously maximize each f_i, i = 1, .. , q, is usually very difficult.

Example 13.3

To find Mr. Davis' ideal point, we solve the following two linear programs (recall Example 13.1).

For y^*_1, we maximize $y_1 = f_1(\mathbf{x}) = 10x_1$ subject to $\mathbf{x} \in \mathbf{X}$ defined the constraints of model (M13.1). The solution is $x^*_1 = 14$, $x^*_2 = 0$. Thus, $y^*_1 = f_1(x^*) = 140$. Similarly, for y^*_2, we maximize $y^*_2 = f_2(\mathbf{x}) = x_1 + 5x_2$ subject to $\mathbf{x} \in \mathbf{X}$. The solution is $x^*_1 = 6$, $x^*_2 = 8$. Thus, $y^*_2 = f_2(x^*) = 46$. Therefore, Davis' ideal point is $\mathbf{y}^* = (140, 46)$, which is specified in Figure 13.4.

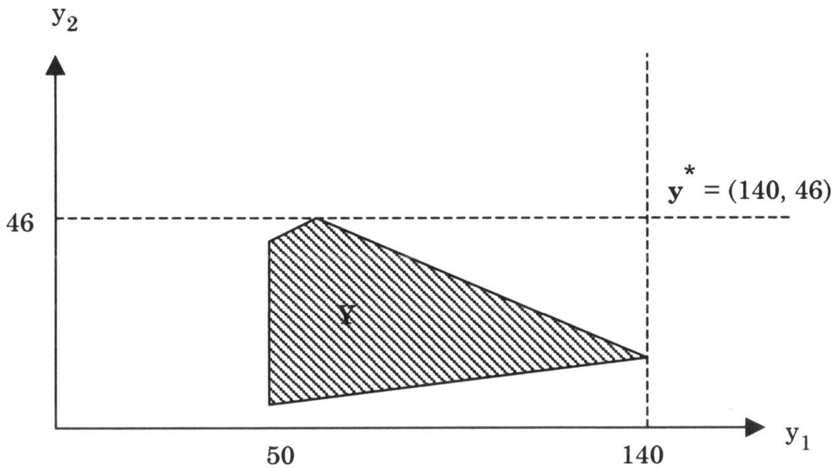

Figure13.4

In group decision problems, if each criterion represents a player's payoff, then $\mathbf{y}^*$, if obtainable, would make each player happy because it simultaneously maximizes each player's payoff. Even if he is the dictator he cannot do better than y^* for himself. As $\mathbf{y}^*$ is usually not attainable, to dissolve a group conflict, a compromise is needed if no other alternative is available. This offers a natural explanation of why the solution to be introduced is called a *compromise solution* (Yu 1973). The reader can extend this explanation easily to multiple criteria problems.

Now, given $\mathbf{y} \in \mathbf{Y}$, the *regret* of using y instead of obtaining the ideal point $\mathbf{y}^*$ may be approximated by the distance between $\mathbf{y}$ and $\mathbf{y}^*$. Thus we define the (group) regret of using y by

$$r(\mathbf{y}) = \|\mathbf{y} - \mathbf{y}^*\|, \tag{13.14}$$

where $\|\mathbf{y} - \mathbf{y}^*\|$ is the distance from $\mathbf{y}$ to $\mathbf{y}^*$ according to some specified norm. Typically, the l_p–norm will be used in our discussion because of its ease in understanding, unless otherwise specified. To make this more specific, define

$$\text{for } p \geq 1,\ r(\mathbf{y}; p) = \|\mathbf{y} - \mathbf{y}^*\|_p = [\sum_i |y_i - y_i^*|^p]^{1/p} \tag{13.14}$$

and

$$\text{for } p = \infty,\ r(\mathbf{y}; \infty) = \max\{|y_i - y_i^*|, i = 1, \ldots, q\} \tag{13.15}$$

Then $r(\mathbf{y};\, \mathrm{p})$ is a measurement of regret from $\mathbf{y}$ to $\mathbf{y}^*$ according to the l_p–norm. Note that the isovalued curves of $r(\mathbf{y};\, \mathrm{p})$ centered at $\mathbf{y}^*$ have some special forms as in Figure 13.5 for the case q = 2. They change shape from a "square-diamond" for p = 1, to a circle for p = 2 and to a square for p = ∞.

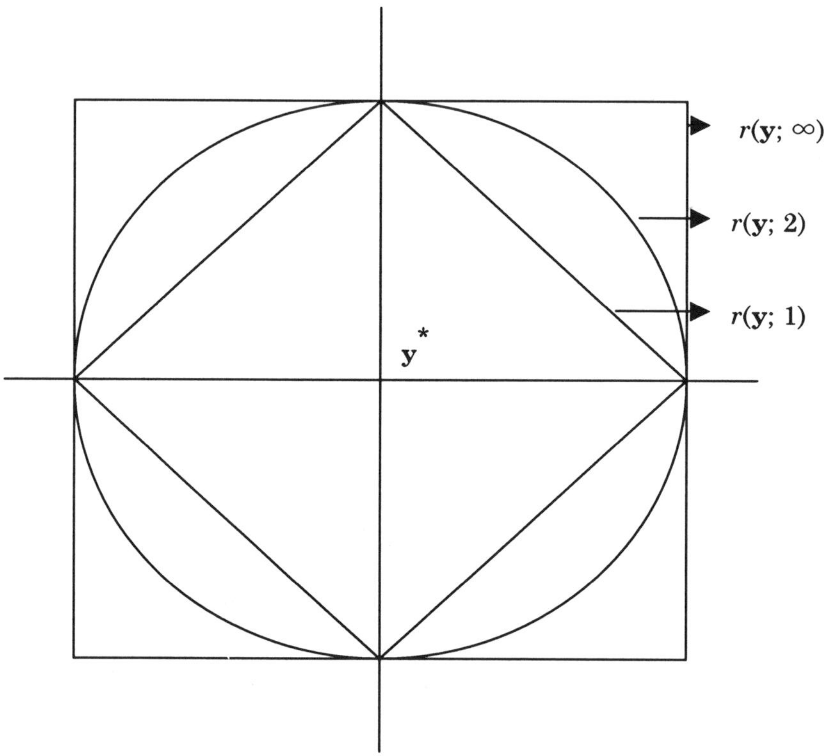

Figure 13.5

Definition 13.2 The compromise solution with respect to the l_p–norm is $\mathbf{y}^p \in \mathbf{Y}$, which minimizes $r(\mathbf{y}; p)$ over $\mathbf{Y}$; or is $\mathbf{x}^p \in \mathbf{X}$, which minimizes $r(f(\mathbf{x}); p)$ over $\mathbf{X}$.

Example 13.4

Using simple isovalued curves of Figure 13.5, we find compromise solutions $\mathbf{y}^1 = (140, 14)$, $\mathbf{y}^2 = (128, 18)$, and $\mathbf{y}^\infty = (117.5, 23)$ for Mr. Davis in Figure 13.6, where $\mathbf{y}^1$ minimizes $r(\mathbf{y}; 1)$, $\mathbf{y}^2$ minimizes $r(\mathbf{y}; 2)$, and $\mathbf{y}^\infty$ minimizes $r(\mathbf{y}; \infty)$. The mathematical solution for finding compromise solutions will be discussed in Section 13.2.2.2.

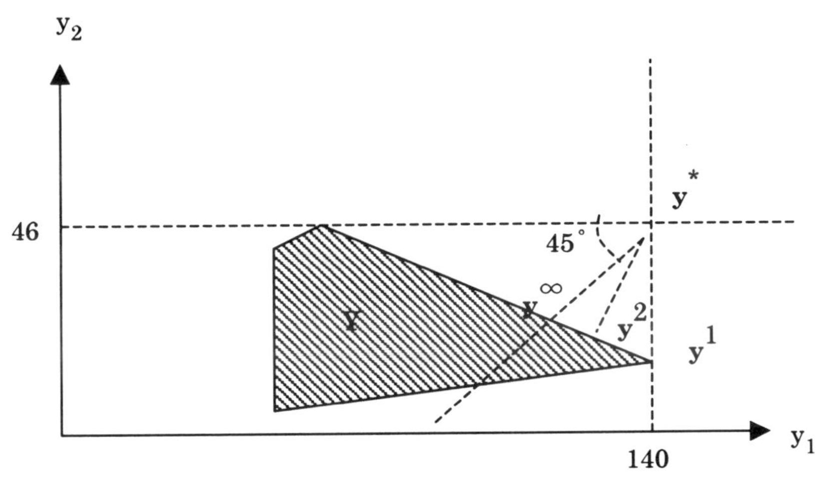

Figure 13.6

Remark 13.7 In group decision problems, $r(\mathbf{y}; p)$ may be interpreted as group regret and the compromise solution $\mathbf{y}^p$ is the one which minimizes the group regret in order to maintain a "cooperative group spirit". As the parameter p varies, the solution $\mathbf{y}^p$ can change.

Note that $r(\mathbf{y}; p)$ treats each $|y_i - y_i^*|$ as having the same importance in forming group regret. In multiple criteria problems, if the criteria have different degrees of importance, then a weight vector $\mathbf{w} = (w_1, w_2, \ldots, w_q) \geq \mathbf{0}$ may be assigned to signal the different degrees of importance. In this case, we define

$$r(\mathbf{y}; p, \mathbf{w}) = \|\mathbf{y} - \mathbf{y}^*\|_{p,w} = [\sum_i w_i^p |y_i - y_i^*|^p]^{1/p} \tag{13.16}$$

The concept of compromise solutions can then naturally be extended to include the regret function of (13.16).

Definition 13.3 The compromise solution with respect to $r(\mathbf{y}; p, \mathbf{w})$ of (13.16) is $\mathbf{y}^{pw} \in \mathbf{Y}$, which minimizes $r(\mathbf{y}; p, \mathbf{w})$ over $\mathbf{Y}$; or is $\mathbf{x}^{pw} \in \mathbf{X}$, which minimizes $r(f(\mathbf{x}); p, \mathbf{w})$ over $\mathbf{X}$.

Remark 13.8 Observe that the weight vector w as in (13.16) has the effect of changing the scale of each criterion. Once the scale is adjusted, (i.e. by $w_i y_i = w_i f_i(\mathbf{x})$ for all $i = 1, \ldots, q$), the regret function is reduced to that of equal weight. Thus, in studying the properties of compromise solutions, without loss of generality, one may focus on the equal weight case. We shall assume from now on unless specified otherwise.

Remark 13.9 Observe that the compromise solution is *not scale dependent*. Scale independence, an important criterion in group decision problems, can prevent players from artificially changing the scale to obtain a better arbitration for themselves. Note that when the scale of $f_i(\mathbf{x})$ or y_i is changed, so is the compromise solution.

Definition 13.4 Given $r(\mathbf{y}; p)$, we define the preference using $r(\mathbf{y}; p)$ by $\mathbf{y}^1 \succ \mathbf{y}^2$ if and only if $r(\mathbf{y}^1; p) < r(\mathbf{y}^2; p)$.

Theorem 13.3 The preference using $r(\mathbf{y}; p)$ is a weak order and, for any $\mathbf{y}^1$ and $\mathbf{y}^2$, $\mathbf{y}^1 \sim \mathbf{y}^2$ if and only if $r(\mathbf{y}^1; p) = \mathbf{r}(\mathbf{y}^2; p)$.

Remark 13.10 Theorem 13.3 reveals that the concept of compromise solutions implicitly has assumed a very strong preference assumption. There are no ambiguities in preference comparison. For any two points $\mathbf{y}^1$ and $\mathbf{y}^2$, either $\mathbf{y}^1$ is better or worse or equivalent to $\mathbf{y}^2$. Exactly one of the three cases must happen. There are no such things as "indefinite" in the preference of compromise solutions. Before we finally use it, precaution is needed to verify that the assumption is fulfilled; otherwise, a misleading solution may result.

13.2.2 Computing Compromise Solutions

To find a compromise solution, we need to know the *target point*, either the ideal point or not, and the parameter p for the regret function $r(f(\mathbf{x}); p)$. When the target point is unknown, we may start with the ideal point as an approximation of the target. Hopefully, by an interactive process, we can gradually locate the target point and the compromise solution. When the ideal point is the target point, computing for the compromise solution is not very complicated, at least theoretically. This will be discussed in Section 13.2.2.1. For general target points the computation can be slightly more complex. We shall discuss this in Section 13.2.2.2.

13.2.2.1 The Ideal Point as the Target Point

To find a compromise solution when the ideal point is the target point, we have to solve q + 1 mathematical programming problems. The first q problems are to find the utopia or ideal point, $\mathbf{y}^* = (y^*_1, \ldots, y^*_q)$ where $y^*_i = \max\{f_i(\mathbf{x}) \mid \mathbf{x} \in \mathbf{X}\}$. The last one is to find the compromise solution $\mathbf{y}^p$. Suppose that $\mathbf{X}$ contains only countable points. We will have q + 1 integer programming problems. Otherwise, if $\mathbf{X}$ is a region, we will have q + 1 non-linear programming problems.

Suppose $\mathbf{X}$ and $f_i(\mathbf{x})$, $i = 1,\ldots,q$, have some special structure. More efficient computational techniques are available. For instance, if $\mathbf{X}$ is a convex set and each $f_i(\mathbf{x})$ is concave, then we have first q concave programming and then a convex programming (because $r(f_i(\mathbf{x}); p)$ is convex under the assumptions). If $\mathbf{X}$ is a polyhedron defined by a system of linear inequalities and each $f_i(\mathbf{x})$ is linear, then the ideal points $\mathbf{y}^*$ can be found by q simple linear programming problems. Furthermore, the compromise solutions of $\mathbf{y}^1$ and $\mathbf{y}^\infty$ can be found by a linear programming problem (the other compromise solution $\mathbf{y}^p$, $1 < p < \infty$ can be found by convex programming). That $\mathbf{y}^1$ can be found by linear programming is obvious. That $\mathbf{y}^\infty$ can be found by linear programming follows directly from the following well-known result:

Theorem 13.4 $\mathbf{x}^0$ solves

$$\text{Model (M13.4): } \min_x \max_i \{| y_i^* - f_i(\mathbf{x}^0) |, i = 1,\ldots,q, \mathbf{x} \in \mathbf{X}\}$$

if and only if $\mathbf{x}^0$ and $\mathbf{v}^0 = \max_i \{y_i^* - f_i(\mathbf{x}^0)\}$ solve

Model (M13.5): $\min \{\mathbf{v} \mid \mathbf{v} \geq y_i^* - f_i(\mathbf{x}^0), i = 1,\ldots,q, \mathbf{x} \in \mathbf{X}\}$.

Example 13.5 Let us use mathematical programs to find the compromise solutions $\mathbf{y}^1$, $\mathbf{y}^2$ and $\mathbf{y}^\infty$ for Mr. Davis when the target point is the ideal point.

From Example 13.3, we know that the ideal point $\mathbf{y}^* = (140, 46)$. Thus,

(i) for the $\mathbf{y}^1$ compromise solution, we solve

$$\min \{[140 - 10x_1] + [46 - (x_1 + 5x_2)]\}, \mathbf{x} = (x_1, x_2) \in \mathbf{X}.$$

We obtain $x^*_1 = 14$, $x^*_2 = 0$, and $\mathbf{y}^1 = (140, 14)$.

(ii) for the $\mathbf{y}^2$ compromise solution, we solve

$$\min \{[140 - 10x_1]^2 + [46 - (x_1 + 5x_2)]^2\}, \mathbf{x} = (x_1, x_2) \in \mathbf{X}.$$

The solution is $x^*_1 = 12.8$, $x^*_2 = 1.04$, and $\mathbf{y}^2 = (128, 18)$.

(iii) for the $\mathbf{y}^\infty$ compromise solution, we solve

$$\min \max [140 - 10x_1, 46 - (x_1 + 5x_2)], \mathbf{x} = (x_1, x_2) \in \mathbf{X}.$$

According to Theorem 13.4, the above problem is simplified as follows:

$$\begin{aligned} &\min \mathbf{v} \\ &\text{Subject to } \mathbf{v} \geq 140 - 10x_1 \\ &\qquad\qquad \mathbf{v} \geq 46 - (x_1 + 5x_2) \\ &\qquad\qquad \mathbf{x} \in \mathbf{X}. \end{aligned}$$

We obtain $x^*_1 = 11.75$, $x^*_2 = 2.25$, and $\mathbf{y}^\infty = (117.5, 23)$.

Table 13.1 is a summary of compromise solutions, when p = 1, 2, and ∞, with respect to the ideal point $\mathbf{y}^* = (140, 46)$.

Table 13.1 Compromise Solutions

	(x_1, x_2)	(y_1, y_2)
p = 1	(14, 0)	$\mathbf{y}^1 = (140, 14)$
p = 2	(12.8, 1.04)	$\mathbf{y}^2 = (128, 18)$
p = ∞	(11.75, 2.25)	$\mathbf{y}^\infty = (117.5, 23)$

13.2.2.2 General Target Points and Goal Programming

Suppose that the target point for compromise solutions is known but is not the ideal point. Then each f_i may not be the type of "more is better". For instance, if the target for f_i is 10 and f_i [**x**] = [8, 20], then $f_i = 15$ is attainable but not as good as $f_i = 10$. Thus the preference is not monotone over f_i. However, moving as close to the target as possible is still the same. Note the resulting compromise solution for the general point, may not be Pareto optimality over **Y**.

The computation for compromise solution for general point can be slightly more complicated. This is due to the possibility that $y^*_i \geq y_i$ for all $y \in \mathbf{Y}$, and i = 1, ..., q may no longer hold. The following method of changing variables can be of help.

Theorem 13.5 Let

$$d_i^+ = \begin{cases} y_i - y_i^*, \text{ if } y_i > y_i^* \\ 0, \text{ otherwise;} \end{cases}$$

$$d_i^- = \begin{cases} y_i^* - y_i, \text{ if } y_i < y_i^* \\ 0, \text{ otherwise.} \end{cases}$$

Then

(i) $y_i^* - y_i = d_i^- - d_i^+$

(ii) $|y_i^* - y_i| = d_i^- + d_i^+$

(iii) $d_i^-,\ d_i^+ \geq 0.$

Note that d_i^+ is the value of y_i exceeding y^*_i (like a surplus variable) and d_i^- is the value of y_i below y^*_i (like a slack variable). The proof of Theorem 13.5 is straightforward.

Theorem 13.6 The definition of d_i^- and d_i^+ in Theorem 13.5 is satisfied if and only if d_i^- and d_i^+ are non-negative and satisfy:

$$y_i^* - y_i = d_i^- - d_i^+ \text{ and } d_i^- d_i^+ = 0$$

Now given a target point **y***, the compromise solution with l_p-distance can be found by solving Model (M13.4):

$$Min \sum_i | y_i^* - f_i(\mathbf{x}) |^p = Min[r(f_i(\mathbf{x}); p)]^p$$

Subject to $g(\mathbf{x}) \leq 0$

Note that since $[r(f_i(\mathbf{x}); p)]^p$ is a strictly increasing function of $r(f_i(\mathbf{x}); p)$, $\mathbf{x}^0$ maximizes $[r(f_i(\mathbf{x}); p)]^p$ if and only if it minimizes $r(f_i(\mathbf{x}); p)$. This makes it easier to find the compromise solution.

We now can apply Thereoms 13.5-13.6 to remove the sign of the absolute values in model (M13.4) and get model (M13.5):

$$Min \sum_{i=1}^{q} (d_i^- + d_i^+)^p$$

$$\begin{aligned} \textit{Subject to} \quad & y_i^* - f(\mathbf{x})_i = d_i^- - d_i^+ \\ & d_i^-, d_i^+ \geq 0 \\ & g(\mathbf{x}) \leq 0 \end{aligned}$$

where i=1, 2,..., q.

Theorem 13.7 Given a target point **y***, $\mathbf{x}^0$ is the compromise solution with respect to the l_p-distance if and only if $\mathbf{x}^0$ solves model (M13.5).

Remark 13.11

(i) Suppose that all $f_i(\mathbf{x})$, i = 1,...,q, and $\mathbf{g}(\mathbf{x})$ are linear. Then model (M13.5) for l_1 -compromise solutions reduces to a linear program typically known as (linear) *goal programming.* By adding weights to or imposing lexicographical ordering on the criteria one can generalize the concept discussed here to a variety of goal programming formats. We note that goal programming is a special class in the domain of compromise solutions. The major advantages of goal programming and l_1-compromise solutions are that they are easily understood and that they can be easily computed by linear programs.

(ii) When all $f_i(\mathbf{x})$, i=1,...,q, and $\mathbf{g}(\mathbf{x})$ are linear, model (M13.5) for l_2-compromise solutions becomes a quadratic program. The program can be solved without major difficulty.

(iii) When all $f_i(\mathbf{x})$, i=1,...,q, and $\mathbf{g}(\mathbf{x})$ are linear, model (M13.5) for l_∞-compromise solutions becomes a linear program by applying Theorem 13.4. This greatly simplifies the computations.

Example 13.6

Suppose instead of the ideal point $\mathbf{y}^* = (140,46)$, Mr. Davis sets (130,46) as his target point. We can find the compromise solutions by using model (M13.5) as follows:

(i) For the $\mathbf{y}^1$-compromise solution, we solve

$$\begin{aligned} &\textit{Min} \quad (d_1^- + d_1^+) + (d_2^- + d_2^+) \\ &\textit{Subject to} \quad 130 - 10x_1 = d_1^- - d_1^+ \\ &\qquad 46 - (x_1 + 5x_2) = d_2^- - d_2^+ \\ &\qquad x_1 + x_2 \le 14 \\ &\qquad x_1 \ge 5 \\ &\qquad x_2 \le 8 \\ &\qquad x_1, x_2 \ge 0,\ d_1^-, d_1^+, d_2^-, d_2^+ \ge 0. \end{aligned}$$

The solution is $d_1^- = d_1^+ = 0$, $d_2^- = 28.5$, $d_2^+ = 0$, $x_1 = 13$, and $x_2 = 0.9$.

Thus, the $\mathbf{y}^1$-compromise solution is $\mathbf{y}^1 = (130,17.5)$.

(ii) For the $\mathbf{y}^2$-compromise solution, we solve

$$\begin{aligned} &\textit{Min} \quad (d_1^- + d_1^+)^2 + (d_2^- + d_2^+)^2 \\ &\textit{Subject to} \quad 130 - 10x_1 = d_1^- - d_1^+ \\ &\qquad 46 - (x_1 + 5x_2) = d_2^- - d_2^+ \\ &\qquad x_1 + x_2 \le 14 \end{aligned}$$

$$x_1 \geq 5$$
$$x_2 \leq 8$$
$$x_1, x_2 \geq 0, d_1^-, d_1^+, d_2^-, d_2^+ \geq 0.$$

We get $d_1^- = 11$, $d_1^+ = 0$, $d_2^- = 23.5$, $d_2^+ = 11.9$, and $x_2 = 2.12$. Therefore, the $\mathbf{y}^2$ -compromise solution is $\mathbf{y}^2 = (119, 22.5)$.

(iii) For the $\mathbf{y}^\infty$ -compromise solution, we have

Min Max $[d_1^- + d_1^+, d_2^- + d_2^+]$

Subject to $130 - 10x_1 = d_1^- - d_1^+$

$$46 - (x_1 + 5x_2) = d_2^- - d_2^+$$
$$x_1 + x_2 \leq 14$$
$$x_1 \geq 5$$
$$x_2 \leq 8$$
$$x_1, x_2 \geq 0, d_1^-, d_1^+, d_2^-, d_2^+ \geq 0.$$

Using Theorem 13.4, we simplify the above program as follows:

Min v

Subject to $v \geq d_1^- + d_1^+$

$$v \geq d_2^- + d_2^+$$
$$130 - 10x_1 = d_1^- + d_1^+$$
$$46 - (x_1 + 5x_2) = d_2^- + d_2^+$$
$$x_1 + x_2 \leq 14$$
$$x_1 \geq 5$$
$$x_2 \leq 8$$
$$x_1, x_2 \geq 0, d_1^-, d_1^+, d_2^-, d_2^+ \geq 0.$$

Solving the program, we get $d_1^- = 20$, $d_1^+ = 0$, $d_2^- = 20$, $d_2^+ = 0$, $x_1 = 11$, $x_2 = 3$, and $v = 20$. The $\mathbf{y}^\infty$ -compromise solution is $\mathbf{y}^\infty = (110, 26)$.

We summarize the $\mathbf{y}^p$-compromise solutions, when p = 1, 2, and ∞, with respect to the target point $\mathbf{y}^* = (130, 46)$ as Table 13.2.

Table 13.2 Compromise Solutions

	(x_1, x_2)	(y_1, y_2)
p = 1	(13, .09)	$\mathbf{y}^1 = (130, 17.5)$
p = 2	(11.9, 2.12)	$\mathbf{y}^2 = (119, 22.5)$
p = ∞	(11, 3)	$\mathbf{y}^\infty = (110, 26)$

13.2.3 Interactive Methods of Compromise Models

For non-trivial decision problems, an optimal decision usually cannot be reached by applying the compromise solution only once. The following interactive method, according to Figure 13.7, may be helpful.

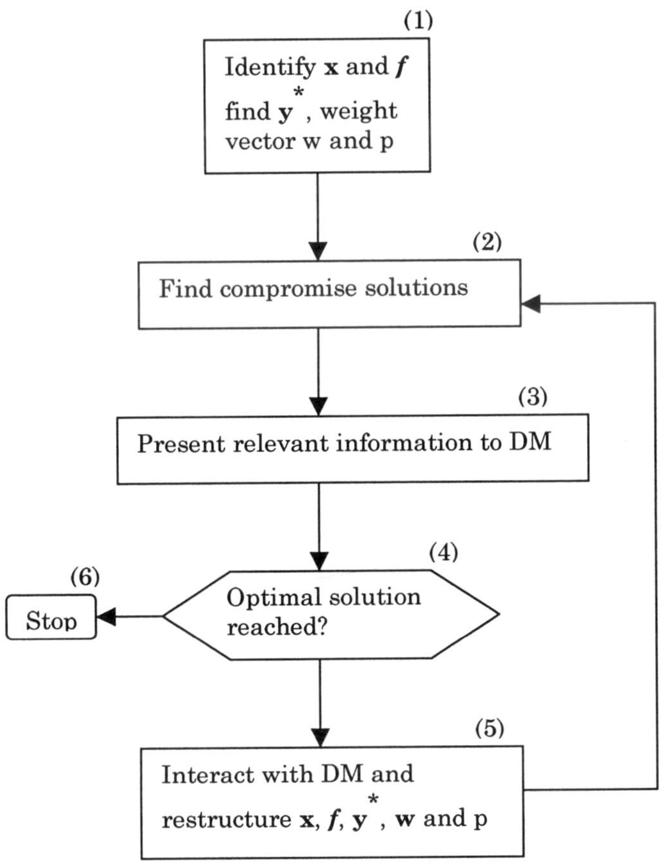

Figure 13.7

Box (1) Identifying **x**, *f* and **y*** is an art that requires careful observation and conversation with the decision maker. The target point **y***, weight vector **w** and p for l_p-distance may be more difficult to specify precisely. However, we can use the ideal point as the first step of approximation for **y***, select some representative weights for **w** and let p = l, 2, ∞ to begin.

Box (2) One can locate compromise solutions according to model (M13.5) for the specifier values of **y***, **w** and p.

Box (3) All relevant information obtained in Box (l) and (2) can be presented to the decision maker. These include **X**, *f*, **Y**, ideal points $\mathbf{y}^{*}_{i}$, i=1, ...,q, the optimal vector value which maximizes the ith criterion, the compromise solutions $\mathbf{y}^{l}$, $\mathbf{y}^{2}$, $\mathbf{y}^{\infty}$ with their weights and norms.

Box (4) and (6) If an optimal solution is reached in Box (3) and the decision maker is satisfied, the process is terminated at Box (6), otherwise, we go to Box (5) to obtain more information.

Box (5) Through conversation, we may obtain new information on **x**, *f*, **y***, weight vector w and norm parameter p. Note that each of these five elements may change with time. To help the decision maker locate or change the target point, we may use the "one at a time method." That is, only one value of f_i or y_i, i=1,..., q, is to be changed and the rest remain unchanged. We may ask the decision maker: "Since no satisfactory solution is obtained with the current target point **y***, would it be possible to decrease (or increase) the value of $\mathbf{y}^{*}_{i}$? And, if so, by how much?" This kind of suggestive question may help the decision maker think hard and carefully of possible changes in the target points. The other method is called the "pairwise trade-off method." We first select two criteria, say f_i and f_j, and then ask: "To maintain the same degree of satisfaction for the target point, everything else being equal, how many units of f_j must be increased in order to compensate a one-unit decrease in f_i? Again, this kind of suggestive question can help the decision maker think hard and carefully of the possible changes in the target points. Finally, we recall that the target point may be regarded as the satisficing set containing a single satisficing solution. Thus, by locating the satisficing set, we may locate the target point.

Remark 13.12

Comparing Section 13.1.1 with Section 13.1.2, we know that satisficing models are more general and flexible in applications, while compromise models are more specific and rigid. The former is loose while the latter makes sense only when strong assumptions are satisfied. However, with flexible interpretation and artful applications in interaction with the decision maker, the two methods can complement each other. After all, both concepts stem from the goal seeking assumption and try to minimize the distance from the criteria space to a set of satisficing or a target point. In a broader perspective, they are the same.

Example 13.7

Assume that Mr. Davis understands the above interactive method. Our interactive process with him to reach a final solution can be as follows:

Step 1. Identify $\mathbf{X}$, f, $\mathbf{y}^*$, $\mathbf{w}$, and p. The decision space $\mathbf{X}$, is given by

$$g_1(\mathbf{x}) = x_1 + x_2 \leq 14$$
$$g_2(\mathbf{x}) = x_1 \geq 5$$
$$g_3(\mathbf{x}) = x_2 \leq 8$$
$$x_1, x_2 \geq 0.$$

and the objective function f_1 and f_2 are

$$y_1 = f_1(\mathbf{x}) = 10x_1$$
$$y_2 = f_2(\mathbf{x}) = x_1 + 5x_2$$

Through conversation with Mr. Davis, it is reasonable to assume that the ideal point is a good approximation for the target point. The ideal point (y^*_1, y^*_2) is (146,46) (see Example 13.3). Mr. Davis agrees that each criterion carries the same weight (i.e., $w_1 = w_2$). We use p =1, 2, ∞ to find the compromise solutions.

Step 2. Find compromise solutions. Refering to Example 13.4 or 13.5, we know that the compromise solutions are $\mathbf{y}^1 = (140, 14)$, $\mathbf{y}^2 = (128,18)$, $\mathbf{y}^\infty = (117.5,23)$.

Step 3. Present relevant information to the decision maker (DM). All relevant information in Steps 1 and 2 is represented to Mr. Davis. Since we are in the two-dimension case, the graphic display in Figure 13.6 can greatly help Mr. Davis to visualize what he can and cannot reach.

Step 4. Optimal Solution Reached? Assume that Mr. Davis *is not satisfied* with the compromise solutions presented to him in Step 3.

Step 5. Restructure new model. Suppose Mr Davis, through conversation with us, prefers to move the target point closer to outcome space **Y** by reducing \$10 from $\mathbf{y}^*_1$ and maintain the equal weights and p values.

Thus, the new target point is $\mathbf{y}^* = (130,46)$ and the new compromise solutions as illustrated in Example 13.6 are $\mathbf{y}^1 = (130, 31.5)$, $\mathbf{y}^2 = (126,33)$, and $\mathbf{y}^\infty = (118,34.5)$ can be located. Once more, suppose that Mr. Davis is still not satisfied with those new compromise solutions, and tells us that his importance weights over the criteria are different, say $\mathbf{w} = (3/4,\ 1/4)$, and **X**, *f*, p, and $\mathbf{y}^*$ as in Example 13.6 remain unchanged. We can again recalculate $\mathbf{y}^1$, $\mathbf{y}^2$ ant $\mathbf{y}^\infty$. The process will continue until Mr. Davis is satisfied with some solutions or quits.

13.3 Further Remarks

This chapter has demonstrated two kinds of models to capture human goal-seeking behavior in decision making processes. The satisficing models seek the intersection of the outcome (objective or criteria) space and the satificing set that can change according to the decision makers' desire. The compromise models are the extensions of the well-known goal linear programming with different norms. Because these models can be flexibly used to interpret the conflict criteria, the results of this chapter have a great impact on many practical research projects such as information management (Chapter 18) and rural telecommunications management (Chapter 19), and credit card portfolio management (Chapter 20).

Exercises

13.1 Given the following MC problem:

Max $f_1(\mathbf{x}) = 10\,x_1 + 12\,x_2$

Max	$f_2(\mathbf{x}) = 15\ x_1 + 8\ x_2$
Subject to	$g_1(\mathbf{x}) = 3\ x_1 + 2x_2 \leq 24$
	$g_2(\mathbf{x}) = 4x_1 + 6x_2 \leq 48$
	$x_1,\ x_2 \geq 0.$

(i) Find the point in the Decision Space **X** that maximize $f_i(\mathbf{x})$, i = 1,2.

(ii) Find the ideal solution $\mathbf{y}^* = (\ y^*_1,\ y^*_2\ )$ that have the maxima of the objective functions.

(iii) Draw both **X** and Outcome Space **Y**.

(iv) Set up your satisficing set **S** and use model (M13.2) or (M13.3) to find the satisfiscing solutions.

(v) If the satisficing solution is not the desired one, use the interactive process in Figure 13.1 to find the preferred solution.

(vi) Compare this solution with the nondominated solution in Problem 3.1 of Chapter 3.

13.2 Find the compromise solutions $\mathbf{y}^1$, $\mathbf{y}^2$ and $\mathbf{y}^\infty$ for Problem 13.1 when the target point.

13.3 Consider the following MC problem:

Max	$f_1(\mathbf{x}) = 4\ x_1 - 3\ x_2$
Max	$f_2(\mathbf{x}) = 5\ x_1 + 9\ x_2$
Subject to	$g_1(\mathbf{x}) = 2x_1 + 4x_2 \leq 20$
	$g_2(\mathbf{x}) = x_1 + \ x_2 \leq 10$
	$x_1,\ x_2 \geq 0.$

(i) Find the point in the Decision Space **X** that maximize $f_i(\mathbf{x})$, i = 1,2.

(ii) Find the ideal solution $\mathbf{y}^* = (\ y^*_1,\ y^*_2\ )$ that have the maxima of the objective functions.

(iii) Draw both **X** and Outcome Space **Y**.

(iv) Set up your satisficing set **S** and use model (M13.2) or (M13.3) to find the satisficing solutions.

(v) If the satisficing solution is not the desired one, use the interactive process in Figure 13.1 to find the preferred solution.

(vi) Find the compromise solutions $\mathbf{y}^1$, $\mathbf{y}^2$ and $\mathbf{y}^\infty$ for the MC problem when the target point.

(vii) Compare the compromise solution with the satisficing solution.

Chapter 14
DE NOVO PROGRAMMING WITH MULTIPLE CRITERIA

The multiple criteria De Novo programming, initiated by Zeleny (1975), is a practical approach to design optimal linear systems for a given decision problems with multiple criteria. In the traditional linear multiple criteria (MC) programming, given a resource availability level (right hand side) for a number of constraints, a set of decision variables is selected to optimize multiple objective functions. Since the resource availability level is determined a *priori*, the result of MC programming is the optimal given system. In contrast, with an unknown resource availability level, the multiple criteria De Novo programming seeks a portfolio of resource availability level to optimize multiple objective functions by allocating a budget according to the resource price. This approach results in an optimal system design. Zeleny (1990) proposed a basic method to construct the optimal system design for solving a De Novo problem via an ideal system design. Li and Lee (1990) then extended Zeleny's basic method to identify fuzzy system designs for De Novo problems.

One of the important issues in the multiple criteria De Novo programming is to determine an optimum-path ration for enforcing a particular budget level of resources so as to establish the optimal system design. In Zeleny's basic method, the optimum-path ratio for achieving the multicriteria performance of the ideal system design related to a given budget level was used to determine the optimal system design. However, the possibility of setting other optimum-path ratios for enforcing different budget levels of resources, derived from linear systems to find alternative optimal system designs in not explored.

In this chapter, we first present several optimum-path ratios for finding optimal system designs in the multiple criteria De Novo programming in Section 14.1. Each optimum-path ratio is chosen to enforce a given budget level to reach a realized performance of the system design. As optimum-path ratios are different, we can establish corresponding different optimal system designs as options for the decision maker. This idea may also be used to formulate and solve optimal pattern matching problems (Zeleny 1990). In Section 14.2, we propose an approach of multiple criteria De Novo programming to dealing with system design problems with multiple decision makers and possible debt. In the system design model, each involved decision maker has his or her own preference for the budget availability level with respect to the criteria (objectives) under consideration. If possible debt occurs in the design time, the model allows flexibility to borrow additional money from a bank with a fixed interest rate to keep the production process feasible. The contingency plan therefore must be prepared to cope with the debt situation. The further remarks will be given in Section 14.3.

14.1 Optimum-path Ratios

14.1.1 A Model of De Novo Programming

In this section, the basic concepts of the De Novo programming with multiple criteria are sketched to facilitate studies on optimum-path ratios (see Zeleny 1990 for further details).

A De Novo programming with multiple criteria is formulated as model (M14.1):

$$\begin{aligned} &\textit{Max} && \mathbf{C}\,\mathbf{x} \\ &\textit{Subject to} && \mathbf{A}\,\mathbf{x} \le \mathbf{b} \\ & && \mathbf{p}\,\mathbf{b} \le B \\ & && \mathbf{x} \ge 0, \end{aligned}$$

where $\mathbf{C} \in \mathbf{R}^{q\times n}$ and $\mathbf{A} \in \mathbf{R}^{m\times n}$ are matrices of dimensions $(q \times n)$ and $(m \times n)$, respectively; and $\mathbf{b} \in \mathbf{R}^m$ is the m-dimensional *unknown resource vector*, $\mathbf{p} \in \mathbf{R}^m$ is the vector of the *unit prices* of m resources, and B is the given total available *budget*.

Solving model (M14.1) means finding the optimal allocation of the budget B so that the resulting portfolio of resources $\mathbf{b}$ maximizes the values of the product mixes, $\mathbf{Cx}$ simultaneously. Such an optimal allocation is referred as the *optimal system design* (Zeleny 1990). According to Zeleny (1990), a basic method of solving model (M14.1) is first to transform it into model (M14.2):

$$
\begin{aligned}
&Max && \mathbf{Z} = \mathbf{C}\,\mathbf{x} \\
&Subject\ to && \mathbf{V}\,\mathbf{x} \le B \\
& && \mathbf{x} \ge 0,
\end{aligned}
$$

where $\mathbf{Z} = (z_1, ..., z_q) \in \mathbf{R}^q$ and $\mathbf{V} = (v_1, ..., v_n) = \mathbf{pA} \in \mathbf{R}^n$.

Let $z_k^* = Max\ z_k$, k = 1, ..., q, be the optimal objective value for the kth criterion of model (14.2) subject to $\mathbf{Vx} \le B$, $\mathbf{x} \ge \mathbf{0}$. Let $\mathbf{Z}^* = (z_1^*, ..., z_q^*)$ be the q criteria performance (objective value) for the ideal systems design with respect to B. Then, a meta-optimal problem is constructed as model (M14.3):

$$
\begin{aligned}
&Min && \mathbf{V}\,\mathbf{x} \\
&Subject\ to && \mathbf{C}\,\mathbf{x} = \mathbf{Z}^* \\
& && \mathbf{x} \ge 0.
\end{aligned}
$$

Solving model (M14.3), we can obtain $\mathbf{x}^*$, B^* (= $\mathbf{Vx}^*$) and $\mathbf{b}^*$ (= $\mathbf{Ax}^*$). The value B^*, called the meta-optimum budget, is the minimum budget to achieve $\mathbf{Z}^*$ through $\mathbf{x}^*$ and $\mathbf{b}^*$. Since $B^* \ge B$ (see Theorem 14.1), the optimum-path ratio for achieving the ideal performance $\mathbf{Z}^*$ related to a given budget level B is defined as $r^* = B/B^*$. Finally, the optimal system design is constructed as ($\mathbf{x}$, $\mathbf{b}$, $\mathbf{Z}$), where $\mathbf{x} = r^*\mathbf{x}^*$, $\mathbf{b} = r^*\mathbf{b}^*$, and $\mathbf{Z} = r^*\mathbf{Z}^*$. To distinguish other optimum-path ratios discussed in the next section from r^*, we shall call r^* the basic *optimum-path ratio.*

14.1.2 Several Optimum-path Ratios in De Novo Programming

In this section, based on the basic optimum-path ratio r^*, we study several optimum-path ratios as alternatives for selecting the optimal system designs to solve De Novo programming problems.

Given a De Novo problem (M14.1), we transform it into model (M14.2), which is a "knapsack" linear program with q criteria. Let **C** = $(c_j^k)_{qxn}$ be the criteria coefficient matrix, where c_j^k is the coefficient for the jth variable on the kth criterion. Without loss of generality, we assume $c_j^k > 0$, $v_j > 0$ and $B > 0$. Consider the kth criterion of model (M14.2) subject to $\mathbf{Vx} \le B$, $\mathbf{x} \ge 0$, which is a linear knapsack program. Let $c_a^k/v_a = Max \{c_j^k/v_j \mid 1 \le j \le n\}$. Then, the optimal solution is $x_j^k = B/v_a$ if j = a; 0, if j ≠ a. Using x_j^k, we have $b_j^k = \mathbf{A}x_j^k$ and $B_j^k = \mathbf{V}\, x_j^k$, k = 1, ..., q. Note that b_j^k is the budget level of production x_j^k with respect to the kth criterion. The design (x_j^k, b_j^k, z_k*) is an optimal system design for a De Novo programming problem with a single criterion (Zeleny 1990). Applying it to model (M14.1), some optimum-path ratios can be constructed for identifying various optimal system designs.

Given a problem of model (M14.1), we assume q ≤ n (i.e., the number of criteria is less than or equal to the number of variables under consideration). Suppose we individually solve k single criterion De Novo programs and obtain solutions x_{j1}^1, x_{j2}^2, ..., x_{jq}^q. Without loss of generality, suppose j1 ≤ j2 ≤ ... ≤ jq. Then, we define $\mathbf{x}^{**} = (x_{j1}^1, x_{j2}^2, ..., x_{jq}^q, 0, 0, ..., 0)$ as a *synthetic optimal solution* from the solutions x_{j1}^1, x_{j2}^2, ..., x_{jq}^q. Thus, we can obtain $\mathbf{b}^{**} = \mathbf{Ax}^{**}$, $\mathbf{Z}^{**} = \mathbf{Cx}^{**}$ and $B^{**} = \mathbf{Vx}^{**}$. To utilize the design ($\mathbf{x}^{**}$, $\mathbf{b}^{**}$, $\mathbf{Z}^{**}$) in our study, we propose a flexible-constraint meta-opitmal problem as model (M14.4):

$$
\begin{aligned}
&\mathit{Min} && \mathbf{V\,x} \\
&\mathit{Subject\ to} && \mathbf{C\,x} \ge \mathbf{Z}^* \\
& && \mathbf{x} \ge 0.
\end{aligned}
$$

Note that comparing model (M14.3) with model (M14.4), the former has the equality constraint $\mathbf{Cx} = \mathbf{Z}^*$, while the latter has the larger than or equal constraint $\mathbf{Cx} \le \mathbf{Z}^*$. We observe that the optimal solution $\mathbf{x}^*$ of model (M14.3) is feasible for model (M14.4). The relationships between the budgets B^{**}, B^*, B and B_j^k are summarized in the following theorem that serves as the basis of the optimum-path study.

Theorem 14.1 $B^{**} \ge B^* \ge B \ge B_j^k$, for k = 1,..., q.

Proof. (i) We first show $B \geq B_j^k$: Since x_j^k is feasible for model (M14.2), $\mathbf{V}x_j^k \leq B$. This implies $B_j^k \leq B$, for k = 1 ,..., q.

(ii) We then show $B^* \geq B$: by contradiction, assume that $B^* < B$. We see that $\mathbf{Vx}^* < B$. That is, the meta-optimum solution $\mathbf{x}^*$ is feasible for model (M14.2). Let $\mathbf{x}^{nd}$ be the nondominated solution and $\mathbf{Z}^{nd} = \mathbf{Cx}^{nd}$ be the corresponding nondominated objective value of model (M14.2). Hence, the ideal objective value $\mathbf{Z}^* = \mathbf{Cx}^* \leq \mathbf{Z}^{nd} = \mathbf{Cx}^{nd}$, which is obviously a contradiction.

(iii) We finally show $B^{**} \geq B^*$: Note that both solutions $\mathbf{x}^{**}$ and $\mathbf{x}^*$ are feasible for model (M14.4). Thus, $B^{**} = \mathbf{Vx}^{**} \geq Min\ \mathbf{Vx} = \mathbf{Vx}^* = B^*$ for model (M14.4).

Recall that B_j^k is the budget level of producing x_j^k with respect to the kth criterion. If we know the weight of importance of utilizing each B_j^k to produce x_j^k in terms of the kth criterion, then the combination of $\{B_{j1}^1, ..., B_{jq}^q\}$ can be considered as a budget level for the system designs with q criteria. Denote such a combination by $\sum_{k=1}^{q} \lambda_k B_{jk}^k$ with $\sum_{k=1}^{q} \lambda_k = 1$ and $0 \leq \lambda_k \leq 1$. From Theorem 14.1, we have:

Theorem 14.2 $B \geq \sum_{k=1}^{q} \lambda_k B_{jk}^k$.

Proof. Since $\sum_{k=1}^{q} \lambda_k = 1$ and $0 \leq \lambda_k \leq 1$, this simply follows from $B \geq B_j^k$, for k = 1, ..., q in Theorem 14.1.

By Theorems 14.1 and 14.2, we can define six types of optimum-path ratios as follows:

(i) $r^1 = B^*/B^{**}$; (ii) $r^2 = B/B^{**}$; (iii) $r^3 = \sum_{k=1}^{q} \lambda_k B_{jk}^k / B^{**}$;

(iv) $r^4 = B/B^*$; (v) $r^5 = \sum_{k=1}^{q} \lambda_k B_{jk}^k / B^*$; and (vi) $r^6 = \sum_{k=1}^{q} \lambda_k B_{jk}^k / B$.

In the above optimum-path ratios, $r^4 = r^*$, which is the basic optimum-path ratio studied by Zeleny (1990). The ratio r^1 is the optimum-path ratio for achieving the synthetic-optimum performance $\mathbf{Z}^{**}$ related to a given meta-optimum budget level B^*; r^2 is the optimum-path ratio for achieving the synthetic-optimum performance $\mathbf{Z}^{**}$ related to a given initial budget level B; r^3 is the optimum-path ratio for achieving the synthetic-optimum performance Z^{**} related to a given combined budget level $\sum_{k=1}^{q} \lambda_k B_{jk}^k$; and r^5 is the optimum-path ratio for achieving the ideal performance $\mathbf{Z}^*$ related to a given combined budget level $\sum_{k=1}^{q} \lambda_k B_{jk}^k$; and r^6 is the optimum-path ratio for achieving the optimum performance $\mathbf{Z}^c$ related to a given combined budget level $\sum_{k=1}^{q} \lambda_k B_{jk}^k$. Assume the given initial budget level B in model (M14.1) or (M14.2) can be replaced by either B^* or $\sum_{k=1}^{q} \lambda_k B_{jk}^k$. By using these optimum-path ratios, the following optimal system designs can be established:

(i) $\mathbf{x}^1 = r^1\mathbf{x}^{**}$, $\mathbf{b}^1 = r^1\mathbf{b}^{**}$ and $\mathbf{Z}^1 = r^1\mathbf{Z}^{**}$,
(ii) $\mathbf{x}^2 = r^2\mathbf{x}^{**}$, $\mathbf{b}^2 = r^2\mathbf{b}^{**}$ and $\mathbf{Z}^2 = r^2\mathbf{Z}^{**}$,
(iii) $\mathbf{x}^3 = r^3\mathbf{x}^{**}$, $\mathbf{b}^3 = r^3\mathbf{b}^{**}$ and $\mathbf{Z}^3 = r^3\mathbf{Z}^{**}$,
(iv) $\mathbf{x}^4 = r^4\mathbf{x}^{**}$, $\mathbf{b}^4 = r^4\mathbf{b}^{**}$ and $\mathbf{Z}^4 = r^4\mathbf{Z}^{**}$,
(v) $\mathbf{x}^5 = r^5\mathbf{x}^{**}$, $\mathbf{b}^5 = r^1\mathbf{b}^{**}$ and $\mathbf{Z}^5 = r^5\mathbf{Z}^{**}$, and
(vi) $\mathbf{x}^6 = r^6\mathbf{x}^{nd}$, $\mathbf{b}^6 = r^6\mathbf{b}^{nd}$ and $\mathbf{Z}^6 = r^6\mathbf{Z}^{nd}$.

The meaning of the above optimal systems design ($\mathbf{x}^i$, $\mathbf{b}^i$, $\mathbf{Z}^i$), i = 1,...,6 is that $\mathbf{b}^i$, the optimum portfolio of resources to be acquired at the current market prices p, allows to produce $\mathbf{x}^i$ and realize the multiple criteria performance $\mathbf{Z}^i$. When model (M14.1) or (M14.2) is actually applied to solve real world problems, these designs may be presented to the decision maker as candidates for the final optimal system design.

Example 14.1

Consider the numerical problem of Zeleny (1990):

$$
\begin{aligned}
\textit{Max} \quad & z_1 = 50\,x_1 + 100\,x_2 + 17.5\,x_3 \\
& z_2 = 92\,x_1 + 75\,x_2 + 50\,x_3 \\
& z_3 = 25\,x_1 + 100\,x_2 + 75\,x_3 \\
\textit{Subject to} \quad & 12\,x_1 + 17\,x_2 \le b_1 \\
& 3\,x_1 + 9\,x_2 + 8\,x_3 \le b_2 \\
& 10\,x_1 + 13\,x_2 + 15\,x_3 \le b_3 \\
& 6\,x_1 + 16\,x_3 \le b_4 \\
& 12\,x_1 + 7\,x_3 \le b_5 \\
& 9.5\,x_1 + 9.5\,x_2 + 4\,x_3 \le b_6, \\
& x_j \ge 0,\ j = 1, 2, 3.
\end{aligned}
$$

with the current market prices of resources $p_1 = .75$, $p_2 = .6$, $p_3 = .35$, $p_4 = .5$, $p_5 = 1.15$ and $p_6 = .65$, and the initial budget level $B = 4658.75$.

Since this problem is a special multiple criteria programming problem, we can use the computer software of Hao and Shi (1996) to compute optimal system designs. For the sake of simplicity, we assume the weights of importance of objective functions z_1, z_2, and z_3 are identical. Then, we first find designs (x_j^k, b_j^k, z_k^*), $(\mathbf{x}^*, \mathbf{b}^*, \mathbf{Z}^*)$ and $(\mathbf{x}^{**}, \mathbf{b}^{**}, \mathbf{Z}^{**})$ and their corresponding budget levels as in Table 14.1. We shall call these designs the structure designs, because they are found from the mathematical structure of the given De Novo problem.

For illustrative purposes, let $\lambda_1 = \lambda_2 = \lambda_3 = 1/3$ and $\lambda_1 B_2^1 + \lambda_2 B_1^2 + \lambda_3 B_3^3 = 4658.20$. Then, from Table 14.1, we get optimum-path ratios:

(i) $r^1 = 47.35\%$; (ii) $r^2 = 33.34\%$; (iii) $r^3 = 33.337\%$;
(ii) $r^4 = 70.41\%$; (v) $r^5 = 70.40\%$; and (vi) $r^6 = 99.99\%$.

Finally, based on the above six optimum-path ratios, we construct the corresponding optimal systems designs, which are shown in Table 14.2.

Table 14.1 Structure Designs

$x_2^1 = 109.14$
$b_2^1 = (1855.38, 982.26, 1418.82, 0, 1309.68, 1036.83)$
$z_1^* = 10914$
$B_2^1 = 4658.10$
$x_1^2 = 198.39$
$b_1^2 = (2380.68, 595.17, 1091.40, 1190.34, 0, 1884.71)$
$z_2^* = 18251.88$
$B_1^2 = 4658.20$
$x_3^3 = 162.31$
$b_3^3 = (0, 1298.48, 2434.65, 2596.96, 1136.17, 649.24)$
$z_3^* = 12173.13$
$B_3^3 = 4658.30$
$\mathbf{x}^* = (131.27, 29.69, 78.97)$
$\mathbf{b}^* = (2079.97, 1292.78, 2883.22, 2051.14, 909.07, 1845)$
$\mathbf{Z}^* = (18251.88, 10914, 12173.13)$
$B^* = 6614.92$
$\mathbf{x}^{**} = (198.39, 109.14, 162.31)$
$\mathbf{b}^{**} = (4236.06, 2875.91, 5837.37, 3787.30, 2445.85, 3570.78)$
$\mathbf{Z}^{**} = (23673.93, 34552.88, 28047)$
$B^{**} = 13973.06$

Table 14.2 Optimal Systems Designs

$\mathbf{x}^1 = (93.94, 51.68, 76.85)$ $\mathbf{b}^1 = (2005.77, 1361.74, 2764, 1793.29, 1158.11, 1690.76)$ $\mathbf{Z}^1 = (11209.6, 16360.79, 13280.28)$ $r^1 = 47.35\%$ with using $B^* = 6614.92$ out of $B^{**} = 13973.06$
$\mathbf{x}^2 = (66.14, 36.39, 54.11)$ $\mathbf{b}^2 = (1412.30, 958.83, 1946.18, 1262.69, 815.45, 1190.50)$ $\mathbf{Z}^2 = (7892.89, 11519.93, 9350.87)$ $r^2 = 33.34\%$ with using $B = 4658.75$ out of $B^{**} = 13973.06$
$\mathbf{x}^3 = (66.14, 36.38, 54.10)$ $\mathbf{b}^3 = (1412.18, 958.74, 1945.82, 1262.57, 815.37, 1190.39)$ $\mathbf{Z}^3 = (7892.18, 11519.89, 9350.03)$ $r^3 = 33.337\%$ with using $\lambda_1 B_2^1 + \lambda_2 B_1^2 + \lambda_3 B_3^3 = 4658.20$ out of $B^{**} = 13973.06$
$\mathbf{x}^4 = (92.43, 20.90, 55.60)$ $\mathbf{b}^4 = (1464.51, 910.25, 2030.08, 1444.21, 640.08, 1299.07)$ $\mathbf{Z}^4 = (7684.80, 12851.30, 8571.10)$ $r^4 = 70.41\%$ with using $B = 4658.75$ out of $B^* = 6614.92$
$\mathbf{x}^5 = (92.41, 20.90, 55.59)$ $\mathbf{b}^5 = (1464.30, 910.16, 2029.79, 1444, 640.08, 1299.07)$ $\mathbf{Z}^5 = (7683.71, 12849.47, 8569.88)$ $r^5 = 70.40\%$ with using $\lambda_1 B_2^1 + \lambda_2 B_1^2 + \lambda_3 B_3^3 = 4658.20$ out of $B^* = 6614.92$
$\mathbf{x}^6 = (198.37, 0, 0)$ $\mathbf{b}^6 = (2380.39, 595.10, 1983.66, 1190.66, 0, 1884.48)$ $\mathbf{Z}^6 = (9918.42, 18249.90, 4959.21)$ $r^6 = 99.99\%$ with using $\lambda_1 B_2^1 + \lambda_2 B_1^2 + \lambda_3 B_3^3 = 4658.20$ out of $B = 4658.75$

14.1.3 An Optimal Pattern Matching Problem

As an important observation, we see that all optimal system designs discussed in the above section from the mathematical structure of the De Novo problem (M14.1). The basic idea is that instead of fixing the budget level B in model (M14.1), we use B^{**}, B^*, B and B_j^k to generate the optimum portfolio of resources $\mathbf{b}^i$, i = 1,..., 6, which "reshape" the system boundaries. The optimum-path ratios r^i, in fact, show the degree of enforcing a particular budget level (B^*, B or B_j^k) to optimize the reshaped system. This idea can be applied to formulate and solve an optimal pattern matching problem. The concept of optimal pattern matching problems with multiple criteria was proposed by Zeleny (1990). Given a proposed multiple criteria problem formulation and an optimal problem formulation (or Pattern) which is preferred by the decision maker and related to the proposed problem, the "optimal matching" means to move the former to the latter as close as possible by reshaping the former, if necessary, so that the decision maker is satisfied. In the following, we apply the De Novo problem formulation (M14.1) and the study of optimum-path ratios discussed in the above section to set up a matching problem with a fixed optimal pattern.

Suppose the decision maker accepts the formulation (M14.1) with the initial available budget level B. But, he or she has the preferred solution ($\mathbf{x}^a$, $\mathbf{Z}^a$), with $\mathbf{Z}^a = \mathbf{C}\mathbf{x}^a$, as the fixed optimal pattern. An *optimal pattern problem* can be formulated as model (M14.5):

Matching the formulation (M14.1) to ($\mathbf{x}^a$, $\mathbf{Z}^a$) *as close as possible*

The evaluation of closeness between model (M14.1) and ($\mathbf{x}^a$, $\mathbf{Z}^a$) can be measured by the optimum-path ratio of a chosen budget level divided by the actual budget level for producing $\mathbf{x}^a$.

We note that solving the matching problem (M14.5) involves an interactive process with the decision maker. Whenever we reshape model (M14.1) towards ($\mathbf{x}^a$, $\mathbf{Z}^a$), we should ask the decision maker to justify the resulting optimal system design and the corresponding optimum-path ratio. If the decision maker is satisfied with the result, then the interactive matching process with the budget levels B^{**}, B^*, and B_j^k, derived from the mathematical structure of model (14.1), the following algorithm is developed for solving problem (M14.5).

Algorithm 14.1

Step 1. Given a problem (M14.5) with ($\mathbf{x}^a$, $\mathbf{Z}^a$) and the initial available budget level B for model (M14.1), find $\mathbf{b}^a = \mathbf{A}\mathbf{x}^a$ and $B^a = \mathbf{V}\mathbf{x}^a$, where $\mathbf{b}^a$ is called the actual needed resource level to produce $\mathbf{x}^a$, and B^a is the actual budget level to purchase $\mathbf{b}^a$ for producing $\mathbf{x}^a$.

Step 2. For model (9.6), find budget levels B^{**}, B^{*} and B_j^k, for k = 1..., q. Go to either one of Steps 3-7.

Step 3. If $B^a \leq B_j^k$, for k = 1, ..., q, then set $r^{a1} = \sum_{k=1}^{q} \lambda_k B_{jk}^k / B^a \geq 1$. The resulting optimal system design is $\mathbf{x}^{a1} = r^{a1}\mathbf{x}^a$, $\mathbf{b}^{a1} = r^{a1}\mathbf{b}^a$, and $\mathbf{Z}^{a1} = r^{a1}\mathbf{Z}^a$. The preferred pattern ($\mathbf{x}^a$, $\mathbf{Z}^a$) is matched by ($\mathbf{x}^{a1}$, $\mathbf{b}^{a1}$, $\mathbf{Z}^{a1}$). Go to Step 8.

Step 4. If $B \geq B^a > B_j^k$, for k = 1,..., q, then set $r^{a2} = B/B^a \geq 1$. The optimal system design is determined as $\mathbf{x}^{a2} = r^{a2}\mathbf{x}^a$, $\mathbf{b}^{a2} = r^{a2}\mathbf{b}^a$, and $\mathbf{Z}^{a2} = r^{a2}\mathbf{Z}^a$. The preferred pattern ($\mathbf{x}^a$, $\mathbf{Z}^a$) is matched by ($\mathbf{x}^{a2}$, $\mathbf{b}^{a2}$, $\mathbf{Z}^{a2}$). Go to Step 8.

Step 5. If $B^* \geq B^a > B$, for k = 1,..., q, then set $r^{a3} = B^*/B^a \geq 1$. The resulting optimal system design is $\mathbf{x}^{a3} = r^{a3}\mathbf{x}^a$, $\mathbf{b}^{a3} = r^{a3}\mathbf{b}^a$, and $\mathbf{Z}^{a3} = r^{a3}\mathbf{Z}^a$. The preferred pattern ($\mathbf{x}^a$, $\mathbf{Z}^a$) is matched by ($\mathbf{x}^{a3}$, $\mathbf{b}^{a3}$, $\mathbf{Z}^{a3}$). Go to Step 8.

Step 6. If $B^{**} \geq B^a > B$, for k = 1,..., q, then set $r^{a4} = B^{**}/B^a \geq 1$. The resulting optimal system design is $\mathbf{x}^{a4} = r^{a4}\mathbf{x}^a$, $\mathbf{b}^{a4} = r^{a4}\mathbf{b}^a$, and $\mathbf{Z}^{a4} = r^{a4}\mathbf{Z}^a$. The preferred pattern ($\mathbf{x}^a$, $\mathbf{Z}^a$) is matched by ($\mathbf{x}^{a4}$, $\mathbf{b}^{a4}$, $\mathbf{Z}^{a4}$). Go to Step 8.

Step 7. If $B^a > B^{**}$, then we ask the decision maker to give another budget level B^t such that $B^{**} < B^t \leq B^a$. Set $r^{at} = B^t/B^a \leq 1$. The resulting optimal system design is determined as $\mathbf{x}^{at} = r^{at}\mathbf{x}^a$, $\mathbf{b}^{at} = r^{at}\mathbf{b}^a$, and $\mathbf{Z}^{at} = r^{at}\mathbf{Z}^a$. The preferred pattern is *approximated* by ($\mathbf{x}^{at}$, $\mathbf{b}^{at}$, $\mathbf{Z}^{at}$) with the optimum-path ratio r^{at}. If the decision maker accepts ($\mathbf{x}^{at}$, $\mathbf{b}^{at}$, $\mathbf{Z}^{at}$), go to Step 8; otherwise, repeat this step.

Step 8. Problem (M14.5) is solved and the process terminates.

We use the following example to illustrate how to use Algorithm 14.1 to solve model (M14.5).

Example 14.2

Suppose the decision maker wants to have the problem of Example 14.1 match the preferred pattern (solution): $\mathbf{x}^a = (200, 120, 180)$ and $\mathbf{Z}^a = \mathbf{C}\mathbf{x}^a = (25150, 36400, 30500)$. Then by Algorithm 14.1, we have:

Step 1. $\mathbf{b}^a = \mathbf{A}\mathbf{x}^a = (4440, 3120, 6260, 4080, 2700, 3760)$ and $B^a = \mathbf{V}\mathbf{x}^a = 14982$.

Step 2. From Table 14.1, we know $B^{**} = 13973.06$, $B^* = 6614.92$, $B_2{}^1 = 4658.10$, $B_1{}^2 = 4658.20$ and $B_3{}^3 = 4658.30$. Since all of these budget levels are smaller than $B^a = 14982$, we go to Step 7.

Step 7. We ask the decision maker to give another budget level B^t such that $13973.06 < B^t \leq 14982$. Suppose the decision maker says $B^t = 14000$. Then, we set $r^{at} = B^t/B^a = 93.45\%$ (≤ 1). The resulting optimal design is

$\mathbf{x}^{at} = (186.9, 112.14, 168.21)$
$\mathbf{b}^{at} = (4149.18, 2915.64, 5849.97, 3812.76, 2523.15, 3513.72)$ and
$\mathbf{Z}^{at} = (23502.68, 34015.80, 28502.25)$.

Thus, if we spend exactly $B^t = 14000$, which replaces $B = 4658.75$ in model (M14.1), then the portfolio of resources $\mathbf{b}^{at}$ allows the decision maker to produce $\mathbf{x}^{at}$ and achieve the performance $\mathbf{Z}^{at} = (23502.68, 34015.80, 28502.25)$ so that the preferred pattern $(\mathbf{x}^a, \mathbf{Z}^a)$ is matched with $r^{at} = 93.45\%$. If the decision maker accepts $(\mathbf{x}^{at}, \mathbf{b}^{at}, \mathbf{Z}^{at})$, go to Step 8 and the given problem is solved; otherwise, repeat this step until the decision maker is satisfied.

Remark 14.1 The fixed optimal pattern and the optimal pattern problem (M14.5) differ from concepts of satisficing solutions or compromise solutions in Chapter 13. In a satisficing solution problem, we want to minimize the distance between the multiple objective function value and the given aspiration level of those objectives; while in a compromise solution problem, the distance between the multiple objective function value and the ideal value of objectives is minimized. Given a resource availability level, both approaches produce a solution which can only minimize the distance from the given system to the goal setting of the decision maker. By matching the De Novo formulation to the fixed optimal pattern, we actively reshape the feasible region of the De Novo problem by selecting different portfolios of resource availability level towards the fixed optimal pattern. Thus, the applicability of the optimal pattern problem could be more likely higher than that of satisficing or compromise solution problems.

14.2 De Novo Design for Multiple Decision Makers and Possible Debt

We observe that this De Novo design model (M14.1) implicitly assumes that (i) there is only a single decision maker involving in the design activity, and (ii) the production process does not allow budget deficit or debt. In many real-world situations, important decision has to be made by a group of responsible people such as the members of the board of directors or trustees (Hwang and Lin 1987). A framework of multiple decision makers thus should be considered in this kind of system design processes. Due to lack of investment and cash flow limitation, a firm may not always have an available budget for production. When a possible budget deficit or debt occurs at the design time, the designed system is not feasible. This needs to be overcome by preparing the corresponding contingency plans (Chapters 9-12).

This section explores an approach of multiple criteria De Novo programming to dealing with system design problems with multi-decision makers and possible debt. In the system design model, each involved decision maker has his or her own preference for the budget availability level with respect to the criteria (objectives) under consideration. If possible debt occurs in the design time, the model allows flexibility to borrow additional money from a bank with a fixed interest rate to keep the production process feasible. The contingency plan therefore must be prepared to cope with the debt situation.

The section is organized as follows. In Section 14.2.1, the formulation of Zeleny's model (M14.1) is extended to reflect the multiple decision makers and the respective budget (or debt) levels for the criteria. Section 14.2.2 discusses how to solve the proposed model. First, the model is transformed into a multi-criteria knapsack problem with a weighted budget level. Second, all single-criterion performances for the ideal system with respect to initial weighted budget level and their contingency plans are identified. Since such a performance is related to a weight change of initial budget level, a multi-criteria performance for the ideal system must be found for a certain change range of the budget weights. Then, with each weight range the optimal system design is constructed by an optimum-path ratio for achieving the ideal system performance. In Section 14.2.3, a solution procedure is summarized and then a numerical example is used to illustrate applicability of the solution procedure.

14.2.1 A Model of System Designs with Multi-Decision Makers and Possible Debt

In this section, model (M14.1) is extended to the system design environment where multi-decision makers participate the design activity and possible debt may occur. Let $\mathbf{b}_i \in \mathbf{R}^m$ be the m-dimensional *unknown resource level* believed by the ith decision maker to achieve the q objectives, i = 1, 2, . . . , s. Then B_i represents the budget availability level preferred by the ith decision maker to achieve the q objectives. If $B_i < 0$, it means the amount of debt that the ith decision maker believes to achieve the q objectives. If $B_i \geq 0$, it is the ith decision maker's preferred budget for the q objectives. The unit prices of m resources $\mathbf{p} \geq 0$, $\mathbf{p} \in \mathbf{R}^m$ are assumed to be the same for the p decision makers. Suppose that there exist $w_i \geq 0$, $\sum_{i=1}^{p} w_i = 1, w_i \in \mathbf{R}$, which is the weight of the ith decision maker associated with the resource $\mathbf{b}_i$ and budget (or debt) B_i. Then, the new formulation of the De Novo programming approach to system designs becomes (M14.6):

$$\begin{array}{ll} \textit{Max} & \mathbf{C}\,\mathbf{x} \\ \textit{Subject to} & \mathbf{A}\,\mathbf{x} \leq \sum_{i=1}^{p} w_i \mathbf{b}_i \\ & \mathbf{p}(\sum_{i=1}^{p} w_i \mathbf{b}_i) \leq \sum_{i=1}^{p} w_i B_i \\ & \mathbf{x} \geq 0, \end{array}$$

where $C \in R^{q \times n}$ is the q objective vector; $A \in R^{m \times n}$ is the unit consumption matrix of resources.

Solving model (M14.6) means finding the optimal allocation of an aggregating budget $\sum_{i=1}^{p} w_i B_i$ so that the resulting portfolio of resources $\sum_{i=1}^{p} w_i \mathbf{b}_i$ maximizes the values of the product mixes, $\mathbf{Cx}$, simultaneously under a certain range of w_i changes. Such an optimal allocation is called the De Novo *optimal system design*. Note that this design differs from the designs in Chapters 9-12 where multiple discrete resource levels are given.

Observe that if some $B_i < 0$, i.e., the ith decision maker thinks that there is a debt B_i to achieve the q objectives, model (M14.6) may have no feasible solution. In order to overcome this situation, model (M14.6) needs to be rebuilt by borrowing additional money from a bank with a fixed interest rate. The solution of the new model can show how much money should be borrowed to keep the production process feasible under some range of w_i changes. This solution is called the *contingency plan* for the De novo optimal system design.

In summary, the problem of de novo system designs with multiple decision makers and possible debt can be solved by two stages. The first stage is to construct all single-criterion performances for the ideal system with respect to a certain weight range. If possible debt occurs, this stage also prepares the contingency plan with respect to a weight range to keep production feasibility. The second stage is to build a meta-optimal model for obtaining the meta-optimal budget for the multi-criteria ideal system. Then, an optimum-path ratio for achieving the multi-criteria ideal system is used to construct a structure of the optimal system design. Note that because the resulting structure is associated with the weights of decision makers' preferences on the resource availability, it can be essentially comparable to a set of nondominated solutions in multi-resource problems (Yu 1985) or a set of efficient agreements in bargaining problems (Young 1991). This comprehensive scenario of the optimal-system-design structure can help each decision maker to realize the role of his or her preference in the undertaking production process. Once the decision makers' weights are determined by some method, such as decision making under uncertainty (Yu 1985 and Shi 1991) or group negotiation games (Brams 1990), the De Novo optimal system design and its contingency plan can be implemented.

14.2.2 Contingency Plan and De Novo Optimal System Design

This section elaborates details of constructing the De Novo optimal system design and preparing its contingency plan for model (M14.6). An assumption of the section is that once the design problem encounters possible debt from some decision maker's preference, it can borrow additional money from a bank with a fixed interest rate for the perceived production process. Without loss of generality, suppose that from some i0th decision maker's point of view, there is a debt $B_{i0} < 0$ for producing $\mathbf{Cx}$. If there exist $w_i \geq 0$, $\sum_{i=1}^{p} w_i = 1, w_i \in \mathbf{R}$, according to Zeleny (1990), model (M14.6) can be transformed into model (M14.7):

$$\begin{array}{ll} \mathit{Max} & z_1 = c^1\,\mathbf{x} \\ & \vdots \\ & z_q = c^q\,\mathbf{x} \\ \mathit{Subject\ to} & \mathbf{V}\,\mathbf{x} \leq \sum_{i=1}^{p} w_i B_i \\ & \mathbf{x} \geq 0, \end{array}$$

where $\mathbf{Z} = (z_1, \ldots, z_q) \in \mathbf{R}^q$, $\mathbf{C} = (c^1, \ldots, c^q) \in \mathbf{R}^{q \times n}$, and $\mathbf{V} = (v_1, \ldots, v_n) = \mathbf{pA} \in \mathbf{R}^n$.

Given $w_i \geq 0$, model (M14.7) is a multiple criteria knapsack problem (or the linear program with only one constraint). Because $B_{i0} < 0$, the aggregating budget $\sum_{i=1}^{p} w_i B_i$ can lead to infeasible solutions for some changes of w_i. To overcome this situation, let y be the amount of money to be borrowed with a fixed interest rate $\alpha > 0$, $\alpha \in \mathbf{R}$. Then, model (M14.7) can be rebuilt as model (M14.8):

$$\begin{array}{ll} \mathit{Max} & z_1 = c^1\,\mathbf{x} - (\alpha + 1)y \\ & \vdots \\ & z_q = c^q\,\mathbf{x} - (\alpha + 1)y \\ \mathit{Subject\ to} & \mathbf{V}\,\mathbf{x} \leq \sum_{i=1}^{p} w_i B_i + y \\ & \mathbf{x}, y \geq 0. \end{array}$$

Model (M14.8) is also a multiple criteria knapsack problem, but with (n+1)-dimensional decision variable. Let $u_j = x_j$, for j = 1, . . . , n; and $u_{n+1} = y$. Then $\mathbf{u} \in \mathbf{R}^{n+1}$. Coordinately, let $f^k = (c_1^{\ k}, \ldots, c_n^{\ k}, -\alpha-1)$, for k = 1, . . . , q, and $\mathbf{g} = (v_1, \ldots, v_n, -1)$. Then, model (M14.8) is rewritten as model (M14.9):

$$
\begin{aligned}
\textit{Max} \qquad & z_1 = f^1 \mathbf{u} \\
& \vdots \\
& z_q = f^q \mathbf{u} \\
\textit{Subject to} \quad & \mathbf{g}\,\mathbf{u} \le \sum_{i=1}^{p} w_i B_i \\
& \mathbf{u} \ge 0.
\end{aligned}
$$

Suppose $u_s \in \mathbf{R}$, s = 1, . . ., n+, is a basic variable of model (M14.9). It can be expressed as $u_s = (\sum_{i=1}^{p} w_i B_i)/g_s$, $g_s \neq 0$. This indicates that a feasible solution of model (M14.9) depends on the range of weight w_i. The following definition specifies such a restriction:

Definition 14.1 The feasible weight set of a basic variable of model (M14.9) u_s is defined as

$$W(u_s) = \{ \mathbf{w} \ge 0 \mid (\sum_{i=1}^{p} w_i B_i)/g_s \, , \; g_s \neq 0 \text{ and } \sum_{i=1}^{p} w_i = 1 \}.$$

Theorem 14.3 The basic variable u_s is a feasible solution of model (M14.9) if and only if $W(u_s) \neq \varnothing$.

From Theorem 14.3, it can be seen that whenever their corresponding feasible weight set shift, one feasible solution of model (M14.9) will shift to another feasible solution.

To prepare the contingency plan for model (M14.7), we solve model (M14.9) (or model (M14.8)) as a special MC^2 problem by using the computer software of Hao and Shi (1996). The resulting feasible solution

$u_{n+1} = y = -\sum_{i=1}^{p} w_i B_i$, $\mathbf{w} \in W(u_{n+1})$, is the amount of money that should be borrowed to keep the production process feasible.

To construct the De Novo optimal system design, we first solve the q single-criterion problems of model (M14.9) with $y = \mathbf{0}$ (i.e., model (M14.7)), respectively. This can result in the q single-criterion performances. Since each single-criterion problem is a linear single-constraint program, let $c_a^k/v_a = Max\ \{c_j^k/v_j\}$ for j = 1, . . . , n+1, where c_j^k is the coefficient for the jth variable on the kth criterion. Then, the optimal solution of the kth criterion problem is $u_a^k = x_a^k = (\sum_{i=1}^{p} w_i B_i)/v_a$, $v_a > 0$, $\mathbf{w} \in W(u_a^k)$, k = 1, . . . , q.

Remark 14.2

If model (M14.9) (or (M14.7)) has some minimization objectives, the corresponding negative ideal solution of the single-criterion problems can be similarly discussed for the purpose of designing optimal systems (Li and Lee 1990).

Now a meta-optimality approach (Zeleny 1990) is used to identify the optimal system design for the q criteria. Let $z_k^*(\mathbf{w}) = Max\ z_k(\mathbf{w})$, k = 1, . . . , q, be the optimal objective function for the kth criteria of model (M14.7). Then $z_k^*(\mathbf{w}) = c_a^k(\sum_{i=1}^{p} w_i B_i)/v_a$, which is the function of weight $\mathbf{w} \in W(u_a^k)$. $\mathbf{Z}^*(\mathbf{w}) = (z_1^*(\mathbf{w}), \ldots, z_q^*(\mathbf{w})) \in \mathbf{R}^q$ is called the *ideal system design* with respect to the aggregating budget $\sum_{i=1}^{p} w_i B_i$. To see how much budget level is needed to achieve the ideal system design $\mathbf{Z}^*(\mathbf{w})$, a meta-optimal problem is constructed as (M14.10):

$$\begin{aligned} &Min && B = \mathbf{V}\,\mathbf{x} \\ &Subject\ to && \mathbf{C}\,\mathbf{x} \geq \mathbf{Z}^*(\mathbf{w}) \\ & && \mathbf{x} \geq 0. \end{aligned}$$

Model (M14.10) is a linear program with a single-objective and multi-constraint levels. Its feasible and optimal solutions are also functions of weight w_i. Let J be the index of basic variables (without confusion, J is also called a *basis*). Given a J with the basic variables denoted by $\mathbf{x}(J(\mathbf{w}))$, we can define the *associated basis matrix* $\mathbf{C}_J$ as the submatrix of $\mathbf{C}$ in model (M14.10) with column index in J and the *associated objective function coefficient* $\mathbf{V}_J$ as the components of $\mathbf{V}$ in model (M14.10) with column index in J. Then, $\mathbf{C}$ and $\mathbf{V}$ can be decomposed as $[\mathbf{C}_J, \mathbf{C}_{J'}]$ and $(\mathbf{V}_J, \mathbf{V}_{J'})$, respectively. Therefore, $\mathbf{x}(J(\mathbf{w})) = \mathbf{C}_J^{-1}\mathbf{Z}^*(\mathbf{w})$ is a basic solution associated with $(J, \mathbf{w})$ and its objective value is represented by

$$B(\mathbf{w}) = \mathbf{V}_J\mathbf{C}_J^{-1}\mathbf{Z}^*(\mathbf{w})$$

when weights $\mathbf{w}$ are specified.

Definition 14.2 The feasible weight set of the basic solution $\mathbf{x}(J(\mathbf{w}))$ is defined as

$$W^*(\mathbf{x}) = \{ \mathbf{w} \geq 0 \mid \mathbf{C}_J^{-1}\mathbf{Z}^*(\mathbf{w}) \geq 0 \text{ and } \sum_{i=1}^{p} w_i = 1 \}.$$

Theorem 14.4

(i) The basic solution $\mathbf{x}(J(\mathbf{w}))$ is feasible if and only if $\mathbf{w} \in W^*(\mathbf{x})$.

(ii) $\mathbf{x}(J(\mathbf{w}))$is the optimal solution if and only if $\mathbf{w} \in W^*(\mathbf{x})$ and the vector $\mathbf{V}_J\mathbf{C}_J^{-1}\mathbf{Z}^*(\mathbf{w}) - \mathbf{V}_{J'} \leq 0$.

The proof of Theorem 14.4 can be easily shown by applying the results in Chapter 8 of Yu (1985). Let the optimal solution of model (M14.10) be $\mathbf{x}^*(\mathbf{w}) = (x_1^*(\mathbf{w}), \ldots, x_n^*(\mathbf{w}))$. Then, the following result regarding the changes of w_i holds.

Theorem 14.5 $W^*(\mathbf{x}^*) \subseteq \cap\{W(x_a^k) \mid k = 1, \ldots, q\}$.

Proof. Because each $z_k^*(\mathbf{w}) = c_a^k(\sum_{i=1}^{p} w_i B_i)/v_a$ is the function of weight $\mathbf{w} \in W(x_a^k)$, the constraint level of model (M14.10), $\mathbf{Z}^*(\mathbf{w}) = (z_1^*(\mathbf{w}), \ldots, z_q^*(\mathbf{w}))$ is the function of $\mathbf{w} \in \cap\{W(x_a^k) \mid k = 1, \ldots, q\}$. Hence, the weight set of the optimal solution, $W^*(\mathbf{x}^*)$, must be contained in $\cap\{W(x_a{}^k) \mid k = 1, \ldots, q\}$.

Theorem 14.5 provides a way to locate the weight set of the optimal solution for model (M14.10) if the weight sets for the optimal solutions of single-criterion problems from model (M14.7) are found.

Given $\mathbf{x}^*(\mathbf{w})$ of model (M14.10), we can obtain $B^*(\mathbf{w})$ $(= \mathbf{V}\ \mathbf{x}^*(\mathbf{w}))$ and $\mathbf{b}^*(\mathbf{w})$ $(= \mathbf{A}\mathbf{x}^*(\mathbf{w}))$. The value $B^*(\mathbf{w})$ represents the minimum budget to achieve $\mathbf{Z}^*(\mathbf{w})$ through $\mathbf{x}^*(\mathbf{w})$ and $\mathbf{b}^*(\mathbf{w})$. The comparison of two budget levels $B^*(\mathbf{w})$ and $\sum_{i=1}^{p} w_i B_i$ is given by:

Theorem 14.6 $B^* \geq \sum_{i=1}^{p} w_i B_i$ for any $\mathbf{w} \in W^*(\mathbf{x}^*)$:

Proof. For every $\mathbf{w}^0 \in W^*(\mathbf{x}^*)$, it follows by applying the Theorem 14.1.

Utilizing the above results, an optimum-path ratio for achieving the ideal performance $\mathbf{Z}^*(\mathbf{w})$ related to the initial budget $\sum_{i=1}^{p} w_i B_i$ can be defined as

$$r^* = \left(\sum_{i=1}^{p} w_i B_i\right) / B^*(\mathbf{w}).$$

Finally, the optimal system design is constructed as $(\mathbf{x}(\mathbf{w}), \mathbf{b}(\mathbf{w}), \mathbf{Z}(\mathbf{w}))$, where $\mathbf{x}^*(\mathbf{w}) = r^*\ \mathbf{x}^*(\mathbf{w})$, $\mathbf{b}(\mathbf{w}) = r^*\ \mathbf{b}^*(\mathbf{w})$, and $\mathbf{Z} = r^*\ \mathbf{Z}^*(\mathbf{w})$.

Remark 14.3

(i) The optimal system design $(\mathbf{x}(\mathbf{w}), \mathbf{b}(\mathbf{w}), \mathbf{Z}(\mathbf{w}))$, the functions of $\mathbf{w} \in W^*(\mathbf{x}^*)$, is the extension of Zeleny's (1990) design. Since $\mathbf{w}$ is unknown at the design time, the implementation of the design $(\mathbf{x}(\mathbf{w}), \mathbf{b}(\mathbf{w}), \mathbf{Z}(\mathbf{w}))$ depends on the multiple decision makers. If they specify the values of $\mathbf{w}$ after reviewing the optimal weight set $W^*(\mathbf{x}^*)$, then the values of $(\mathbf{x}(\mathbf{w}), \mathbf{b}(\mathbf{w}), \mathbf{Z}(\mathbf{w}))$ can be readily calculated. If they use a known probability distribution to represent their preferences, the expected values of $(\mathbf{x}(\mathbf{w}), \mathbf{b}(\mathbf{w}), \mathbf{Z}(\mathbf{w}))$ can also be obtained.

(ii) In addition to budget levels ($\sum_{i=1}^{p} w_i B_i$) and $B^*(\mathbf{w})$, there are two more budget levels may be considered. The first is $B^k(\mathbf{w})$, the budget level of producing x^k with respect to the kth criterion in model (M14.7). The other $B^{**}(\mathbf{w})$, for the case $q \le n$ (the number of criteria is less than or equal to that of variables), comes from a meta-optimal problem: *Min* $\mathbf{Vx}$ subject to $\mathbf{Cx} \le \mathbf{Z}^*(\mathbf{w})$, $\mathbf{x} \ge \mathbf{0}$. Approaches of Section 14.1.1 can be applied to study various optimum-path ratios for model (M14.6). As a result, different optimal system designs and contingency plans with changes of w_i can be constructed to solve model (M14.6).

Algorithm 14.2

Step 1. Transform a given problem (M14.6) into problem (M14.7). Then, problem (M14.7) is rebuilt as problem (M14.8) (or (M14.9)), where $y \in \mathbf{R}$ is the necessary amount of money to be borrowed with a fixed (given) interest rate α.

Step 2. Solve problem (M14.9) to find the feasible solution $u_{n+1} = y = -\sum_{i=1}^{p} w_i B_i$, $\mathbf{w} \in W^*(u_{n+1})$, as a contingency plan to cope with the possible debt.

Step 3. Solve the q single-criterion problems of (M14.9) with $y = \mathbf{0}$ (i.e., problem (M14.7)) respectively. This can result in the q single-criterion performances $z_k^*(\mathbf{w}) = c_a^k(\sum_{i=1}^{p} w_i B_i)/v_a$, where $\mathbf{w} \in W(x_a^k)$, $k = 1, \ldots, q$.

Step 4. Using $\mathbf{Z}^*(\mathbf{w}) = (z_1^*(\mathbf{w}), \ldots, z_q^*(\mathbf{w}))$, the ideal system design with respect to the aggregating budget $\sum_{i=1}^{p} w_i B_i$ from the meta-optimal problem (M14.10), find the optimal solution $\mathbf{x}^*(\mathbf{w}) = \mathbf{C}_J^{-1}\mathbf{Z}^*(\mathbf{w})$ with $\mathbf{w} \in W^*(\mathbf{x}^*)$.

Step 5. For given $\mathbf{x}^*(\mathbf{w})$ of model (M14.10), obtain $B^*(\mathbf{w})$ (= $\mathbf{V}\ \mathbf{x}^*(\mathbf{w})$) and $\mathbf{b}^*(\mathbf{w})$ (= $\mathbf{A}\mathbf{x}^*(\mathbf{w})$). $B^*(\mathbf{w})$ is the minimum budget to achieve $\mathbf{Z}^*(\mathbf{w})$ through $\mathbf{x}^*(\mathbf{w})$ and $\mathbf{b}^*(\mathbf{w})$.

Step 6. Set the optimum-path ratio $r^* = (\sum_{i=1}^{p} w_i B_i)/B^*(\mathbf{w})$. The De Novo optimal system design is constructed as ($\mathbf{x}(\mathbf{w})$, $\mathbf{b}(\mathbf{w})$, $\mathbf{Z}(\mathbf{w})$), where $\mathbf{x}^*(\mathbf{w}) = r^*\ \mathbf{x}^*(\mathbf{w})$, $\mathbf{b}(\mathbf{w}) = r^*\ \mathbf{b}^*(\mathbf{w})$, and $\mathbf{Z} = r^*\ \mathbf{Z}^*(\mathbf{w})$.

Step 7. If the multi-decision makers specify the values of $\mathbf{w}$ after reviewing the optimal weight set $W^*(\mathbf{x}^*)$, then the values of ($\mathbf{x}(\mathbf{w})$, $\mathbf{b}(\mathbf{w})$, $\mathbf{Z}(\mathbf{w})$) can be readily calculated. If they use a known probability distribution on $W^*(\mathbf{x}^*)$ for their preferences, the expected values of ($\mathbf{x}(\mathbf{w})$, $\mathbf{b}(\mathbf{w})$, $\mathbf{Z}(\mathbf{w})$) can also be obtained.

The following example, adopted from Example 14.1, is used to illustrate Algorithm 14.2 for solving problem (M14.6).

Example 14.3

Consider a problem of the system design with three criteria and two decision makers. One of decision maker believes the budget level is $B_1 = 4658.75$. The other thinks there is a debt $B_2 = -100$. The current market prices of resources are given as $p_1 = 0.75$, $p_2 = 0.6$, $p_3 = 0.35$, $p_4 = 0.5$, $p_5 = 1.15$, and $p_6 = 0.65$. The formulation with unknown resource allocations (b_1, b_2) is:

$$
\begin{array}{ll}
\textit{Max} & 50\ x_1 + 100\ x_2 + 17.5\ x_3 \\
 & 92\ x_1 + 75\ x_2 + 50\ x_3 \\
 & 25\ x_1 + 100\ x_2 + 75\ x_3 \\
\textit{Subject to} & 12\ x_1 + 17\ x_2 \le w_1\, b_{11} + w_2\, b_{12} \\
 & 3\ x_1 + 9\ x_2 + 8\ x_3 \le w_1\, b_{21} + w_2\, b_{22} \\
 & 10\ x_1 + 13\ x_2 + 15\ x_3 \le w_1\, b_{31} + w_2\, b_{32} \\
 & 6\ x_1 + 16\ x_3 \le w_1\, b_{41} + w_2\, b_{42} \\
 & 12\ x_1 + 7\ x_3 \le w_1\, b_{51} + w_2\, b_{52}, \\
 & 9.5\ x_1 + 9.5\ x_2 + 4\ x_3 \le w_1\, b_{61} + w_2\, b_{62}
\end{array}
$$

where $x_j \ge 0$, $j = 1, 2, 3$, and $w_1, w_2 \ge 0$, $w_1 + w_2 = 1$, are the unknown weights for the decision makers' preferences.

Step 1. The above problem is first transformed as:

$$\begin{aligned} \textit{Max} \quad & 50\,x_1 + 100\,x_2 + 17.5\,x_3 \\ & 92\,x_1 + 75\,x_2 + 50\,x_3 \\ & 25\,x_1 + 100\,x_2 + 75\,x_3 \end{aligned}$$

$$\begin{aligned} \textit{Subject to} \quad & 23.475\,x_1 + 42.675\,x_2 + 28.000\,x_3 \leq 4658.75\mathrm{w}_1 - 100\,\mathrm{w}_2 \\ & \mathrm{x}_j \geq 0,\ j = 1, 2, 3. \end{aligned}$$

Since there is -100 debt, let y be the necessary amount of money to be borrowed with a fixed (given) interest rate $\alpha = 6\%$. Then, this problem is rebuilt as:

$$\begin{aligned} \textit{Max} \quad & 50\,x_1 + 100\,x_2 + 17.5\,x_3 - 1.06\,y \\ & 92\,x_1 + 75\,x_2 + 50\,x_3 - 1.06\,y \\ & 25\,x_1 + 100\,x_2 + 75\,x_3 - 1.06\,y \\ \textit{Subject to} \quad & 23.475\,x_1 + 42.675\,x_2 + 28.000\,x_3 \leq 4658.75\mathrm{w}_1 - 100\,\mathrm{w}_2 + y \\ & \mathrm{x}_j,\ y \geq 0,\ j = 1, 2, 3. \end{aligned}$$

Step 2. Using the computer software of Hao and Shi (1996), we solve the above problem and find $y = -4658.75\mathrm{w}_1 + 100\,\mathrm{w}_2$ with $0 < \mathrm{w}_1 \leq 0.021$, as the contingency plan in Table 14.3.

Step 3. We solve the above problem and find three single-criterion optimal solutions: x_2, x_1, and x_3 with $0.021 \leq \mathrm{w}_1 \leq 1$ as in Table 14.3.

Step 4. Because the ideal system design associated with the aggregating budget $4658.75\mathrm{w}_1 - 100\,\mathrm{w}_2$ is $\mathbf{Z}^*(\mathbf{w}) = (z_1^*(\mathbf{w}), z_2^*(\mathbf{w}), z_3^*(\mathbf{w}))$ of Table 14.3, the meta-optimal problem (M14.10) is constructed as:

$$\begin{aligned} \textit{Min} \quad & 23.475\,x_1 + 42.675\,x_2 + 28.000\,x_3 - y \\ \textit{Subject to} \quad & 50\,x_1 + 100\,x_2 + 17.5\,x_3 - 1.06\,y = 10916.81\,\mathrm{w}_1 - 234.33\,\mathrm{w}_2 \\ & 92\,x_1 + 75\,x_2 + 50\,x_3 - 1.06\,y = 18257.93\,\mathrm{w}_1 - 391.91\,\mathrm{w}_2 \\ & 25\,x_1 + 100\,x_2 + 75\,x_3 - 1.06\,y = 12174.43\,\mathrm{w}_1 - 261.32\,\mathrm{w}_2 \\ & \mathrm{x}_j,\ y \geq 0,\ j = 1, 2, 3. \end{aligned}$$

This problem is a linear program with one criterion and two constraint levels. Solving it by using Hao and Shi (1996) software, the optimal solution $\mathbf{x}^*(J)$ and the weight set $W^*(\mathbf{x}^*)$ are listed in Table 14.4. Note that from Table 14.3 and Table 14.4, $W^*(\mathbf{x}^*) = W(x_2) \cap W(x_1) \cap W(x_3)$ which verifies Theorem 14.5.

Step 5. For given $\mathbf{x}^*(J)$, we obtain:

$$B^*(\mathbf{w}) = 23.475\, x_1^* + 42.675\, x_2^* + 28.000\, x_3^*$$
$$= 6616.53\, w_1 - 142.3\, w_2; \text{ and}$$
$$\mathbf{b}^*(\mathbf{w}) = \mathbf{A}\ \mathbf{x}^*(\mathbf{w}) = (b_1^*(\mathbf{w}),\ b_2^*(\mathbf{w}),\ b_3^*(\mathbf{w}),\ b_4^*(\mathbf{w}),\ b_5^*(\mathbf{w}),\ b_6^*(\mathbf{w})),$$

where $b_1^*(\mathbf{w}) = 2080.56\, w_1 - 44.72\, w_2$,

$b_2^*(\mathbf{w}) = 1292.98\, w_1 - 27.82\, w_2$,

$b_3^*(\mathbf{w}) = 2883.94\, w_1 - 87.52\, w_2$,

$b_4^*(\mathbf{w}) = 2051.72\, w_1 - 44.12\, w_2$,

$b_5^*(\mathbf{w}) = 909.02\, w_1 - 45.74\, w_2$, and

$b_6^*(\mathbf{w}) = 1845.61\, w_1 - 39.67\, w_2$.

Step 6. The optimum-path ratio can be set as:

$$r^*(\mathbf{w}) = (4658.75\, w_1 - 100\, w_2)/(6616.53\, w_1 - 142.3\, w_2) \text{ for } 0.021 \le w_1 \le 1.$$

The sketch of $r^*(\mathbf{w})$ is depicted in Figure 14.1. Then the optimal system design is constructed as $(\mathbf{x}(\mathbf{w}), \mathbf{b}(\mathbf{w}), \mathbf{Z}(\mathbf{w}))$, where $\mathbf{x}(\mathbf{w}) = r^*(\mathbf{w})\ \mathbf{x}^*(\mathbf{w})$, $\mathbf{b}(\mathbf{w}) = r^*(\mathbf{w})\ \mathbf{b}^*(\mathbf{w})$, and $\mathbf{Z}(\mathbf{w}) = r^*(\mathbf{w})\mathbf{Z}^*(\mathbf{w})$. Note that each component of the design is a function of (w_1, w_2) for $0.021 \le w_1 \le 1$, $w_1 + w_2 = 1$.

Step 7. Suppose after reviewing the optimal weight set $W^*(\mathbf{x}^*) = \{(w_1, w_2) \ge 0 \mid 0.021 \le w_1 \le 1, w_1 + w_2 = 1\}$, the two decision makers agree that $w_1 = 0.75$ and $w_2 = 0.25$. Then, the values of r^* and $(\mathbf{x}, \mathbf{b}, \mathbf{Z})$ are readily calculated as:

$r^* = 70.41\%$,

$\mathbf{x} = r^*\ \mathbf{x}^* = (97.795, 22.1, 58.513)$,

$\mathbf{b} = r^*\ \mathbf{b}^* = (1090.82, 677.89, 1507.53, 1075.70, 471.98, 967.63)$, and

$$\mathbf{Z} = r^* \mathbf{Z}^* = (5723.65, 9572.57, 6383.01).$$

Table 14.3 All Single-Criterion Designs and Contingency Plan

$\mathbf{u}^k$	$z_k^*(\mathbf{w})$	$W(\mathbf{u}^k)$
$u_a^1 = x_a^1$ $= x_2$ $= 109.17\ w_1 - 2.34\ w_2$	$z_1^* = 10916.81\ w_1 - 234.33\ w_2$	$0.021 \le w_1 \le 1$
$u_a^2 = x_a^2$ $= x_1$ $= 189.46\ w_1 - 4.26\ w_2$	$z_2^* = 18257.93\ w_1 - 391.91\ w_2$	$0.021 \le w_1 \le 1$
$u_a^3 = x_a^3$ $= x_3$ $= 162.33\ w_1 - 3.48\ w_2$	$z_3^* = 12174.43\ w_1 - 261.32\ w_2$	$0.021 \le w_1 \le 1$
$u_4 = y$ $= -4658.75\ w_1 + 100\ w_2$		$0 < w_1 < 0.021$

Table 14.4 Meta-Optimal Solution

$\mathbf{x}^*(J)$	$W^*(\mathbf{x}^*)$
$x_1^* = 131.34\ w_1 - 2.82\ w_2$, $x_2^* = 29.68\ w_1 - 0.64\ w_2$, $x_3^* = 78.98\ w_1 - 1.70\ w_2$,	$0.021 \le w_1 \le 1$

However, if two decision makers agree that $w_1 = 0.02$ and $w_2 = 0.98$, then $(w_1, w_2) \in W(y) = \{(w_1, w_2) \ge 0 \mid 0 < w_1 < 0.021, w_1 + w_2 = 1\}$. The problem needs the additional budget $y = 4.825$ as the contingency plan to make the production process feasible.

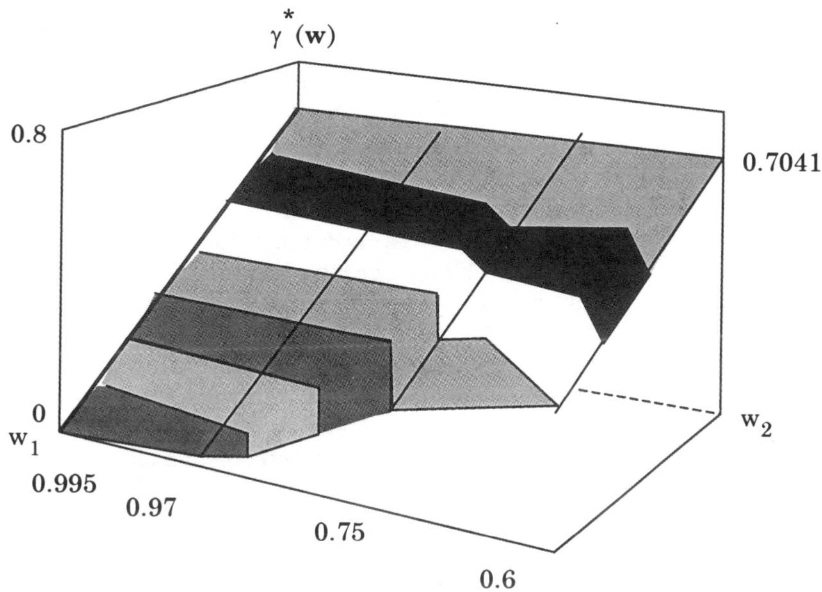

Figure 14.1 The optimum-path ratio $\gamma^*(\mathbf{w})$

As a comparison with Zeleny's (1990) approach, one can see that if only the first decision maker involves the design activity and the second decision maker does not exist, i.e., $w_1 = 1$ and $w_2 = 0$, then Zeleny's optimal design is obtained as:

$r^0 = 70.41\%$,

$\mathbf{x}^0 = (92.43, 20.91, 55.60)$,

$\mathbf{b}^0 = (1464.51, 910.25, 2030.08, 1444.21, 640.08, 1299.07)$, and

$\mathbf{Z}^0 = (7684.80, 12851.30, 8571.10)$.

Note that $r^0 = 70.41\%$ is identical to $r^* = 70.41\%$ with $w_1 = 0.75$ and $w_2 = 0.25$. This is because r^*, the function of $W^*(\mathbf{x}^*)$, contains r^0 (see Figure 14.1). Therefore, Zeleny's optimal design is the special one in the design domain of this approach.

14.3 Further Remarks

The concept of multiple criteria De Novo programming differs from that of MC² linear programming. The former assumes the resource level is an unknown variable and the purpose of solving the problem is to design a best system for a given budget level to purchase the resource. The latter has given multiple discrete resources levels and the purpose of solving the problem is to allocate an optimal solution structure (i.e., the set of potential solutions). The approach of multiple criteria De Novo programming to dealing with system design problems with multiple decision makers and possible debt in Section 14.2 is the application of MC² linear programming in the context of De Novo programming. That is, if the discrete resource levels (related to multiple decision makers) are unknown, how to design the optimal system under given prices of the resources. The approach indeed demonstrates an inner connection between De Novo and MC² linear programming.

Exercises

14.1 Given the following MC problem:

$$\begin{array}{ll} \textit{Max} & z_1 = 10\,x_1 + 12\,x_2 + 8\,x_3 \\ \textit{Max} & z_2 = 15\,x_1 + 8\,x_2 + 15\,x_3 \\ \textit{Max} & z_3 = 6\,x_1 + 2\,x_2 + 21\,x_3 \\ \textit{Subject to} & 3\,x_1 + 2x_2 + 12\,x_3 \le b_1 \\ & 4x_1 + 6x_2 + 5\,x_3 \le b_2 \\ & 7x_1 + x_2 + 4\,x_3 \le b_3 \\ & x_1, x_2, x_3 \ge 0, \end{array}$$

with the market prices of resources $p_1 = .50$, $p_2 = .70$, and $p_3 = .35$, and the initial budget level $B = 5000$.

(i) Find the optimum-path ratios r^k, k = 1, 2, 3, 4, 5, 6.
(ii) Find the structure designs like Table 14.1 for the problem.
(iii) Find various De novo optimal designs like Table 14.2 for the problem.

14.2 Consider a problem of the system design with three criteria and two decision makers. One of decision maker believes the budget level is B_1 = 5000. The other thinks there is a debt B_2 = - 300. The current market prices of resources are given as $p_1 = 0.50$, $p_2 = 0.70$, and $p_3 = 0.35$. The formulation with unknown resource allocations (b_1, b_2) is:

Max $z_1 = 10\,x_1 + 12\,x_2 + 8\,x_3$

Max $z_2 = 15\,x_1 + 8\,x_2 + 15\,x_3$

Max $z_3 = 6\,x_1 + 2\,x_2 + 21\,x_3$

Subject to $3\,x_1 + 2x_2 + 12\,x_3 \le w_1\,b_{11} + w_2\,b_{12}$

$4x_1 + 6x_2 + 5\,x_3 \le w_1\,b_{21} + w_2\,b_{22}$

$7x_1 + x_2 + 4\,x_3 \le w_1\,b_{31} + w_2\,b_{32}$

$x_1,\ x_2,\ x_3 \ge 0,$

where $x_j \ge 0$, j = 1, 2, 3, and $w_1, w_2 \ge 0$, $w_1 + w_2 = 1$, are the unknown weights for the decision makers' preferences.

Apply Algorithm 14.2 to identify the De Novo optimal designs and contingency plans.

Chapter 15
ELIMINATING PERMANENTLY DOMINATED OPPORTUNITY

In LP problems, let x_j be a nonbasic variable of a basic feasible solution for a given primal problem. If x_j remains a non-basic variable for all optimal basic solutions of the problem, then x_j is called a "permanent nonbasic variable" and can be eliminated from further consideration in the process of finding the optimal basic solution. For the techniques of identifying and eliminating permanent non-basic variables in linear programming problems, the reader is referred to Zionts (1965), Thompson, Tonge, and Zionts (1966), Cheng (1980 and 1985), and Ye (1989). In this chapter, we *extend* the elimination techniques for LP problems to the procedure of solving MC^2 problems (Shi, Yu and Zhang 1994). The technique is to eliminate *permanently dominated opportunities* that can never have a chance to be in a potential solution J of a given model (M4.1).

Given an MC^2 problem (M4.1), we assume that there is the set of all potential solutions $\vartheta = \{J_1, J_2, \ldots, J_r\}$ identified by the MC^2-simplex method. In Section 15.1, we describe a method to eliminate the permanently dominated opportunities from further consideration in the process of finding the subset of ϑ which covers a subset of $\Gamma \times \Lambda$. In Section 15.2, we assume that for any $(\gamma, \lambda) \in \Gamma \times \Lambda$, (M4.1) always offers a bounded optimal solution. With this assumption ϑ covers the space $\Gamma \times \Lambda$. Then, based on the method described in Section 15.1, we develop a method of eliminating all permanently dominated opportunities from further consideration of identifying the entire ϑ. In Section 15.3, suppose that for some $(\gamma, \lambda) \in \Gamma \times \Lambda$, (M4.1) has no feasible solution. Then ϑ covers only a subset of $\Gamma \times \Lambda$. We discuss how to apply the elimination technique described in Sections 15.1 and 15.2 to eliminate permanently dominated opportunities (PDOs) for the subset of $\Gamma \times \Lambda$ that is covered by ϑ. Finally, Section 15.4 provides the further remarks.

15.1 Eliminating Permanently Dominated Opportunities for Subsets of $\Gamma \times \Lambda$

Given an MC² problem (M4.1), from Section 4.2, we can define $N^* = \{1, \ldots, n+m\}$ be the set of all possible decision and slack opportunities under consideration. Suppose we solve (M4.1) by using the MC²-simplex method and obtain $\vartheta = \{J_1, J_2, \ldots, J_r\}$. We observe that the set of all selected opportunities involved in ϑ is a subset of N^*. We denote such a set by $N^*(\vartheta)$ and write $N^* = N^*(\vartheta) \cup N'(\vartheta)$, where $N'(\vartheta)$ is the complement set of $N^*(\vartheta)$. Then, $N'(\vartheta)$ contains all PDOs that are not selected in any potential solution J of ϑ. We call $N'(\vartheta)$ *the set of all PDOs* with respect to (M4.1).

Now we discuss how to recognize and eliminate the set $N'(\vartheta)$ before identifying the entire ϑ for a given MC² problem (M4.1) as follows:

For convenience, according to Section 4.5, we shall call the model (M4.1) the *primal (or P) problem*. Then, the *dual problem* of (M4.1) is defined as (M15.1):

$$\begin{array}{ll} \textit{Min} & \mathbf{u}^t\, \mathbf{D}\gamma \\ \textit{Subject to} & \mathbf{u}^t\, \mathbf{A} \geq \boldsymbol{\lambda}^t\, \mathbf{C} \\ & \mathbf{u}^t \geq 0. \end{array}$$

Let $\Gamma = \{\gamma > 0 \mid \gamma = (\gamma_1, \gamma_2, \ldots, \gamma_p)^t$ with $\gamma_k > 0$ and $\sum_{k=1}^{p} \gamma_k = 1\}$ be the space of the primal parameter γ; and $\Lambda = \{\lambda > 0 \mid \lambda^t = (\lambda_1, \lambda_2, \ldots, \lambda_q)$ with $\lambda_k > 0$ and $\sum_{k=1}^{q} \lambda_k = 1\}$ be the dual parameter λ; $\Gamma \times \Lambda$ be the Cartesian product space of parameters of (γ, λ).

Definition 15.1 Given J as a basis for (M4.1),

(i) the space Γ is said to be covered by J if $\Gamma \subseteq \Gamma(J)$;

(ii) the space Λ is said to be covered by J if $\Lambda \subseteq \Lambda(J)$;

(iii) the space the space $\Gamma \times \Lambda$ is said to be covered J if $\Gamma \times \Lambda \subseteq \Gamma(J) \times \Lambda(J)$;

(iv) the spaces Γ, Λ and $\Gamma \times \Lambda$ are said to be covered by the set of potential solutions, $\vartheta = \{J_1, J_2, \ldots, J_r\}$ for (M4.1) can be similarly defined. (For instance, Γ is said to be covered by ϑ if $\Gamma \subseteq \cup \{\Gamma(J_i) \mid i = 1, \ldots, r\}$.)

Given a problem (M4.1), instead of using Tables 4.1-4.3 to represent the MC² simplex method, we write Table 15.1-Table 15.2. Table 15.1 is Table 4.1:

Table 15.1 The Initial Simplex Tableau

x	**s**	RHS	
$\mathbf{A}$	$\mathbf{I}$	$\mathbf{D}\gamma$	(15.1)
$-\lambda' \mathbf{C}$	$\mathbf{0}$	$\mathbf{0}$	(15.2)

Given a basis J, we use a row linear transformation on Table 15.1 by:

(15.3) = $\mathbf{B}_J^{-1}$ (15.1) (i.e., premultiply (4.3) by $\mathbf{B}_J^{-1}$), and

(15.4) = $\lambda' \mathbf{C}_J$ (15.3) + (15.2).

This results in Table 15.2:

Table 15.2 The Simplex Tableau

x	S	RHS	
$\mathbf{B}_J^{-1}\mathbf{A}$	$\mathbf{B}_J^{-1}$	$\mathbf{B}_J^{-1}\mathbf{D}\gamma$	(15.3)
$\lambda' \mathbf{C}_J \mathbf{B}_J^{-1}\mathbf{A}-\lambda\mathbf{C}$	$\lambda' \mathbf{C}_J \mathbf{B}_J^{-1}$	$\lambda' \mathbf{C}_J \mathbf{B}_J^{-1}\mathbf{D}\gamma$	(15.4)

For a given basis J for (M4.1), if there is $\gamma > 0$ such that $\mathbf{x}(J) = \mathbf{B}_J^{-1}\mathbf{D}\gamma > 0$, we call J is *a nondegenerate feasible basis*, and if for all possible $\gamma > 0$ at least one component of $\mathbf{x}(J)$ is zero, we call J *a degenerate feasible basis*. Note that a degenerate feasible basis is dependent on $\gamma > 0$. For the sake of simplicity, in the following we concentrate on the nondegerate feasible bases (see Remark 15.2).

Assumption 15.1 All feasible bases under discussion for Section 15.1 are nondegenerate.

Now Definition 4.4 is restated as:

Definition 15.2 Given a basis J for an MC[2] problem,

(i) $\Gamma(J) = \{\gamma > 0 \mid \mathbf{B}_J^{-1}\mathbf{D}\gamma \geq 0\}$ or $\Gamma^*(J) = \{\gamma > 0 \mid \mathbf{B}_J^{-1}\mathbf{D}\gamma > 0\}$ is called the primal parameter set; and

(ii) $\Lambda(J) = \{\lambda > 0 \mid \lambda^t (\mathbf{C}_J\mathbf{B}_J^{-1}\mathbf{A}\text{-}\mathbf{C},\ \mathbf{C}_J\mathbf{B}_J^{-1}) \geq 0\}$ is called the dual parameter set.

By Definition (i) of 15.2, given any primal potential basis K for the primal problem (P) (or (M4.1)), its primal parameter set is $\Gamma(K) = \{\gamma > 0 \mid \mathbf{B}_K^{-1}\mathbf{D}\gamma \geq 0\}$, where $\mathbf{B}_K$ is the basis matrix for K. Given any dual potential basis Q for the problem (P), its dual parameter set is $\Lambda(Q) = \{\lambda > 0 \mid \lambda^t\ (\mathbf{C}_Q\mathbf{B}_Q^{-1}\mathbf{A}\text{-}\mathbf{C},\ \mathbf{C}_Q\mathbf{B}_Q^{-1}) \geq 0\}$ (see (ii) of Definition 15.2).

Using Definitions 4.10-4.11 and Remark 4.5, given a primal potential basis K and a dual potential basis Q, we call (K, Q) a primal-dual pair. Then, we have the following result:

Theorem 15.1 Given a primal-dual pair (K, Q), for any $(\gamma^0, \lambda^0) \in \Gamma(K) \times \Lambda(Q)$, there is a $J(\gamma^0, \lambda^0)$ which is an optimal basis of the resulting LP of Problem (P) with fixed (γ^0, λ^0).

The above theorem can be proved by Theorem 8.25 of Yu (1985). Note that $J(\gamma^0, \lambda^0)$ is also a potential basis for Problem (P) since $(\gamma^0, \lambda^0) \in \Gamma(J(\gamma^0, \lambda^0)) \times \Lambda(J(\gamma^0, \lambda^0))$. Such a potential basis may not be unique.

If (γ^0, λ^0) is fixed, then Problem (P) can be viewed as a LP problem. To facilitate our discussion, in the following we introduce a theorem of complementary slackness, which can be found in any standard LP text (For example, see Bazaraa, Jarvis and Sherali 1990).

Theorem 15.2 Consider the LP problem resulting from Problem (P) with a fixed $(\gamma^0, \lambda^0) > 0$. Let $(\mathbf{x}; \gamma^0)$ be (P)-feasible and $(\mathbf{u}; \lambda^0)$ be (D)-feasible (Remark 4.5). Then they are respectively optimal if and only if $\mathbf{u}^t (\mathbf{Ax} - \mathbf{D}\gamma^0) = 0$ and $(\mathbf{u}^t \mathbf{A} - (\lambda^0)^t \mathbf{C})\, \mathbf{x} = 0$.

Remark 15.1 By Theorem 15.2, we see that given a fixed $(\gamma^0, \lambda^0) > 0$, if $(\mathbf{x}^*; \gamma^0)$ and $(\mathbf{u}^*; \lambda^0)$ are the optimal solutions to the LP problem (P) and the LP problem (D), respectively, then $((\mathbf{u}^*)'\,\mathbf{A}_j - (\lambda^0)'\,\mathbf{C}_j)\,x^*_j = \mathbf{0}$, j $\in$N*, where $\mathbf{A}_j$ denotes the jth column of matrix **A** and $\mathbf{C}_j$ denotes the jth column of matrix **C**. Therefore, $(\mathbf{u}^*)'\,\mathbf{A}_j - (\lambda^0)'\,\mathbf{C}_j) > \mathbf{0}$ implies that x^*_j = 0.

Given a dual potential basis Q for Problem (P), let $H_j^>(Q) = \{\lambda > 0 \mid \lambda'\,\mathbf{C}_Q\mathbf{B}_Q^{-1}\mathbf{A}_j - \lambda'\,\mathbf{C}_j > 0\}$. Then, we see that a sufficient condition for $\lambda'\,\mathbf{C}_Q\mathbf{B}_Q^{-1}\mathbf{A}_j - \lambda'\,\mathbf{C}_j > \mathbf{0}$, for all $\lambda \in \Lambda(Q)$ is that $\Lambda(Q) \subseteq H_j^>(Q)$.

Theorem 15.3. Given a primal-dual pair (K, Q) and a fixed $(\gamma^0, \lambda^0) \in \Gamma(K) \times \Lambda(Q)$, let $J(\gamma^0, \lambda^0)$ be a potential basis. If for given j $\in$ N, $\Lambda(Q) \subseteq H_j^>(Q)$ and $\mathbf{B}_Q^{-1}\mathbf{A}_j \geq \mathbf{0}$, then x_j is not a basic variable in $J(\gamma^0, \lambda^0)$. That is, the jth product is not (chosen) in $J(\gamma^0, \lambda^0)$.

Proof. Note that for the fixed $(\gamma^0, \lambda^0) \in \Gamma(K) \times \Lambda(Q)$, $(\gamma^0, \lambda^0) \in \Gamma(J(\gamma^0, \lambda^0)) \times \Lambda(J(\gamma^0, \lambda^0))$. Let $\mathbf{x}^*(J(\gamma^0, \lambda^0), \gamma^0) = \mathbf{B}_J^{-1}\mathbf{D}\gamma^0$ be the optimal solution associated with the basis $J(\gamma^0, \lambda^0)$ with the fixed (γ^0, λ^0) for Problem (P) and $(\mathbf{u}^*(J(\gamma^0, \lambda^0), \gamma^0))' = (\lambda^0)^t\mathbf{C}_J\,\mathbf{B}_J^{-1}$ be the optimal solution associated with the basis $J(\gamma^0, \lambda^0)$ with the fixed (γ^0, λ^0) for Problem (D). By Remark 15.1, as $(\gamma^0, \lambda^0) \in \Gamma(J(\gamma^0, \lambda^0)) \times \Lambda(J(\gamma^0, \lambda^0))$, we have that

(i) $[(\mathbf{u}^*(J(\gamma^0, \lambda^0), \gamma^0))^t\,\mathbf{A}_j - (\lambda^0)^t\mathbf{C}_j]\;x^*_j\,(J(\gamma^0, \lambda^0), \gamma^0) = \mathbf{0}$, j $\in$N.

Since $\Lambda(Q) \subseteq H_j^>(Q)$, we see that $\lambda'\,\mathbf{C}_Q\mathbf{B}_Q^{-1}\mathbf{A}_j - \lambda'\,\mathbf{C}_j > 0$, for all $\lambda \in \Lambda(Q)$. Thus, we have that $(\lambda^0)^t\,\mathbf{C}_Q\mathbf{B}_Q^{-1}\mathbf{A}_j - (\lambda^0)^t\,\mathbf{C}_j > \mathbf{0}$, since $\lambda^0 \in \Lambda(Q)$.

Because $\mathbf{B}_J$ is an optimal basis matrix, we have that $(\lambda^0)^t\,\mathbf{C}_J\mathbf{B}_J^{-1}\mathbf{B}_Q \geq (\lambda^0)^t\,\mathbf{C}_Q$. Then, we have that

$$
\begin{aligned}
(\mathbf{u}^*(J(\gamma^0, \lambda^0), \gamma^0))'\,\mathbf{A}_j &= (\lambda^0)^t\,\mathbf{C}_J\mathbf{B}_J^{-1}\mathbf{A}_j \\
&= (\lambda^0)^t\,\mathbf{C}_J\mathbf{B}_J^{-1}\mathbf{B}_Q\,(\mathbf{B}_Q^{-1}\mathbf{A}_j) \\
&\geq (\lambda^0)^t\mathbf{C}_Q\mathbf{B}_Q^{-1}\mathbf{A}_j,
\end{aligned}
$$

since $\mathbf{B}_Q^{-1}\mathbf{A}_j \geq 0$, by the assumption.

Thus, $(\mathbf{u}^*(J(\gamma^0, \lambda^0), \gamma^0))'\,\mathbf{A}_j - (\lambda^0)^t\mathbf{C}_j \geq (\lambda^0)^t\mathbf{C}_Q\mathbf{B}_Q^{-1}\mathbf{A}_j - (\lambda^0)^t\mathbf{C}_j > \mathbf{0}$. This implies that

(ii) $(\mathbf{u}^*(J(\gamma^0, \lambda^0), \gamma^0))'\,\mathbf{A}_j - (\lambda^0)^t\mathbf{C}_j > \mathbf{0}$, where $\lambda^0 \in \Lambda(Q) \cap \Lambda(J(\gamma^0, \lambda^0))$.

From (i) and (ii), we see that $x^*_j(J(\gamma^0, \lambda^0), \gamma^0) = \mathbf{0}$. Hence, $x^*_j(J(\gamma^0, \lambda^0), \gamma^0)$ does not appear in $\mathbf{x}^*(J(\gamma^0, \lambda^0), \gamma^0)$, the basic variables associated with $J(\gamma^0, \lambda^0)$ for Problem (P). This is because for any $\gamma \in \Gamma(J(\gamma^0, \lambda^0))$, x^*_j

$(J(\gamma^0, \lambda^0), \gamma) > \mathbf{0}$ whenever $x^*_j(J(\gamma^0, \lambda^0), \gamma)$ appears in $\mathbf{x}^*(J(\gamma^0, \lambda^0))$. Therefore, the jth product is not in the potential basis $J(\gamma^0, \lambda^0)$.

Given a primal-dual pair (K, Q), note that for any $(\gamma, \lambda) \in \Gamma(K) \times \Lambda(Q)$, by Theorem 15.1, there is *at least* one potential basis $J(\gamma, \lambda)$ such that $(\gamma, \lambda) \in \Gamma(K) \times \Lambda(Q)$.

We define $\Pi(K, Q) = \{J(\gamma, \lambda) \mid J(\gamma, \lambda)$ is an optimal basis of the resulting linear programming of Problem (P) with fixed (γ, λ), $(\gamma, \lambda) \in \Gamma(K) \times \Lambda(Q)\}$.

Theorem 15.4 $\Gamma(K) \times \Lambda(Q)$ is covered by $\Pi(K, Q)$.

Proof. For any $(\gamma, \lambda) \in \Gamma(K) \times \Lambda(Q)$, there is $J(\gamma, \lambda) \in \Pi(K, Q)$ such that $(\gamma, \lambda) \in \Gamma(J(\gamma, \lambda)) \times \Lambda(J(\gamma, \lambda))$. Thus $(\gamma, \lambda) \in \cup\{\Gamma(J(\gamma, \lambda)) \times \Lambda(J(\gamma, \lambda)) \mid (\gamma, \lambda) \in \Gamma(K) \times \Lambda(Q)\}$, which is the union of parameter sets of $\Pi(K, Q)$.

Note that the condition that $\Lambda(Q) \subseteq H_j^{>}(Q)$ and $\mathbf{B}_Q^{-1}\mathbf{A}_j \geq \mathbf{0}$ is independent of $J(\gamma^0, \lambda^0)$ in Theorem 15.3. Thus, we have the following:

Theorem 15.5 Given a primal-dual pair (K, Q), if for given $j \in N$, $\Lambda(Q) \subseteq H_j^{>}(Q)$ and $\mathbf{B}_Q^{-1}\mathbf{A}_j \geq 0$, then the jth product is not in $J(\gamma^0, \lambda^0)$ for any $(\gamma, \lambda) \in \Gamma(K) \times \Lambda(Q)$.

Example 15.1

Consider the following MC² problem:

$$Max\begin{pmatrix}\lambda_1 & \lambda_2\end{pmatrix}\begin{pmatrix}3 & 1 & 2 & 0\\ -1 & 5 & 1 & 2\end{pmatrix}\begin{pmatrix}x_1\\ x_2\\ x_3\\ x_4\end{pmatrix}$$

$$Subject\ to \begin{pmatrix}1 & 0 & 2 & 1\\ 0 & 1 & 1 & 2\\ 1 & 1 & 0 & 0\end{pmatrix}\begin{pmatrix}x_1\\ x_2\\ x_3\\ x_4\end{pmatrix} \leq \begin{pmatrix}20 & 50\\ 40 & 20\\ 30 & 40\end{pmatrix}\begin{pmatrix}\gamma_1\\ \gamma_2\end{pmatrix}$$

$$x_i \geq 0, i = 1, 2, 3, 4$$

Let x_5, x_6, and x_7 be slack variables for constraints 1, 2, and 3 respectively.

Using the computer software of the MC^2-simplex method (Chien, Shi, and Yu 1989), we obtain the set of all potential solutions $\vartheta = \{J_1, J_2, J_3, J_4, J_5\}$ are in Table 15.3 and the corresponding optimal situation sets $\{S_1, S_2, S_3, S_4, S_5\}$ as in Fig 9.1. Note that in Table 15.3 the primal parameter sets of J_i are denoted by $\Lambda(J_i)$, i = 1,...,5. The detailed explanation for the potential solutions can be referred to Example 9.11 of Lee, Shi and Yu (1990). From Table 15.3, we see that $N(\vartheta) = \{1,2,3,5,6,7\}$ and $N'(\vartheta) = \{4\}$.

Now suppose that we have not identified the entire set of potential solutions (bases) and that we want to use the elimination technique (i.e., Theorem 15.3 and Theorem 15.5) to eliminate some PDOs.

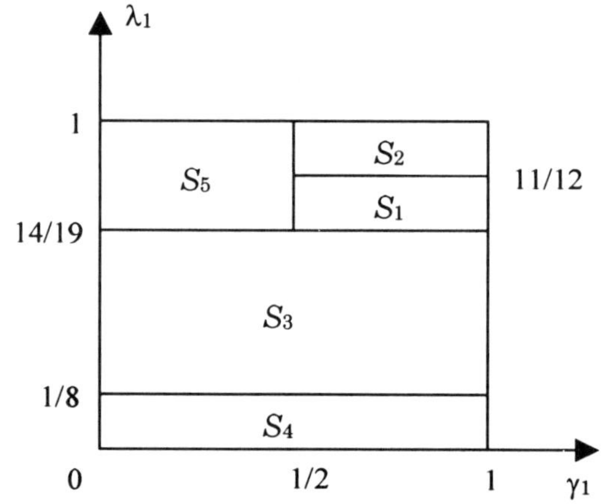

Figure 15.1

Table 15.3

Bases	Basic Variables	$\Gamma(J_i)$	$\Lambda(J_i)$
J_1	(x_1, x_2, x_3)	$1/2 \le \gamma_1 \le 1$	$14/19 \le \lambda_1 \le 11/12$
J_2	(x_1, x_3, x_6)	$1/2 \le \gamma_1 \le 1$	$11/12 \le \lambda_1 \le 1$
J_3	(x_2, x_3, x_7)	$0 < \gamma_1 \le 1$	$1/8 \le \lambda_1 \le 14/19$
J_4	(x_2, x_5, x_7)	$0 < \gamma_1 \le 1$	$0 < \lambda_1 \le 1/8$
J_5	(x_1, x_3, x_7)	$0 < \gamma_1 \le 1/2$	$14/19 \le \lambda_1 \le 1$

Assume that by the MC^2–simplex method, we have obtained a primal potential basis K_1 and a dual potential basis Q_1 in the processes of locating the set of all potential solutions. (See Tables 15.4 and 15.5)

From Table 15.4, we have $\mathbf{x}(K_1) = (x_3, x_6, x_7)$, $0 < \gamma_1 \le 1$.
From Table 15.5, we have $\mathbf{x}(Q_1) = (x_2, x_3, x_5)$, $0 < \gamma_1 \le 1/8$, and

$$(\mathbf{u}(Q_1, \lambda))^t = \lambda^t \mathbf{C}_{Q_1} \mathbf{B}_{Q_1}^{-1} = \lambda^t \begin{pmatrix} 0 & 0.6 & -1.4 \\ 0 & 1.2 & 0.2 \end{pmatrix}$$

Let j = 4, from the given problem, we have that $\boldsymbol{A}_4 = (3, 2, 3)^t$ and $\mathbf{C}_4 = (1, 2)^t$. Then, $\lambda^t \mathbf{C}_{Q_1} \mathbf{B}_{Q_1}^{-1} \boldsymbol{A}_4 - \lambda^t \mathbf{C}_4 = \lambda^t \begin{pmatrix} -4 \\ 1 \end{pmatrix}$.

Therefore,

$$\Lambda(Q_1) = \{\lambda > 0 \mid 0 < \lambda_1 \le 1/8,\ \lambda_1 + \lambda_2 = 1\}$$

$$\subseteq H_4^{>}(Q_1) = \{\lambda > 0 \mid \lambda^t \begin{pmatrix} -4 \\ 1 \end{pmatrix} > 0\}$$

$$= \{\lambda > 0 \mid 0 < \lambda_1 < 1/5,\ \lambda_1 + \lambda_2 = 1\}.$$

Also, from Table 15.5, we have that $\mathbf{B}_{Q_1}^{-1} \boldsymbol{A}_4 = (1, 10, 0)^t \ge 0$.

By Theorem 15.3 and Theorem 15.5, we know that the 4th opportunity is not in any potential basis of $\Pi(K_1, Q_1)$.

Table 15.4 Simplex Tableau of K_1

x_1	x_2	x_3	x_4	x_5	x_6	x_7	*RHS*	
0.5	0.25	1	0.75	0.25	0	0	7.5	30
2.5	3.75	0	1.25	-0.25	1	0	52.5	60
2.5	1.25	0	3.75	0.25	0	1	22.5	90
-2	-0.5	0	0.5	0.5	0	0	15	60
1.5	-4.75	0	-1.25	0.25	0	0	7.5	30

Table 15.5 Simplex Tableau of Q_1

x_1	x_2	x_3	x_4	x_5	x_6	x_7	*RHS*	
1	1	0	1	0	0.2	0.2	15	30
5	0	0	10	1	-1	3	15	210
-1	0	1	0	0	-0.2	-0.8	0	-30
-4	0	0	-4	0	0.6	-1.4	15	-30
5	0	0	1	0	1.2	0.2	75	120

15.2 Eliminating Permanently Dominated Opportunities for The Entire Space of Parameters

Based on Theorem 15.3 and Theorem 15.5, in this section, we describe a method of eliminating the PDOs from further consideration in the process of identifying the set of all potential solutions, $\vartheta = \{J_1, \ldots, J_g\}$, for the MC[2] problem (M4.1).

Assume that for any (γ, λ), (M4.1) always has a bounded optimal solution. With this assumption ϑ covers the space $\Gamma \times \Lambda$. By (iv) of Definition 15.1, this means that $\Gamma \times \Lambda \subseteq (\Gamma(J_1) \times \Lambda(J_1)) \cup \ldots \cup (\Gamma(J_g) \times \Lambda(J_g))$. Therefore, without loss of generality, we can assume that there is a group of primal-dual pairs denoted by $\Upsilon = \{(K^i, Q^i) \mid i = 1, \ldots, d\}$ *such that* $\Gamma \times \Lambda \subseteq \cup \{\Gamma(K^i) \times \Lambda(Q^i) \mid (K^i, Q^i) \in \Upsilon\}$. Note that some K^i and/or Q^i in such a Υ may be potential bases; and there may be a case where $K^i = K^j$ (or $Q^i = Q^j$) $i \neq j$. By Definition 15.2, a potential solution satisfies that both the primal and dual parameter sets are non-empty. But, a primal (dual) potential solution satisfies only that the primal (dual) parameter set is non-empty. Thus, to find such a Υ may be much easier than to find the entire ϑ.

Recall that $\Pi(K,Q) = \{J(\gamma, \lambda) \mid J(\gamma, \lambda)$ is an optimal basis of the resulting linear programming of Problem (M4.1) with fixed(γ, λ), $(\gamma, \lambda) \in \Gamma(K) \times \Lambda(Q)\}$. Let $\Pi(\Upsilon) = \cup \{\Pi(K^i, Q^i) \mid (K^i, Q^i) \in \Upsilon\}$. Then, we have:

Theorem 15.6 $\Gamma \times \Lambda$ is covered by $\Pi(\Upsilon)$.

Proof. By Theorem 15.4, for any $(K^i, Q^i) \in \Upsilon$, $\Gamma(K^i) \times \Lambda(Q^i) \subseteq \cup \{(\Gamma(J(\gamma_i, \lambda_i)) \times \Lambda(J(\gamma_i, \lambda_i)) \mid (\gamma_i, \lambda_i) \in \Gamma\Gamma(K^i) \times \Lambda(Q^i)\}$, which is the union of the optimal situation sets of $\Pi(K^i, Q^i)$. Then,

$\Gamma \times \Lambda \subseteq \cup \{\Gamma\Gamma(K^i) \times \Lambda(Q^i) \mid (K^i, Q^i) \in \Upsilon\} \subseteq \cup \{(\cup \{\Gamma(J(\gamma_i, \lambda_i)) \times \Lambda(J(\gamma_i, \lambda_i)) \mid (\gamma_i, \lambda_i) \in \Gamma\Gamma(K^i) \times \Lambda(Q^i)\} \mid (K^i, Q^i) \in \Upsilon)$, which is the union of the optimal situation sets of $\Pi(\Upsilon)$. Therefore, $\Gamma \times \Lambda$ is covered by $\Pi(\Upsilon)$ by (iii) of Definition 15.1.

Theorem 15.7. $\Pi(\Upsilon) = \vartheta$.

Proof. Since $\Pi(\Upsilon) \in \vartheta$, we need to show that $\vartheta \in \Pi(\Upsilon)$.

Given any $J \in \vartheta$, note that $\Gamma(J) \times \Lambda(J) \subseteq \Gamma \times \Lambda \subseteq \cup \{\Gamma(K^i) \times \Lambda(Q^i) \mid (K^i, Q^i) \in \Upsilon\}$. Then, for any $(\gamma, \lambda) \in \Gamma(J) \times \Lambda(J)$, there is (K^i, Q^i) such that $(\gamma, \lambda) \in \Gamma(K^i) \times \Lambda(Q^i)$. Thus, J is an optimal basis of the resulting linear programming of Problem (M4.1) with fixed (γ, λ) related to (K^i, Q^i), since $(\gamma, \lambda) \in (\Gamma(K^i) \times \Lambda(Q^i)) \cap (\Gamma(J) \times \Lambda(J))$. By the definition of $\Pi(K^i, Q^i)$, $J \in \Pi(K^i, Q^i)$. Thus, $J \in \Pi(\Upsilon)$.

Theorem 15.8. Suppose that Assumption 15.1 holds. Let $\Upsilon = \{(K^i, Q^i) \mid i = 1, \ldots, d\}$ be a group of primal-dual pairs such that $\Gamma \times \Lambda \subseteq \cup \{\Gamma(K^i) \times \Lambda(Q^i) \mid (K^i, Q^i) \in \Upsilon\}$. If the jth opportunity is not in any potential basis of $\Pi(K^i, Q^i)$ for any i, then the jth opportunity is not in any potential basis of ϑ and, consequently, the jth column corresponding to variable x_j can be eliminated from further consideration in the process of identifying the entire ϑ.

Proof. If for each (K^i, Q^i), the jth opportunity is not in any potential basis of $\Pi(K^i, Q^i)$, then the jth opportunity is not in any potential basis of $\Pi(\Upsilon)$. Note that by Theorem 15.6, $\Pi(\Upsilon)$ covers the space $\Gamma\times\Lambda$ and by Theorem 15.7, $\Pi(\Upsilon) = \vartheta$. We conclude that the jth opportunity is not in any potential basis of ϑ. Therefore, the jth column corresponding to variable x_j can be eliminated from further consideration in the process of identifying the entire ϑ.

Since it is easier to cover the space $\Gamma\times\Lambda$ by a group of primal-dual pairs Υ, based on Theorem 15.3, Theorem 15.5, and Theorem 15.8, we propose the following procedure to eliminate all PDOs from further consideration in the process of identifying the entire ϑ as follows:

Algorithm 15.1 For the given MC² problem (M4.1), suppose that ϑ covers the space $\Gamma\times\Lambda$.

Step 1. We find a group of primal-dual pairs, $\Upsilon = \{(K^i, Q^i) \mid i = 1, \ldots, d\}$, such that $\Gamma\times\Lambda \subseteq \cup\{\Gamma(K^i)\times\Lambda(Q^i) \mid (K^i, Q^i) \in \Upsilon\}$.

Step 2. For each pair (K^i, Q^i), we use Theorem 15.3 and Theorem 15.5 to check whether or not the jth opportunity, $j\in N$, is in any potential basis of $\Pi(K^i, Q^i)$.

Step 3. By Theorem 15.8, we check whether or not the jth opportunity, $j \in N$, is in any potential basis of ϑ. Then, we can identify and eliminate $N'(\vartheta)$, the set of all PDOs, from further consideration in the process of finding the entire ϑ.

Remark 15.2 Recall that Theorem 15.3 is true if and only if $J(\gamma^0, \lambda^0)$ is non-degenerate potential basis. If $J(\gamma^0, \lambda^0)$ is a degenerate potential basis and $x^*_j(J(\gamma^0, \lambda^0), \gamma^0) = 0$, for some i, then Theorem 15.3 fails. That ith opportunity is in $J(\gamma^0, \lambda^0)$ or not is unknown. Consequently, Algorithm 15.1 cannot apply to the degenerate cases.

Example 15.2 (continuation of Example 15.1)

We show how Algorithm 15.1 can be used to eliminate $N'(\vartheta)$ from further consideration in the process of locating $\vartheta = \{J_1, \ldots, J_5\}$.

Step 1. By the MC²-simplex method, we obtain bases K_1, J_3,, J_5 and Q_1 as listed in Table 15.6.

Table 15.6

Bases	Basic Variables	$\Gamma(\,.\,)$	$\Lambda(\,.\,)$
K_1	$(x_3,\ x_6,\ x_7)$	$0 < \gamma_1 \leq 1$	
J_3	$(x_2,\ x_3,\ x_7)$	$0 < \gamma_1 \leq 1$	$1/8 \leq \lambda \leq 14/19$
J_5	$(x_1,\ x_3,\ x_7)$	$0 < \gamma_1 \leq 1/2$	$14/19 \leq \lambda \leq 1$
Q_1	$(x_2,\ x_3,\ x_5)$		$0 \leq \lambda \leq 1/8$

Note that in Table 15.6, K_1 is a primal potential basis, J_3, and J_5 are potential bases, and Q_1 is a dual potential basis. We assign the group of primal-dual pairs as: $\Upsilon = \{(K_1,\ Q_1),\ (K_1,\ J_3),\ (K_1,\ J_5)\}$. (In the group, we have one primal potential basis and three dual potential bases where we treat J_3 and J_5 as two dual potential bases.)

Then, we see that $\Gamma \times \Lambda \subseteq (\Gamma(K_1) \times \Lambda(Q_1)) \cup (\Gamma(K_1) \times \Lambda(J_3)) \cup (\Gamma(K_1) \times \Lambda(J_5))$ (see Figure 15.2).

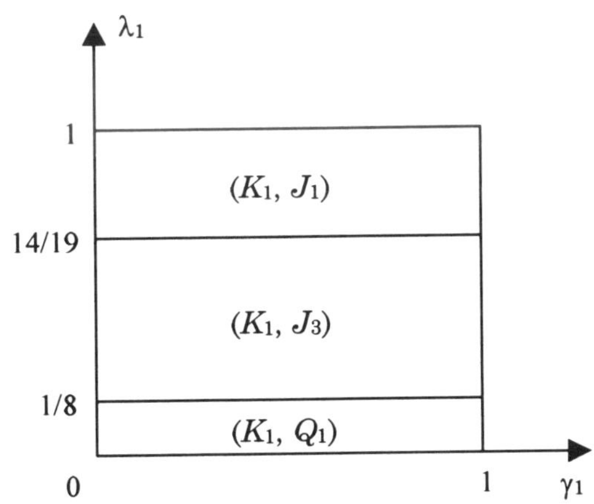

Figure 15.2

Note that in Figure 15.2, for simplicity, we use (K_1, Q_1), (K_1, J_3), and (K_1, J_5) to represent $\Gamma(K_1)\times\Lambda(Q_1)$, $\Gamma(K_1)\times\Lambda(J_3)$, and $\Gamma(K_1)\times\Lambda(J_5)$, respectively.

Table 15.7

x_1	x_2	x_3	x_4	x_5	x_6	x_7	RHS	
0	- 2.5	0	2.5	0.5	-1	1	-30	30
1	1.5	0	0.5	-0.1	0.4	0	21	24
0	-0.5	1	0.5	0.3	-0.2	0	-3	18
0	2.5	0	1.5	0.3	0.8	0	57	108
0	-7	0	-2	0.4	-0.6	0	-24	-6

Step 2. For primal-dual pair (K_1, J_5), we have the tableau of J_5 in Table 15.7 and the tableau of K_1in Table 15.4.

From the given problem, we know that N = {1, 2, 3, 4}. Note that only nonbasic column j with $\mathbf{B}_{J_5}^{-1}\mathbf{A}_j \geq 0$, $j \in N$, in Table 15.7 needs to be checked.

Let j = 4, from Table 15.7, we obtain that

$$H_4^{>}(J_5) = \{\lambda > 0 \mid \lambda^t \begin{pmatrix} 1.5 \\ -2 \end{pmatrix} > 0\} = \{\lambda > 0 \mid 4/7 < \lambda_1 \leq 1, \lambda_1 + \lambda_2 = 1\}.$$

Thus, $\Lambda(J_5) \subseteq H_4^{>}(J_5)$.

Also, from Table 15.7, we have that $\mathbf{B}_{J_5}^{-1}\mathbf{A}_4 = (2.5, 0.5, 0.5)^t > 0$.

By Theorem 15.3 and Theorem 15.5, we know that the 4th opportunity is not in any potential basis of $\Pi(K_1, J_5)$.

Similarly, we know that the 1st and 4th opportunities are not in any potential basis of $\Pi(K_1, J_3)$; and the 4th opportunity is not in any potential basis of $\Pi(K_1, Q_1)$.

Step 3. By Theorem 15.8, we conclude that only the 4th opportunity is not in any potential basis of $\Pi(\Upsilon) = \vartheta$ that covers the space $\Gamma\times\Lambda$. Thus, the 4th product is not in any potential bases J_1, J_2, J_3, J_4, and J_5. That is, $N'(\vartheta) = \{4\}$. We can eliminate the column corresponding x_4 from further consideration in the process of locating $\vartheta = \{J_1, J_2, J_3, J_4, J_5\}$. In fact, x_4 is not a basic variable for J_1, J_2, J_3, J_4, and J_5 (see Table 15.3).

Remark 15.3 We may be interested in knowing the surplus cost of a computation with the elimination technique, Algorithm 15.1, and comparing it with the MC^2-simplex method without Algorithm 15.1. For a given primal-dual pair (K, Q), Theorem 15.3 and Theorem 15.5 in Step 2 of Algorithm 15.1 provide a check for each nonbasic opportunity j in a total of O((q+m)n) arithmetic operations. Suppose there exists a potential basis J such that $\Gamma(K)\times \Lambda(Q)$ covers $\Gamma(J)\times \Lambda(J)$. By using the MC^2-simplex method to identify J, we first pivot the basis J from an adjacent basis that needs O((q+m)(n+m+p)) arithmetic operations. Then, we identify both the primal and dual parameter sets by using (Yu, 1985) which needs O(m(p+q)) arithmetic operations. Thus, the total of O((q+m)(n+m+p)+m(p+q)) arithmetic operations is needed to identify J. The cost of this computation is much higher than that of the computation by using Theorem 15.3 and Theorem 15.5 to eliminate the PDOs. Therefore, comparing this with the MC^2-simplex method with Algorithm 15.1 would have a significant cost saving in computation for identifying ϑ and may greatly reduce the number of iterations required to obtain ϑ. However, the precise analysis about the density of a large number of PDOs in a given MC^2 problem and the comparison study between the standard MC^2-simplex method with Algorithm 15.1 are ongoing research problems.

15.3 Elimination Techniques for Multiple Criteria Linear Programming Problems

Recall that if we replace Υ in the MC^2 problem (M4.1) by an m-dimensional vector $\boldsymbol{d} \in \mathbf{R}^m$, then (M4.1) reduces to an MC linear programming problem. An MC linear programming problem can be written as (M3.1):

$$\begin{array}{ll} \textit{Max} & \mathbf{C\,x} \\ \textit{Subject to} & \mathbf{A\,x} \le \mathbf{d} \\ & \mathbf{x} \ge 0 \end{array}$$

where $\mathbf{C} \in \mathbf{R}^{q\times n}$, $\mathbf{A} \in \mathbf{R}^{m\times n}$, and $\boldsymbol{d} \in \mathbf{R}^m$.

In this section a given feasible basis J for (M3.1) is called nondegenerate if $\mathbf{x}(J) = \mathbf{B}_J^{-1}\mathbf{d} > 0$.

Assumption 15.2 All feasible bases for Problem (M3.1) under discussion are nondegenerate.

For a given MC problem (M3.1), we have the following well-known results (chapter 8 of Yu (1985)).

Definition 15.3 A feasible basis J is a nondominated basis (solution) for the MC problem (M3.1) if there exists $\lambda > \mathbf{0}$ such that J is an optimal basis for

$$\begin{array}{ll} \text{Max} & \boldsymbol{\lambda}'\mathbf{C}\mathbf{x} \\ \text{Subject to} & \mathbf{A}\mathbf{x} \leq \mathbf{d} \\ & \mathbf{x} \geq 0. \end{array}$$

Theorem 15.9 A feasible basis J is a nondominated basis for the MC problem (M3.1) if and only if $\boldsymbol{\Lambda}(J) \neq \phi$.

The following theoretical results can be derived by using analyses simslar to Theorems 15.3, 15,5, and 15.8. We list them without providing the proofs.

Theorem 15.10 Given a dual basis Q and a fixed $\lambda^0 \in \Lambda$ (Q), let $J(\lambda^0)$ be an optimal basis. Suppose that Assumption 15.2 holds. If for given j $\in$ N, Λ (Q)$\subseteq H_j^{>}$(Q) and $\mathbf{B}_Q^{-1}\mathbf{A}_j \geq \mathbf{0}$, then x_j is not a basic variable in $J(\lambda^0)$. That is, the jth opportunity is not in $J(\lambda^0)$.

Define $\Pi'(Q)$ = $\{J(\lambda) \mid J(\lambda)$ is an optimal basis of the resulting linear programming of the problem (M4.1) with fixed $\lambda \in \Lambda(Q)\}$.

Theorem 15.11 Suppose that Assumption 15.2 holds. Given a dual basis Q, if for given j $\in$ N, $\Lambda(Q) \subseteq H_j^{>}(Q)$ and $\mathbf{B}_Q^{-1}\mathbf{A}_j \geq \mathbf{0}$, then the jth opportunity is not in $J(\lambda) \in \Pi'(Q)$ for any $\lambda \in \Lambda(Q)$.

Let $\vartheta' = \{J_1, \ldots, J_h\}$ be the set of all nondominated solutions identified by the MC-simplex method (Yu, 1985) for the MC problem (M3.1). We assume that for any $\lambda > \mathbf{0}$, (M3.1) has a bounded optimal solution. Then ϑ' covers the space Λ . This means that $\Lambda \subseteq \Lambda($ $J_1) \cup \ldots \cup \Lambda(J_h)$. Note that there may be a case where $\Lambda(J_i) \cap \boldsymbol{\Lambda}(J) \neq \phi$, $i \neq j$. Without loss of generality, we can assume that there is a group of dual bases denoted by $\Upsilon' = \{Q_i \mid i = 1, ..., d\}$ be a group of dual bases such that $\Lambda \subseteq \cup\{\Lambda(Q_i) \mid (Q_i) \in \Upsilon'\}$. Let $\Pi(\Upsilon') = \cup\{\Pi'(Q_i) \mid (Q_i) \in \Upsilon'\}$. Then, we have:

Theorem 15.12 Let $\Upsilon' = \{Q_I \mid i = 1, ..., d\}$ be a group of dual bases such that $\Lambda \subseteq \cup\{\Lambda(Q_i) \mid (Q_i) \in \Upsilon'\}$. Suppose that Assumption 15.2 holds. If the jth opportunity is not in any optimal basis of $\Pi'(Q_i)$ for any i, then the jth opportunity is not in any nondominated basis of ϑ' and, consequently, the jth column corresponding to variable x_j can be eliminated from further consideration in the processes of identifying the entire ϑ'.

A procedure to eliminate all PDOs from further consideration in the process of solving an MC problem is as follows:

Algorithm 15.2

Step 1. We find a group of dual bases, $\Upsilon' = \{Q_i \mid i = 1, ..., d\}$, such that $\Lambda \subseteq \cup\{\Lambda(Q_i) \mid (Q_i) \in \Upsilon'\}$.
Step 2. For each basis Q_i, we use Theorem 15.10 and Theorem 15.12 to check whether or not the jth opportunity, $j \in N$, is in any optimal basis of $\Pi'(Q_i)$.
Step 3. By Theorem 15.11, we check whether or not the jth opportunity, $j \in N$, is in any nondominated basis of ϑ'. Thus, we can identify all opportunities that are not in any nondominated basis of ϑ'. Those PDOs can be eliminated from further consideration in the process of finding the entire ϑ'.

Remark 15.4 Similar to Remark 15.2, Algorithm 15.2 cannot apply to a nondominated basis of a given MC problem if the basis is degenerate.

Example 15.3 Let $\gamma_1 = 2/3$ and $\gamma_2 = 1/3$ in the MC^2 problem of Example 15.1. Then the MC^2 problem reduces to the MC problem

$$Max \begin{pmatrix}\lambda_1 & \lambda_2\end{pmatrix} \begin{pmatrix} 3 & 1 & 2 & 1 \\ -1 & 5 & 1 & 2 \end{pmatrix} \begin{pmatrix} x_1 \\ x_2 \\ x_3 \\ x_4 \end{pmatrix}$$

$$Subject\ to \begin{pmatrix} 2 & 1 & 2 & 1 \\ 3 & 4 & 1 & 2 \\ 2 & 1 & -1 & 3 \end{pmatrix} \begin{pmatrix} x_1 \\ x_2 \\ x_3 \\ x_4 \end{pmatrix} \le \begin{pmatrix} 60 \\ 70 \\ 30 \end{pmatrix}$$

$$x_i \ge 0, i = 1,\ 2,\ 3,\ 4$$

By applying the computer program of (Chien, Shi and Yu 1989) we obtain the set of all nondominated solutions $\vartheta' = \{J_1, \ldots, J_2\}$as in Table 15.8. Note that when we use $\lambda_1 + \lambda_2 = 1$ in Table 9.14, λ_2 is a uniquely specified when λ_1 is given.

The meaning of the nondominated solution $J_1$with the basic variables (x_1, x_2, x_3) is that if the feasible basis J_1 is chosen, then {1,2,3} are the opportunities that we are committed to undertake. Thus, J_1 is an optimal basis whenever $14/19 \le \lambda_1 \le 11/12$. The nondominated solutions J_2, J_3, and J_4 in Table 15.8 can be similarly explained.

Now suppose before we find $\vartheta' = \{J_1, \ldots, J_4\}$ of Table 15.8, we want to use Algorithm 15.2 to eliminate all PDOs from further consideration in the process of locating $\vartheta' = \{J_1, \ldots, J_4\}$.

Step 1. By the MC-simplex method, assume that we obtain a group of dual bases $\Upsilon' = \{Q_1, J_3, J_4\}$ as listed in Table 15.9. (Note that we treat nondominated bases, J_3 and J_4 as the dual bases).

Table 15.8

Feasible bases	Basic Variables	$\Lambda(J_1)$
J_1	(x_1, x_2, x_3)	$14/19 \le \lambda_1 \le 11/12$
J_2	(x_1, x_3, x_6)	$11/12 \le \lambda_1 \le 1$
J_3	(x_2, x_3, x_7)	$1/8 \le \lambda_1 \le 14/19$
J_4	(x_2, x_5, x_7)	$0 \le \lambda_1 \le 1/8$

Table 15.9

Bases	Basic Variables	$\Lambda(\,.\,)$
Q_1	(x_1, x_3, x_7)	$14/19 \le \lambda_1 \le 1$
J_3	(x_2, x_3, x_7)	$1/8 \le \lambda_1 \le 14/19$
J_4	(x_2, x_5, x_7)	$0 \le \lambda_1 \le 1/8$

Then, we see that $\Lambda \subseteq \Lambda(Q_1) \times \Lambda(J_3) \times \Lambda(J_4)$.

Step 2. For the basis Q_1, we have the tableau of Q_1 in Table 15.10.

Since the set of possible opportunities is $N = \{1,2,3,4,5,6,7\}$, we check only the nonbasic column j with $\mathbf{B}_Q^{-1}\mathbf{A}_j \ge 0$, $j \in N$, in Table 15.10.

Checking j = 4, from Table 15.10, we obtain that

$$H_4^{>}(Q_1) = \{\lambda > 0 \mid \lambda^t \begin{pmatrix} 1.5 \\ -2 \end{pmatrix} > 0\} = \{\lambda > 0 \mid 4/7 < \lambda_1 \le 1,\ \lambda_1 + \lambda_2 = 1\}.$$

Thus, $\Lambda(Q_1) \subseteq H_4^{>}(Q_1)$.

Also, from Table 15.10, we see that $\mathbf{B}_Q^{-1}\mathbf{A}_j = (0.5, 0.5, 2.5)^t > 0$.

By Theorem 15.10 and Theorem 15.12, we know that the 4th opportunity is not in any optimal basis of $\Pi'(Q_1)$.

Similarly, for the bases J_3 and J_4, the 1th and 4th opportunities are neither in any optimal basis of $\Pi'(J_3)$, nor any optimal basis of $\Pi'(J_4)$.

Step 3. By Theorem 15.11, we know that only the 4th opportunity is not in any optimal basis of $\Pi'(\Upsilon') = \vartheta'$ which covers the space Λ. In other words, only the 4th opportunity is not in any nondominated bases J_1, J_2, J_3, and J_4. Thus we can eliminate the column corresponding to x_4 from further consideration in the process of locating $\vartheta' = \{J_1, J_2, J_3, J_4\}$. As the verification, from Table 15.8, we see that x_4 is not a basic variable for J_1, J_2, J_3, or J_4.

Table 15.10 Simplex Tableau of Q_1

x_1	x_2	x_3	x_4	x_5	x_6	x_7	*RHS*
0	1.5	0	0.5	-0.1	-0.4	1	22
1	-0.5	0	0.5	0.3	0.2	0	4
0	-2.5	1	2.5	0.5	-1	0	-10
0	2.5	0	1.5	0.3	0.8	0	74
0	-7	0	-2	0.4	-0.6	0	-18

15.4 Further Remarks

In this chapter we have proposed a set of techniques to eliminate permanently dominated opportunities in solving MC^2 linear programming problems. If a given MC^2 problem involves a large number of opportunities, then the elimination techniques described in Algorithm 15.1 can be useful to reduce much computation in identifying the set of all potential solutions. We have also discussed the corresponding elimination technique (Algorithm 15.2) for solving MC linear programming problems.

Some other techniques to eliminate the permanently dominated opportunities for a given MC^2 linear problem can be developed based on our discussion in Section 15.2 and 15.3. For instance, to relax the assumption in Section 15.3 that for any (γ, λ), the MC^2 problem (M4.1) has a bounded optimal solution, we may assume that for any (γ, λ), (M4.1) either has no feasible solution or has a bounded optimal solution. With this assumption, the set of all potential solutions may not cover the whole parameter space $\Gamma \times \Lambda$. The elimination techniques can be similarly developed. We shall not do so (see (Shi, 1991) for details).

Exercises

15.1 Given the following MC^2 problem:

$$\begin{array}{ll} Max & 3x_1 + 2x_2 + x_3 - x_4 + 4x_5 \\ Max & 2x_1 - x_2 + 2x_3 + 5x_4 + 3x_5 \end{array}$$

Subject to $x_1 - 2x_3 + x_4 \leq (10, 20)$
$- x_2 + x_3 + 2x_4 + 2x_5 \leq (40, 20)$
$x_1 + x_2 - x_5 \leq (30, 80)$
$x_1, x_2, x_3 \geq 0,$

By the computer software of the MC²-simplex method (Hao and Shi 1996), we know the set of potential solutions are $\mathbf{x}(J_1) = (x_1, x_3, x_5)$, $\mathbf{x}(J_2) = (x_1, x_2, x_5)$, $\mathbf{x}(J_3) = (x_2, x_5, x_6)$, and $\mathbf{x}(J_4) = (x_1, x_2, x_3)$. Use Algorithm 15.1 to show the variable x_4 is not in any of these solutions before we find them.

15.2 Consider the following MC² problem:

Max $10x_1 - 5x_2 + 10x_3 + 20x_4 + 8x_5$
Max $5x_1 + 9x_2 + 12x_3 + 3x_4 - 3x_5$

Subject to $4x_1 + 3x_2 - 2x_3 + 7x_4 + 2x_5 \leq (50, 20)$
$3x_1 - 1x_2 + 5x_3 + 2x_4 + x_5 \leq (90, 40)$
$5x_1 + x_2 + 2x_3 - 2x_4 + 8x_5 \leq (50, 70)$
$x_1, x_2, x_3 \geq 0,$

Suppose that we know the set of potential solutions are $\mathbf{x}(J_1) = (x_4, x_3, x_8)$, $\mathbf{x}(J_2) = (x_4, x_3, x_5)$, $\mathbf{x}(J_3) = (x_4, x_3, x_2)$, and $\mathbf{x}(J_4) = (x_2, x_3, x_8)$, use Algorithm 15.1 to show the variable x_1 is not in any of these solutions.

Part Six: Applications

Chapter 16
OPTIMAL TRADE-OFF ANALYSIS IN TRANSFER PRICING

Ever since Du Pont and General Motors Corporation of USA initiated transfer pricing systems for the interdivisional transfer of resources among their divisions, many large organizations, with the creation of profit centers, have used a transfer pricing system in one way or the other. Recently, transfer pricing problems have become more important because most corporations increase transfer of goods or services dramatically among their divisions as a result of restructuring or downsizing their organizations (Tang 1992). Therefore, designing a good transfer pricing strategy should be a major concern for both top management and divisional managers.

Transfer pricing problems have been extensively studied by a number of scholars. Many of them have recognized that a successful transfer pricing strategy should consider multiple criteria (objectives), such as overall profit, total market share, divisional autonomy, performance evaluation, and utilized production capacity (Abdel-khalik and Lusk 1974, Bailey and Boe 1976, Merville and Petty 1978, Yunker 1983, Lecraw 1985 and Lin *et al.* 1993). However, few developed methods have a capability of dealing with all possible optimal trade-offs of multiple criteria in optimal solutions of the models.

In this chapter, we propose a multiple factor model to provide managers, who are involved in transfer price decision making of a multi-divisional corporation, with a systematic and comprehensive scenario about all possible optimal transfer prices depending on both multiple criteria and multiple constraint levels (in short, multiple factors) (Shi, Kwak and Lee 1998). The trade-offs of the optimal transfer prices, which have rarely been considered in the literature, can be used as a basis for managers of a corporation to make a high-quality decision in selecting their transfer pricing systems for business competition. First, existing transfer pricing models will be reviewed in Section 16.1. Then, the methodology of formulating and solving a transfer pricing model with multiple factors will be described in Section 16.2. A prototype of a transfer pricing problem in a corporation will be illustrated in Section 16.3 to explain the implications of the multiple factor transfer pricing model. Finally, further remarks will be presented in Section 16.4.

16.1 Existing Transfer Pricing Models

In the literature of transfer pricing problems, the various approaches can be categorized into four groups: (i) market-based pricing, (ii) accounting-based pricing, (iii) marginal cost pricing (or opportunity cost pricing), and (iv) negotiation-based pricing. Market-based prices are ideal for transfer prices when external market prices are available. Even though empirical research has found that some corporations prefer cost-based prices to market-based prices, market-based pricing method is recommended when the emphasis is on the motivation of divisional managers (Borkowski 1990).

However, if there is no outside market for intermediate goods or services, then accounting-based pricing, marginal cost pricing (economic models and mathematical programming techniques), or negotiation-based pricing (behavioral approach) is commonly recommended for finding a transfer pricing system.

In the accounting-based pricing approach, the divisional managers simply use accounting measurements of the divisions, such as full costs or variable costs, as their transfer prices. Thus, the transfer price of one division may differ from that of another division. These transfer prices may not be globally optimal for the corporation as a whole (Abdel-khalik and Lusk 1974 and Eccles 1983).

In marginal cost pricing approaches, Hirschleifer (1956) recommended use of an economic model to set transfer pricing at a supplying division's marginal cost to achieve the global optimal output. A problem of this economic model is that it can destroy the divisional manager's autonomy, and the supplying division may not get the benefit of efficiencies. Moreover, the manager of supplying division in this case should be evaluated based on cost, not on profit.

As an alternative method to identify marginal costs as transfer prices, a mathematical programming approach becomes more attractive for the transfer pricing problem because it handles complex situations in a trade setting (Dopuch and Drake 1964). The application of linear programming to the transfer pricing problem is based on the relationship between the primal and dual solutions in the linear programming problem. Shadow prices, which reflect the input values of scarce resources (or opportunity cost) implied in the primal problem, can be used as the basis for a transfer price system. However, these transfer prices have the following limitations for decision making: (i) those transfer prices based on dual values of a solution tend to reward divisions with scarce resources, (ii) the linear formulation requires a great deal of local information, and (iii) transfer prices based on shadow prices do not provide a guide for performance evaluation of divisional managers.

Ruefli (1971) proposed a generalized goal decomposition model for incorporating multiple criteria (objectives) and some behavioral aspects within a three-level hierarchical organization into the mathematical formulation of transfer pricing problems. Bailey and Boe (1976) suggested another goal programming model as a supplement of Ruefli's model. Both models overcome the shortcoming of linear programming in dealing with multiple criteria and the organizational hierarchy. Merville and Petty (1978) used a dual formulation of goal programming to directly find the shadow price as the optimal transfer price for multinational corporations. However, the optimal solution of these models that results in the transfer prices is determined by a particular distance (norm) function because of the mathematical structure of goal programming. Thus, the optimal solution represents only a single optimal trade-off of the multiple criteria, not all possible optimal trade-offs.

Watson and Baumler (1975) and Thomas (1980) criticized the lack of behavioral considerations in the mathematical programming approaches. They suggested a negotiated transfer pricing to further the integration of differentiation within the organization. The differentiation of the organization exists because, for example, different managers can interpret the same organizational problem differently. Generally, the negotiated transfer pricing will be successful under conditions such as the existence of some form of outside markets for the intermediate product, freedom to buy or sell outside, and sharing all market information among the negotiators. Negotiated prices may require iterative exchanges of information with the central office as a part of a mathematical programming algorithm.

16.2 Multiple Factor Transfer Pricing Model

The review of existing transfer pricing models shows that two major shortcomings, from a management point of view, need to be overcome in the previous mathematical models of transfer pricing problems. First, neither the linear programming approach nor the goal programming approach can provide a comprehensive scenario of all possible optimal trade-offs between multiple objectives under consideration for a given transfer pricing problem, such as maximizing the overall profits for a corporation and minimizing the under utilization of production capacity (Tang 1992). Second, none of the previous models can deal with the organizational differentiation problems, as Watson and Baumler (1975) pointed out. In real-life cases, when a corporation designs its transfer prices for the divisions, the involved decision makers (executives or the members of the task force) can give different opinions on the same issue, such as production capacity and customer's demand. In mathematical models, these different interpretations can be represented by different "constraint levels." Because both linear programming and goal programming presume a fixed single constraint level, they fail to mathematically describe such an organizational differentiation problem.

The MC^2 linear programming framework can resolve the above shortcomings inherent in the previous transfer pricing models. Based on Al-Eryani, Alamand and Akhter (1990), Yunker (1983) and Tang (1992), the four important objectives of transfer pricing problems in most corporations are considered: (i) maximizing the overall profit, (ii) maximizing the total market share, (iii) maximizing the subsidiary profit (note that the subsidiary profit maximization is used to reflect the degree of the subsidiary autonomy in decision making. It may differ from the overall profit), and (iv) maximizing the utilized production capacity. Even though the following model contains only these four specific objectives, the generality of the modelling process fits in all transfer pricing problems with multiple criteria and multiple constraint levels.

Let k be the number of divisions in a corporation under consideration, and t be the index number of the products that each division of the corporation produces.

Define x_{ij} as the units of the jth product made by the ith division, $i = 1, \ldots, k; j = 1, \ldots, t$. For the coefficients of the objectives, let p_{ij} be the unit overall profit generated from the jth product made by the ith division; m_{ij} be the market share value for the jth product made by the ith division in the market; s_{ij} be the unit subsidiary profit generated from the jth product made by the ith division; and c_{ij} be the unit utilized production capacity of the ith division to produce the jth product. For the coefficients of the constraints, let b_{ij} be the budget allocation rate for producing the jth product by the ith division. For the coefficients of the constraint levels, let b_{ij}^s be the budget availability level believed by the sth manager (or executive) for producing the jth product by the ith division, s = 1, . . . , h; let d_i^s be the production capacity level believed by the sth manager for the ith division; d_{ij}^s be the production capacity level believed by the sth manager for the ith division to produce the jth product; e_{ij}^s be the initial inventory level believed by the sth manager for the ith division to hold the jth product. Then, the multiple factor transfer pricing model (M16.1) is:

$$Max \sum_{i=1}^{k}\sum_{j=1}^{t} p_{ij}x_{ij}$$

$$Max \sum_{i=1}^{k}\sum_{j=1}^{t} m_{ij}x_{ij}$$

$$Max \sum_{i=1}^{k}\sum_{j=1}^{t} s_{ij}x_{ij}$$

$$Max \sum_{i=1}^{k}\sum_{j=1}^{t} c_{ij}x_{ij}$$

$$\text{Subject to} \qquad \sum_{i=1}^{k}\sum_{j=1}^{t} b_{ij}x_{ij} \le (b_{ij}^1,\dots,b_{ij}^h)$$

$$\sum_{j=1}^{t} x_{ij} \le (d_i^1,\dots,d_i^h)$$

$$x_{ij} \le (d_{ij}^1,\dots,d_{ij}^h)$$

$$-x_{ij} + x_{(i+1)j} \le (e_{ij}^1,\dots,e_{ij}^h)$$

$$x_{ij} \ge 0, i = 1,\dots,k,\ j = 1,\dots,t$$

In the next section, a prototype of the transfer pricing model in a corporation will be illustrated to demonstrate the implications for decision makers.

16.3 Model Implications

16.3.1 Numerical Example

As an illustration of the multiple factor transfer pricing model, the United Chemical Corporation has two divisions that process raw materials into intermediate or final products. Division 1, which is located in Kansas City, Missouri, manufactures two kinds of chemicals, called Products 1 and 2. Product 1 in Division 1 is intermediate product and cannot be sold externally. It can, however, be processed further by Division 2 into a final product. Division 2, which is located in Atlanta, Georgia, manufactures Product 2 and finalizes Product 1 in Division 1. The executives (president, vice president for production, and vice president for finance) and all divisional managers agree on the following multiple objectives:

(i) Maximize the overall company's profit.

(ii) Maximize the market share goal of Product 2 in Division 1, and Products 1 and 2 in Division 2.

(iii) Maximize the utilized production capacity of the company so that each division manager can avoid any underutilization of normal production capacity.

The data related to these objectives is given in Table 16.1.

Table 16.1 Data of Objectives in United Chemical Corporation

	Division 1		Division 2	
	Product 1	Product 2	Product 1	Product 2
Unit profit ($)	-4	8	13	5
Market price	0	40	42.6	38
Unit utilizing production capacity (h)	4	4	3	2

Table 16.2 Constraint Levels of Vice-Presidents

	Vice-president for production	Vice-president for finance
Transfer product constraint	0	100
Budget constraint ($)	45000	40000
Production capacity of Product 2 in Division 1	38000	12000
Production capacity of Product 1 in Division 2	45000	50000
Production capacity of Product 2 in Division 2	36000	10000

In the Table 16.1, Product 1 of Division 1 is a by-product which has no sale value in Division 1 at all. The unit profit -$4 of this product means its production cost. All other products generate profits. We use a market price of each product to maximize the market share goal. Note that the market prices of three products in Table 16.1 are different. In this company, besides the president, the vice presidents for production and for finance are the main decision makers and may have different interpretations of the same resource availability across the divisions. The vice president for production views the constraint level based on the material, manpower, and equipment under control, while the vice president for finance views the constraint level based on the available cash flow. The president will make the final decision for the company's transfer price setting on the basis of the compromises of both vice presidents. All divisional managers will carry out the president's decision, although they have their autonomy to provide information about the divisional profits and costs for the executives. The interpretations of the vice presidents for the resource constraint levels are summarized in Table 16.2. All executives and divisional managers agree on the units of resources consumed to produce the products. The data is given in Table 16.3.

Table 16.3 Unit Consumptions of Resources

	Division 1		Division 2	
	Product 1	Product 2	Product 1	Product
Transfer product constraint	1.0	0	1.4	0
Budget constraint($)	0	0.4	0.4	0.4
Production capacity of Product 2 in Division 1		1.0		
Production capacity of Product 1 in Division 2			1.0	
Production capacity of Product 2 in Division 2				1.0

Let x_{ij} be the units of the jth product produced by the ith division, i = 1, 2; j = 1, 2. Using the information in Tables 16.1-16.3, the multiple factor transfer pricing model is formulated as:

$$
\begin{aligned}
&\mathit{Max} && -4\, x_{11} + 8\, x_{12} + 13\, x_{21} + 5\, x_{22} \\
&\mathit{Max} && 40\, x_{12} + 46.2\, x_{21} + 38\, x_{22} \\
&\mathit{Max} && 4\, x_{11} + 4\, x_{12} + 3\, x_{21} + 2\, x_{22} \\
&\mathit{Subject\ to} && -\, x_{11} + x_{21} \leq (0,\ 100) \\
& && 0.4\, x_{12} + 0.4\, x_{21} + 0.4\, x_{22} \leq (45{,}000,\ 40{,}000) \\
& && x_{12} \leq (38{,}000,\ 12{,}000) \\
& && x_{21} \leq (45{,}000,\ 50{,}000) \\
& && x_{22} \leq (36{,}000,\ 10{,}000) \\
& && x_{ij} \geq 0,\ i = 1,\ 2;\ j = 1,\ 2.
\end{aligned}
$$

Since this multiple factor transfer pricing problem is a typical MC^2 problem, the MC^2 software of Chien *et al.* (1989) can be used to solve the problem. Let $\lambda = (\lambda_1, \lambda_2, \lambda_3)$ be the weight parameter for the objectives, where $\lambda_1 + \lambda_2 + \lambda_3 = 1$ and $\lambda_1, \lambda_2, \lambda_3 \geq 0$. Let $\gamma = (\gamma_1, \gamma_2)$ be the weight parameter for the constraint levels, where $\gamma_1 + \gamma_2 = 1$ and $\gamma_1, \gamma_2 \geq 0$.

Because both weight parameters (γ, λ) are unknown before design time, the solution procedure of MC^2 linear programming must be used to locate all possible potentially optimal solutions as (γ, λ) vary. The implications of the potentially optimal solutions for accounting decision makers will be explained in the next subsection. After putting (γ, λ) into the above model, it becomes:

$$\begin{array}{ll} \textit{Max} & \lambda_1 (-4\, x_{11} + 8\, x_{12} + 13\, x_{21} + 5\, x_{22}) + \\ & \lambda_2 (40\, x_{12} + 46.2\, x_{21} + 38\, x_{22}) + \\ & \lambda_3 (4\, x_{11} + 4\, x_{12} + 3\, x_{21} + 2\, x_{22}) \\ \textit{Subject to} & -x_{11} + x_{21} \leq 100\, \gamma_2 \\ & 0.4\, x_{12} + 0.4\, x_{21} + 0.4\, x_{22} \leq 45{,}000\, \gamma_1 + 40{,}000\, \gamma_2 \\ & x_{12} \leq 38{,}000\, \gamma_1 + 12{,}000\, \gamma_2 \\ & x_{21} \leq 45{,}000\, \gamma_1 + 50{,}000\, \gamma_2 \\ & x_{22} \leq 36{,}000\, \gamma_1 + 10{,}000\, \gamma_2 \\ & x_{ij} \geq 0,\ i = 1, 2;\ j = 1, 2. \end{array}$$

Let s_q, q = 1, . . ., 5, be the slack variables corresponding to the constraints. The MC^2 software yields two potentially optimal solutions $\{J_1, J_2\}$ and their associated values of (γ, λ) as shown in Table 16.4. Here, $V(J_i)$ denotes the objective value of potentially optimal solution J_i, which is a function of (γ, λ). When (γ, λ) are specified, $V(J_i)$ is the payoff of using J_i.

Table 16.4 shows that if the γ takes the value from $\Gamma(J_1)$ and the λ takes the value from $\Lambda(J_1)$, J_1 is the optimal solution for the transfer pricing problem. In this case, products $\{x_{11}, x_{12}, x_{21}, x_{22}\}$ will be produced to achieve the objective payoff $V(J_1)$ and the resource of x_{22} has the amount of s_5 unused. Similarly, the potentially optimal solution J_2 can be interpreted. However, J_2 is different from J_1 because there is the unused budget s_2 for J_2 and the parameter set $\Gamma(J_1) \neq (J_2)$. Because Product 1 in Division 1 is a byproduct and its unit profit is -\$4, whenever $\lambda_1 < \lambda_3$, both J_1 and J_2 are not optimal.

Let $p_q(J_i)$, q = 1, . . ., 5; i = 1, 2, be the shadow price of J_i for the qth constraint. According to the marginal cost pricing approach, the optimal transfer prices of $\{J_1, J_2\}$ are designated as the shadow prices of $\{J_1, J_2\}$. These optimal transfer prices are found in Table 16.5. Table 16.5 shows that (i) the relative transfer price between x_{11} and x_{21} is $p_1(J_1) = 4\lambda_1 - 4\lambda_3$; (ii) the transfer price for budget across the divisions is $p_2(J_1) = 12.5\lambda_1 + 95\lambda_2 + 5\lambda_3$; (iii) the transfer price for x_{12} is $p_3(J_1) = 3\lambda_1 +$

$2\lambda_2 + 2\lambda_3$; (iv) the transfer price for x_{21} is $p_4(J_1) = 4\lambda_1 + 8.2\lambda_2 + 5\lambda_3$, and (v) the transfer price for x_{22} is $p_5(J_1) = 0$.

Table 16.4 All Potentially Optimal Solutions

J_i	$\Gamma(J_i)$	$\Lambda(J_i)$	$V(J_i)$
J_1	$\gamma_1 + \gamma_2 = 1$ $65\gamma_1 - 280\gamma_2 \geq 0$ $\gamma_1, \gamma_2 \geq 0$	$\lambda_1 + \lambda_2 + \lambda_3 = 1$ $\lambda_1 \geq \lambda_3$ $\lambda_1, \lambda_2, \lambda_3 \geq 0$	$\lambda^t \begin{pmatrix} 856{,}000 & 736{,}400 \\ 4720{,}000 & 4234{,}000 \\ 526{,}000 & 47444{,}000 \end{pmatrix} \gamma$
J_2	$\gamma_1 + \gamma_2 = 1$ $-65\gamma_1 + 280\gamma_2 \geq 0$ $\gamma_1, \gamma_2 \geq 0$	$\lambda_1 + \lambda_2 + \lambda_3 = 1$ $\lambda_1 \geq \lambda_3$ $\lambda_1, \lambda_2, \lambda_3 \geq 0$	$\lambda^t \begin{pmatrix} 889{,}000 & 596{,}400 \\ 4967{,}000 & 3170{,}000 \\ 539{,}000 & 41184{,}000 \end{pmatrix} \gamma$

Table 16.5 All Optimal Transfer Prices

J_i	$p_q(J_i)$
$J_1 = (x_{11}, x_{12}, x_{21}, x_{22}, s_5)$	$p_1(J_1) = 4\lambda_1 - 4\lambda_3$ $p_2(J_1) = 12.5\lambda_1 + 95\lambda_2 + 5\lambda_3$ $p_3(J_1) = 3\lambda_1 + 2\lambda_2 + 2\lambda_3$ $p_4(J_1) = 4\lambda_1 + 8.2\lambda_2 + 5\lambda_3$ $p_5(J_1) = 0$
$J_2 = (x_{11}, x_{12}, x_{21}, x_{22}, s_2)$	$p_1(J_1) = 4\lambda_1 - 4\lambda_3$ $p_2(J_1) = 0$ $p_3(J_1) = 8\lambda_1 + 40\lambda_2 + 4\lambda_3$ $p_4(J_1) = 9\lambda_1 + 46.2\lambda_2 + 7\lambda_3$ $p_5(J_1) = 5\lambda_1 + 38\lambda_2 + 2\lambda_3$

16.3.2 Optimal Trade-offs and Their Accounting Implications

Optimal trade-offs related to transfer prices of the multiple factor model consist of three components: (i) trade-offs among multiple objectives, (ii) trade-offs among multiple constraint levels, and (iii) trade-offs between multiple objectives and multiple constraint levels. The trade-offs among multiple objectives imply that all possible optimal compromises of the multiple objectives are determined by locating all possible weights of importance of these objectives. Similarly, the trade-offs among multiple constraint levels imply that all possible optimal compromises of the multiple constraint levels which represent the executives' different opinions are determined by locating all possible weights of importance of these opinions. The trade-offs between multiple objectives and multiple constraint levels measure the effects on the transfer pricing problem by the interaction of the objectives and constraint levels. These three cases of the optimal trade-offs can be analyzed through the potentially optimal solutions of the multiple factor model. The optimal transfer prices also result from those solutions. In the following, three trade-off cases and accounting implications of the corresponding optimal transfer prices are explored in detail by using the above numerical example.

Because there are two potentially optimal solutions $\{J_1, J_2\}$ in the example, the optimal trade-offs should be studied in terms of both J_1 and J_2. For the trade-offs among three objectives that are the overall company's profit, the market share of the company, and the utilized production capacity of the company:

(i) If the overall company's profit is not considered (this implies $\lambda_1 = 0$ and $\lambda_3 \neq 0$), then either J_1 or J_2 is not optimal since $\lambda_3 < 0$ (see Table 4).

(ii) If the market share of the company is not considered (i.e., $\lambda_2 = 0$), then both J_1 and J_2 are optimal for which $\lambda_1 + \lambda_3 = 1$, $\lambda_1 \geq \lambda_3$, and λ_1, $\lambda_3 \geq 0$. The graphical representation is shown in Figure 16.1. From Figure 16.1, any weighting values of λ_1 and λ_3 taken from the feasible (dark) segment guarantees that the utility values of both the overall company's profit and the utilized production capacity of the company are maximized. The resulting optimal transfer prices associated with J_1 are $p_1(J_1) = 4\lambda_1 - 4\lambda_3$, $p_2(J_1) = 12.5\lambda_1 + 5\lambda_3$, $p_3(J_1) = 3\lambda_1 + 2\lambda_3$, $p_4(J_1) = 4\lambda_1 + 5\lambda_3$, and $p_5(J_1) = 0$, respectively. The resulting optimal transfer prices associated with J_2 are $p_1(J_2) = 4\lambda_1 - 4\lambda_3$, $p_2(J_2) = 0$, $p_3(J_2) = 8\lambda_1 + 4\lambda_3$, $p_4(J_2) = 9\lambda_1 + 7\lambda_3$, and $p_5(J_2) = 5\lambda_1 + 2\lambda_3$, respectively. The weighting values of λ_1 and λ_3 in the transfer prices for J_1 and J_2 are given in Figure 16.1, because $\Lambda(J_1) = \Lambda(J_2)$.

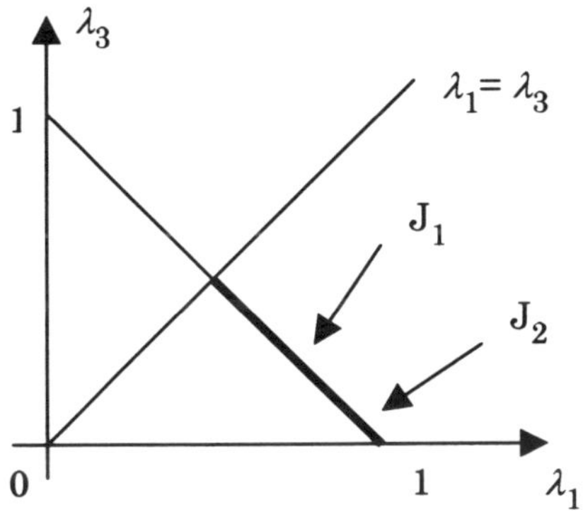

Figure 16.1 Optimal Trade-offs between Profit (λ_1) and Production Capacity (λ_3)

(iii) If the utilized production capacity of the company is not considered (i.e., $\lambda_3 = 0$), then both J_1 or J_2 are optimal for which $\lambda_1 + \lambda_2 = 1$ and $\lambda_1, \lambda_2 \geq 0$, where the graphical representation is shown in Figure 16.2. Figure 16.2 implies that any weighted combination of λ_1 and λ_1 taken from the indicated feasible segment maximizes the utility values of both the overall company's profit and the market share of the company. The resulting optimal transfer prices of J_1 are $p_1(J_1) = 4\lambda_1$, $p_2(J_1) = 12.5\lambda_1 + 95\lambda_2$, $p_3(J_1) = 3\lambda_1 + 2\lambda_2$, $p_4(J_1) = 4\lambda_1 + 8.2\lambda_2$, and $p_5(J_1) = 0$, while the resulting optimal transfer prices of J_2 are $p_1(J_2) = 4\lambda_1$, $p_2(J_2) = 0$, $p_3(J_2) = 8\lambda_1 + 40\lambda_2$, $p_4(J_2) = 9\lambda_1 + 46.2\lambda_2$, and $p_5(J_2) = 5\lambda_1 + 38\lambda_2$, respectively.

For the trade-offs among two constraint levels (the different opinions of two vice presidents), the weight of the vice president for production is γ_1 while that of the vice president for finance is γ_2 such that $\gamma_1 + \gamma_2 = 1$, $\gamma_1, \gamma_2 \geq 0$. The range of γ_1 and γ_1 are decomposed into two subsets: $\Gamma(J_1) = \{\gamma_1, \gamma_2 \geq 0 \mid 65\gamma_1 - 280\gamma_2 \geq 0 \text{ and } \gamma_1 + \gamma_2 = 1\}$ and $\Gamma(J_2) = \{\gamma_1, \gamma_2 \geq 0 \mid -65\gamma_1 + 280\gamma_2 \geq 0 \text{ and } \gamma_1 + \gamma_2 = 1\}$ (see Table 16.4). The graphical representation of $\Gamma(J_1)$ and $\Gamma(J_2)$ is shown in Figure 16.3. Whenever the weighting values of γ_1 and γ_2 are taken from $\Gamma(J_1)$, the corresponding compromise of two vice presidents will result in the optimal transfer prices of J_1 in Table 16.5. Similarly, the decision situation of constraint levels for J_1 can be explained.

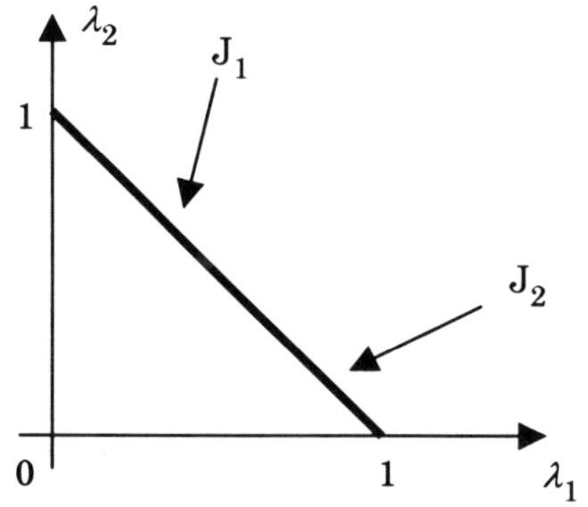

Figure 16.2 Optimal Trade-offs between Profit (λ_1) and Market Share (λ_2)

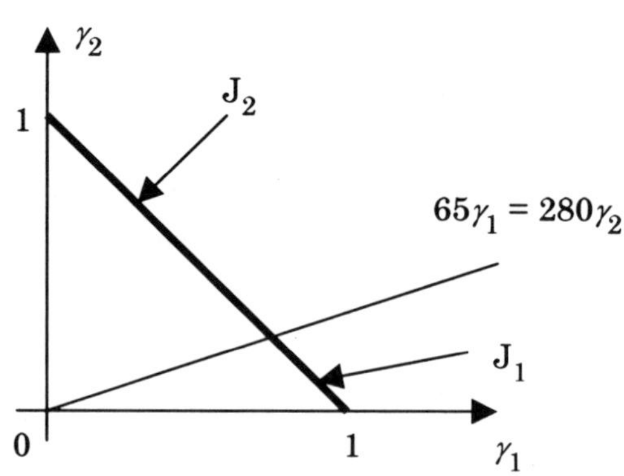

Figure 16.3 Optimal Trade-offs between Vice-President for Production (γ_1) and Vice-President for Finance (γ_2)

Finally, there are many optimal trade-off situations between three objectives and two constraint levels involved with the transfer pricing problem. For example, two cases are illustrated as follows:

(i) In the case that the utilized production capacity of the company is not considered (i.e., $\lambda_3 = 0$) in contrast to the weighting value of the market share and that of the vice president for finance (i.e., $0 \leq \lambda_2 \leq 1$ and $0 \leq \gamma_2 \leq 1$), the optimal trade-offs between the overall company's profit and the constraint level believed by the vice president for production for J_1 and J_2 are shown in Figure 4. Here, if the values of (γ_1, λ_1) are taken from $0 \leq \lambda_1 \leq 1$ and $.81 \leq \gamma_1 \leq 1$, then the optimal transfer prices associated with J_1 will be chosen. They are $p_1(J_1) = 4\lambda_1$, $p_2(J_1) = 12.5\lambda_1 + 95\lambda_2$, $p_3(J_1) = 3\lambda_1 + 2\lambda_2$, $p_4(J_1) = 4\lambda_1 + 8.2\lambda_2$, and $p_5(J_1) = 0$. Otherwise, the values of (γ_1, λ_1) are taken from $0 \leq \lambda_1 \leq 1$ and $0 \leq \gamma_1 \leq .81$, and the optimal transfer prices associated with J_2, $p_1(J_2) = 4\lambda_1$, $p_2(J_2) = 0$, $p_3(J_2) = 8\lambda_1 + 40\lambda_2$, $p_4(J_2) = 9\lambda_1 + 46.2\lambda_2$, and $p_5(J_2) = 5\lambda_1 + 38\lambda_2$, will be chosen.

(ii) In the case that the weighting value of the overall company's profit is fixed at .5 (i.e., $\lambda_1 = .5$) and the weighting value of the market share and that of the vice president for finance are any of $0 \leq \lambda_2 \leq 1$ and $0 \leq \gamma_2 \leq 1$, respectively, Figure 16.5 shows the optimal trade-offs between the utilized production capacity of the company and the constraint level believed by the vice president for production for J_1 and J_2. In Figure 16.5, if the values of (γ_1, λ_3) are taken from $0 \leq \lambda_3 \leq .5$ and $.81 \leq \gamma_1 \leq 1$, then the optimal transfer prices associated with J_1 are $p_1(J_1) = 2 - 4\lambda_3$, $p_2(J_1) = 6.25 + 95\lambda_2 + 5\lambda_3$, $p_3(J_1) = 1.5 + 2\lambda_2 + 2\lambda_3$, $p_4(J_1) = 2 + 8.2\lambda_2 + 5\lambda_3$, and $p_5(J_1) = 0$, respectively. If the values of (γ_1, λ_3) are taken from $0 \leq \lambda_3 \leq .5$ and $0 \leq \gamma_1 \leq .81$, then the optimal transfer prices associated with J_2 are $p_1(J_2) = 2 - 4\lambda_3$, $p_2(J_2) = 0$, $p_3(J_2) = 4 + 40\lambda_2 + 4\lambda_3$, $p_4(J_2) = 4.5 + 46.2\lambda_2 + 7\lambda_3$, $p_5(J_2) = 2.5 + 38\lambda_2 + 2\lambda_3$, respectively. Note that when the weighting values fall in the range of $0.5 < \lambda_3 \leq 1$ and $0 \leq \gamma_1 \leq 1$, there is not any optimal trade-off because both J_1 and J_2 are not optimal solutions (recall that this is caused by the byproduct 1 in Division 1 which has -\$4 as the unit profit).

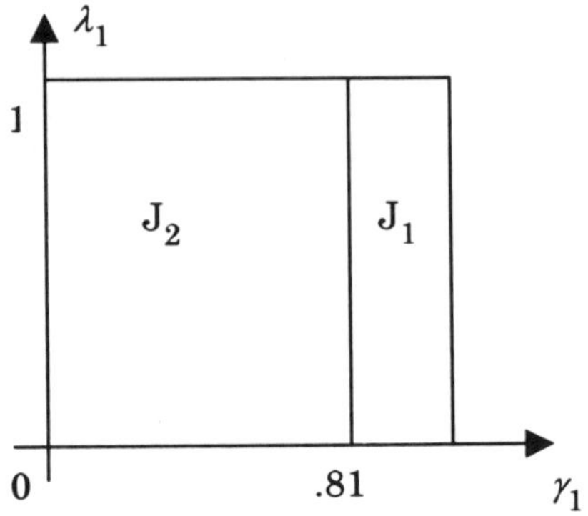

Figure 16.4 Optimal Trade-offs between Vice-President for Production(λ_1) and Total Profit (λ_1)

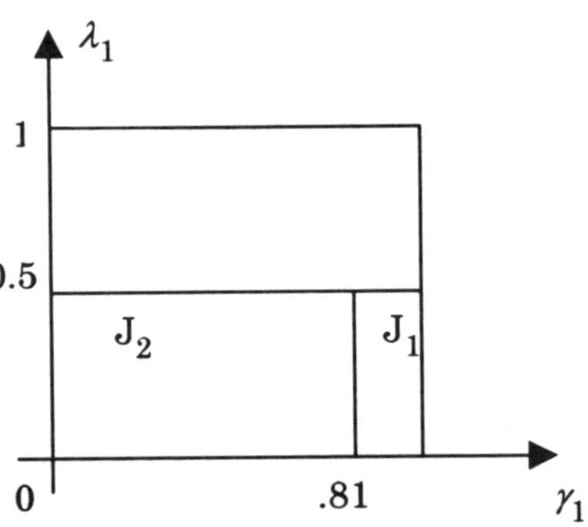

Figure 16.5 Optimal Trade-offs between Vice-President for Production(γ_1) and Total Profit (γ_3)

It is worth noting some important implications for accounting decision makers from the above trade-off analyses. First, the multiple factor transfer pricing model has a capability of systematically locating all possible optimal transfer prices through the optimal trade-offs of multiple objectives and multiple constraint levels. Since the set of all possible optimal transfer prices found by this model describes every possible decision situation within the model framework, the optimal transfer price obtained from either linear programming or goal programming models is included in the subset as a special case.

Second, the multiple factor transfer pricing model can be applied to solve transfer pricing problems not only with complex-structure, but also with some organizational behavior contents, such as the organizational differentiation (see Table 16.2 for different constraint levels). Consequently, this model can foster more autonomous flexibility than any other mathematical programming models by allowing central management or local managers to express their own preference structures and weights.

Third, the proposed method facilitates decision makers' participation that may make a positive management achievement of organizational goals (Locke *et al.* 1981). The model can aid coordination and synthesis of multiple conflicting views. This may be quite effective in a transfer pricing situation in which many objectives are contradictory to each other and these objectives are measured differently by a number of decision participants.

Fourth, in most multiple criteria solution techniques, including the goal programming approach, if the decision makers are not satisfied with the optimal solution obtained by using their preferred weights of importance for the multiple objectives, then an iterative process has to be conducted for incorporating the decision makers' new preference on the weights and finding new solutions until decision makers are satisfied. However, the proposed model eliminates such a time and cost consuming iterative process since it already considers all possible optimal solutions with respect to the changes of parameter (γ, λ). Whenever decision makers want to change their preferences on the weights, the corresponding optimal transfer prices can be immediately identified from the results like in Table 16.5. This model, in turn, allows the performance evaluation of optimal transfer prices.

Finally, the optimal transfer prices obtained by the multiple factor model have a two-fold significance in terms of decision characteristics. If the problem is viewed as a deterministic decision problem, whenever the preferred weighting values of objectives and constraint levels are known, the resulting optimal solutions can be identified from the potentially optimal solutions of the model. Then, the corresponding optimal transfer prices can be adopted to handle the business situation (recall the above trade-off analyses). If the problem is viewed as a probabilistic decision problem, it involves the assessment of the likelihood of (γ, λ) to occur at the various points of the range. With proper assumptions, the uncertainty may be represented by random variables with some known probability distribution. A number of known criteria, such as maximizing expected payoff, minimizing the variance of the payoff, maximin payoff, maximizing the probability of achieving a targeted payoff, stochastic dominance, probability dominance, and mean-variance dominance can be used to choose the optimal transfer prices (see Shi 1991). In summary, the multiple factor transfer pricing model fosters flexibility in designing the optimal transfer prices for a corporation to cope with all possible changes of business competition. This model is more likely to be a better aid for executives or managers to understand and deal with their current or future transfer pricing problems.

16.4 Further Remarks

A multiple factor transfer pricing model has been developed to solve the transfer pricing problems in a multi-divisional corporation. This model can provide a systematic and comprehensive scenario about all possible optimal transfer prices depending on multiple criteria and multiple constraint levels. The trade-offs of optimal transfer prices offer a broad basis for managers of a corporation to flexibly implement the optimal transfer pricing strategy and cope with various business situations. Furthermore, this method also aids global optimization, division autonomy, and performance evaluation.

There are some research problems remaining to be explored. From a practical point of view, the framework of this model can be applied to other accounting areas, such as capital budgeting, cost allocation, audit sampling objectives, and personnel planning for an audit firm if the decision variables and formulation are expressed appropriately. From a theoretical point of view, the decomposition algorithm of linear programming (Dantzig and Wolfe 1960) can be incorporated into the MC^2-simplex method (Seiford and Yu 1979) to sharpen the multiple factor transfer pricing model's capability of solving the large-scale transfer pricing problem. Thus, this incorporation may result in the development of more effective solution procedures. How to incorporate other trade-off techniques, such as the satisficing trade-off method (Nakayama 1994), into the MC^2 framework for designing optimal transfer pricing strategies is another interesting research problem.

Chapter 17
CAPITAL BUDGETING WITH MULTIPLE CRITERIA AND MULTIPLE DECISION MAKERS

Capital budgeting is not a trivial task if a firm is to maintain competitive advantages adopting new information or manufacturing systems. A firm may implement innovative accounting systems such as activity-based costing (ABC) to generate more useful information for better economic decision making in the ever-changing business environment. ABC can provide value-adding and non-value-adding activity information about new capital investments. Investment justification in the new manufacturing environment, however, requires a comprehensive decision-making process that involves competitive analysis, overall strategy, and evaluation of uncertain cash flows (Howell and Schwartz 1994). The chalk here is to measure cash flows as well as intangible benefits that these new systems bring. Furthermore, conflicts of goals, limited resources, and uncertain risk factors may complicate the capital budgeting problem (see, Hillier (1963) and Lee (1993) for details). These problems of conflicts of goals among decision makers and limited resources in a typical organization support the use of MC^2 linear programming.

This chapter attempts to incorporate multiple decision makers' preferences using (1) The analytic hierarchy process (AHP) approach; and (2) MC^2 linear programming in a capital budgeting situation (Kwak, Shi, Lee and Lee 1996). Our approach shows its strength in quantifying strategic and nonfinancial factors that are important in the current competitive business environment. The problem of incommensurable units in the selection criteria, because of nonfinancial and qualitative measures, can be resolved by using AHP approach. The MC^2 approach fosters modeling flexibility by incorporating decision makers' preferences as multiple constraint levels. Section 17.1 reviews the background of capital budgeting. Section 17.2 introduces the framework of AHP. Section 17.3 discusses the proposed MC^2 model. Section 17.4 demonstrates the managerial significance and implications of capital budgeting problems by illustrating an example. Section 17.5 concludes the chapter with several future research avenues.

17.1 Background of Capital Budgeting

While traditional techniques such as payback or accounting rate of return are use a secondary method, discounted cash flow (DCF) methods, including net present value (NPV) and internal rate of return (IRR), are the primary quantitative methods in ca budgeting (Kim and Farragher 1981). The payback method estimates how long it will to recover the original investment. However this method incorporates neither the cash flows after the payback period nor the variability of those cash flows (Boardman and Baker 1982). The accounting rate of return method measures a return on the original cost of the investment. Both of the above methods ignore the time value of money. DCF methods ma} be adequate to evaluate new manufacturing or information systems, because of a bias in favor of short-term investments with quantifiable benefits (Mensah and Miranti 1989).

A current trend in capital budgeting methods utilizes mathematical programming higher discount rates to incorporate higher risk factors (Pike 1983). Hillier (1963) suggested useful ways to evaluate risky investments by estimating expected values and standard deviations of net cash flows for each alternative investment. He showed that the standard deviation of cash flows is easily obtainable. With this information a complete description o risky investment is possible via probability distribution of the IRR, NPV, or annual of the proposed investment under the assumption of the net cash flows from the investment are normally distributed. Similarly, Ouederni and Sullivan (1991) used a semi-variance model to incorporate risk in capital budgeting. Turney (1990) suggested a stochastic dynamic adjustment model to incorporate greater risk premiums when significant additional funds are required in multiple time periods. Lin (1993) also proposed a multiple-criteria ca budgeting model under risk. His model used chance constraints of uncertain cash flows accounting earnings as risk factors. Pike (1988) empirically tested the correlation between sophisticated capital budgeting techniques and decision-making effectiveness and found management believed that sophisticated investment techniques improve effectiveness u evaluation and control of large capital projects. Weingartner (1963) introduced a mathematical programming approach in the capital budgeting problem. Other researchers have extended Weingartner's work with different directions (Baumol and Quandt 1965,

Bernard 1969, and Howe and Patterson 1985). The development of chance-constrained programming (CCP) by Charnes and Cooper (1961) also enriched applications of mathematical programming models in the capital budgeting problem. Hillier (1969) used the CC solve the capital budgeting problem. He also developed an approximation solution me to the CCP with zero-one variables using a linear constraint.

A typical mathematical capital budgeting approach maximizes DCFs that measure a project's desirability on the basis of its expected net present value as a primary goal. DCF analysis, however, ignores strategic factors such as future growth opportunities (Cheng 1993). Furthermore, management can change their plans if operating conditions change. For example, they can change input and output mixes or abandon the project in a multiperiod situation. The increasing involvement of stakeholders, other than shareholders, in a business organization supports a multiple-objective approach (Bhaskar 1979). Other empirical studies also found that firms used multiple criteria in their capital budgeting problems (Bhaskar and McNamee 1983 and Thanassoulis 1985). The goal programming approach has used to handle multiple-objective problems. It emphasizes weights of importance of the multiple objectives with respect to the decision makers (DM) preference (Bhaskar 1979). Within the framework of hierarchical goal optimization, several goal programming models have been suggested to unravel multiple-objective capital budgeting problems (Ignizio 1976 and Lee and Lerro 1974). Similarly, Santhanam, Muralidhar and Schniederjans (1989) used a zero goal programming approach for information system project selection, but their article lacks a multiple time horizon (learning curve) effect. Choi and Levary (1989) investigated use of a chance-constrained goal programming approach to reflect multiple goals for a multinational capital budgeting problem. Reeves and Hedin (1993) suggested interactive programming (IGP) which allows more flexibility for DMs in considering trade-offs adjusting goal target levels. However, because of the difficulties of measuring the preferences and priorities of decision makers. Reeves and Franz (1985) developed a simplified interactive multiple objective linear programming (SIMOLP) method which uses an interactive procedure for a DM to identify a preferred solution and Gonzalez *et al.* (1987) applied this procedure in a capital budgeting problem (Comer, Deckno and Spahr 1993 and Reeves *et al.* 1988). Interactive multiple-objective techniques such as SIMOLP or can reduce the difficulty of the solution process. Such studies suggest that most capital budgeting problems require the analysis of multiple criteria that better reflect a real-world situation. Thanassoulis (1985) addressed multiple objectives such as the maximization shareholder wealth, maximization of firm growth, minimization of financial risk, maximization of the firm liquidity, and minimization of environmental pollution.

However, these existing multiple-criteria approaches, including goal programming, implicitly assume that there is only one decision maker setting up the constraint (budget availability) level of a capital budgeting problem. This assumption is not realistic because in most real capital budgeting problems, such as constructing a major highway, or shopping mall, multiple decision makers must involve the decision of the constraint levels. We use the proposed MC^2 approach to remove this assumption.

17.2 AHP Framework

AHP is a practical measurement technique that has been widely applied in modeling human judgment process (Saaty 1980 and Zahedi 1986). AHP enables decision makers to structure a complex problem in the form of a hierarchy of its elements according to an organization's structure or ranks of management levels. It captures managerial decision preferences through a series of comparisons of relevant criteria. This feature of the AHP minimizes the risk of inconsistent decisions due to incommensurable units in the selection criteria. Recently, AHP has been applied to several accounting problems (e.g., divisional performance evaluation (Chan and Lynn 1991), capital budgeting (Liberatore, Mohnhan and Stout 1992), and real estate investment (Kamath and Khaksari 1991).

The preference of each manager may be analyzed through the use of AHP or multi attribute utility technique (MAUT). Both AHP and MAUT have their own strengths and weaknesses. For a recent debate regarding the two methods, readers are referred to Dyer and Forman (1990). In this chapter, the AHP is employed mainly due to its capability reducing computational complexity and availability of software.

17.3 A Model of Multi-Criteria and Multiple Decision Makers Capital Budgeting

The linear programming approach has limitations for solving real-world problems multiple objectives. The goal programming or

MOLP approach provides only one optimal trade-off among several objectives of the firm. None of the past mathematical models supports multiple decision makers in a capital budgeting decision-making process. As discussed before, in a real-world situation the decision makers (DMs) can have different opinions on the same issue. These different interpretations can be represented by different constraint levels in mathematical models. Because both linear programming and goal programming a fixed single constraint level, they fail to mathematically describe such an organizational differentiation problem.
The MC^2 linear programming framework can resolve the above shortcomings in current capital budgeting models. The framework is flexible enough to include any objective pending on problem situation. However, for the sake of clear presentation, we address groups of objectives that are common in capital budgeting: (1) Maximization of net present value; (2) profitability; (3) growth; and (4) flexibility of financing. Some group objectives may have multitime periods. Net present value measures the expected net monetary or loss from a project by discounting all expected future cash flows to the present point in time, using the desired rate of return. If we want to incorporate risk factors, we estimate the standard deviation of net cash flows as Hillier (1963) suggested. Profitability can be measured by return on investment (ROI). Growth can be measured by sales growth or a nonfinancial measure of market share growth each of which focuses on the long-success of a firm. Flexibility of financing (leverage) can be measured by the debt to equity ratio. As a firm's debt to equity ratio goes up, the firm's cost of borrowing becomes more expensive. Even though our model contains only these four specific objective groups flexibility of the modeling process fits in all capital budgeting problems with multiple criteria and multiple constraint levels.

This model integrates AHP with a mathematical model. The strengths of such an integration have been shown m several areas (e.g., advertisement media choice (Dyer and Forman 1990), R&D portfolio selection (Suh, Suh and Choi 1993), and telecommunication hub design (Lee, Nazem and Shi 1995)). The use of both AHP and MC^2, as we will explore in this article, is the first in the decision-making literature.

In general, let i be the time period for a firm to consider capital budgeting, and the index number of the projects that the firm can select. Define x_j as jth project that can be selected by the firm, j = 1, ..., t. For the coefficients of the objectives, let v_j be the net present value generated by the jth project; g_{ij} the net sales increase generated by the jth project in the ith time period to measure sales growth, i = 1, . .., S; r_{ij} be the ROI generated from the jth project in the ith time period; and l_{ij} be the firm's equity to debt ratio in the ith time period after jth project is selected to measure liquidity. Here, the optimum debt to equity ratio for the firm is assumed to and the inverse of debt to equity ratio is used to measure leverage. For the coefficients the constraints, let b_{ij} be the cash outlay required by the jth project in the ith time period. For the coefficients of the constraint levels, let c_{ki} be the budget availability level believed by the kth manager (or executive) for the firm in the ith time period, k = 1, . ..u.

In this chapter, for illustration, if we treat t = 10, i = 5, and u = 5, then the model (M17.1) is:

$$\text{Max} \quad \sum_{j=1}^{10} v_j x_j$$

$$\text{Max} \quad \sum_{j=1}^{10} g_{ij} x_j \forall i = 1,\ldots,5$$

$$\text{Max} \quad \sum_{j=1}^{10} r_{ij} x_j \forall i = 1,\ldots,5$$

$$\text{Max} \quad \sum_{j=1}^{10} l_{ij} x_j \forall i = 1,\ldots,5$$

$$\text{Subject to:} \quad \sum_{j=1}^{10} b_{ij} x_j \leq (c_{1i},\ldots,c_{5i}) \forall i = 1,\ldots,5$$

$$\text{And} \quad x_j = \{0,1\}$$

where k = 1, . . ., 5, for C_{ki}, represents the five possible DMs (i.e., president, controller, production manager, marketing manager, and engineering manager). The weights of 16 total objectives can be expressed by $(\lambda_1,\ldots,\lambda_{16})$ with $\sum \lambda_q = 1$, q = 1, ..., 16, $0 < \lambda_q < 1$) and the weights of five DMs can be expressed by $(\gamma_1,\ldots,\gamma_5)$ with $\sum \gamma_k = 1$, k = 1, . .., 5,

$0 < \gamma_k < 1$.

The above model is an integer MC² problem with 16 objectives and five constraint levels. Solving this problem by using a currently available solution technique (Seiford Yu 1979) is not a trivial task because of its substantial computational complexity. To overcome this difficulty, we propose a two-phased solution procedure as depicted in Figure 17.1.

The first phase is referred to as the AHP phase in the sense that AHP is applied to de weights of λ and γ, which reduces the model's complexity. The computation of weights is effectively handled by Expert Choice (Forman et al. 1985), commercial software AHP. In this phase, we first induce preferences about objectives from all five DMs involved. Note that each DM may have different preferences about the four goals. The president think maximization of ROI is the most important goal, while the controller may think maximization of NPV is the most important goal. AHP generates relative weights for each goal. Then, multiple constraint levels are incorporated using the MC² framework. For instance, DMs may estimate different budget availability levels for each time period. The controller may believe that \$40 million is available for the first year, while the production manager may believe that the company can spend \$50 million for this time period. This scenario is more realistic if we consider the characteristics of a capital budgeting problem from the perspective of the different utility functions or risk factors of each DM. An example using AHP to generate relative weights for the five DMs' constraint levels (preferences) can be found in the Appendix.

The second phase is called the integer programming (IP) phase. After applying the AHP phase to the model, the model is reduced to a linear IP problem, in which the potentially optimal projects must be selected. To solve this problem, we employ the ZOOM software (Singhal, Marsten and Morin 1989) that was originally introduced for solving zero-one integer linear programming problems.

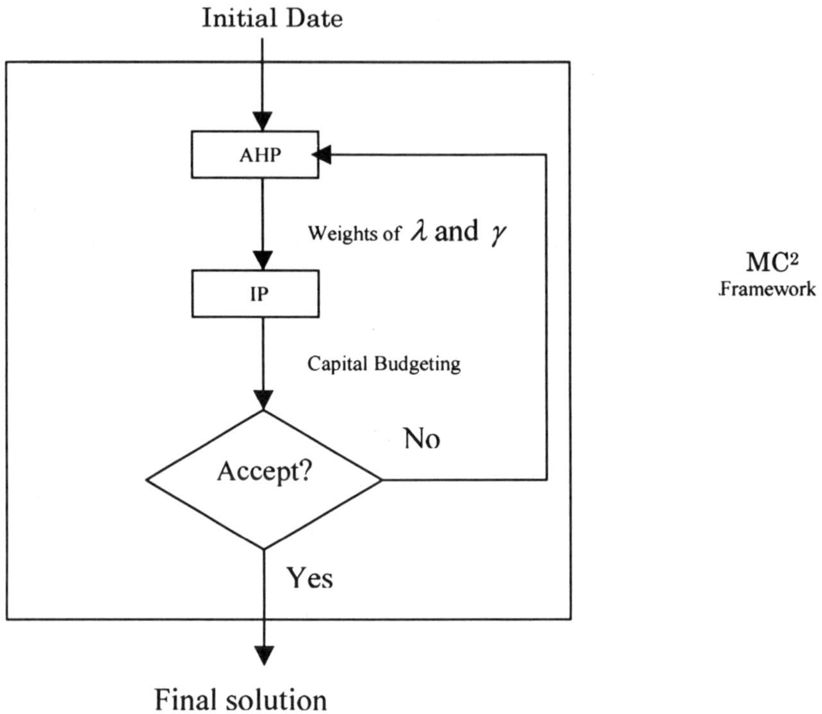

Figure 17.1 A Two-phased Capital Budgeting Model

17.4 Model Implications

17.4.1 Numerical Example

A prototype of our capital budgeting model demonstrates the implications for multiple criteria and multiple decision makers. We use a Lorie and Savage (1955) type problem as follow Lone-Savage Corporation has ten projects under consideration. All projects have five-year time periods.

The executives (president, controller, production manager, marketing manager, and engineering manager) agree on the following multiple objectives:

1. Maximize net present value of each project. Net present value is computed as the sum of all the discounted, estimated future cash flows, using the desired rate of return, minus the initial investment.
2. Maximize the growth of the firm due to each project. Growth can be measured by net sales increase from each project. This can measure a strategic success factor of a firm.
3. Maximize the profitability of the company. Profitability can be measured by ROI for each time period generated by each project.
4. Maximize the flexibility of financing. The flexibility of financing (leverage) can be measured by debt to equity ratio. If a firm's debt to equity ratio is higher than the industry average or the optimum level, the firm's cost of debt financing will become more expensive. In this example, we use the inverse of the debt to equity ratio and assume I as the optimum debt to equity ratio.

The data related to these objectives is given in Table 17.1. In this example, all projects are assumed to be independent. In table 1, the firm's cost of capital is assumed to be known a priori and to be independent of the investment decisions. Based on these assumptions, the net present value of each project can be defined as the sum of the cash flows discounted by the cost of capital. Cash outlay is the amount of expenditure required for project j, j = 1, 2, 3, ..., 10, in each time period. To measure growth, the net sales increase for each time period for each project is estimated. These data are provided in Table 17.2.

To measure the profitability of each project, ROI is estimated after reflecting additional income and capital expenditures from each investment for each time period. These date are provided in Table 17.3.

To measure leverage, the inverse of debt to equity ratio is used after adopting each project. Here, the optimum debt to equity ratio is assumed to be 1 for the Lorie-Savage Corporation. These data are provided in Table 17.4.

The five key DMs in this company (president, controller, production manager, marketing manager, and engineering manager) have different beliefs regarding resource availability. For example, for budget availability levels, each DM may have a different opinion. Of course, the president will make the final decision based on the opinions of other managers. However, the DMs' preferences collection process should improve the quality of the final decision. Budget availability level data are provided in Table 17.5.

Table 17.1 Net Present Value Data for the Lorie-Savage Corporation (in millions)

Present Project Value	Net	Cash Outlays for Each Period				
		Period 1	Period 2	Period 3	Period 4	Period 5
1	$20	$24	$16	$ 6	$ 6	$ 8
2	16	12	10	2	5	4
3	11	8	6	6	6	4
4	4	6	4	7	5	3
5	4	1	6	9	2	3
6	18	18	18	20	15	15
7	7	13	8	10	8	8
8	19	14	8	12	10	8
9	24	16	20	24	16	16
10	4	4	6	8	6	4

Table 17.2 Net Sales Increase Data for the Lorie-Savage Corporation (in millions)

Project	Cash Outlays for Each Period				
	Period 1	Period 2	Period 3	Period 4	Period 5
1	$124	$130	$145	$150	$170
2	100	120	140	150	160
3	80	90	95	95	100
4	40	50	50	55	60
5	40	45	50	55	60
6	110	120	140	150	165
7	60	70	65	70	80
8	110	120	100	110	120
9	150	170	180	190	200
10	35	40	40	50	50

Table 17.3 Return on Investment Data for the Lorie-Savage Corporation (in millions)

Project	Cash Outlays for Each Period				
	Period 1	Period 2	Period 3	Period 4	Period 5
1	$10	$12	$14	$15	$17
2	10	12	18	16	17
3	12	15	15	15	18
4	8	15	10	8	12
5	15	10	8	20	18
6	12	12	10	15	15
7	8	12	10	12	12
8	14	16	13	15	16
9	12	10	9	12	12
10	10	8	9	8	12

Table 17.4 Debt to Equity Data for the Lorie-Savage Corporation (in millions)

Project	Cash Outlays for Each Period				
	Period 1	Period 2	Period 3	Period 4	Period 5
1	0.85	0.90	0.98	0.95	0.95
2	0.90	0.98	0.95	0.95	0.96
3	0.96	0.97	0.98	0.92	0.92
4	0.98	0.95	0.96	0.92	0.95
5	0.90	0.95	0.95	0.98	0.95
6	0.90	0.95	0.94	0.95	0.95
7	0.96	0.98	0.98	0.98	0.98
8	0.96	0.95	0.90	0.92	0.95
9	0.90	0.88	0.85	0.95	0.95
10	0.98	0.95	0.95	0.98	0.95

Table 17.5 Budget Availability Level Data for the Lorie-Savage Corporation (in millions)

Decision Maker	Estimate Budget Availability				
	Period 1	Period 2	Period 3	Period 4	Period 5
President	\$50	\$30	\$30	\$35	\$30
Controller	40	45	30	30	20
Production Manager	55	40	20	30	35
Marketing Manager	45	30	40	45	30
Engineering Manager	50	40	45	30	35

The parameters for budget availability levels are derived by using AHP. Here, one level of an AHP hierarchy is used. The aggregated objective function also can be obtained by using AHP. The difference here is that two levels of an AHP hierarchy must be used to provide preferences for the objective functions. The first level of this AHP hierarchy corresponds to the four groups of objectives. The second level corresponds to time periods each objective group. Note that the second level is not required for the objective of maximizing the net present value. Hence, four pairwise comparison matrixes are required for eigenvalue computation. More detailed information about the formulation of this problem is presented in the appendix. In sum, the objective function aggregated from the 16 objectives is written as:

Maximize $56.34\, x_1 + 51.74\, x_2 + 40.54\, x_3 + 25.43\, x_4 + 25.44\, x_5 + 53.63\, x_6 + 31.98\, x_7 + 50.59\, x_8 + 67.62\, x_9 + 23.17\, x_{10}$.

The optimum IP solution for this numerical example is selecting projects 2, 3, 4, and 8. This solution reflects the preferences of multiple DMs and satisfies budget constraints. If the DMs are not satisfied with the solution, we have to use AHP to assess and compute new values of weights γ and λ, according to Figure 17.1. This process will terminate whenever all DMs accept the compromise solution.

17.4.2 Managerial Implications

The model and solution procedure proposed in this study have several managerial implications. First, the model integrates the multiple-objective capital budgeting problem multiple decision makers. The most common problem in a capital investment decision-making situation is how to estimate cash flows and to incorporate the risk factors of decision makers (Hillier 1963 and Lee 1993). In our model, we allow multiple constraint levels to incorporate each DM's preference about budget availability. Neither linear programming nor goal programming models can handle these characteristics.

Second, our model can be applied to solve capital budgeting problems not only with the representation of a complex structure, but also with motivational implications. Our method facilitates DMs' participation in order to minimize sub-optimization of overall company goals (Locke *et al.* 1981). The model can aid coordination and synthesis of multiple objectives, some in conflict. This feature can be quite effective in a capital budgeting situation, in which many objectives are contradictory to each other and these objectives can be measured differently by a number of decision participants. This model can foster more autonomous control than any other mathematical programming models by allowing DMs to express their own priority structures and weights.

Third, adoption of the AHP reduces the solution complexity of the resulting MC^2 IP program. The MC^2 IP is a very complex problem to solve (see chapter 5). Real life-size MC^2 IP problems are intractable by using current computational technology. Furthermore, the computer software for MC^2 IP is still under development (Shi and Lee 1997). Even though the use of AHP may lose some possible trade-offs among objectives and/or DMs' preferences, the AHP reduces MC^2 IP to a traditional zero-one IP that is much easier to be solved using currently available IP software.

Lastly, our two-phased framework (Figure 17.1) for capital budgeting problem is highly flexible. The DMs can reach an agreement interactively. For example, alternative budgeting policy can be obtained by providing a different preference matrix of objectives and/or resource availability levels. Furthermore, the two-phased framework may be attempted iteratively until all of the DMs are satisfied with the final budgeting policy.

17.5 Further Remarks

A decision-making process with a multiple-criteria model has been addressed to solve capital budgeting problems. This model can foster a high-quality decision in capital budgeting under multiple criteria and multiple decision makers. This decision-making strategy reflects each decision maker's preference and limits sub-optimization of overall company goals. This method can also better handle real-world problems that may include uncertain factor. By incorporating information from each influential manager, our model is more likely to provide better budgeting solutions than the previous linear programming or goal programming approaches.

There are other research problems remaining to be explored. From a practical point of view, the estimation of future cash flows, determining a firm's cost of capital, measuring intangible benefits, and measuring residual value of assets are still important issues. Models like ours can reduce the chance of an ineffective decision making by incorporating multiple decision makers' preferences. This framework can be applied to other accounting problems, such as cost allocation, audit sampling objectives, and personnel planning for an audit if the decision variables and formulation are expressed appropriately.

Appendix

For illustrative purposes, we show how to use AHP to induce the five DMs' preferences of budget availability and to compute the relative weights. Generally, AHP collects input judgments of DMs in the form of a matrix by pairwise comparisons of criteria (i.e., their budget availability levels). An eigenvalue method is then used to scale weights of such criteria. That is, the relative importance of each criterion is computed. The result from all of pairwise comparison is stored in an input matrix as follows:

President	Controller	Production Manager	Marketing Manager	Engineering Manager
1	3	4	5	6
	1	2	5	5
		1	3	4
			1	2
				1

Applying an eigenvalue method to the above input matrix results in a vector W_i = (0.477, 0.251, 0.154, 0.070, 0.048). In addition to the vector, the inconsistency ratio (γ) is obtained to estimate the degree of inconsistency in pairwise comparisons. In this example, the inconsistency ratio is 0.047. A common guideline is that if the ratio surpass 0.1, a new input matrix must be generated. Therefore, this input matrix is acceptable

A similar computing process can be applied for the 16 objective functions. However, two hierarchical levels are required for this case. The first level of AHP hierarchy corresponds to the four groups of objectives and the second level corresponds to the time periods within each objective group.

Let x_j be the jth project which can be selected by the firm. Using data in Tables 17.1-17.4, the model for capital budgeting with multiple criteria and multiple DMs is formulated as:

Maximize $20\,x_1 + 16\,x_2 + 11\,x_3 + 4\,x_4 + 4\,x_5 + 18\,x_6 + 7\,x_7 + 19\,x_8 + 24\,x_9 + 4\,x_{10}$.

Maximize $120\,x_1 + 100\,x_2 + 80\,x_3 + 40\,x_4 + 40\,x_5 + 120\,x_6 + 70\,x_7 + 110\,x_8 + 150\,x_9 + 35\,x_{10}$

Maximize $130\,x_1 + 120\,x_2 + 90\,x_3 + 50\,x_4 + 45\,x_5 + 120\,x_6 + 70\,x_7 + 120\,x_8 + 170\,x_9 + 40\,x_{10}$

Maximize $145\,x_1 + 140\,x_2 + 95\,x_3 + 50\,x_4 + 50\,x_5 + 140\,x_6 + 65\,x_7 + 100\,x_8 + 180\,x_9 + 40\,x_{10}$

Maximize $150\,x_1 + 150\,x_2 + 95\,x_3 + 55\,x_4 + 55\,x_5 + 150\,x_6 + 70\,x_7 + 110\,x_8 + 190\,x_9 + 50\,x_{10}$

Maximize $170\,x_1 + 160\,x_2 + 100\,x_3 + 60\,x_4 + 60\,x_5 + 165\,x_6 + 80\,x_7 + 120\,x_8 + 200\,x_9 + 50\,x_{10}$

Maximize $10\,x_1 + 10\,x_2 + 12\,x_3 + 8\,x_4 + 15\,x_5 + 12\,x_6 + 8\,x_7 + 14\,x_8 + 12\,x_9 + 10\,x_{10}$

Maximize $12\,x_1 + 12\,x_2 + 15\,x_3 + 15\,x_4 + 10\,x_5 + 12\,x_6 + 12\,x_7 + 16\,x_8 + 10\,x_9 + 8\,x_{10}$

Maximize $14\,x_1 + 18\,x_2 + 15\,x_3 + 10\,x_4 + 8\,x_5 + 10\,x_6 + 10\,x_7 + 13\,x_8 + 9\,x_9 + 9\,x_{10}$

Maximize $15\,x_1 + 16\,x_2 + 15\,x_3 + 8\,x_4 + 20\,x_5 + 15\,x_6 + 12\,x_7 + 15\,x_8 + 12\,x_9 + 8\,x_{10}$

Maximize $17\,x_1 + 17\,x_2 + 8\,x_3 + 12\,x_4 + 18\,x_5 + 15\,x_6 + 12\,x_7 + 16\,x_8 + 12\,x_9 + 12\,x_{10}$

Maximize $0.85\,x_1 + 0.90\,x_2 + 0.96\,x_3 + 0.98\,x_4 + 0.90\,x_5 + 0.90\,x_6 + 0.96\,x_7 + 0.96\,x_8 + 0.90\,x_9 + 0.98\,x_{10}$

Maximize $0.90\,x_1 + 0.98\,x_2 + 0.97\,x_3 + 0.95\,x_4 + 0.95\,x_5 + 0.95\,x_6 + 0.98\,x_7 + 0.95\,x_8 + 0.88\,x_9 + 0.95\,x_{10}$

Maximize $098\,x_1 + 0.\,95\,x_2 + 0.98\,x_3 + 0.96\,x_4 + 0.95\,x_5 + 0.94\,x_6 + 0.98\,x_7 + 0.90\,x_8 + 0.85\,x_9 + 40\,x_{10}$

Maximize $\quad 0.95\,x_1 + 0.95\,x_2 + 0.92\,x_3 + 0.92\,x_4 + 0.98\,x_5 + 0.95\,x_6 + 0.98\,x_7 + 0.92\,x_8 + 0.95\,x_9 + 0.98\,x_{10}$

Maximize $\quad 0.95\,x_1 + 0.96\,x_2 + 0.92\,x_3 + 0.95\,x_4 + 0.95\,x_5 + 0.95\,x_6 + 0.98\,x_7 + 0.95\,x_8 + 0.95\,x_9 + 0.95\,x_{10}$

subject to $\quad 24\,x_1 + 12\,x_2 + 8\,x_3 + 6\,x_4 + x_5 + 18\,x_6 + 13\,x_7 + 14\,x_8 + 16\,x_9 + 4\,x_{10} \leq 49.37$

$16\,x_1 + 10\,x_2 + 6\,x_3 + 4\,x_4 + 6\,x_5 + 18\,x_6 + 8\,x_7 + 8\,x_8 + 20\,x_9 + 6\,x_{10} \leq 35.30$

$6\,x_1 + 2\,x_2 + 6\,x_3 + 7\,x_4 + 9\,x_5 + 20\,x_6 + 10\,x_7 + 12\,x_8 + 24\,x_9 + 8\,x_{10} \leq 28.91$

$6\,x_1 + 5\,x_2 + 6\,x_3 + 5\,x_4 + 2\,x_5 + 15\,x_6 + 8\,x_7 + 10\,x_8 + 16\,x_9 + 6\,x_{10} \leq 33.44$

$8\,x_1 + 4\,x_2 + 4\,x_3 + 3\,x_4 + 3\,x_5 + 15\,x_6 + 8\,x_7 + 8\,x_8 + 16\,x_9 + 4\,x_{10} \leq 29.96$

and $\quad x_j = \{0, 1\}$, for j = 1,…, 10.

Chapter 18
DATA FILE ALLOCATIONS

Distributed information technology is one the most important computing developments. Many organizations from industry, academia and government have used distributed databases to support their information needs. Their use has been accelerated by two driving forces: (i) technological advances in telecommunication systems and (ii) difficulties in satisfying organizational need for geographically dispersed information. A telecommunication infrastructure has great potential for reducing the cost of information systems. Indeed, most database systems are being distributed (Nolan 1984 and Stonebraker 1989).

A distributed database is a collection of data that logically belongs to the same organization or system but is physically spread over the sites of a data communication network. The two important aspects in the design of distributed databases are fragmentation and allocation (Ceri and Pelagatti 1984, and Ozsu and Valdureiz 1991). Fragmentation is the partitioning of an entire database into a set of different databases. Allocation concentrates on the (possibly replicated) distribution of these local fragments across geographically dispersed sites. An integration of both designs is referred to as distribution design (or data allocation design). A general method for designing centralized databases includes four phases: requirements collection and analysis, conceptual design, data model mapping, and physical design (Elmasri and Navathe 1989). The design of distributed databases adds to the above phases an additional one, distribution design.

Solving these two designs simultaneously is not a trivial task because of its complexity. Each design problem has been shown to be NP-complete (Apers 1988 and Eswaran 1974). Therefore, a common approach is to solve one design in isolation, assuming that the other is determined a priori. This chapter concentrates on allocation decisions, i.e., it studies how to allocate database fragments, that were previously computed according to any fragmentation method, to dispersed database locations. These database fragments are referred to as (data) files.

Data communications is of particular importance in distribution design. Broadly speaking, telecommunication networks can be categorized into local and wide area networks (LANs and WANs, respectively). Sheng and Lee (1992) demonstrated how the distribution design depends upon the supporting data communication network: the primary incentive for distributed database implementations in LANs is load balancing, while the development of distributed databases via WANs is motivated by locality of reference. Today, LANs typically use cable, whereas WANs use telephone lines or satellites. Several approaches to file allocation on LANs include (Lee 1993, Lien, Chang and Wah 1988, Wah and Lien 1985, Yu, Siu, and Lam and Chen 1985). However, most file allocation research has been based on WANs. The operation of distributed databases on WANs has been considered important (Ozsu and Valdureiz 1991).

The investigation of the file allocation problem on WANs dates to a pioneering work by (Chu 1969). Since then, the file allocation design problem via wide area networks has been extensively studied in the literature (Casey 1972, Dowdy and Foster 1982, Gavash and Pirkul 1986, Ram 1989, and Ramesh and Ryan 1990). A common objective of these models is to minimize telecommunication costs. However, in a distributed information system, further economy can be achieved: the cost of allocating data files over geographically dispersed sites may be lower than the cost of frequently accessing a remote file over the network.

Furthermore, planning for distributed information systems involves other multiple design criteria, such as system response time and data availability. Solving for file allocation under multiple criteria in conflict is an open research area, even though a number of studies have been reported (Jain and Dutta 1986, and Jain 1987). In a conflicting situation, designers typically set their goals or targets, and iteratively modify them to find more effective designs under given constraints until satisfied. Therefore, it is of particular interest to address the goal setting behavior of designers of distributed information systems.

An analytical technique known as a compromise solutions approach supports a policy maker's goal setting process when multiple goals are in conflict. This chapter applies a compromise modeling technique to file allocation design (Lee, Shi and Stolen 1994). Because this involves zero-one decision variables, the resulting model is termed a zero-one compromise model. This fits the designers' interactive decision making process very well, since target value is adjusted until a compromise design is finally reached.

In this chapter, Section 18.1 outlines the framework of file allocation via wide area networks. Section 18.3 proposes a zero-one compromise model to solve the file allocation problem. Section 18.3 illustrates a sample distribution information system that is implemented by using the proposed modeling approach. Finally, Section 18.4 gives some further remarks regarding to this practical project.

18.1 File Allocation Via Wide Area Networks

A WAN typically has a point-to-point topology, with any two points connected by one or more communication channels. Let the total number of channels be m. A typical principle for data communications over WANs is that the data to be transmitted should be divided into smaller units termed packets. These traverse and queue for channels. User requests for data in distributed information systems include queries and updates. A query requests data, whereas updates require modification of current data content in addition to retrieval. Each site is assumed to have its own central processor and its own software. A data file may have to be allocated to a local site because remote user access to it would be too slow.

We define a zero-one integer variable x^f_{ik}, to 1 when a copy of file i is allocated to site k; otherwise, 0. Denote a file allocation decision vector by $\mathrm{x}^f = (x^f_{11},..., x^f_{ln})$, where l is the total number of files and n is the total number of sites. For multiple file copies, a query must be transmitted to the site that allows the most economical data retrieval. In order to describe query routing decisions, we define a zero-one integer variable k^q_{ijk} that is 1 when a site j requests file i to be routed through site k; otherwise, 0. Denoting this query routing decision vector by $x^q = (x^q_{111},..., x^q_{lnn})$, the decision variable becomes the vector $\mathbf{x} = (\mathrm{x}^f, \mathrm{x}^q)$.

The following system parameters are used to represent the file allocation design problem.

λ = total rate of data requests entering the system

γ = total data request rate within the system

α = total query request rate entering the system

β = total update request rate entering the system

α_{ik} = the query request rate at site k for file i

β_{ik} = the update request rate at site k for file i

q_{ijk} = the telecommunication cost of the query for file i between site j and site k

u_{ijk} = the telecommunication cost of the update for file i between site j and site k

s_{ik} = the storage cost of file i at site k

p_{jk} = the set of channels that are contained in the path taken by data request (from site j to site k)

Alternatively, variable x_{ijk}^{q} may be defined as the fraction of site j requests for file i that are routed via site k, i.e., $0 \le x_{ijk}^{q} \le 1$. In this case, the design problem can be solved relatively easily, because it involves less complexity, i.e., there are fewer zero-one variables. However, this design incurs further communication overhead due to its probabilistic scheduling nature. Thus, we confine ourselves to zero-one variables of x_{ijk}^{q}.

r_{jk} = the probability that site j can communicate with site k

r_k = the reliability of site k

T = the average site delay of processing a packet

l/μ = the average length of a packet

σ_i = the size of file i

θ_k = the processing capacity of channel k

ϕ_k = the storage capacity of site k.

The goal of building a distributed information system is a statement of the degree to which the managerial purpose must be realized. For example, if the purpose of building a distributed system is to minimize the telecommunication cost, a goal can be "Build the system while spending less than fifty thousand dollars per month". Goal setting is an important aspect of human management practice and is a critical phase in the human decision making process.

In the design of distributed information systems, the following three goals are important:

1. Minimization of operating costs (y_c): Operating costs include both telecommunication and site storage costs. Telecommunication costs occur due to data transfers for queries and updates. Storage costs occur because of file storage.

2. Minimization of system response time (y_t): system response time is the expected time that either a query or an update spends in the system. It is estimated as the delays experienced in site processing and channel transfers. If data files are appropriately allocated, data can be accessed more effectively, i.e., the system response time can be reduced.

3. Maximization of data availability (y_d): E availability is defined as the fraction of data requests that are successfully executed (with respect to requests submitted to the system) over a given time interval (Lee 1993). Certain data requests may be lost due to site and/or channel failures; from these the source-destination reliabilities r_{jk} are computed (Ball 1979). However, if the system supports multiple file copies, a data request can still be satisfied, i.e., the system can improve data availability.

These goals and the constraints are represented in terms of decision variables and system parameters (Lee and Sheng 1992). Operating costs consist of query telecommunication, update telecommunication, and data storage costs:

$$y_c = \sum_{i=1}^{l} \sum_{k=1}^{n} \left[\sum_{\substack{j \neq k \\ j=1}}^{n} a_{ij} q_{ijk} x_{ijk}^{q} + \left(\sum_{\substack{j \neq k \\ j=1}}^{n} \beta_{ij} \mu_{ijk} + s_{ik} \right) x_{ik}^{f} \right]$$

Representing each channel as an M/M/1 queue, the average time that a data request spends within the system is:

$$y_t = \tau + \frac{1}{\gamma} \sum_{c=1}^{m} \sum_{\substack{j,k \\ c \in p_{jk}}} \sum_{i=1}^{l} \left[\tau + \frac{1}{\mu \theta_c} \right] \times (a_{ij} x_{ijk}^{q} + \beta_{ij} x_{ik}^{f}) .$$

For query-intensive distributed information systems, γ is approximated by λ ($\gamma \approx \lambda$). In this case, y_t is a linear function of decision variable. In contrast, system-wide data availability is:

$$y_d = \sum_{i=1}^{l} \sum_{j=1}^{n} \left[\sum_{\substack{k \neq j \\ k=1}}^{n} w_\alpha \frac{\alpha_{ij}}{\alpha} (r_{jk} x_{ijk}^{q} + r_j x_{ij}^{f}) + \prod_{\substack{k=1 \\ x_{ik}^{f}=1}}^{n} w_\beta \frac{\beta_{ij}}{\beta} r_{jk} \right]$$

The designer assigns relative importance w_α and w_β respectively, so that $w_\alpha + w_\beta = 1$. For convenience, the set of system constraints is denoted by s_c. First, because a data file must be allocated to at least one site,

$$\sum_{j=1}^{n} x_{ij}^{f} \geq 1, \forall i = 1,\ldots,l .$$

The site j request of file i must be transferred to one of the remote sites if file / is not placed at site j. Therefore,

$$\sum_{\substack{k \neq j \\ k=1}}^{n} x_{ijk}^{q} + x_{ij}^{f} = 1, \forall i = 1,\ldots,l, j = 1,\ldots,n$$

Because the site j request of file i is routed to site k only if file i is placed at site k,

$$x_{ijk}^{q} \leq x_{ik}^{f}, \forall i = 1,\ldots,l, j = 1,\ldots,n, k = 1,\ldots,n, j \neq k$$

Because updates are transmitted to all of the sites containing the data to be updated, the total request rate within the system depends on file allocation:

$$\sum_{i=1}^{l} \sum_{j=1}^{n} \sum_{\substack{k \neq j \\ k=1}}^{n} \beta_{ij} x_{ik}^{f} = \gamma - \alpha$$

In order for the communication network remain stable,

$$\sum_{i=1}^{l} \sum_{\substack{j,k \\ c \in p_{jk}}} (\alpha_{ij} x^{\overset{f}{ijk}} + \beta_{ij} x_{ik}^{f}) \leq \mu \theta_c, \forall c = 1,\ldots,m .$$

For the storage requirements of each site,

$$\sum_{i=1}^{l} \sigma_i x_{ik}^f \leq \phi_k, \forall k = 1,\ldots,n.$$

18.2 Zero-One Compromise Model

This file allocation formulation is incorporated into the framework of a compromise model (Yu 1973, Freimer and Yu 1976, Yu 1985 and Shi and Yu 1989). One of its advantages is its applicability in the representation of human goal setting behavior. Furthermore, a compromise model is comprehensive: it includes goal programming models (Hillier and Lieberman 1986, Ignizio 1982, Lee 1972, and Romero 1991), which have been widely applied in multiple criteria environments.

In practice, target values for goals are set several times until the best design is reached. The flexibility of the compromise solutions accommodates a creeping commitment approach to goal setting. The trade-off among the three goals for file allocation requires zero-one integer decision variables. These variables have not been explored in the compromise models of Chapter 13 (Yu 1973, Zeleny 1974 and Yu 1985).

In the compromise model, the objective is to minimize a regret function. For given K goals define a vector

$y = (y_1, y_2, \ldots y_k)$ as a set of goal functions. Let **y*** be a target vector that is initially set by designers. The regret of achieving y instead of achieving the target **y*** is represented by the distance between y and **y***. Thus the regret function is defined by

$$r(\mathbf{y}) = \| \mathbf{y} - \mathbf{y}^* \|$$

Typically, the l_p norm is used to denote the distance. Therefore, for $p \geq 1$,

$$r(\mathbf{y} \mid p) = \left[\sum_{i=1}^{k} | y_i - y_i^* |^p \right]^{1/p}$$

Because the goals have different degrees of importance, the weighting vector w is applied. Applying weighting factors,

$$\| \mathbf{y} - \mathbf{y}^* \|_{p.w.}$$

The expression can be rewritten as

$$r\ (\mathbf{y} \mid p,\ \mathbf{w}) = \left[\sum_{i=1}^{k} w_i^p \mid y_i - y_i^* \mid^p \right]^{1/p}$$

Applying Theorem 13.5 in Chapter 13, we remove the absolute value sign of regret function and yield:

$$r\ (\mathbf{y} \mid p,\ \mathbf{w}) = \left[\sum_{i=1}^{k} w_i^p (d_i^- + d_i^+)^p \right]^{1/p}$$

Minimization of this is equivalent to minimization of

$$r'\ (\mathbf{y} \mid p,\ \mathbf{w}) = \left[\sum_{i=1}^{k} w_i^p (d_i^- + d_i^+)^p \right]^{p}$$

Without loss of generality, we will use the above function r' for the regret function.

The goal function y_i is a function of x such that $y_i = f_i(\mathbf{x})$ for i = 1,...,k. In consequence, the file allocation design problem can be presented in the form of a compromise model as

Minimize

$$r\ (\mathbf{y} \mid p,\ \mathbf{w}) = \sum_{i=1}^{k} w_i^p (d_i^- + d_i^+)^p$$

subject to

$$y_i = f_i(\mathbf{x}) = d_i^- - d_i^+, \forall i = 1, \ldots, k$$
$$x \in s_c$$
$$x \in \{0,1\}$$

The constraint, $d_i^+ \times d_i^- = 0$ is redundant and thus not incorporated into the formulation. In the case of one-sided goals, where

$y_i \geq y_i^*$ (a lower bound goal) or $y_i \leq y_i^*$ (an upper bound goal), r does not need both d_i^- and d_i^+. For example, the case of a lower bound goal, d_i^+ would deleted from r because only a d_i^- deviation may be penalized. In other words, the goal is to maximize y_i, only when y_i is less than y_i^* i.e., it is not necessary to consider d_i^+. In the same sense, d_i^- would be deleted from r in the case of an upper bound goal.

To solve the resulting zero-one integer programming problem, we use the ZOOM system as developed by Singhal *et al.* (1989). The total number of zero-one variables in the model is $1 \times n \times (n + 1)$. For example, allocating five files over ten sites requires 550 binary variables. This size is consistent with real-life file allocation designs that can be solved by the ZOOM. For example, Ram and Marsten (1991) adopted the ZOOM to solve file allocations that require 6500 binary variables.

The weighting vector w must be estimated once the goals are identified. Either the MAUT (Multiple Attribute Utility Technique) (Fishburn 1982 and Keeney and Raiffa 1976) or the AHP (Analytic Hierarchy Process) (Saaty 1977 and Saaty 1980) may be adopted on the basis of the designers' preference with respect to each goal. Historically, researchers have maintained both methods as sepa- rate areas of application. For a current debate between exponents and critics of these methods, see (Dyer 1990, Forman 1990, Harker and Vargas 1990, Saaty 1990, Schoner, Wedley and Choo 1992).

In this project, we illustrate the weighting process by using the AHP for its simplicity. The method starts by pairwise comparisons. A computer software package called Expert Choice (Buede 1992, Fonnan, Saaty, Selly and Waldron 1985) may be used for this estimation. A pairwise comparison is likely to generate the best estimation for simplicity and convenience. The basic premise is that two goals are compared at a time until all combinations of comparison have been considered. Thus, the total number of pairwise comparisons for given K goals is:

$$\binom{k}{2} = \frac{k(k-1)}{2}$$

For our file allocation model, the total number is three because K = 3. An example of a possible question during pairwise comparison is "In the file allocation, what is the relative importance of goal 1 (e.g., minimization of operating costs) versus goal 2 (e.g., minimization of system response time)?" The result from all of the pairwise comparisons is stored in a matrix. A =[a_{ij}], a k×k matrix. Typically, the goals are sorted in descending order of importance. The element a_{ij} gives the relative importance of goal i to goal j. Matrix A is arranged so that all diagonal elements are one. Its lower triangle is the reciprocal of the upper one. The relative weight of each goal may be calculated by an eigenvalue method and then stored in theform of a vector **w** = ($w_1,\ldots,w_k$). Another measuring method, based on shadow prices instead of pairwise comparisons, can be found in (Ballestero and Romero 1994).

Designers often update target values several times throughout the file allocation design process. This iterative process is often used by systems designers (Whitten, Bentley and Barlow 1989). The target value that is feasible at one stage of database systems design may become less feasible or even infeasible later. At same time, the compromise design may provide a better opportunity to assess model parameters (i.e., target values, weighing factors, etc.) for the designers; i.e., they may review the resulting file allocation design and value of the target function, and then interactively adjust parameters.

Within the framework of the creeping commitment, an iterative generation of compromise solutions is inevitable. The process continues until the best design is identified. This process is summarized in Figure 18.1.

18.3 A Sample Distributed Information System

Consider a credit card transaction company that maintains a distributed information system over six sites (n = 6). To process transaction requests, the company needs to operate databases over these geographically dispersed sites. The sites are connected by eight channels (m = 8) via a wide area network as depicted in Figure 18.2. The company uses a single logical data file (l = 1). This prototype illustrates the applicability of the zero-one compromise model in file allocation design. Table 18.1 shows the transaction request rates, data storage costs, and site reliabilities. The average packet size is 256 bytes ($1/\mu = 256$). Table 18.2 shows query communication costs and capacities for communication channels.

For example, c_{12} is the channel between Seattle (Site 1) and Los Angeles (Site 2). The communication costs between these sites are the summation of costs for channels traversed by data requests. An expansion factor for updates is 20%, i.e., $u_{1jk} = 1.2 \times q_{1jk}$, for j = k = 1,..., n, j ≠ k.

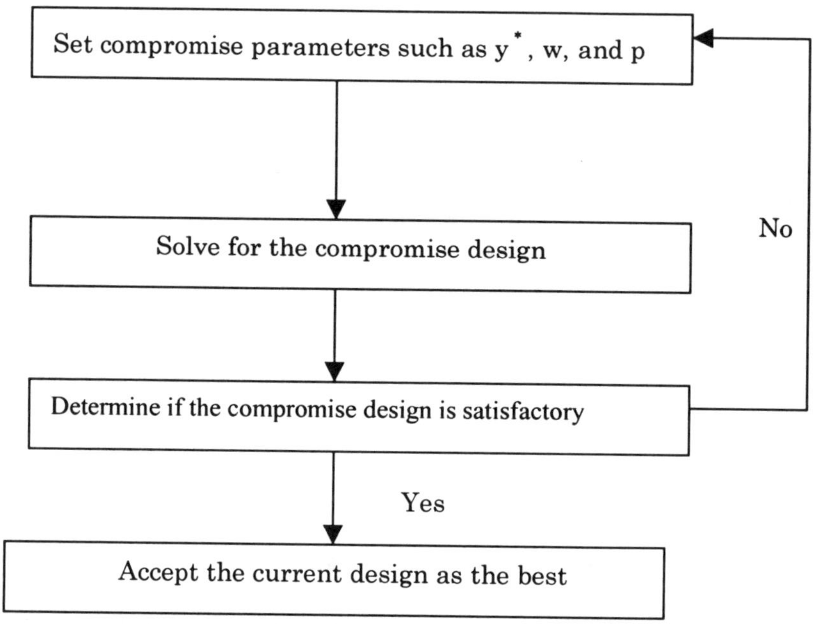

Figure.18.1 Creeping Commitment Design Process in Compromise Model

Table 18.3 shows source-destination reliabilities r_{jk}. The system operates 100 hours per week and 4.2 weeks per month. Sufficient storage capacity is assumed at each site. In this system, the nodal delay is dominated by the telecommunication delay and thus ignored, i.e., T = 0.

The emphasis is on query transactions so that $y \approx \lambda = 45.5, w_a = 1$, and $w_\beta = 0$.

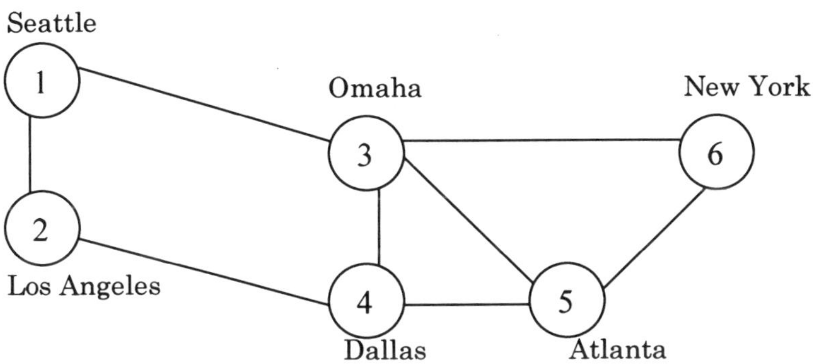

Figure. 18.2 A Network Topology for Sample Distributed System

Table 18.1 Site Related Data for A Distributed Information System

System parameter	Specification
Query request rate, sites 1..6(a_{1k})	6.0 9.0 3.0 6.0 3.0 8.0 (request/sec)
Update request rate, sites 1..6(β_{1k})	2.02.01.52.01.02.0 (request/sec)
Storage cost, sites 1..6(s_{ik})	16.67 3.33 20.00 30.00 6.67 3.33 (\$/day)
Site reliability, sites 1..6(r_k)	98.5 95.5 98.0 99.0 97.5 99.0 (%)

Table 18.2 Query Communication Costs and Channel Capacities

C_{12}	1.8	76.8	C_{35}	2.5	19.2
C_{13}	2.7	76.8	C_{45}	1.5	76.8
C_{24}	0.5	200.0	C_{36}	3.1	38.4
C_{34}	1.1	76.8	C_{56}	1.3	76.8

The designer first has to determine the relative importance of goals by pairwise comparison. For this purpose, the designer must define one unit of each goal. This must be determined so that the designer can make comparison in a consistent fashion. The units used are $1,000 / month for operating costs, 1 second for system response time, and 1% for data availability. With respect to these units, goals are compared. For instance, a question is asked to systems users: What is more important, and by how much, a $1,000 / month decrease of operating costs or a 1 second decrease of system response time? If the answer is that the first is five times more important than the second, then $a_{12} = 5$. The result of all such pairwise comparisons is then represented in matrix A:

$$A = \begin{pmatrix} 1 & 5 & 7 \\ \frac{1}{5} & 1 & 3 \\ \frac{1}{7} & \frac{1}{3} & 1 \end{pmatrix}$$

Table 18.3 Source-destination Reliability (r_{jk}) of the Communication Network (unit: %)

Source	Destination					
	1	2	3	4	5	6
1	-	97.0	95.5	92.1	92.6	91.5
2	97.0	-	92.0	94.3	93.1	91.5
3	95.5	92.0	-	96.3	96.5	95.1
4	92.1	94.3	96.3	-	94.6	93.6
5	92.6	93.1	96.5	94.6	-	96.5
6	91.5	91.5	95.1	93.6	96.5	-

The eigenvalues that represent the estimated normalized weights are then:

$\mathbf{w} = (0.731, 0.188, 0.081)$.

The inconsistency ratio is 0.130, which is accepted by the designer. Cost and response time goals are upper bounded, whereas the availability goal is lower bounded. The designer feels that his regret is linearly proportional to the distance between the target point and goal achievement, i.e., p = 1. Hence, the objective function for the zero-one compromise model is:

$$\textit{Minimize } \mathrm{r}(\mathbf{y}) = 0.731\, d_c^+ + 0.188\, d_t^+ + 0.081\, d_d^- .$$

The total number of zero-one variables for the compromise model is 30. The designer sets an initial target value as:

$\mathbf{y}_1^* = (\$10{,}000,\ 45.0\text{ sec},\ 99.0\%)$.

To meet this target value, a compromise design is obtained by recommending that a file be allocated to Dallas in isolation($x_4^f = 1$). Therefore, only Dallas can serve query services, i.e., $x_{14}^q = x_{24}^q = x_{34}^q = x_{54}^q = x_{64}^q = 1$. Each goal function is then achieved at the level of

$\mathbf{y}_1 = (\$10{,}840,\ 46.5\text{ sec},\ 96.0\%)$.

The designer is not satisfied with such low achievements of system response time and data availability and is willing to pay up to $12,000 in exchange for reducing the system response time to 45 seconds while increasing data availability to 97.5%. Therefore, the second target is:

$$\mathbf{y}_2^* = (\$12{,}000,\ 45.0\text{ sec},\ 97.5\%).$$

Another compromise design for this target is obtained by recommending that file copies be allocated to both Los Angeles and New York, i.e., $x_2^f = x_6^f = 1$. Query requests from Seattle, Omaha, and Dallas are routed to Los Angeles, whereas query requests from Atlanta are routed to New York, i.e., $x_{12}^q = x_{32}^q = x_{42}^q = x_{56}^q = 1$. Goal functions are then:

$$\mathbf{y}_2 = (\$12{,}150,\ 46.2\text{ sec},\ 96.7\%).$$

Actually, the response time and data availability are improved by 0.65% (= {46.5 - 46.2} / {46.5} 100) and 0.73% (= {96.7 - 96.0} / {96.0} x 100), respectively, with the detriment of 12.1% (= {12,150 - 10,840} / {10,840} x 100) of operating costs. The designer still attempts to explore the possibility of improving response time and data availability by allowing $12,500 of operating costs (this is the maximum allowable costs). Therefore, new target is set as

$$\mathbf{y}_3^* = (\$12{,}500,\ 45.0\text{ sec},\ 97.0\%).$$

Even under this target, the compromise design remains unchanged. Therefore, the designer accepts this as the best one and quits the iterative process, i.e., file copies are allocated to Los Angles and New York. The monthly operating cost is $12,150, the system response time is 46.2 seconds, and data availability is 96.7%.

18.4 Further Remarks

This chapter presents a methodology for incorporating human goal seeking behavior into file allocation decisions in a distributed information system using WANs. The applicability of a zero-one compromise modeling in file allocation was demonstrated with a sample-distributed system. The iterative target setting process is likely to better represent designers' compromising behavior while trading off multiple goals.

From a practical perspective, the design of distributed information systems involves processor location, program location, and network topology as well as file allocation. Modeling and optimization of this comprehensive model under multiple goals is not a trivial task. One promising approach would be to find a method that can be applied in real-life projects by incorporating some major system parameters into the model. Incorporating an exchange heuristic (Ignizio 1980) into the proposed design procedure may be a good candidate.

Chapter 19
RURAL TELECOMMUNICATIONS SYSTEMS

In spite of rapid advances in telecommunications technology, in comparison to urban areas most rural communities show symptoms that raise concerns for their future. Rural communities experience inferior school systems, sparse population, insufficient medical services, and low employment levels, in addition to poor economies. If rural areas are not to experience further decline, some measures must be taken to ensure that they have access to the needed infrastructures in economically viable ways (Walzer 1991).

It is clear that telecommunications technologies can play a major role in fostering infrastructures needed by rural communities. Telecommunications infrastructure in rural areas facilitates job creation through the support of service-sector growth. At the same time, rural inhabitants can regain the competitive advantage of the comfort level that urban dwellers enjoy because information can travel instantaneously via telecommunications. Telecommunication has the ability to provide needed access to the human resource development by bridging distances. Among possible telecommunications structures, a "hub" architecture has been widely accepted for implementing rural telecommunications. Hubs are somewhat larger than their neighboring rural communities. The sparsely distributed population in surrounding areas can rely upon available facilities in the hub cities (Nazem 1992). Hub cities should foster a substantial level of basic infrastructures, such as business, health care, education, and transportation. As a result, determining hub cities is an important design issue in rural telecommunications networks.

Nazem and Liu (1992) addressed several important economic criteria, and proposed a multi-criteria set covering model. Lee *et al.* (1994) proposed a two-phased decision support system, and built a prototype for an IBM-PC compatible system. Furthermore, selecting hub cities is likely to raise many policy issues. Policy makers typically set goals for their criteria, and proceed to find a trade-off among goals. The analytical technique called compromise programming in Chapter 13 (Yu 1973, Zeleny 1974, Freimer and Yu 1976, Yu 1985, Shi and Yu 1989) is a well-established method used to support policy maker's goal-setting behavior. Lee *et al.* (1993) recently explored the possibility of employing the compromise programming technique to estimate the relative importance of each individual criterion. The main objective of this chapter is to sharpen the compromise model further to aid in hub selection for rural telecommunications. The proposed compromise design incorporates normalizing factors based on economic shadow prices, in addition to weighting factors depending on the relative importance of each criterion. The answer proposed in this chapter is more likely to provide better hub decisions, in an economic sense, than previous approaches (Lee, Shi and Nazem 1996). Section 19.1 discusses the rural telecommunications system. Section 19.2 describes the compromise model for selecting hub cities of the system. Section 19.3 uses the rural telecommunications system of the State of Nebraska as a real-world example to show the applicable potential of the proposed model. Finally, the further remarks are given in Section 19.4.

19.1 Needs for Rural Telecommunications

Among emerging global technologies, telecommunications infrastructure is becoming an increasingly important weapon in maintaining competitive advantages for international or regional economies. In spite of rapid changes in technological, innovations, most rural areas are still isolated and remote. One of the most significant problematic factors in rural areas is the widely dispersed market demand. Consequently, advanced telecommunications networks have concentrated on dense urban areas. The telecommunications infrastructure in rural areas lags far behind that of urban areas in its capabilities (Adams and Stephens 1991). However, telecommunications technology can foster the development of needed infrastructures for rural areas, in an economically and technically feasible manner. Several quantitative studies have demonstrated a strong positive correlation between investment in telecommunications and economic development (Cronin *et al.* 1991 and Parker *et al.* 1992).

Rural residents can immediately grasp the benefits of investments in telecommunications. Telecommunications infrastructures have a great potential of enhancing the economies of rural regions. Many rural policy makers increasingly recognize the crucial importance of telecommunications infrastructures in improving productivity and efficiency in business, and enhancing the quality of social services throughout rural regions. Parker *et al.* (1989) noted that, through telecommunications, the overall breadth of beneficial changes can occur in virtually all areas of rural business and social services, such as manufacturing, agriculture, health care, education, retirement, etc.

Rural areas must be distinguished from isolated areas where access to any telecommunication network is non-existent because of physical remoteness of either distance or terrain. The emphasis of telecommunication policy must be on upgrading communication infrastructures for reaching residents who are not only rural but also physically remote, because there are few truly remote customers (Egan 1992). The rural telecommunications is developed through the use of existing public or private communication networks. This architecture is one of the most economically viable ways because of lower initial investment costs.

19.2 A Compromise Model

Each policy maker has a set of goals to achieve. Any management action starts with this goal-seeking behavior. The behavior typically moves the perceived states to the ideal states. Under multiple criteria, determining the appropriate action is not a trivial task. For example, assume that two goals of selecting hubs for a rural telecommunication network are (i) to reach at least one million rural residents via telecommunications, and (ii) to build the network by spending at most 0.2 million dollars. Achieving these two goals simultaneously is typically not possible. The policy maker must trade off one goal in favor of the other. One of the key issues in this trade-off process is determining the relative importance of each goal.

A compromise model is a well-established mathematical model used to define the human goal-seeking behavior. A fundamental advantage of the compromise model is its conceptual simplicity in computing. Furthermore, the compromise model includes a goal programming model (Lee 1972) that has been widely applied in multiple-criteria environments; i.e.. a goal-programming model is a special case of a compromise model. We apply the compromise model to telecommunication network design so as to seek the trade-off among several goals. This trade-off requires the use of zero-one integer decision variables, an approach not explored in previous linear compromise model research. In contrast, our model is referred to as a zero-one compromise model.

In order to illustrate the basic skeleton of the zero-one compromise model, consider n criteria for rural telecommunications establishment. Define a vector $\boldsymbol{y} = (y_1, y_2, \ldots, y_n)$ as a set of goal functions. Let **y*** be an ideal value (also called an anchor value). Each component *of* $\boldsymbol{y}^*$ may be obtained by optimizing one criterion in isolation. The regret of using $\boldsymbol{y}$ instead of using the target $\boldsymbol{y}^*$ is approximated by the distance between $\boldsymbol{y}$ and $\boldsymbol{y}^*$. Thus, we define the regret of using y by

$$r(\mathbf{y}) = \| \mathbf{y}^* - \mathbf{y} \|$$

where $\| \mathbf{y}^* - \mathbf{y} \|$ is the distance from $\boldsymbol{y}$ to $\boldsymbol{y}^*$ according to some specified norm.

In a multicriteria environment, goals may have different degrees of importance. One of the practical difficulties with a compromise model is finding a way to assign weights to the goals. We adopt a scaling method, called the Analytic Hierarchy Process (AHP) (Saaty 1977), which has been widely used for practical human judgmental decision making. The effectiveness of using the AHP in weighting the goals within the framework of mathematical programming is highlighted in a number of sources (Dyer and Forman 1991, and Suh *et al.* 1993). The availability of commercial software called *Expert Choice* (Forman *et al.* 1985) further facilitates its use. The AHP employs a pairwise comparison of criteria, and the result is stored in an input matrix A. The relative weight of each criterion is calculated by the application of an eigenvalue method to A, and then storing it in the form of a vector $\boldsymbol{w} = (w_1, \ldots, w_n)$.

In addition, because goals are measured by different scales, the measures of the criteria must be normalized to avoid a meaningless summation. A common solution to this problem is to use a normalizing factor like the following one (Romero 1991):

$$\delta_i = \frac{1}{| y_i^* - y_i^o |}, \forall i = 1, \ldots, n$$

Here, y_i^o is the anti-ideal (nadir value) for the *i*th criterion, i.e., the worst value possible. The anti-ideal value is obtained by solving the single objective problem in an opposite direction. For example, if the *i*th goal is to be maximized, y_i^o is obtained by minimizing the problem. Recently, Ballestero and Romero (1991) showed that this kind of normalizer provides a good economic interpretation; i.e., they are good shadow prices in economic scenarios.

The regret function is then

$$\mathbf{r}(\mathbf{y} \mid \mathrm{p}, \mathbf{w}, \delta) = \| \mathbf{y}^* - \mathbf{y} \|_{p,w,\delta}.$$

The function is further explained as

$$\mathbf{r}(\mathbf{y} \mid \mathrm{p}, \mathbf{w}, \delta) = \left[\sum_{i=1}^{n} \frac{w_i^p}{| y_i^* - y_i^o |^p} | y_i^* - y_i |^p \right]^{1/p}$$

Applying Theorem 13.5, we remove the absolute value sign of the regret function and obtain:

$$\mathbf{r}(\mathbf{y} \mid \mathrm{p}, \mathbf{w}, \delta) = \left[\sum_{i=1}^{n} \frac{w_i^p}{| y_i^* - y_i^o |^p} (d_i^- + d_i^+)^p \right]^{1/p}$$

Minimizing this r is equivalent to minimizing r'. The third condition, i.e., $d_i^+ \times d_i^- = 0$, is redundant and does not have to be incorporated into the model.

Then it is seen that minimization of the regret function is equivalent to minimization of

$$\mathbf{r}'(\mathbf{y} \mid \mathrm{p}, \mathbf{w}, \delta) = \sum_{i=1}^{n} \frac{w_i^p}{| y_i^* - y_i^o |^p} (d_i^- + d_i^+)^p$$

For convenience, we will denote r by the above regret function *r'*. Hence, the objective of our compromise model is reduced to minimizing this function.

In order to incorporate hub decisions into the compromise programming modeling of the telecommunications design, let $L = \{1, \ldots, l\}$ be a set of hub candidate cities under consideration. Hub decision requires determining a subset S as optimal hub cities so that the regret function is minimized. Define zero-one integer decision variables: $x_k = 1$ if the candidate hub location k is to be selected as the hub city, otherwise

$x_k = 0$. Denote a decision vector $\mathbf{x} = (x_1,...,x_l)$, where l is the number of hub candidates. Also note that the goal function y_i is a function of $\mathbf{x}$ such that

$$y_i = f_i(\mathbf{x}).$$

Telecommunications services must be accessed by rural residents within a certain distance, which is denoted by θ. Let d_{ik} be the distance between locations i and k. The variable e_{ik} has the value of one if $d_{ik} \leq \theta$; otherwise, e_{ik} has the value of zero. Thus, the constraint common to the kind of set covering problem can be derived, i.e., $\sum_{k=1}^{l} e_{ik} x_k \geq 1$, for $i \neq k$. Consequently, the design problem can be presented in the form of zero-one compromise programming as

$$\text{Minimize} \qquad r(y \mid p, w, \delta) = \sum_{i=1}^{n} \frac{w_i^p}{| y_i^* - y_i^o |^p} (d_i^- + d_i^+)^p$$

$$\text{Subject to} \qquad y_i - f_i(\mathbf{x}) = d_i^- - d_i^+, \qquad \forall i = 1,...,n,$$

$$\sum_{k=1}^{l} e_{ik} x_k \geq 1, \qquad \forall i = 1,...,l, \; i \neq k,$$

$$x_k \in \{0,1\}, \qquad \forall i = 1,...,l.$$

For the sake of convenience, the first set of the constraints is called the regret distance constraints and the second set of constraints the hub distance constraints.

When p = 1, the computational effort of solving the corresponding compromise model is minimum. The ZOOM (Zero/One Optimization Methods) software, as developed by Singhal *et al.* (1989), was used to solve this kind of zero-one compromise model. The software used was originally designed for up to approximately 200 zero-one variables. This capability is sufficient for real-life telecommunications hub city decisions. For instance, the case for the state of Nebraska studied in this research considers twenty-four zero-one variables.

Several merits of our methodology are worth noting. It is reasonable to expect that telecommunications systems will be established by a group rather than an individual, because of its political nature. Our model enables the decision makers to facilitate a group decision. Policy makers, along with network designers, can be involved in hub city selection committee that is responsible for this group decision-making process. This involvement is more likely to reduce the communication gap between system designers and policy makers. Furthermore, concentrating on well-defined criteria mitigates human biases in assessing the overall utility of the telecommunications establishment.

One should also keep in mind that our design methodology is highly flexible. Both designers and policy makers may use the model as a stepping stone and perform a parametric study. This model is likely to better serve as a decision support tool because of this iterative potential.

19.3 Nebraska as A Case Study

Our design methodology is applied to one of the Midwest states of America, Nebraska. A Nebraska rural telecommunication network is to be built on the basis of this study. We intend to compare our results with the actual network when the Nebraska rural network takes its full shape in the near future. However, the current prototype, as proposed in this study, can be easily applied to any rural area in America.

Such diverse factors as strong dependence on agriculture, poor comforts level, inferior schools, and limited access to medical facilities have reduced the attractiveness of the rural Midwest. The Midwest rural areas are increasingly becoming troubled regions because of inferior economies and extensive emigration (Sofranko 1991).

In the early 1970s, the future looked bright for rural areas in the Midwest. The gap between the rural and urban standard of living was lessened because of slow emigration and an improved employment level. However, in the 1980s the situation reversed. An article entitled "No housing, no growth for small town" of August 1, 1993, published by the Sunday World Herald (Omaha, Nebraska), highlights the crisis this way: During the 1980s, the population of 58 rural Nebraska towns with 1000 to 2500 residents declined by an average of 6.6%, while habitable housing in the same communities dropped by 1.6%, according to the 1990 US Census. For example, a small rural town, Crawford, lost 15.2% of its population in the 1980s, while the supply of habitable housing there fell 9.4%. The reason for decline in small-town population is a common story across the Midlands. Population loss results in a lowered overall income level, which reduces consumer spending and eliminates more jobs. In contrast, the population growth in the U.S. during the same period was 10%.

The success of telecommunications infrastructure in rural areas of Idaho and Iowa indicates its potential (Bahl *et al.* 1991). From these perspectives, we apply the zero-one compromise model in the establishment of telecommunication networks for rural Nebraska.

Telecommunications policy makers first solicit hub candidates for rural telecommunications. Nebraska created a fully deregulated telecommunication environment, and thus the telecommunication network is developed through a partnership between the public and private sectors, i.e., the State Directorate of Telecommunication and private sectors cooperate for mutual benefits. Hub candidates must be somewhat larger than their neighboring rural communities; these candidate cities function as service centers. Twenty-four major population centers are chosen as hub candidates, comprising twenty-two communities, each having a population in excess of 5,000, in addition to the two major metropolitan areas of Omaha and Lincoln.

Next, policy makers need to determine the criteria for measures of effectiveness of telecommunications via hubs for rural communities. The following criteria are chosen, as suggested by Nazem (1992). A policy study of rural telecommunications undertaken by the US Office of Technology Assessment has also recommended this approach.

1. *Economic activity.* Local economic activity is one of the most significant components to consider when evaluating the effectiveness of rural telecommunications. Economic activity was measured by the number of employees in the areas surrounding a hub candidate city.

2. *Population.* The primary aim of rural telecommunications is to allow as many rural residents as possible to access needed infrastructure. Population is the total population of the counties for which the hub candidate serves as the center.

3. *Budget.* To establish a hub city requires construction and operation costs. The telecommunication cost is also related to the transmission facilities for data, voice, and videotext traffic; it includes both hardware and software components. These costs must be estimated in the form of present value for each individual hub candidate.

4. *Health care.* The health care level is measured by health care professionals. We develop a measure called the health care index on the basis of two factors, namely the number of hospital beds and the number of practicing physicians in the city. For instance, if the number of beds is 100 or less, and the number of physicians is 10 or less, the health care index of that hub city is assigned the value of 1. If the number of hospital beds is 400 or greater and the number of physicians is 40 or greater, the index is 10. Indexes 2 through 6 are assigned for hub cities in between.

5. *Education.* An index is assigned to each hub candidate on the basis of the level of educational opportunity. For instance, cities with only a high-school-level education program receive an index of 2. Cities with comprehensive higher learning more than a state-supported four-year college program receive an index of 10. Hub candidates are assigned education indexes from 2 through 10.

6. *Transportation.* An index is assigned to each hub candidate on the basis of the type of highways to support transportation services. For instance, the hub candidate having only a state highway is assigned a transportation index of 3. Hub candidates are assigned transportation indexes from 3 through 10.

Measures for twenty-four hub candidates are summarized in Table 19.1. The following are the units of each criterion. These units are used as the basis for the pairwise comparison of criteria.

1. Economy: Increase of 10,000 employees supported via rural telecommunications.
2. Population: Increase of 10,000 rural residents supported via rural telecommunications.
3. Budget: Decrease of $ 10,000 of costs spent on rural telecommunications establishment.
4. Health care: Increase of one health care index of any rural community.
5. Education: Increase of one education index of any rural community.
6. Transportation: Increase of one transportation index of any rural community.

The relative importance of the criteria in the telecommunications establishment can thus be determined within the framework of paired comparison through the use of the AHP.

For instance, in justifying rural telecommunications, policy makers are asked "Which is more important, and by how much: economy or budget?" If the decision maker slightly favors the first over the second, then $a_{13} = 3$. If the first is strongly favored over the second, then $a_{13} = 7$. Practically, a ratio scale between 1 and 9 is used. The experience of many users of the AHP implies that the l-to-9 scale captures fairly well the preferences of the decision makers (Harker and Vargas 1987).

The result of all of the fifteen-paired comparisons is represented in the form of the matrix A:

$$A = \begin{pmatrix} 1 & 2 & 3 & 5 & 6 & 7 \\ & 1 & 1 & 3 & 5 & 5 \\ & & 1 & 3 & 4 & 6 \\ & & & 1 & 3 & 5 \\ & & & & 1 & 2 \\ & & & & & 1 \end{pmatrix}$$

The lower triangular is not listed because the lower triangle is simply the reciprocal of the upper triangle. According to the eigenvalue method of computing, the estimated normalized weights are:

$$\mathbf{y}^* = (43.62, 111.06, 236.0, 77.0, 106.5, 0.5)$$

and

$$\mathbf{y}^o = (14.51, 43.92, 413.0, 33.0, 56.0. 58.5).$$

For example, minimizing the population goal in isolation results in the objective value of 14.51. Normalizers are then

$$\delta = (0.03435, 0.01489, 0.005650, 0.02273, 0.01980, 0.01923).$$

Table 19.1 Telecommunications Hub Candidate Cities for the State of Nebraska

Hub candidate	Number of employees (unit:10,000)	Counties population (unit:10,000)	Budget (unit:$10, 000)	Health Care index	Educt'n index	Transp. index
1.Alliance	0.27	1.42	25	1	4	3
2.Blair	0.32	1.55	20	1	4	3
3.Beatrice	0.55	3.58	20	5	2.5	3
4.Chadron	0.18	0.94	40	1	6	3
5.Columbus	1.01	2.96	15	1	4	4
6.Falls City	0.19	1.10	10	1	2	3
7.Fremont	1.04	3.58	8	4	4	4
8.Gering/ Scottsbluff	1.07	3.84	20	5	4	3
9.Grand Island	1.94	4.99	8	5	4	4
10.Hstings	1.06	3.11	10	5	5	4
11.Holdredge	0.30	1.01	10	1	2	4
12.Kearney	1.17	3.79	5	5	7	6
13.Lexington	0.53	2.23	20	5	2	6
14.Lincoln	8.18	20.30	7.5	10	10	9
15.McCook	0.34	1.30	45	1	4	3
16.Nevraska City	0.36	1.50	18	1	2	4
17.Norfolk	1.24	3.23	20	5	4	3.5
18.North Platte	0.74	3.47	10	4	4	6
19.Ogallala	0.25	0.92	10	1	2	6
20.Omaha	21.32	41.03	2	10	9	10
21.Sydney	0.22	1.01	50	1	4	6
22.South Sioux City	0.69	1.72	7.5	1	4	5
23.Wayne	0.19	0.98	22	1	6	3
24.York	0.46	1.50	10	2	4	5
Total	43.62	111.06	413	77	106.5	110.5

The distance d_{ik}, between hub candidates can be found through the Nebraska map prepared by the Department of Roads (the rough picture is Figure 19.1). Policy makers suggest that rural residents should be within 30 miles from any hub city, i.e. $\theta = 30$. Policy makers decide to use an absolute distance penalty for their regrets, i.e., $p = 1$.

All of the goals are lower bounded, except for the budget goal, which is upper bounded. A detailed formulation of the resulting zero-one compromise model is given in the appendix. Solving the model yields the total of twenty-two hub cities such that

$$\mathbf{x}^* = (1,0,1,0,\ 1,1,1,1,1,1,1,1,1,1,1,1,1,1,1,1,1,1,1,1).$$

Only Chadron and Falls City in Table 19.1 are not selected. Each goal function is achieved at the level of

$$y = (43.12, 108.57. 353.0. 75.0, 96.5, 104.5).$$

For example, the optimal hub design can cover 1.0857 million rural residents. For a parametric study, decision makers may change the values *of p* and/or *w* to reach other designs.

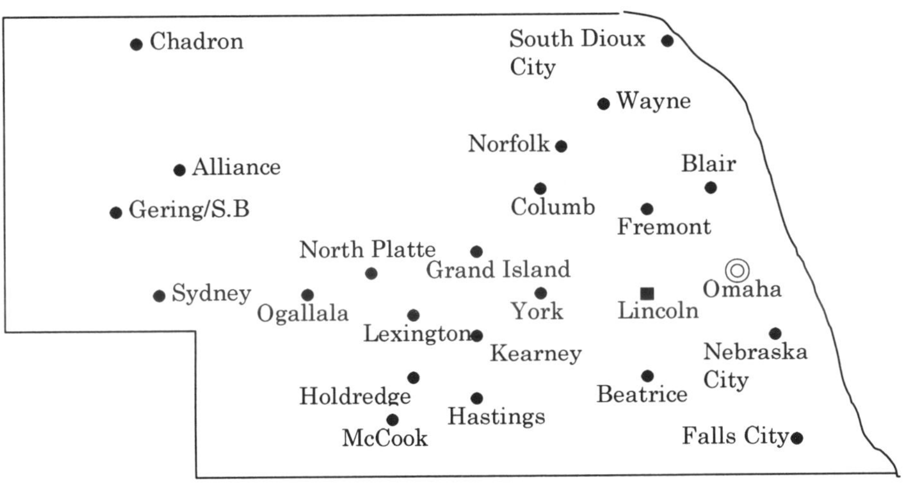

Figure 19.1 Hub candidate cities in Nebraska

19.4 Further Remarks

Telecommunications has been increasingly recognized as an infrastructure rural communities need in order to foster services in an economically viable way. Thanks to emerging broadband fiber optics, rural area networks can be built via hub cities that function as centers for neighboring rural communities.

This chapter investigates a design methodology for building rural telecommunication networks via hub cities in order to support rural policy makers goal-setting behavior. A compromise among conflicting targets is obtained through the use of a zero-one compromise model. A prototype model for the state of Nebraska can be readily applied to other rural regions.

Several research areas still remain to be explored. A possible extension is to consider multiple resource availability levels by using linear design models developed by Lee, Shi and Yu (1990). The policy maker can prepare the optimal contingency plans to deal with changes of all possible decision situations. Another possibility is the incorporation of other telecommunication design criteria, such as system response time and/or availability, into a set of goals (Lee and Sheng 1992). This issue is not a trivial task, because of the nonlinearity of these criteria. An exchange heuristic (Ignizio 1980) can be iteratively adopted until a compromise solution is reached. From a practical point of view, an important contribution is to sharpen parameter measures so that this work may be readily applicable to other cases.

Appendix: Compromise programming formulation

A. *Objective function*

Minimize

$$13.60d_1^- + 3.202d_2^- + 1.141d_3^+ + 2.341d_4^- + 0.990d_5^- + 0.654d_6^-.$$

For simplicity, the objective function is multiplied by 1000 and minimized. It is also noted that both d_i^- and d_i^+ do not need to be optimized for lower bound or upper bound goals. For example, in the case of Economy, only a d_i^- may be minimized because Economy is a lower bound goal. In other words, the goal is to maximize y_1 as long as y_1 is less than y_1^* i.e., we do not need to consider d_1^+ In contrast, in the case of Budget, only d_3^+ is minimized because it is an upper bound goal.

B. Regret distance constraints

$46.62 - (0.27\ x_1 + 0.32\ x_2 + 0.55\ x_3 + 0.18\ x_4 + 1.01\ x_5 + 0.19\ x_6 + 1.04\ x_7 + 1.07\ x_8 + 1.94\ x_9 + 1.06\ x_{10} + 0.30\ x_{11} + 1.17\ x_{12} + 0.53\ x_{13} + 8.18\ x_{14} + 0.34\ x_{15} + 0.36\ x_{16} + 1.24\ x_{17} + 0.74\ x_{18} + 0.25\ x_{19} + 21.32\ x_{20} + 0.22\ x_{21} + 0.69\ x_{22} + 0.19\ x_{23} + 0.46\ x_{24}) = d_1^- - d_1^+$;

$111.06 - (1.42\ x_1 + 1.55\ x_2 + 3.58\ x_3 + 0.94\ x_4 + 2.96\ x_5 + 1.10\ x_6 + 3.58\ x_7 + 3.84\ x_8 + 4.99\ x_9 + 3.11\ x_{10} + 1.01\ x_{11} + 3.79\ x_{12} + 2.23\ x_{13} + 20.30\ x_{14} + 1.39\ x_{15} + 1.50\ x_{16} + 3.23\ x_{17} + 3.47\ x_{18} + 0.92\ x_{19} + 41.03\ x_{20} + 1.01\ x_{21} + 1.72\ x_{22} + 0.98\ x_{23} + 1.50\ x_{24}) = d_2^- - d_2^+$;

$236.0 - (25\ x_1 + 20\ x_2 + 20\ x_3 + 40\ x_4 + 15\ x_5 + 10\ x_6 + 8\ x_7 + 20\ x_8 + 8\ x_9 + 10\ x_{10} + 10\ x_{11} + 5\ x_{12} + 20\ x_{13} + 7.5\ x_{14} + 45\ x_{15} + 18\ x_{16} + 20\ x_{17} + 10\ x_{18} + 10\ x_{19} + 2\ x_{20} + 50\ x_{22} + 7.5\ x_{22} + 22\ x_{23} + 10\ x_{24}) = d_3^- - d_3^+$;

$77.0 - (1\ x_1 + 1\ x_2 + 5\ x_3 + 1\ x_4 + 1\ x_5 + 1\ x_6 + 4\ x_7 + 5\ x_8 + 5\ x_9 + 5\ x_{10} + 1\ x_{11} + 5\ x_{12} + 5\ x_{13} + 10\ x_{14} + 1\ x_{15} + 1\ x_{16} + 5\ x_{17} + 4\ x_{18} + 1\ x_{19} + 10\ x_{20} + 1\ x_{21} + 1\ x_{22} + 1\ x_{23} + 2\ x_{24}) = d_4^- - d_4^+$;

$106.5 - (4\ x_1 + 4\ x_2 + 2.5\ x_3 + 6\ x_4 + 4\ x_5 + 2\ x_6 + 4\ x_7 + 4\ x_8 + 4\ x_9 + 5\ x_{10} + 2\ x_{11} + 7\ x_{12} + 2\ x_{13} + 10\ x_{14} + 4\ x_{15} + 2\ x_{16} + 4\ x_{17} + 4\ x_{18} + 2\ x_{19} + 9\ x_{20} + 4\ x_{21} + 4\ x_{22} + 6\ x_{23} + 4\ x_{24}) = d_5^- - d_5^+$;

$110.5 - (3\ x_1 + 3\ x_2 + 3\ x_3 + 3\ x_4 + 4\ x_5 + 3\ x_6 + 4\ x_7 + 3\ x_8 + 4\ x_9 + 4\ x_{10} + 4\ x_{11} + 6\ x_{12} + 6\ x_{13} + 9\ x_{14} + 3\ x_{15} + 4\ x_{16} + 3.5\ x_{17} + 6\ x_{18} + 6\ x_{19} + 10\ x_{20} + 6\ x_{21} + 5\ x_{22} + 3\ x_{23} + 5\ x_{24}) = d_6^- - d_6^+$;

Chapter 20
CREDIT CARD PORTFOLIO MANAGEMENT

Many data mining problems in credit card portfolio management decisions can be carried out through a large-size real-life data. One of the typical credit card portfolio management decisions is to predict the cardholders' behavior (e.g., either "Bads" for bankruptcy or "Goods" for otherwise). In the past decades, academic research has successfully developed a number of techniques, such as the linear and logistic regression (Breiman *et al.* 1984, Joachimsthaler and Stam 1988), decision tree induction (Quilan 1986), linear programming (Freed and Glover 1981) and neural network (Guo and Gelfand 1992), to deal with classification or discriminant problems. In financial business, the practitioners have applied some of these techniques to conduct the credit card portfolio management. The examples of current approaches are (1) Behavior Score developed by Fair Isaac Corporation (FICO); (2) Credit Bureau Score most of which are also developed by FICO; (3) First Data Resource (FDR)'s Proprietary Bankruptcy Score; and (4) a new competent score using Set Enumeration (SE) decision tree (Rymon 1993). Among them, (1)-(3) are linear while (4) is a nonlinear model. A comparison of these methods can be demonstrated by the Kolmogorov-Smirnov (KS) value that measures the largest separation of these two cumulative distributions of Goods and Bads (Conover 1999). The KS values of the Behavior Score, Credit Bureau Score, FDR Bankruptcy Score, and SE Score on a sample of 1000 cases are 55.26, 45.55, 59.16, and 60.22, respectively.

The objective of this chapter is to outline the research project on searching an alternative data mining approach (preferred by an MCLP approach) that could outperform the current approaches for business practice (Shi, Wise, Luo and Lin 2000). Note that the reason of why to choose the MCLP discriminant analysis instead of a cluster analysis is that the behavior of credit card holders can be classified as a problem given appropriate groups (eg., Good and Bad). A cluster analysis, however, does not require a *prior* grouping. Section 20.1 proposes a new MCLP model for credit card portfolio management decisions. Section 20.2 presents the results of computational experiences in real-world data and compares the variations of different approaches. Finally, Section 20.3 summarizes this chapter with further research directions.

20.1 Methodology for Credit Card Portfolio Decisions

Data mining for credit card portfolio management decisions is to classify the different cardholder behaviors in terms of their payment to the credit card companies, such as banks and mortgage loan firms. In realty, while all credit card companies share some common variables, some credit card company has different variables to describe the cardholders' behavior. The examples of common categories of the variables are balance, purchase, payment and cash advance. Some credit card company may consider residence state category and job security as special variables. In the case of FDR, there are 38 original variables from the common variables over the past seven months. Then, a set of 65 derived variables is internally generated from the 38 variables to perform the precise data mining.

Given a set of r variables about the cardholders $A = (\mathbf{a}_1, \ldots, \mathbf{a}_r)$, let $\boldsymbol{A}_i = (a_{i1}, \ldots, a_{ir})$ be the development sample of data for the variables, where $i = 1, \ldots, n$ and n is the sample size. We want to determine the coefficients for an appropriate subset of the variables, denoted by $\boldsymbol{X} = (x_1, \ldots, x_r)$, and a boundary value b to separate two groups: G (Goods) and B (Bads); that is,

(a) $\boldsymbol{A}_i\, \mathbf{x} \leq \mathrm{b}$, $\boldsymbol{A}_i \in \mathrm{G}$ and $\boldsymbol{A}_i\, \mathbf{x} \geq \mathrm{b}$, $\boldsymbol{A}_i \in \mathrm{B}$,
where $\boldsymbol{A}_i$ are the vector values of the subset of the variables.

We consider two kinds of measurements for better separation of Goods and Bads. Let α_i be the overlapping degree with respect to $\boldsymbol{A}_i$, and β_i be the distance from $\boldsymbol{A}_i$ to its adjusted boundary (Freed and Glover 1981 and Glover 1990) (for example, see Figure 20.1). We want to minimize the sum of α_i and maximize the sum of β_i simultaneously. By putting α_i into (a), we have:

(b) $\boldsymbol{A}_i\, \mathbf{x} \leq \mathrm{b} + \alpha_i$, $\boldsymbol{A}_i \in \mathrm{G}$ and $\boldsymbol{A}_i\, \mathbf{x} \geq \mathrm{b} - \alpha_i$, $\boldsymbol{A}_i \in \mathrm{B}$.

However, considering β_i, we can further rewrite (b) as

(c) $\boldsymbol{A}_i\, \mathbf{x} = \mathrm{b} + \alpha_i - \beta_i$, $\boldsymbol{A}_i \in \mathrm{G}$ and $\boldsymbol{A}_i\, \mathbf{x} = \mathrm{b} - \alpha_i + \beta_i$, $\boldsymbol{A}_i \in \mathrm{B}$.

Our two-criteria linear programming model is stated as

(d) *Minimize* $\Sigma_i\alpha_i$ and *Maximize* $\Sigma_i\beta_i$

Subject to

$$\boldsymbol{A}_i\,\mathbf{x} = b + \alpha_i - \beta_i,\ \boldsymbol{A}_i \in G,$$
$$\boldsymbol{A}_i\,\mathbf{x} = b - \alpha_i + \beta_i,\ \boldsymbol{A}_i \in B,$$

where $\boldsymbol{A}_i$ are given, $\mathbf{x}$ and b are unrestricted, and α_i and $\beta_i \geq \mathbf{0}$.

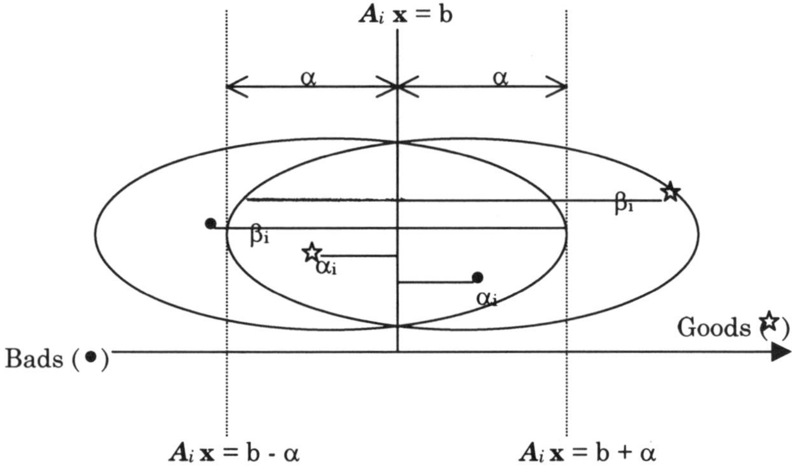

Figure 20.1 An Illustrative Example

In order to apply real-life data for business practice in FDR, we choose the SAS LP (linear programming) package to facilitate the computation (SAS/OR 1990). In addition, the compromise solution approach (Yu 1973, Yu 1985 and Shi and Yu 1989) has been employed to reform the model (d) so that the SAS LP can systematically identify the best trade-off between $\Sigma_i\alpha_i$ and $\Sigma_i\beta_i$ for an optimal solution. To explain this, we assume the "ideal value" of $-\Sigma_i\alpha_i$ be α^* and the "ideal value" of $\Sigma_i\beta_i$ be β^*. Then, if $-\Sigma_i\alpha_i > \alpha^*$, we define the regret measure as $d_\alpha^+ = \Sigma_i\alpha_i$ +

α^*; otherwise, it is 0. If $-\Sigma_i\alpha_i < \alpha^*$, the regret measure is defined as $d_\alpha^- = \alpha^* + \Sigma_i\alpha_i$; otherwise, it is 0. Thus, we have (i) $\alpha^* + \Sigma_i\alpha_i = d_\alpha^- - d_\alpha^+$, (ii) $|\alpha^* + \Sigma_i\alpha_i| = d_\alpha^- + d_\alpha^+$, and (iii) $d_\alpha^-, d_\alpha^+ \geq 0$. Similarly, we derive $\beta^* - \Sigma_i\beta_i = d_\beta^- - d_\beta^+$, $|\beta^* - \Sigma_i\beta_i| = d_\beta^- + d_\beta^+$, and $d_\beta^-, d_\beta^+ \geq 0$ (Figure 20.2). An MCLP model has been gradually evolved as

(e) *Minimize* $d_\alpha^- + d_\alpha^+ + d_\beta^- + d_\beta^+$

Subject to
$\alpha^* + \Sigma_i\alpha_i = d_\alpha^- - d_\alpha^+$,
$\beta^* - \Sigma_i\beta_i = d_\beta^- - d_\beta^+$,
$\boldsymbol{A_i}\,\mathbf{x} = b + \alpha_i - \beta_i$, $\boldsymbol{A_i} \in G$,
$\boldsymbol{A_i}\,\mathbf{x} = b - \alpha_i + \beta_i$, $\boldsymbol{A_i} \in B$,

where $\boldsymbol{A_i}$, α^*, and β^* are given, $\mathbf{x}$ and b are unrestricted, and α_i, β_i, d_α^-, d_α^+, d_β^-, $d_\beta^+ \geq 0$.

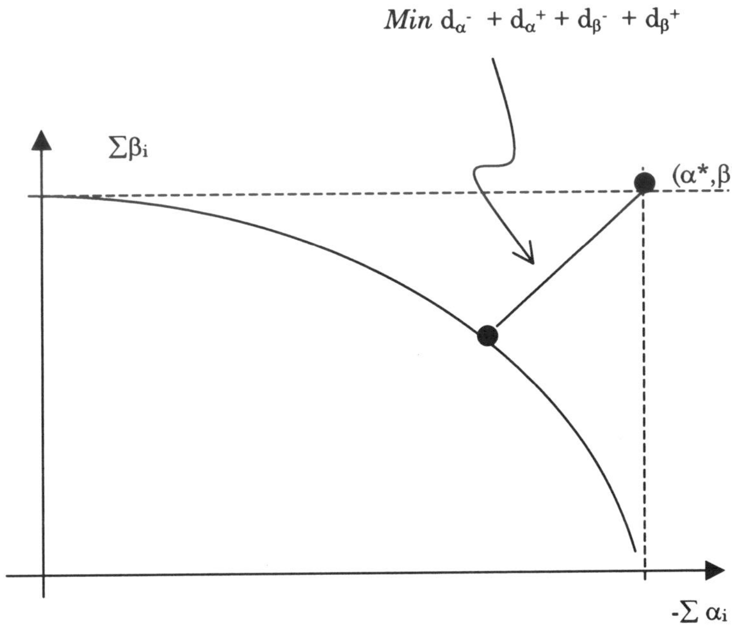

Figure 20.2 The Concept of Minimization

The best separation can be interpreted in terms of many different criteria. For example, to measure the separation in terms of the pair of the center and the dispersion, we can use information value, which is the quotient of the difference between means and standard deviation. To measure the separation in terms of matching pair of Good and Bad, we can use the concordance. To measure the separation in terms of the cumulative distribution, we can use the KS, which measures the largest separation of these two cumulative distributions of Good and Bad. Comparing these available measurements of separation, the KS measurement has empirically demonstrated the high correlation to identify Bads. Thus, it has been widely accepted in the credit risk management area. Furthermore, it is also a common sense in the credit risk area that the bigger value of KS, the better separation.

For the problem of the two-class classification in this paper, the KS separation is given as:

$$KS = \max \mid \text{Cum. distribution of Good} - \text{Cum. distribution of Bad} \mid .$$

Although its asymptotic distribution can be obtained, such as some statistical testing procedures may be applied, we did not use this feature here. What we used is that KS is an indicator to measure the maximum separation between the two classes: Goods and Bads. The code of KS test is also implemented by SAS for the convenience of analyzing the large real-life data.

20.2 Computational Experience and MCLP Model Performance

Prior to the development of the MCLP model (e), we constructed seven special models of (e) (M1-M7) and tested the total of 42 LP or MCLP solutions (see Appendix). 14 out of those 42 solutions in model (M1)-(M4) used a set of 65 variables in FDR on a sample of 6000 cases. The KS value ranged from 6.19 to 47.55. The rest 28 solutions in model (M3)-(M7) used a set of 64 variables on a sample of 1000 cases. The highest KS value was 49.88.

The model (e) has four versions (M8-M11) on different sample data and variable sizes in which 9 MCLP solutions have been tested. A summary is given in Table 20.1. The explanation is as follows:

Table 20.1 Summary of MCLP Model (e)

Versions	# of Variables	Sample Size	KS Value	Note
(M8)	64	1000	59.49	Optimal
(M8)	64	6000	51.97	Cross-Validation
(M9)	30 (Log.Reg.)	2000	60.10	Optimal
(M9)	30 (Log. Reg.)	6000	56.43	Cross-Validation
(M10)	29 (Log. Reg.)	3000	56.19	Optimal
(M10)	29 (Log. Reg.)	6000	55.81	Cross-Validation
(M11)	27 (Expert)	3000	55	Optimal
(M11)	27 (Expert)	6000	53.63	Cross-Validation

Model (M8) with 64 variables, $\alpha^* = 0.001$ and $\beta^* = 10{,}000{,}000$ has been tested on a 1000-sample (KS = 59.49) and then used in the statistical cross-validation study on a 6000-sample (KS = 51.97). The deviation 7.52 (59.49-51.97) shows the model is not robust. However, The KS value of the Behavior Score, Credit Bureau Score, FDR Bankruptcy Score, and SE Decision Tree Score on this 1000-sample are 55.26, 45.55, 59.16, and 60.22, respectively. This indicates that the MCLP model has achieved a better KS value in this particular case. To test the robustness of the model (e) with regard to sample sizes, model (M9) employing 30 variables chosen by the logistic regression, $\alpha^* = 0.001$ and $\beta^* = 10{,}000{,}000$ has been applied on a 2000-sample. The KS value is 60.1. The corresponding KS value of the cross-validation study on the 6000-sample is 56.43. This shows that while the KS deviation is reduced from 7.52 to 3.67 (60.1-56.43), the KS value 60.1 on the 2000-sample has improved from that 59.49 on the 1000-sample. Because of this encouragement, we further applied model (M10) with 29 variables, $\alpha^* = 0.001$ and $\beta^* = 10{,}000{,}000$ on a sample of 3000 cases. The results are (i) the KS = 56.19 on the 3000-sample, (ii) the KS = 55.81 on the 6000-

sample, and (iii) the percentage of the cases in 15 point interval is less than 5%. Therefore, the good news is that since the KS deviation is only 0.38 (56.19-55.81), the MCLP model (e) starts showing the robustness (or the cross-validation test is successful) from a 3000 sample size. The bad news is that the KS value of the 3000-sample has dropped 3.91 from 60.1 on the 2000-sample to 56.19. In model (M11), instead of using 30 variables from the set of 65 variables by the logistic regression, we applied our working experience to select 28 variables with $\alpha^* = 0.001$ and $\beta^* = 10,000,000$ for the 3000-sample. The test results are KS = 55 on the 3000-sample and KS = 53.63 on the 6000-sample. This implicates that using the logistic regression to choose a subset of variables can result in the better KS value.

20.3 Main Research Findings and Further Research

The current main research findings on the MCLP approach are (i) the MCLP model is fully controlled by the formulation; (ii) the 3000 is a stable size of sample for the robustness in the separation process; and (iii) the MCLP model can be easily adapted to multi-group separation problems. These findings have shown that the MCLP model is an attractive technology for data mining in credit card portfolio management. It is possible to upgrade our model into a production version. We can also observe the following further researches:

First, to improve the KS value and keep the robustness, the model (e) may be revised in two different ways. The first way is to consider a general norm p on the regret functions $|\alpha^* + \Sigma_i\alpha_i|^p = (d_\alpha^- + d_\alpha^+)^p$ and $|\beta^* - \Sigma_i\beta_i|^p = (d_\beta^- + d_\beta^+)^p$, $p \geq 1$. When p =1, we have a linear objective function as in (d). When p > 1, we have a nonlinear objective function for (d). When $p = \infty$, we have a series of linear objective function in (d) (Yu 1973, Yu 1985 and Shi and Yu 1989) (Figure 13.5). Since the theoretical study has shown that a different norm can produce a different regret value, switching p = 1 to other p value may give the MCLP model a better chance for the better separation of Goods and Bads. As a result, the KS value may be improved. The second revision is to define α_{ij} as the degree to which $\boldsymbol{A}_i$ overlap the two-group boundary b with respect to the jth variable and define β_{ij} be the distance of $\boldsymbol{A}_i$ from its adjusted boundary with respect to the jth variable. Then, a more general MCLP model can be formulated and this model can capture more insights of the separation.

Second, according to the known research results (Freed and Glover 1981), the linear programming (LP) approach has advantage to separate multi-group problems. If a given problem has k different groups, then the boundary value b of the MCLP model can be expanded to $(b_1, \ldots, b_{k-1})$. The solving process is straightforward. However, in the logistic regression, the multi-group problems only can be handled by pair-wise comparison approach (i.e., one group vs. the rest groups). We have verified the MCLP approach on multi-group separation problems for a small sample data. Using a large-size of real-life data to handle the credit card problem with several groups is an ongoing project.

Appendix: Summary of tested LP/MCLP models

(M1) *Minimize* α

Subject to

$A_i\, \mathbf{x} \leq b + \alpha,\ A_i \in G,$

$A_i\, \mathbf{x} \geq b - \alpha,\ A_i \in B,$

where A_i are given, $\mathbf{x}$ and b are unrestricted, and $\alpha \geq 0$ is a common degree to which all group observation A_i violate the two-group boundary b.

Sample Size: 6000 The highest KS: 19.70 # of Variables: 65

(M2) *Maximize* β

Subject to

$A_i\, \mathbf{x} \geq b - \beta,\ A_i \in G,$

$A_i\, \mathbf{x} \leq b + \beta,\ A_i \in B,$

where A_i are given, $\mathbf{x}$ and b are unrestricted, and $\beta \geq 0$ is a common the distance of the observation A_i from its adjusted boundary.

Sample Size: 6000 The highest KS: 8.80 # of Variables: 65

(M3) *Minimize* $\Sigma_i \alpha_i$

Subject to

$A_i\, \mathbf{x} \leq b + \alpha_i,\ A_i \in G,$

$A_i\, \mathbf{x} \geq b - \alpha_i,\ A_i \in B,$

where A_i are given, $\mathbf{x}$ and b are unrestricted, and $\alpha_i \geq 0$.

Sample Size: 6000 The highest KS: 43.81 # of Variables: 65

Sample Size: 1000 The highest KS: 48.31 # of Variables: 64

(M4) *Maximize* $\Sigma_i \beta_i$

Subject to

$\boldsymbol{A}_i \mathbf{x} \geq b - \beta_i, \; \boldsymbol{A}_i \in G,$

$\boldsymbol{A}_i \mathbf{x} \leq b + \beta_i, \; \boldsymbol{A}_i \in B,$

where $\boldsymbol{A}_i$ are given, $\mathbf{x}$ and b are unrestricted, and $\beta_i \geq 0$.

Sample Size: 6000 The highest KS: 47.55 # of Variables: 65

(M5) *Minimize* $d_\alpha^- + d_\alpha^+ + d_\beta^- + d_\beta^+$

Subject to

$\alpha^* + \alpha = d_\alpha^- - d_\alpha^+,$

$\beta^* - \beta = d_\beta^- - d_\beta^+,$

$\boldsymbol{A}_i \mathbf{x} = b + \alpha - \beta, \; \boldsymbol{A}_i \in G,$

$\boldsymbol{A}_i \mathbf{x} = b - \alpha + \beta, \; \boldsymbol{A}_i \in B,$

where $\boldsymbol{A}_i$, $\alpha^* = 0.1$, and $\beta^* = 10$ are given, $\mathbf{x}$ and b are unrestricted, and α, β, d_α^-, d_α^+, d_β^-, $d_\beta^+ \geq 0$.

Sample Size: 1000 The highest KS: 36.50 # of Variables: 65

(M6a) *Maximize* $\Sigma_i d_i$

Subject to

$\boldsymbol{A}_i \mathbf{x} \geq b + d_i, \; \boldsymbol{A}_i \in G,$

$\boldsymbol{A}_i \mathbf{x} \leq b - d_i, \; \boldsymbol{A}_i \in B,$

where $\boldsymbol{A}_i$ are given, $\mathbf{x}$, b and d_i are unrestricted, where d_i is a distance from the observation $\boldsymbol{A}_i$ to the two-group boundary b.

Sample Size: 1000 The highest KS: 18.12 # of Variables: 64

(M6b) *Maximize* d

Subject to

$\boldsymbol{A}_i \mathbf{x} \geq b + d, \; \boldsymbol{A}_i \in G,$

$\boldsymbol{A}_i \mathbf{x} \leq b - d, \; \boldsymbol{A}_i \in B,$

where $\boldsymbol{A}_i$ are given, $\mathbf{x}$, b and d are unrestricted, where d is a distance from all observation $\boldsymbol{A}_i$ to the two-group boundary b.

Sample Size: 1000 The highest KS: 32.13 # of Variables: 64

(M7) *Minimize* $\Sigma_i h_i \alpha_i - \Sigma_i k_i \beta_i$

Subject to

$\boldsymbol{A}_i \mathbf{x} = b + \alpha_i - \beta_i, \; \boldsymbol{A}_i \in G,$

$\boldsymbol{A}_i \mathbf{x} = b - \alpha_i + \beta_i$, $\boldsymbol{A}_i \in B$,
where $\boldsymbol{A}_i$, $h_i = 50{,}000{,}000$, and $k_i = -0.01$ are given, $\mathbf{x}$ and b are unrestricted, and α_i and $\beta_i \geq 0$.

Sample Size: 1000 The highest KS: 49.88 # of Variables: 64

(e) *Minimize* $d_\alpha^- + d_\alpha^+ + d_\beta^- + d_\beta^+$

Subject to

$\alpha^* + \Sigma_i\alpha_i = d_\alpha^- - d_\alpha^+$,

$\beta^* - \Sigma_i\beta_i = d_\beta^- - d_\beta^+$,

$\boldsymbol{A}_i \mathbf{x} = b + \alpha_i - \beta_i$, $\boldsymbol{A}_i \in G$,

$\boldsymbol{A}_i \mathbf{x} = b - \alpha_i + \beta_i$, $\boldsymbol{A}_i \in B$,

where $\boldsymbol{A}_i$, $\alpha^* = 0.001$, and $\beta^* = 10{,}000{,}000$, $b = 1$ are given, $\mathbf{x}$ are unrestricted, and α_i , β_i , d_α^-, d_α^+, d_β^-, $d_\beta^+ \geq 0$.

(M8):
Sample Size: 1000 The KS: 59.49 # of Variables: 64
Sample Size: 6000 The KS: 51.97 # of Variables: 64

(M9):
Sample Size: 2000 The KS: 60.10 # of Variables: 30 by logistic regression
Sample Size: 6000 The KS: 56.43 # of Variables: 30 by logistic regression

(M10):
Sample Size: 3000 The KS: 56.19 # of Variables: 29 by logistic regression
Sample Size: 6000 The KS: 55.81 # of Variables: 29 by logistic regression

(M11):
Sample Size: 3000 The KS: 55 # of Variables: 27 by Expert
Sample Size: 6000 The KS: 53.63 # of Variables: 27 by Expert

Note: We used b = 1 in some problems of (M1) – (M4) and all problems of (M5) – (M11).

Chapter 21
AGGREGATE PRODUCTION PLANNING

Production planning is one of the most important functions in the process of production and operations management. In production planning, managers of manufacturing firms need to make crucial decisions on which specific aggregate levels of production, inventory and work force have to be produced to meet possible demands. Such decisions are traditionally interpreted as finding a best combination of the production, inventory and work force quantities that yields a minimum overall cost.

Production planning has drawn a great deal of attention from both practitioners and academia. Historically, economists initiated research on production planning and product development (Phelps 1947). Since then, production planning has been extensively studied in the literature (Brown 1967, Buffa 1965, Churchman 1961, Holt, Modigliani, Muth and Simon 1960, and Menipaz 1984 and Starr 1989). According to Saad (1982) and Heizer and Render (1993), all various approaches to production planning problems may be classified as five categories: (i) linear programming (Bowman 1956, Goodman 1974, Jones 1967 and Singhal and Adlakha 1989), (ii) linear decision rule (Bergstrom and Smith 1970, Holt, Modigliani, Muth and Simon 1960, and Schwarz and Johnson 1978), (iii) management coefficient approaches (Bowman 1963), (iv) simulation (Vergin 1966), and (v) search decision rule (Bitran and Hax 1977, Nagasawa, Nishiyama and Hitomi 1982, Oliff, and Leong 1987, and Taubert 1968). While the linear programming approach has been accepted by some manufacturing firms (Ledbetter and Cox 1977), other models have been recognized to perform well in certain conditions (Lee and Khumawala 1972).

Bowman (Bowman 1956) first used a linear programming approach to formulate and solve production planning problems. By incorporating fluctuations in demand, production and inventory on a seasonal basis, this model regards production periods that consist of inventory, regular and overtime production rates as "resources," and the corresponding demand periods as "destinations." Then, a transportation method of linear programming is applied to identify an optimal solution that yields the minimum total costs of production and inventory in supplying the product units to the demands. The framework of this model has a great impact on many modeling developments, because it provides a useful channel of not only using mathematical modeling approaches to production planning, but also solving a real aggregate production planning problem by any available commercial linear programming software.

Due to the structure of linear programming, Bowman's (1956) model can only handle production planning problems with a single objective (i.e., a sum of production and inventory). A number of authors adopted techniques of goal programming (Charnes and Cooper 1961 and Lee 1972) and multiple objective programming (Yu 1985 and Zeleny 1974) to solve aggregate production planning problems with multiple objectives, for instance, treating production, inventory, work force and shortage as individual objectives (Goodman 1974 and Singhal and Adlakha 1989). The best advantage of multiple objective approaches is to understand trade-offs of the objectives for different policies of production planning (Singhal and Adlakha 1989).

However, all of the current models have explicitly or implicitly assumed that both capacity (supply) and demand levels are fixed. That is, capacity and demand levels in the models are given for a certain situation. It is well known that in reality, capacity and demand fluctuate on both seasonal and situational bases. Keeping in mind that the purpose of designing production planning is to predict and forecast details of production periods in the future, given a future season or period, the corresponding capacity and demand levels may change subject to certain situations, such as economy variations and group decision making. For example, when the economy is in recession, massive layoffs may cause a change in labor availability levels. When global business competition occurs, the availability level of raw materials may decrease at one time and increase at another time as the price of raw material fluctuates. These economy variations can cause change in capacity and demand levels. In the situation of group decision making, production and finance managers may have different interpretations of the capacity level for a future season. Production managers generally view the capacity level based on the material, manpower, and equipment under control, while finance managers may view it based on the available cash flow. For the demand level of the season, marketing managers' opinion may differ from that of either production managers or finance managers.

In this chapter, we propose a model of aggregate production planning with multi-objective and multi-capacity-demand levels (in short, multiple factors) by using the concepts of multi-criteria and multi-constraint level (MC^2) linear programming (Gyetvan and Shi 1992, Seiford and Yu 1979, and Yu 1985). This production planning model provides manufacturing decision makers with a systematic and comprehensive scenario incorporating all possible trade-offs of aggregate production planning depending on the multiple factors. These trade-offs can be used as a basis for the production planning decision making under fluctuations in capacity and demand in terms of production periods and situations. The organization of the paper is as follows.

In Section 21.1, we use the framework of an MC^2 transportation model to formulate aggregate production planning problems with multiple factors (APPMF) (Shi and Haase 1996). Next, the set of potentially optimal solutions of the problem is identified by computer software of the MC^2 transportation model developed by Haase and Shi (1994). Each potentially optimal solution contains a number of optimal trade-offs between multiple factors. In Section 21.2, a prototype of an APPMF problem is illustrated to explain implications of the multiple trade-offs in production planning. Finally, the chapter is concluded with further remarks in Section 21.3.

21.1 A Model of Aggregate Production Planning with Multiple Factors

The existing linear programming-based production planning models have some modeling shortcomings from a management perspective. Neither the linear programming approach (Bowman 1956), nor the goal programming approach (Goodman 1974), nor the multiobjective programming approach (Singhal and Adlakha 1989) can provide a comprehensive scenario of all possible optimal trade-offs between multiple objectives and multiple capacity-demand levels under consideration of a given production planning problem. Production planning by linear programming only reflects a single objective (as the sum of production, inventory and work force costs) of a manufacturing firm. As a result, this approach cannot help the firm find all the optimal trade-offs of several individual objectives. The production plan determined by the goal programming approach is a compromise (i.e., an optimal trade-off) among several objectives of the firm. However, it misses other possible compromises of the objectives which lead to different production plans for different decision situations that the firm may face. Multiple objective production plan lacks the capability of coping with possible multiple supply-demand levels.

The MC^2 linear programming framework can resolve the above shortcomings inherent in the previous production planning models and provide a scenario of all possible optimal trade-offs between multiple objectives and multiple capacity-demand levels. Therefore, we adopt the MC^2 transportation model ((M6.2) of Chapter 6) as a general form of the proposed AAPMF model. However, the interpretation of APPMF models using the above MC^2 transportation model should depend on contexts of aggregate production planning problems. For illustrative purposes, we now adopt some notations of Singhal and Adlakha (1989) to formulate a three-objective and three-capacity-demand level production planning problem with three production periods as follows:

Consider three objectives: (i) production cost, (ii) inventory holding cost, and (iii) shortage cost. Assume that there are three capacity-demand levels: (i) economy boom, (ii) normal economy, and (iii) economy recession.

Let r = regular unit production cost,

v = overtime unit production cost,

h = unit inventory holding cost per period,

s = unit shortage cost,

c_{ik} = production capacity for the i-th planning period at the k-th capacity level, i = 1, . . . , 7, k = 1, 2, 3.

d_{jk} = demand for the j-th planning period at the k-th demand level, j = 1, 2, 3, k = 1, 2, 3, where d_{4k} are slack products.

Using the general form, the production planning problem with 3 objectives and 3 capacity-demand levels is shown in Table 21.1. Row 1 contains the initial inventory over three production periods; rows 2-7 contain the regular and overtime production activities, respectively. The last row contains the demand levels of economic situations. For the columns, the 1st, 2nd and 3rd are production periods, the 4th is slack, and the last is three capacity levels of economic situations. Each cost cell in Table 1 contains three unit objective costs. If the values of r, v, s, c_{ik}, and d_{jk} are estimated or determined by certain techniques, then the problem can be solved by using the TP MC^2 Program, the computer software of Haase and Shi (1994). In the next section, a prototype of the APPMF problem is illustrated to demonstrate the optimal trade-offs and their management implications for decision makers.

Table 21.1 An Aggregate Production Planning with 3×3 Multiple Factors

Periods	P1	P2	P3	Slack	Capacity Levels
Initial Inventory	0,0,0	0,h,0	0,2h,0	0,0,0	c_{11} c_{12} c_{13}
R1	r,0,0	r,h,0	r,2h,0	0,0,0	c_{21} c_{22} c_{23}
01	v,0,0	v,h,0	v,2h,0	0,0,0	c_{31} c_{32} c_{33}
R2	r,0,s	r,0,0	r,h,0	0,0,0	c_{41} c_{42} c_{43}
02	v,0,s	v,0,0	v,h,0	0,0,0	c_{51} c_{52} c_{53}
R3	r,0,2s	r,0,s	r,0,0	0,0,0	c_{61} c_{62} c_{63}
03	v,0,2s	v,0,s	v,0,0	0,0,0	c_{71} c_{72} c_{73}
Demand Levels	d_{11} d_{12} d_{13}	d_{21} d_{22} d_{23}	d_{31} d_{32} d_{33}	d_{41} d_{42} d_{43}	

21.2 Optimal Trade-offs and Management Implications

As a convenient illustration of optimal trade-offs of APPMF problems, we use the TPMC2 Program (Hasse and Shi 1994) for a twelve-period aggregate production planning problem with two objectives, production and inventory, and two supply-demand levels, economy booming and recession. Assume there are two supply sources, regular and overtime for each period. Considering the initial inventory and slack supply, the problem has 13 demand destinations and 25 supply sources (i.e., 325 variables) as shown in Table 21.2. When the economy is booming, the data of regular, overtime, and demand comes from Table 1 of Singhal and Adlakha (1989) and Menipaz (1984). The data of economy in recession is made up by us. Cells in Table 21.2 with "X" are infeasible cells for production, because the shortage cost may occur and it is not under consideration.

Table 21.2 An Aggregate Production Planning with 2×2 Multiple Factors

Production Periods		1	2	3	4	5	6	7	8	9	10	11	12	Slack	Capacity:	
															Eco Boomin	Eco Recession
0	Initial Inventory	0, 0	0, h	0, 2h	0, 3h	0, 4h	0, 5h	0, 6h	0, 7h	0, 8h	0, 9h	0, 10h	0, 11h	0, 0	100	100
1	Regular	r, 0	r, h	r, 2h	r, 3h	r, 4h	r, 5h	r, 6h	r, 7h	r, 8h	r, 9h	r, 10h	r, 11h	0, 0	800	900
1	Overtime	v, 0	v, h	v, 2h	v, 3h	v, 4h	v, 5h	v, 6h	v, 7h	v, 8h	v, 9h	v, 10h	v, 11h	0, 0	320	400
2	Regular	X	r, 0	r, h	r, 2h	r, 3h	r, 4h	r, 5h	r, 6h	r, 7h	r, 8h	r, 9h	r, 10h	0, 0	760	600
2	Overtime	X	v, 0	v, h	v, 2h	v, 3h	v, 4h	v, 5h	v, 6h	v, 7h	v, 8h	v, 9h	v, 10h	0, 0	304	400
3	Regular	X	X	r, 0	r, h	r, 2h	r, 3h	r, 4h	r, 5h	r, 6h	r, 7h	r, 8h	r, 9h	0, 0	840	800
3	Overtime	X	X	v, 0	v, h	v, 2h	v, 3h	v, 4h	v, 5h	v, 6h	v, 7h	v, 8h	v, 9h	0, 0	336	420
4	Regular	X	X	X	r, 0	r, h	r, 2h	r, 3h	r, 4h	r, 5h	r, 6h	r, 7h	r, 8h	0, 0	880	700
4	Overtime	X	X	X	v, 0	v, h	v, 2h	v, 3h	v, 4h	v, 5h	v, 6h	v, 7h	v, 8h	0, 0	352	400
5	Regular	X	X	X	X	r, 0	r, h	r, 2h	r, 3h	r, 4h	r, 5h	r, 6h	r, 7h	0, 0	840	800
5	Overtime	X	X	X	X	v, 0	v, h	v, 2h	v, 3h	v, 4h	v, 5h	v, 6h	v, 7h	0, 0	336	350
6	Regular	X	X	X	X	X	r, 0	r, h	r, 2h	r, 3h	r, 4h	r, 5h	r, 6h	0, 0	800	770
6	Overtime	X	X	X	X	X	v, 0	v, h	v, 2h	v, 3h	v, 4h	v, 5h	v, 6h	0, 0	320	340
7	Regular	X	X	X	X	X	X	r, 0	r, h	r, 2h	r, 3h	r, 4h	r, 5h	0, 0	800	770
7	Overtime	X	X	X	X	X	X	v, 0	v, h	v, 2h	v, 3h	v, 4h	v, 5h	0, 0	320	340
8	Regular	X	X	X	X	X	X	X	r, 0	r, h	r, 2h	r, 3h	r, 4h	0, 0	760	700
8	Overtime	X	X	X	X	X	X	X	v, 0	v, h	v, 2h	v, 3h	v, 4h	0, 0	304	300
9	Regular	X	X	X	X	X	X	X	X	r, 0	r, h	r, 2h	r, 3h	0, 0	840	800
9	Overtime	X	X	X	X	X	X	X	X	v, 0	v, h	v, 2h	v, 3h	0, 0	336	350
10	Regular	X	X	X	X	X	X	X	X	X	r, 0	r, h	r, 2h	0, 0	880	820
10	Overtime	X	X	X	X	X	X	X	X	X	v, 0	v, h	v, 2h	0, 0	352	400
11	Regular	X	X	X	X	X	X	X	X	X	X	r, 0	r, h	0, 0	840	800
11	Overtime	X	X	X	X	X	X	X	X	X	X	v, 0	v, h	0, 0	336	350
12	Regular	X	X	X	X	X	X	X	X	X	X	X	r, 0	0, 0	800	750
12	Overtime	X	X	X	X	X	X	X	X	X	X	X	v, 0	0, 0	320	300
Demand:	Eco Booming	1100	500	1200	1300	1500	1000	800	500	1000	1200	1400	1400	976	13876	
	Eco Recession	1200	400	1500	1000	1600	900	900	450	900	1300	1500	1500	510		13660

According to Menipaz (1984), let r = 120, and v = 180, h = 3, and initial inventory for both economic situations be 100 (see Table 21.2). Then, using the TP MC² Program, running on the UNIX ULTRIX 4.2 operation system in a mainframe, the potentially optimal solutions $\{J_1, J_2, J_3, J_4\}$ are obtained as shown in Table 21.3, Table 21.4 and Table 21.5, respectively. The CPU time is 117.95 seconds because of the problem complexity (Shi and Yu 1992). Table 21.3 contains the supply-demand level weighting set $\Gamma(J_k)$, the criteria weighting set $\Lambda(J_k)$, and the total cost matrix $TC(J_k)$, k = 1, 2, 3, 4. Table 21.4 and Table 21.5 give the units of products to be produced, x_{ij}. Note that $TC(J_k)$ is a function of (γ, λ), where $\gamma_1 + \gamma_2 = 1$ and $\lambda_1 + \lambda_2 = 1$, $0 < \gamma_1, \gamma_2, \lambda_1, \lambda_2 \leq 1$.

Tables 21.3 Potentially Optimal Solutions

J_k	$\Gamma(J_k)$	$\Lambda(J_k)$	$TC(J_k)$
J_1	$0.71 \leq \gamma_1 \leq 1$	$0.05 \leq \lambda_1 \leq 1$	$\lambda\begin{pmatrix} 1713600 & 1796400 \\ 10116 & 14220 \end{pmatrix}\gamma$
J_2	$0.46 \leq \gamma_1 \leq 1$	$0.17 \leq \lambda_1 \leq 1$	$\lambda\begin{pmatrix} 1713600 & 1796400 \\ 15048 & 14520 \end{pmatrix}\gamma$
J_3	$0.60 \leq \gamma_1 \leq 1$	$0.08 \leq \lambda_1 \leq 0.09$	$\lambda\begin{pmatrix} 1730640 & 1786200 \\ 9660 & 15210 \end{pmatrix}\gamma$
J_4	$0.60 \leq \gamma_1 \leq 1$	$0 \leq \lambda_1 \leq 0.02$	$\lambda\begin{pmatrix} 1750800 & 1827100 \\ 8052 & 12570 \end{pmatrix}\gamma$

Table 21.4 Units of Products of Potentially Optimal Solutions J_1 and J_2

J_1	J_2
$x_{11} = 32\,\gamma_1 + 200\,\gamma_2$	$x_{19} = 100\,\gamma_1 + 100\,\gamma_2$
$x_{14} = 68\,\gamma_1 - 100\,\gamma_2$	$x_{21} = 780\,\gamma_1 + 800\,\gamma_2$
$x_{21} = 800\,\gamma_1 + 900\,\gamma_2$	$x_{22} = 20\,\gamma_1 + 100\,\gamma_2$
$x_{31} = 320\,\gamma_1 + 400\,\gamma_2$	$x_{31} = 268\,\gamma_1 + 100\,\gamma_2$
$x_{3,13} = 52\,\gamma_1 + 300\,\gamma_2$	$x_{42} = 176\,\gamma_1 - 100\,\gamma_2$
$x_{42} = 196\,\gamma_1$	$x_{43} = 192\,\gamma_1 + 350\,\gamma_2$
$x_{43} = 240\,\gamma_1 + 150\,\gamma_2$	$x_{44} = 392\,\gamma_1 + 350\,\gamma_2$
$x_{45} = 324\,\gamma_1 + 450\,\gamma_2$	$x_{52} = 304\,\gamma_1 + 400\,\gamma_2$
$x_{52} = 304\,\gamma_1 + 400\,\gamma_2$	$x_{63} = 840\,\gamma_1 + 800\,\gamma_2$
$x_{63} = 840\,\gamma_1 + 800\,\gamma_2$	$x_{73} = 168\,\gamma_1 + 350\,\gamma_2$
$x_{73} = 120\,\gamma_1 + 550\,\gamma_2$	$x_{7,13} = 168\,\gamma_1 + 70\,\gamma_2$
$x_{7,13} = 216\,\gamma_1 - 130\,\gamma_2$	$x_{84} = 556\,\gamma_1 + 250\,\gamma_2$
$x_{84} = 880\,\gamma_1 + 700\,\gamma_2$	$x_{85} = 324\,\gamma_1 + 450\,\gamma_2$
$x_{94} = 352\,\gamma_1 + 400\,\gamma_2$	$x_{94} = 352\,\gamma_1 + 450\,\gamma_2$
$x_{10,5} = 840\,\gamma_1 + 800\,\gamma_2$	$x_{10,5} = 840\,\gamma_1 + 800\,\gamma_2$
$x_{11,5} = 336\,\gamma_1 + 350\,\gamma_2$	$x_{11,5} = 336\,\gamma_1 + 350\,\gamma_2$
$x_{12,6} = 800\,\gamma_1 + 770\,\gamma_2$	$x_{12,6} = 800\,\gamma_1 + 770\,\gamma_2$
$x_{13,6} = 200\,\gamma_1 + 130\,\gamma_2$	$x_{13,6} = 200\,\gamma_1 + 130\,\gamma_2$
$x_{13,13} = 120\,\gamma_1 + 210\,\gamma_2$	$x_{13,13} = 120\,\gamma_1 + 210\,\gamma_2$
$x_{14,7} = 732\,\gamma_1 + 340\,\gamma_2$	$x_{14,7} = 604\,\gamma_1 + 620\,\gamma_2$
$x_{14,12} = 68\,\gamma_1 + 430\,\gamma_2$	$x_{14,8} = 196\,\gamma_1 + 150\,\gamma_2$
$x_{15,7} = 68\,\gamma_1 + 560\,\gamma_2$	$x_{15,7} = 196\,\gamma_1 + 280\,\gamma_2$
$x_{15,13} = 252\,\gamma_1 - 220\,\gamma_2$	$x_{15,13} = 124\,\gamma_1 + 60\,\gamma_2$
$x_{16,8} = 388\,\gamma_1 + 580\,\gamma_2$	$x_{16,9} = 216\,\gamma_1 - 130\,\gamma_2$
$x_{16,9} = 160\,\gamma_1 + 100\,\gamma_2$	$x_{16,10} = 320\,\gamma_1 + 480\,\gamma_2$
$x_{16,12} = 212\,\gamma_1 + 20\,\gamma_2$	$x_{16,11} = 224\,\gamma_1 + 350\,\gamma_2$
$x_{17,8} = 112\,\gamma_1 - 130\,\gamma_2$	$x_{17,8} = 304\,\gamma_1 + 300\,\gamma_2$
$x_{17,11} = 192\,\gamma_1 + 430\,\gamma_2$	$x_{18,9} = 684\,\gamma_1 + 930\,\gamma_2$
$x_{18,9} = 840\,\gamma_1 + 800\,\gamma_2$	$x_{18,12} = 156\,\gamma_1 - 130\,\gamma_2$
$x_{19,13} = 336\,\gamma_1 + 350\,\gamma_2$	$x_{19,13} = 336\,\gamma_1 + 350\,\gamma_2$
$x_{20,10} = 848\,\gamma_1 + 900\,\gamma_2$	$x_{20,10} = 880\,\gamma_1 + 820\,\gamma_2$
$x_{20,11} = 32\,\gamma_1 - 80\,\gamma_2$	$x_{21,12} = 352\,\gamma_1 + 400\,\gamma_2$
$x_{21,10} = 352\,\gamma_1 + 400\,\gamma_2$	$x_{22,11} = 840\,\gamma_1 + 800\,\gamma_2$
$x_{22,11} = 840\,\gamma_1 + 800\,\gamma_2$	$x_{23,11} = 336\,\gamma_1 + 350\,\gamma_2$
$x_{23,11} = 336\,\gamma_1 + 350\,\gamma_2$	$x_{24,12} = 800\,\gamma_1 + 750\,\gamma_2$
$x_{24,12} = 800\,\gamma_1 + 750\,\gamma_2$	$x_{25,12} = 92\,\gamma_1 + 480\,\gamma_2$
$x_{25,12} = 320\,\gamma_1 + 300\,\gamma_2$	$x_{25,13} = 228\,\gamma_1 - 180\,\gamma_2$

Table 21.5 Units of Products of Potentially Optimal Solutions J_3 and J_4

J_3	J_4
$x_{16} = 100\,\gamma_1 + 100\,\gamma_2$	$x_{1,13} = 100\,\gamma_1 + 100\,\gamma_2$
$x_{21} = 800\,\gamma_1 + 900\,\gamma_2$	$x_{21} = 800\,\gamma_1 + 900\,\gamma_2$
$x_{31} = 300\,\gamma_1 + 300\,\gamma_2$	$x_{31} = 300\,\gamma_1 + 300\,\gamma_2$
$x_{3,31} = 20\,\gamma_1 + 100\,\gamma_2$	$x_{3,31} = 20\,\gamma_1 + 100\,\gamma_2$
$x_{42} = 344\,\gamma_1 - 30\,\gamma_2$	$x_{42} = 344\,\gamma_1 - 30\,\gamma_2$
$x_{43} = 24\,\gamma_1 + 280\,\gamma_2$	$x_{43} = 24\,\gamma_1 + 280\,\gamma_2$
$x_{44} = 68\,\gamma_1 - 100\,\gamma_2$	$x_{44} = 68\,\gamma_1 - 100\,\gamma_2$
$x_{45} = 324\,\gamma_1 + 450\,\gamma_2$	$x_{45} = 324\,\gamma_1 + 450\,\gamma_2$
$x_{52} = 156\,\gamma_1 + 430\,\gamma_2$	$x_{52} = 156\,\gamma_1 + 430\,\gamma_2$
$x_{5,13} = 148\,\gamma_1 - 30\,\gamma_2$	$x_{5,13} = 148\,\gamma_1 - 30\,\gamma_2$
$x_{63} = 840\,\gamma_1 + 800\,\gamma_2$	$x_{63} = 840\,\gamma_1 + 800\,\gamma_2$
$x_{73} = 336\,\gamma_1 + 420\,\gamma_2$	$x_{73} = 336\,\gamma_1 + 420\,\gamma_2$
$x_{84} = 880\,\gamma_1 + 700\,\gamma_2$	$x_{84} = 880\,\gamma_1 + 700\,\gamma_2$
$x_{94} = 352\,\gamma_1 + 400\,\gamma_2$	$x_{94} = 352\,\gamma_1 + 400\,\gamma_2$
$x_{10,5} = 840\,\gamma_1 + 800\,\gamma_2$	$x_{10,5} = 840\,\gamma_1 + 800\,\gamma_2$
$x_{11,5} = 336\,\gamma_1 + 350\,\gamma_2$	$x_{11,5} = 336\,\gamma_1 + 350\,\gamma_2$
$x_{12,6} = 800\,\gamma_1 + 770\,\gamma_2$	$x_{12,6} = 800\,\gamma_1 + 770\,\gamma_2$
$x_{13,6} = 100\,\gamma_1 + 30\,\gamma_2$	$x_{13,6} = 200\,\gamma_1 + 130\,\gamma_2$
$x_{13,13} = 220\,\gamma_1 + 310\,\gamma_2$	$x_{13,13} = 120\,\gamma_1 + 210\,\gamma_2$
$x_{14,7} = 480\,\gamma_1 + 560\,\gamma_2$	$x_{14,7} = 480\,\gamma_1 + 560\,\gamma_2$
$x_{14,8} = 36\,\gamma_1 + 380\,\gamma_2$	$x_{14,13} = 320\,\gamma_1 + 210\,\gamma_2$
$x_{14,13} = 284\,\gamma_1 - 170\,\gamma_2$	$x_{15,7} = 320\,\gamma_1 + 340\,\gamma_2$
$x_{15,7} = 320\,\gamma_1 + 340\,\gamma_2$	$x_{16,8} = 500\,\gamma_1 + 450\,\gamma_2$
$x_{16,8} = 464\,\gamma_1 + 70\,\gamma_2$	$x_{16,10} = 72\,\gamma_1 + 280\,\gamma_2$
$x_{16,10} = 72\,\gamma_1 + 280\,\gamma_2$	$x_{16,11} = 188\,\gamma_1 - 30\,\gamma_2$
$x_{16,11} = 224\,\gamma_1 + 350\,\gamma_2$	$x_{17,11} = 36\,\gamma_1 + 380\,\gamma_2$
$x_{17,13} = 304\,\gamma_1 + 300\,\gamma_2$	$x_{17,13} = 268\,\gamma_1 - 80\,\gamma_2$
$x_{18,9} = 664\,\gamma_1 + 550\,\gamma_2$	$x_{18,9} = 664\,\gamma_1 + 550\,\gamma_2$
$x_{18,10} = 176\,\gamma_1 + 250\,\gamma_2$	$x_{18,10} = 176\,\gamma_1 + 250\,\gamma_2$
$x_{19,9} = 336\,\gamma_1 + 350\,\gamma_2$	$x_{19,9} = 336\,\gamma_1 + 350\,\gamma_2$
$x_{20,10} = 800\,\gamma_1 + 370\,\gamma_2$	$x_{20,10} = 800\,\gamma_1 + 370\,\gamma_2$
$x_{20,12} = 280\,\gamma_1 + 450\,\gamma_2$	$x_{20,12} = 280\,\gamma_1 + 450\,\gamma_2$
$x_{21,10} = 352\,\gamma_1 + 400\,\gamma_2$	$x_{21,10} = 352\,\gamma_1 + 400\,\gamma_2$
$x_{22,11} = 840\,\gamma_1 + 800\,\gamma_2$	$x_{22,11} = 840\,\gamma_1 + 800\,\gamma_2$
$x_{23,11} = 336\,\gamma_1 + 350\,\gamma_2$	$x_{24,12} = 800\,\gamma_1 + 750\,\gamma_2$
$x_{24,12} = 336\,\gamma_1 + 350\,\gamma_2$	$x_{24,12} = 336\,\gamma_1 + 350\,\gamma_2$
$x_{25,12} = 320\,\gamma_1 + 300\,\gamma_2$	$x_{25,12} = 320\,\gamma_1 + 300\,\gamma_2$

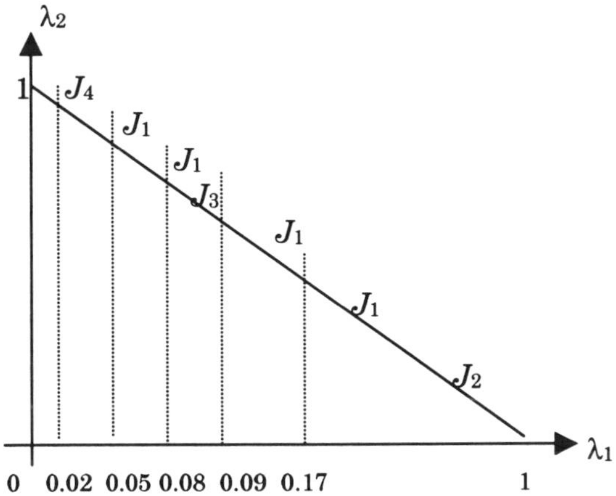

Figure 21.1 Production Cost(λ_1) vs. Inventory Cost(λ_2)

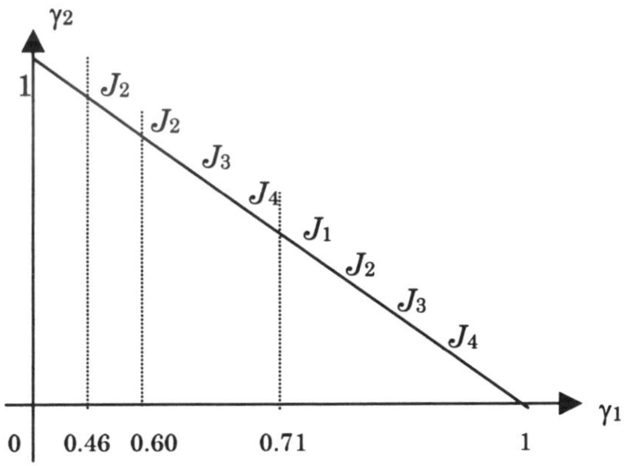

Figure 21.2 Economic Booming(γ_1) vs. Economic Recession(γ_2)

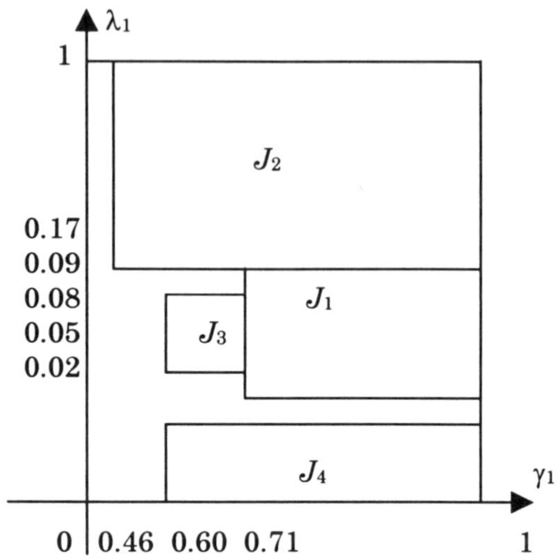

Figure 21.3 Production Cost(λ_1) vs. Economic Booming(γ_1)

Optimal trade-offs of the production planning model with multiple factors consist of three components: (i) trade-offs between the multiple objectives, production and inventory costs; (ii) trade-offs between the multiple supply-demand levels, economy booming and recession; and (iii) trade-offs between multiple objectives and multiple supply-demand levels. The trade-offs between the multiple objectives imply that all possible compromises of production and inventory costs are determined by locating all possible weights of importance of these two objectives. Similarly, the trade-offs between the multiple supply-demand levels imply that all possible economic situations which may occur between the economy booming and recession can be expressed by locating all possible weights of these two situations. The trade-offs between multiple objectives and multiple supply-demand levels show the effects on the APPFM problem by certain combinations of the objective and supply-demand level. These three cases of optimal trade-offs can be analyzed through the potentially optimal solutions of the APPFM model. In the following, three trade-off cases and management implications of the corresponding optimal production planning are explored in detail by using the APPFM example.

Because there are four potentially optimal solutions $\{J_1, J_2, J_3, J_4\}$ in the example, all possible optimal trade-offs should be studied in terms of each J. For the trade-offs between two objectives, production cost (λ_1) and inventory cost (λ_2):

(i) When $0 < \lambda_1 \leq 0.02$, J_4 is the optimal production plan.

(ii) When $0.02 < \lambda_1 < 0.05$, there is no optimal production plan.

(iii) When $0.05 \leq \lambda_1 \leq 0.08$, J_1 is the optimal production plan.

(iv) When $0.08 < \lambda_1 \leq 0.09$, both J_1 and J_3 are the optimal production plans.

(v) When $0.09 < \lambda_1 \leq 0.17$, J_1 is the optimal production plan.

(vi) When $0.17 < \lambda_1 \leq 1$, both J_1 and J_2 are the optimal production plans.

The graphical representation of objective trade-offs is shown in Figure 21.1. From Figure 21.1, any weighting values of λ_1 and λ_2 taken from the dark segment guarantee that the utility values of both production cost and inventory cost are minimized.

For the trade-offs among the economy booming (γ_1) and recession (γ_2):

(i) When $0 < \gamma_1 \leq 0.46$, there is no feasible production plan.

(ii) When $0.46 < \gamma_1 \leq 0.60$, J_2 is the optimal production plan.

(iii) When $0.60 < \gamma_1 \leq 0.71$, J_2, J_3 and J_4 are the optimal production plans.

(iv) When $0.71 < \gamma_1 \leq 1$, all J_1, J_2, J_3 and J_4 are the optimal production plans.

These trade-offs can easily be verified by Tables 21.4-21.5. Their graphical representations are shown in Figure 21.2. Whenever the weighting values of γ_1 and γ_2 are taken from a segment in Figure 21.2, the corresponding possible economic situation will result in either the optimal production plan or no feasible production plan. Finally, there are four trade-off situations between two objectives and two supply-demand levels in this problem. For example, production cost (λ_1) vs. economy booming (γ_1) has been illustrated in Table 21.3. Whenever (γ_1, λ_1) are taken from $\Gamma(J_k) \times \Lambda(J_k)$, k = 1, 2, 3, 4, there is an optimal production plan. Otherwise, there is no optimal production plan. This situation is shown in Figure 21.3. The other situations: production cost (λ_1) vs. the economy recession (γ_2), inventory cost (λ_2) vs. the economy booming (γ_1), and inventory cost (λ_2) vs. the economy recession (γ_2), can be similarly studied.

It is worth noting some important implications for production planning decision makers from the above trade-off analysis. First, the APPMF model has a capability of systematically locating all possible optimal aggregate production policies through the optimal trade-offs of multiple objectives and multiple supply-demand levels (Gardner and Dannenbring 1979). Since the set of all possible optimal production plans found by this model describes every possible decision situation, the optimal production plans obtained from any of linear programming, goal programming and MC linear programming models is included as a subset.

Second, the APPMF method facilitates decision makers' participation that may make a positive management achievement of organizational goals. The model can effectively aid coordination and synthesis of multiple conflicting objectives in the designing process of production planning. These objectives may be measured differently by a number of decision participants.

Third, in most multiple criteria solution techniques, including the goal programming approach, if the decision makers are not satisfied with the optimal solution obtained by using their preferred weights of importance for the multiple objectives, then an iterative process has to be conducted for incorporating the decision makers' new preference on the weights and finding new solutions until decision makers are satisfied. However, the APPMF model eliminates such a time and cost consuming iterative process since it already considers all possible optimal solutions with respect to changes in parameter (γ, λ). Whenever decision makers want to change their preference on weights, the corresponding optimal aggregate production plan can immediately be identified from the results like Table 21.3. This model, in turn, allows the performance evaluation of the optimal aggregate production plans.

Fourth, the optimal aggregate production plans obtained by the APPMF model have a two-fold significance in terms of decision characteristics. If the problem is viewed as a deterministic decision problem, whenever the preferred weighting values of objectives and supply-demand levels are known, the resulting optimal solutions can be identified from the potentially optimal solutions of the model. Then, the corresponding optimal aggregate production plans can be adopted to handle the changeable business situation. If the problem is viewed as a probabilistic decision problem, it involves the assessment of the likelihood of (γ, λ) to occur at the various points of the range. With proper assumptions, the uncertainty may be represented by random variables with some known probability distribution. A number of known criteria, such as maximizing expected payoff, minimizing the variance of the payoff, maximizing the probability of achieving a targeted payoff, stochastic dominance, probability dominance, and mean-variance dominance can be used to choose the optimal aggregate production plans (Lee, Shi and Yu 1990).

21.3 Further Remarks

By using the framework of MC^2 linear programming, we have proposed a multiple factor aggregate production planning model to solve production planning problems of manufacturing firms. This model can provide a systematic and comprehensive scenario about all possible optimal aggregate production plans depending on multiple objective and multiple supply-demand levels. The trade-offs of optimal aggregate production plans offer a broad basis for managers to flexibly implement the optimal production planning strategy and cope with various business competitions. This method also aids multi-situation optimization, decision makers' participation and performance evaluation. Furthermore, the TPMC2 computer software of Haase and Shi (1994) can be used to solve not only production planning problems, but also any application problems with a "transportation model" feature.

There are a number of research problems that remain to be explored. For example, Haase (1994) explored different modifications of the algorithm in Shi (1995). With these modifications, the TP MC^2 computer software can be improved so that few pivoting iterations are required to locate the potential solutions. This, in turn, will greatly reduce the computation time.

Another important research problem is the study of contingency plans for the potential solutions of the multiple factor aggregate production planning model. Recall from Section 21.3, when $0.02 < \gamma_1 < 0.05$, there is no optimal production plan (see Figure 21.1); and when $0 < \lambda_1 \leq 0.46$, there is no feasible production plan (see Figure 21.2). According to Chapters 9-12, these cases indicate that the possible contingency plans should be constructed to recover the feasibility and/or optimality. Similarly, we can construct contingency plans for the infeasibility of Figure 21.3.

Optimal trade-offs of this paper result from the fixed multiple capacity-demand levels. If the capacity-demand level is assumed as a random variable, then our model can be extended to eliminate "optimal trade-off" policies. In this case, Zeleny's (1995) method can be applied to study "trade-off-free" management in aggregate planning problems. These problems are on-going projects and deserve careful studies. We will report any significant results in the near future.

Chapter 22
AGRICULTURAL POLICY MAKING

Agricultural policy making has been recognized as a complicated decision problem involving many decision makers. In the United Sates, the agricultural market system can be described by three types of decision variables (Erickson and House 1985 and Erickson 1985). The first type is policy instruments which represent the preferences of the policy makers, such as the Federal Government and Congress. The second type is called behavioral variables which are controlled by individual producers and consumers. The third type is impact (or target) variables which express the common interest and relationship between the policy makers and individual producers or consumers.

A number of researchers have used mathematical programming approaches to solve agricultural policy making problems. Candler and Norton (1977) built a two-level mathematical programming model for achieving a competitive agricultural market equilibrium. In this model, the lower level submodel handles the relationship between the policy makers and producers or consumers. Then, the upper level submodel optimizes policy targets over the best choices of the lower level. Due to the complexity of computation, it is difficult to determine an optimal solution (or policy) for this model. To cope with this difficulty, Erickson and House (1985) developed a two-level approximation method. This approach uses an experimental design to simulate the data of the lower level submodel. The data is then used to construct an upper level submodel with multiple policy targets.

However, the optimal solution illustration of Erickson and House (1985) did not show how to determine a non-dominated set in terms of all possible combinations of weights for multiple policy targets. In addition, both models implicitly assume that the future economic environment is given as a *priori*. Thus, they have limited usefulness when dealing with future economic changes. For example, if the foreign demand for American agricultural products increases in the future, the budget constraint level and policy instrument bounds may increase. Otherwise, they may decrease. The optimal policy making solutions for different future economic environments can be different. Neither previous model addresses how to deal with such future economic changes.

By considering the multiple upcoming economic environments, this chapter extends the mathematical programming framework into a multiple criteria (policy targets) and multiple economic situation (MCMES) model for agricultural policy making (Shi and Lin 2000). This generalized model can be readily computed by recently developed computer software called the "large-scale MC^2 program" (Hao and Shi 1996). In comparison with the previous mathematical programming approaches to agricultural policy making, this model provides policy makers with a more systematic and comprehensive scenario of all possible optimal values depending on multiple criteria and multiple economic situations. The trade-offs of all possible optimal values can be used as a basis for agricultural policy makers to make high-quality decisions in selecting their optimal pricing policy for a future agricultural commodity program.

Section 22.1 will discuss the nature of agricultural policy making and the previous models for agricultural policy making. Section 22.2 will elaborate details about the proposed model and the implementation of the model. Section 22.3 will present the trade-off analysis and discuss the economic implication of the model. An example adopted from the 1985 U.S. Farm Bill is also used to show the applicability of the proposed model. Section 22.4 concludes the paper with some remarks.

22.1 Agricultural Policy Making

Agricultural activity always plays a key role in Human's life. With a large amount of farming land and advanced technology, the major agricultural goods production countries, such as the U. S. A., Canada and Australia, can produce a huge amount of food. However, because of uncertainty in both demand and supply in the agricultural goods market, these countries face a great challenge in making an effective agricultural policy. The problem of U. S. agricultural policy making is very complicated in its nature. We may use the following characteristics to describe these complications:

(i) The problem involves multiple criteria. These criteria include budget, production level, farm income, consumer prices, inflation rate, natural resources and environmental quality. Some criteria are conflicting, while some are complementary.

(ii) There are multiple decision makers participating in making decisions. In general, we can classify three interest groups. One is the policy makers who are respective officers of the Federal Government or members of Congress. The second group is farmers and the third group is both domestic and international consumers. The process of policy making involves a complex bargaining while one group cannot absolutely dominate another group.

(iii) The problem can be considered as a two-level decision making process. The first level decisions come from farmers who represent the agricultural sector. Their decisions can influence the second level decisions of government policy makers. The government policy makers give policies or regulations to farmers, but do not directly engage in any agricultural activity.

(iv) The future agricultural market is uncertain. Instead of finding an optimal policy for the agricultural market under a certain or given economic situation, the problem formulation should identify a set of optimal policies which can cope with various economic situations. In other words, the formulation should have a predicating capability to handle a changeable future agricultural market.

Samuelson (1952) initiated the linear programming approach to problems of competitive market equilibrium. Since then, a number of researchers have worked on formulating and solving large-scale and multi-product linear and nonlinear programming models for market equilibrium problems (McCarl and Spreen 1980). Candler and Norton (1977) proposed a multi-level mathematical programming, known as USMP, for agricultural economics. This model can be potentially used to deal with the second and third characteristics since it does not incorporate multiple criteria consideration and its solution is obtained from a given economic environment. Therefore, this model has limitations in both discussing possible trade-offs of the multiple criteria mentioned in the first characteristic and preparing the optimal policies for different future economic situation mentioned in the fourth characteristic. In addition, due to the complexity of computing the local optima, applications of this model in large-scale agricultural problems are prohibited (Candler *et al.* 1981).

Erickson and House (1985) and Erickson (1985) developed a two-level mathematical programming model with multiple criteria. The advantage of this model is that it uses a trade-off analysis of involved criteria to show some possible alternative agricultural policies. To cope with the computing complexity, Erickson and House (1984) proposed a heuristic approach to estimating the optimal agricultural policies. At the lower level of this approach, a factor-based experimental design is used to estimate the functional relationship (or equations) between policy instruments and target variables given a predetermined system of behavioral variables. Then, the resulting data of the lower level is used to build an upper level multiple linear programming model to find the non-dominated agricultural policies expressed by the policy instruments. In order to demonstrate the applicability of the model, Erickson and House (1985) identified a number of alternative agricultural policies for the 1985 U. S. Farm Bill.

Even though this model advanced the mathematical modeling tools for agricultural policy making, it still had two shortcomings. First, because the economic situation of agricultural markets is given as a priori, the model cannot respond to the changeable future agricultural markets as mentioned in the fourth characteristic. Second, although Erickson (1985) emphasized that the upper level model would be resolved for all possible combinations of weights of multiple criteria in the non-dominated policies, the example in Erickson and House (1985) only showed a few alternative non-dominated policies under some given combinations of weights of multiple criteria. Thus, the mechanism of solving the upper level model was incomplete. The cause of these shortcomings does not come from the strategic thinking of modeling, but from the technical development of mathematical modeling tools.

The purpose of this chapter is to overcome the disadvantages of the above mathematical programming approaches to problems of agricultural policy making. In the next section, a technical development of mathematical programming will be reviewed. Then, the MC^2 linear programming will be used to formulate a multiple criteria and multiple economic situation (MCMES) model for agricultural policy making. This proposed model intends to incorporate all four above mentioned characteristics into the process of identifying the best agricultural policy.

22.2 Multiple Criteria And Multiple Economic Situation Model

22.2.1 Formulation of MCMES Model

To show the generality of the proposed multiple criteria and multiple economic situation (MCMES) model, we assume that the future agricultural market is "uncertain". This uncertainty can be predictably generated by linear combinations of extreme economic situations, such as "booming", "normal" and "recession." Let $\mathbf{x}_0 \in \mathbf{R}^q$ be the vector of multiple criteria (impacts, or targets), $\mathbf{x}_1 \in \mathbf{R}^m$ be the vector of individual behaviors, and $\mathbf{x}_2 \in \mathbf{R}^n$ be the vector of policy instruments. As discussed before, $\mathbf{x}_0$ represents the goals of policy makers; $\mathbf{x}_1$ is controlled by producers; and $\mathbf{x}_2$ represents the preferences of the government. Since the policy makers involve the data of economic development tendency, we assume that the policy makers are in charge of predicting the changes of the future agricultural market and producers believe the prediction.

A basic framework of the two-level agricultural policy making can be stated as follows (Erickson 1985):

The Lower Level: Determine the relationship between the policy makers and the producers in terms of the goals of policy makers,

$$\mathbf{x}_0 = f(\mathbf{x}_1, \mathbf{x}_2) \tag{22.1}$$

The Upper Level: Optimize the utility of the multiple criteria X_0 which is an explicit function of the policy instruments,

$$Max \quad \mathbf{w}^t \mathbf{x}_0 = \mathbf{w}^t g(\mathbf{x}_2(\mathbf{x}_1)) \tag{22.2}$$

where $\mathbf{w}^t \in \mathbf{R}^q$ is a given weight vector and $g(\mathbf{x}_2(\mathbf{x}_1))$ is the resulting function of the lower level.

Using this basic framework, we propose a multiple criteria and multiple economic situation (MCMES) model for agricultural policy making as:

The Lower Level: Adopt the factorial design of Erickon and House (1984) to determine a linear relationship between the multiple criteria ($\mathbf{x}_0$) and the policy makers ($\mathbf{x}_2$) for a predetermined system of relationships for the producers ($\mathbf{x}_1$). A simulation, if necessary, should be used to conduct this analysis. In general, let x_{2j}, j = 1, . . . , n be the jth policy instrument such as loan rates on wheat. Let $\mathbf{e}_i$, i = 1, . . . , q, be the variance of the ith criterion (target). Then, a factorial design of the lower level formulation of the MCMES model can be:

$$\begin{aligned} x_{01} &= a_{11} x_{21} + \dots + a_{1n} x_{2n} + c_1 e_1 \\ x_{02} &= a_{21} x_{21} + \dots + a_{2n} x_{2n} + c_2 e_2 \\ &\vdots \\ x_{0q} &= a_{q1} x_{21} + \dots + a_{qn} x_{2n} + c_q e_q \end{aligned} \tag{22.3}$$

where $a_{ij} = r(x_{0i}(\mathbf{x}_1), x_{2j}(\mathbf{x}_1))$, $i = 1, \ldots, q$; $j = 1, \ldots, n$, are the coefficients that measure the effect of a unit change in x_{2j} on the impact variable x_{0i}, and $\mathbf{c}_i$ is the coefficient of $\mathbf{e}_i$. The values of a_{ij}, $\mathbf{c}_i$ and $\mathbf{e}_i$ have to be determined according to some analysis, such as regression or simulation through the influence of the behavioral variables $\mathbf{x}_1$. Details of such a factor analysis can be found in Rummel (1970). As an alternative, this lower model also can be run by a mathematical programming model (Erickson 1985).

The Upper Level: Given the data of a_{ij}, c_i and e_i from the lower level model, an upper level formulation is built as:

$$\begin{aligned} &Max \quad \lambda^t (x_{01}, x_{02}, \ldots, x_{0q}) \\ &s.t. \quad B_1 x_{21} + \ldots + B_n x_{2n} \le (b_1, b_2, b_3)\,\gamma \qquad (22.4) \\ &\qquad\; x_{2j} \ge 0,\ j = 1, \ldots, n, \end{aligned}$$

where $\lambda = (\lambda_1, \lambda_2, \ldots, \lambda_q) \in \mathbf{R}^q$ and $\gamma = (\gamma_1, \gamma_2, \gamma_3) \in \mathbf{R}^3$ are the *criteria weights* and the *constraint level weights*, respectively. Both (γ, λ) exist, but are unknown at the time of policy making. Vectors $B_j \in \mathbf{R}^k$ are the effects of policy instruments x_{2j} on the constraints. Vectors $b_1, b_2, b_3 \in \mathbf{R}^k$ are the predicted constraint levels or policy instrument bounds for the "booming", "normal" and "recession" future agricultural markets, respectively.

Note that the way of handling the uncertainty of future agricultural markets in the above MCMES model is different from that of well-known stochastic models. In the stochastic models (Kall 1976), the uncertainty is dealt with by a given probability distribution prior to solving the problem. In the MCMES model, the uncertainty is systematically generated by parameters (λ, γ) without giving a probability distribution. After the model is solved, the optimal expected policy of any suitable distribution is computed or the optimal policy for any preferred values of (λ, γ) is chosen. Thus, the approach to the uncertainty by the MCMES model is completely new and more general than that of the stochastic models (see the implication section for more details).

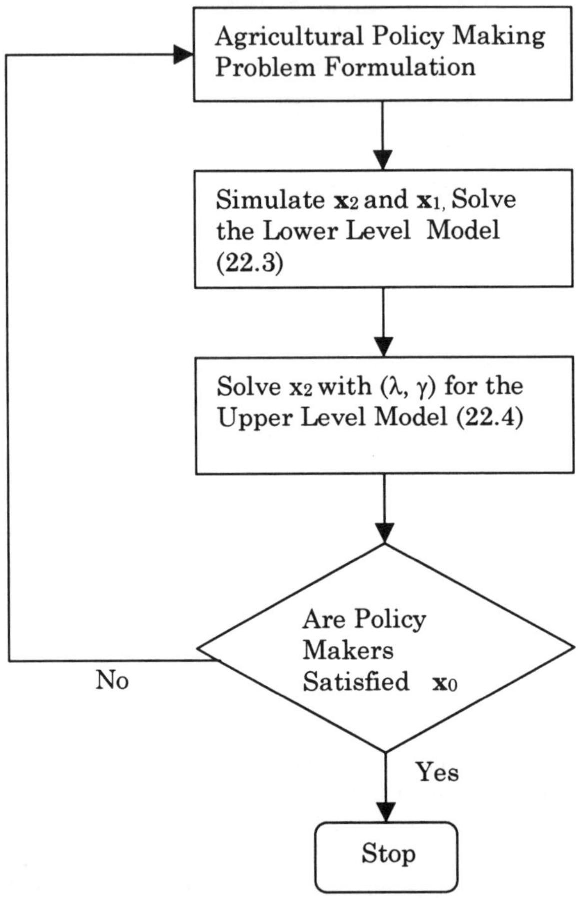

Figure 22.1 An Implementation Procedure of MCMES Model

An implementation procedure of the above MCMES model can be described as follows (see Figure 22.1):

Procedure 22.1

Step 1. Given a problem of agricultural policy making, identify the three groups of variables: $\mathbf{x}_0$ (policy goals), $\mathbf{x}_2$ (policy instruments) and $\mathbf{x}_1$ (producer behaviors).

Step 2. Build the lower level model (22.3) and simulate the relationship between $\mathbf{x}_0$ and $\mathbf{x}_2$ in terms of a given system of $\mathbf{x}_1$ to generate the data a_{ij}, c_i and e_i.

Step 3. Formulate the upper level model (22.4) and use the software of Hao and Shi (1996) to identify a set of $\{\mathbf{x}_0\}$ with all possible ranges of (λ, γ). This provides a comprehensive scenario about all possible optimal values depending on multiple criteria (λ) and multiple economic situations (γ).

Step 4. Present the results of Step 3 to the policy makers. If they are satisfied with the results, the decision making process is complete. Otherwise, interact with the policy makers and the producers to modify the original data of the problem according to their preferences and return to Step 1 again.

22.2.2 Modeling Implications

Using the MCMES model, important policy implications can be obtained through "trade-off" analyses.

The trade-offs related to the MCMES model consist of three components: (i) trade-offs based on multiple criteria, (ii) trade-offs based on multiple economic situations, and (iii) trade-offs between multiple criteria and multiple economic situations. The trade-offs based on multiple criteria imply that all possible optimal compromises of the multiple criteria are determined by locating all possible weights of importance of these criteria (or targets). The "trade-offs" based on multiple economic situations means that all possible economic situations are predicted by generating all possible weights of importance of multiple extreme economic situations. The trade-offs between multiple criteria and multiple economic situations measure the effects on the agricultural policy making problem by the interaction of the criteria and economic situations. These three cases of the optimal trade-offs can be analyzed through the potentially optimal solutions of the MCMES model.

When comparing a single criterion mathematical programming model of Candler and Norton (1977) and multiple criteria model of Erickson and House (1985) with the MCMES model, five major advantages of the latter model become clear:

First, the MCMES model has the capability of systematically locating all possible optimal policies through the optimal trade-offs of multiple criteria and multiple economic situations. Since the set of all possible optimal policies found by this model describes every possible economic situation within the model framework, the optimal policy obtained from either Candler and Norton (1977)'s model or Erickson and House (1985)s' model is included in the subset as a special case.

Second, in most multiple criteria solution techniques, including goal programming, if policy makers are not satisfied with the optimal solution obtained by using their preferred weights of importance for the multiple criteria, then an iterative process must be conducted for incorporating the policy makers' new preference on the weights. New solutions must be tried until the policy makers are satisfied. However, the MCMES model eliminates this time and cost consuming iterative process since it already considers all possible optimal solutions with respect to the changes of parameters (λ, γ). Whenever the policy makers want to change their preferences on the weights, the corresponding optimal policies can be immediately identified from the results (see Table 22.4). This model, in turn, allows the performance evaluation of optimal policies.

Third, if the policy makers are not satisfied with the optimal policy scenario of the MCMES formulation, the procedure shown in Figure 1 allows the analysts to interact with the policy makers to rebuild and resolve the problem. This process will continue until the policy makers agree with the solution.

Fourth, the optimal policies obtained by the MCMES model have a two-fold significance in terms of decision making characteristics. If the agricultural policy making problem is viewed as a deterministic decision problem, whenever the preferred weighting values of criteria and economic situation are known, the resulting optimal solutions can be identified from the potentially optimal solutions of the model. Then, the corresponding optimal policy can be adopted to handle the problem. If the agricultural policy making problem is viewed as a probabilistic decision problem, it involves the assessment of the likelihood of (λ, γ) to occur at the various points of the range. With proper assumptions, the uncertainty may be represented by random variables with some known probability distribution. A number of known criteria, such as maximizing expected payoff, minimizing the variance of the payoff, maximin payoff, maximizing the probability of achieving a targeted payoff, stochastic dominance, probability dominance, and mean-variance dominance can be chosen to determine the optimal policies (Shi 1991).

Finally, using the computer software developed by Hao and Shi (1996) makes the MCMES model capable of dealing with a large and complex agricultural policy system.

In summary, the MCMES model fosters feasibility in designing optimal agricultural policies to cope with all possible changes of criteria and economic situations. This model is a useful aid for policy makers who try to understand and deal with current or future agricultural policy making problems.

22.3 Illustrative Example

The example of the 1985 U. S. Farm Bill in Erickson and House (1985) is adopted to illustrate the application of the proposed MCMES model.

Given the problem of U. S. agricultural policy making, we conduct Procedure 1 as follows:

Step 1. Define the policy instrument variables by

x_{21} = the loan rates on wheat ($/bushel);
x_{22} = the loan rates on corn ($/bushel);
x_{23} = the supply reduction programs (mixes of target prices and land set-aside) for wheat ($/bushel);
x_{24} = the supply reduction programs for corn ($/bushel);
x_{25} = the export demand levels for wheat (percentage);
x_{26} = the export demand levels for corn (percentage); and
x_{27} = the interest rates and exchange rates (percentage).

The data of original unit policy instrument magnitudes was given in Erickson and House (1985). For the sake of convenience, it is reprinted as Table 22.1.

Table 22.1 Data of Original Policy Instrument Magnitudes

Level	x_{21}	x_{22}	x_{23}	x_{24}	x_{25}	x_{26}	x_{27}
High:	3.60	2.35	3.88	2.47	115	115	250
Low:	1.42	1.57	2.56	1.65	85	85	150

Step 2. Define the criteria (target) variables by

x_{01} = the budget cost (million \$);
x_{02} = the net farm revenue (million \$);
x_{03} = the net export value (million \$);
x_{04} = the average wage (\$/hour); and
x_{05} = the land cropped (million acres).

According to Erickson and House (1985), for the lower level model (22.3), the relationship between $\mathbf{x}_0 = (x_{01}, x_{02}, x_{03}, x_{04}, x_{05})$ and $\mathbf{x}_2 = (x_{21}, \ldots, x_{27})$ in terms of a given system of $\mathbf{x}_1$ can be estimated by the equations:

$$x_{0i} = a_{i1}\, x_{21} + \ldots + a_{i7}\, x_{27} + c_i\, , \; i = 1, \ldots, 5.$$

where the data a_{ij} and c_i represent the given system of behavioral variables $\mathbf{x}_1$. One of the simulation results for a_{ij} and c_i in Erickson and House (1985) is given in Table 22.2.

Table 22.2 Simulation Results of Multiple Criteria Values

Criteria:	x_{21}	x_{22}	x_{23}	x_{24}	x_{25}	x_{26}	x_{27}	c_i
x_{01}	770	603	443.99	838.38	-156.11	-468.09	401.67	3216.58
x_{02}	-154.3	161.98	-108.14	-31.8	-66.13	1061.3	-1207.83	24328.35
x_{03}	192.96	47.63	-55.38	16.06	640.17	1986.03	2022.77	13870.14
x_{04}	0.01	0	0	0.01	0	0.01	-0.02	3.02
x_{05}	3.85	1.52	2.96	2.82	2.87	2.49	-4.42	332.92

Step 3. For illustrative purposes, suppose that the policy makers use the "booming" and "recession" to generate the future agricultural markets. The predicted design units of the upper bounds for policy instruments are given as in Table 22.3, whose measurement scale differs from the original data of Table 22.1 (Erickson and House 1985).

Table 22.3 Predicted Design Units of Policy Instruments

Market	x_{21}	x_{22}	x_{23}	x_{24}	x_{25}	x_{26}	x_{27}
Booming:	1.80	1.80	2.00	2.50	2.00	1.50	1.70
Recession:	1.50	1.00	1.50	1.50	1.00	1.00	1.00

From Tables 22.2 and 22.3, an upper level formulation of the MCMES model (22.4) can be stated as:

$$\begin{aligned}
Max\quad & \lambda_1(770x_{21} + 603x_{22} + 443.99x_{23} + 838.38x_{24} - 156.11x_{25} - 468.09x_{26} + 401.67x_{27} + 3216.58) + \\
& \lambda_2(-154.3x_{21} + 161.98x_{22} - 108.14x_{23} - 31.8x_{24} - 66.13x_{25} + 1061.3x_{26} - 1207.83x_{27} + 24328.35) + \\
& \lambda_3(192.96x_{21} + 47.63x_{22} - 55.38x_{23} + 16.06x_{24} + 640.17x_{25} + 1986.03x_{26} + 2022.77x_{27} + 13870.14) + \\
& \lambda_4(0.01x_{21} + 0.01x_{24} + 0.01x_{26} - 0.02x_{27} + 3.02) + \\
& \lambda_5(3.85x_{21} + 1.52x_{22} + 2.96x_{23} + 2.82x_{24} + 2.87x_{25} + 2.49x_{26} - 4.42x_{27} + 332.92)
\end{aligned}$$

$$\begin{aligned}
s.t.\quad & x_{21} \le 1.5\,\gamma_1 + 1.8\,\gamma_2 \\
& x_{22} \le 1.5\,\gamma_1 + 1.8\,\gamma_2 \\
& x_{23} \le 1.5\,\gamma_1 + 2.0\,\gamma_2 \\
& x_{24} \le 1.5\,\gamma_1 + 2.5\,\gamma_2 \\
& x_{25} \le 1.0\,\gamma_1 + 2.0\,\gamma_2 \\
& x_{26} \le 1.0\,\gamma_1 + 1.5\,\gamma_2 \\
& x_{27} \le 1.0\,\gamma_1 + 1.7\,\gamma_2 \\
& x_{2j} \ge 0,\ j = 1, \ldots, 7.
\end{aligned}$$

The above problem has five criteria and two constraint levels. Let $\gamma_1 + \gamma_2 = 1$ and $\lambda_1 + \lambda_2 + \lambda_3 + \lambda_4 + \lambda_5 = 1$. Using the computer software of Hao and Shi (1996) on a PC, eight potentially optimal solutions $\{J_1, \ldots, J_8\}$ are identified in 21 seconds. The union of (λ, γ) ranges associated with these solutions covers all possible values of (λ, γ).

For the sake of space, we only list the potentially optimal solution $J_1(x_{2j}) = (x_{21}, x_{22}, x_{23}, x_{24})$ with the data regarding the constraint level weight set $\Gamma(J_1)$, the criteria weight set $\Lambda(J_1)$ and the objective matrix $V(J_1)$ in Table 22.4.

Table 22.4 Potentially Optimal Solution J_1

$\Gamma(J_1)$	$\Lambda(J_1)$	$V(J_1)$
$\gamma_1 + \gamma_2 = 1$ $\gamma_1, \gamma_2 \geq 0$	$\lambda_1 + \lambda_2 + \lambda_3 + \lambda_4 + \lambda_5 = 1$	$\lambda \begin{pmatrix} 7199.33 & 8671.54 \\ 61218.68 & 62483.23 \\ 14172.05 & 14232.59 \\ 3.05 & 3.06 \\ 349.64 & 355.55 \end{pmatrix} \gamma$
	$770\lambda_1-154\lambda_2+\lambda_3+0.01\lambda_4+3.8\lambda_5 \geq 0$	
	$443\lambda_1-108\lambda_2+2.96\lambda_3 \geq 0$	
	$838\lambda_1-32\lambda_2+0.01\lambda_4+2.8\lambda_5 \geq 0$	
	$156\lambda_1+66\lambda_2-2.8\lambda_5 \geq 0$	
	$468\lambda_1-1060\lambda_2-0.01\lambda_4-2.5\lambda_5 \geq 0$	
	$-402\lambda_1+1207\lambda_2+\lambda_3+0.02\lambda_4+4.42\lambda_5 \geq 0$	
	$\lambda_1, \lambda_2, \lambda_3, \lambda_4, \lambda_5 \geq 0$	

Now we use the potentially optimal solutions $\{J_1, \ldots, J_8\}$ to demonstrate the trade-off analysis in terms of weights and economic implications. Because there are five criteria: ($x_{01}, \ldots x_{05}$), we can consider 10 kinds of any two criteria trade-offs. For illustrative purposes, let us consider the following two criteria trade-offs:

(i) the budget cost (λ_1) versus the net farm revenue (λ_2): Let the other criteria weights $\lambda_3 = \lambda_4 = \lambda_5 = 0$. If $\lambda_1 + \lambda_2 = 1$, the relationship of the budget cost (λ_1) and the net farm revenue (λ_2) is shown in Figure 22.2. The purpose of this figure is identical to that of Figure 22.2 in Erickson and House (1985). Because all the optimal solutions regarding the budget cost and the net farm revenue over either economic booming and recession can be generated by weights (λ_1, λ_2) in Figure 22.2, the solutions of Erickson and House (1985) are special cases. For example, if $\gamma_1 = 1$ and $\gamma_2 = 0$, then $x_{21} = x_{22} = x_{23} = x_{24} = 1.5$ (design units) with the objective function $7199.33\lambda_1 + 61218.68\lambda_2$, which is a number of solutions shown in Table 22.4 of Erickson and House (1985).

(ii) the budget cost (λ_1) versus the net export value (λ_3) with $\lambda_1 + \lambda_3 = 1$ can be represented by Figure 22.3. Note that when $0 \leq \lambda_1 \leq 0.03$, the potentially optimal solutions J_1, J_2, J_3, J_4, and J_5 are alternatives, while J_6, J_7, and J_8 are alternatives for $0.03 < \lambda_1 \leq 1$.

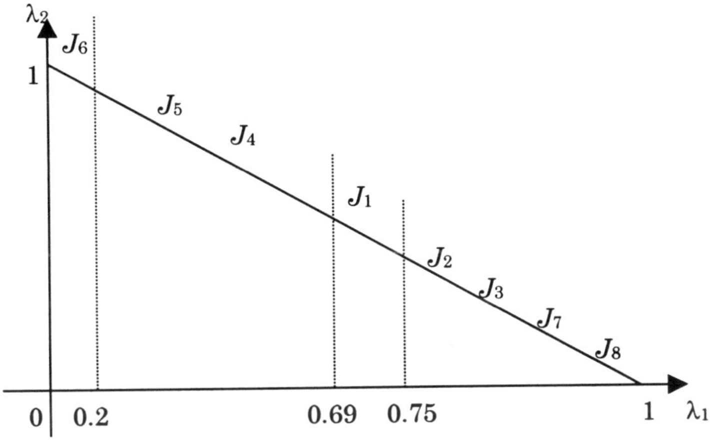

Figure 22.2 Budget Cost (λ_1) vs. Net Farm Revenue (λ_2)

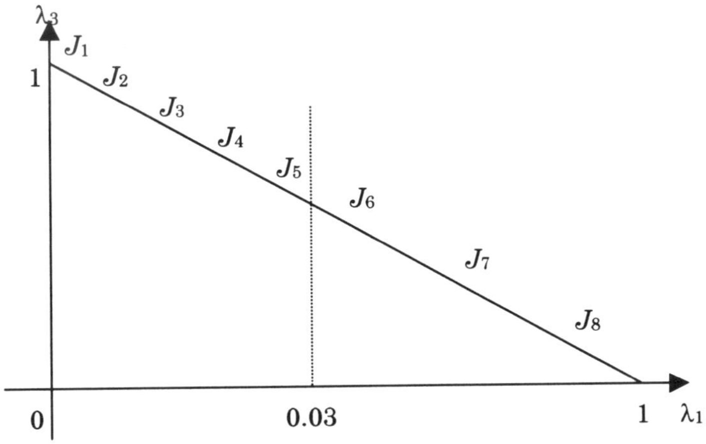

Figure 22.3 Budget Cost (λ_1) vs. Net Export Value (λ_3)

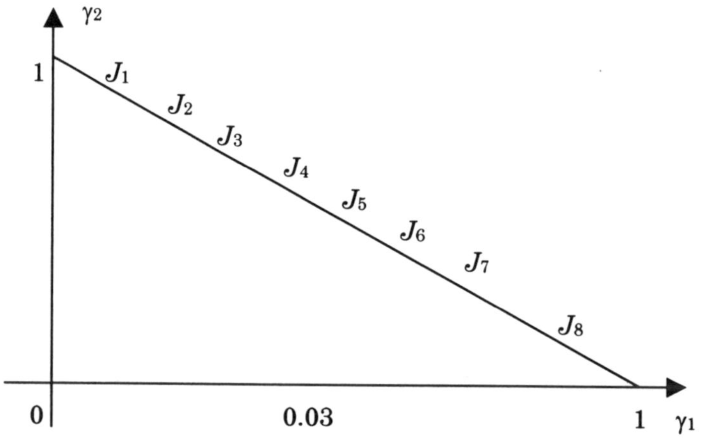

Figure 22.4 Economic Booming (γ_1) vs. Economic Recession (γ_2)

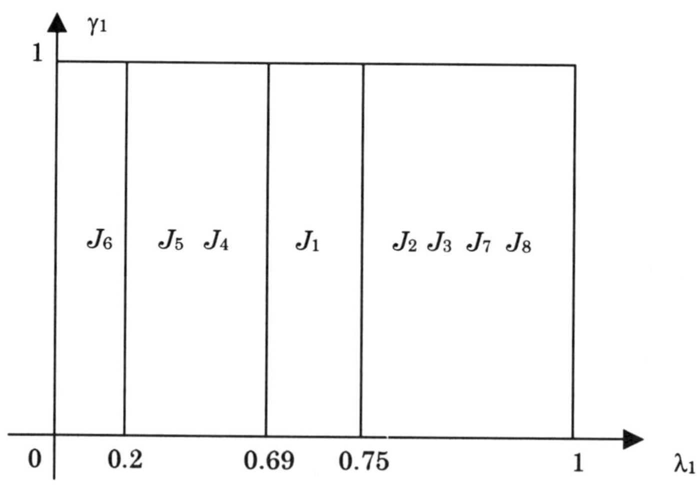

Figure 22.5 Budget Cost (λ_1) vs. Economic Booming (γ_1)

For the constraint level trade-offs: let $\gamma_1 + \gamma_3 = 1$, then the weight trade-offs of the economic booming (γ_1) versus the economic recession (γ_2) are given in Figure 22.4. In this case, all of the potentially optimal solutions $\{J_1, \ldots, J_8\}$ are alternatives. All possible upcoming uncertain situations between the economic booming and the economic recession can be generated by this trade-off graph.

For the criteria and constraint level trade-offs: we only show the weight trade-offs of the budget cost (λ_1) versus the economic booming (γ_1). The other cases can be similarly shown. Assume λ_1 is independent on g_1. Then, their relationship can be depicted as Figure 22.5. When $0.69 \leq \lambda_1 \leq 0.75$ and $0 \leq \gamma_1 \leq 1$, the potentially optimal solution J_1 of Table 22.4 is the optimal policy that we should choose.

22.4 Concluding Remarks

Based on the known mathematical programming approaches to agricultural policy making problems, this chapter considers the multiple upcoming economic environments and proposes a multiple criteria and multiple economic situation model. This model can be readily applied to reality and implemented by large-scale MC^2 computer software (Hao and Shi 1996). In comparison with previous mathematical programming approaches to agricultural policy making, this model provides policy makers with a more systematic and comprehensive scenario of all possible optimal values depending on multiple objectives and multiple economic situations. The trade-offs of all possible optimal values can be used as a basis for agricultural policy makers to make high-quality decisions in selecting their optimal pricing policies for the future agricultural commodity programs.

Bibliography

1. A. R. Abdel-khalik and E. J. Lusk, Transfer pricing - a synthesis, *The Accounting Review* (1974), 8-24.
2. P. Adams and R. Stephens, Challenge to rural economies and telecommunications infrastructure, in *Telecommunications and Rural Developments: A Study of Private and Public Sector Innovation*, J. Schmandt, F. Williams, R.H. Wilson and S. Strover (eds.), Praeger, 1991.
3. A.V. Aho, J. E. Hopcroh and J. D. Ullman, *Data Structures and Algorithms*, Addison-Wesley, Massachusetts, 1987.
4. M. F. Al-Eryani, P. Alam and S. H. Akhter, Transfer pricing determinants of U.S. multinationals, *Journal of International Business Studies* (1990), 409-425.
5. Y. P. Aneja and K. P. K. Nair, Bicriteria transportation problem, *Management Science* **25** (1979), 73-78.
6. P.M.G. Apers, Data allocation in distributed database systems, *ACM Trans. Database Systems* **13** (1988), 236-304.
7. L.D. Bahl, O.M. Coelho, R. Cutler, J.A. Marsh and R. Stephens, Telecommunications and community development, in *Telecommunications and Rural Development: A Study of Private and Public Sector Innovation*, J. Schmandt, R Williams. R.H. Wilson and S. Strover (eds.), Praeger, 1991.
8. A.D. Bailey and W. J. Boe, Goal and resource transfers in the multigoal organization, *The Accounting Review* (1976), 559-573.
9. E. Balas, An additive algorithm for solving linear problems with zero-one variables, *Operations Research* **13** (1965), 517-546.
10. M.O. Ball, Computing network reliabilities *Operation Research*, **27** (1979), 823-838.
11. E. Ballestero and C. Romero, Weighting in compromise programming: A theorem on shadow prices, *Operations Research Letters* **10** (1991), 421-427.
12. W.J. Baumol and R.E. Quandt, Investment and discount fate under capital rational- approach, *Economic Journal* **75** (1965), 173-329.
13. M. S. Bazaraa, J. Jarvis and H. D. Sherali, *Linear Programming and Network Flows*, John Wiley, New York, 1990.
14. G.L. Bergstrom and B.E. Smith, Multi-item production planning - an extension of the HMMS rules, *Management Science* **16** (1970), B614-629.
15. R.H. Bernard, Mathematical programming models for capital budgeting- a survey generalization and critique, *Journal of Financial and Quantitative Analysis* **4**(2) (1969), 111-158.
16. K.A. Bhaskar, Multiple objective approach to capital budgeting, *Accounting and Business Research* **10**(37) (1979), 25-46.
17. K.A. Bhaskar and P. McNamee, Mutiple objectives in accounting and finace, *Journal of Business Finance and Accounting* **10**(4) (1983), 595-621.
18. G.R. Bitran and A.C. Hax, On the design of hierarchical production planning systems, *Decision Sciences* **8** (1977), 28-55.
19. C.M. Boardman, W.J. Reinhart, and S.E. Celec, The role of payback period in the theory and application of duration to capital budgeting, *Journal of Business Finance and Accounting* **9**(4) (1982), 511-522.

20. C. E. Bodington and T. E. Baker, A history of mathematical programming in the petroleum industry, *Interfaces* **20** (1990), 117-127.

21. S. C. Borkowski, Environmental and organizational factors affecting transfer pricing: a survey, *Journal of Management Accounting Research* (1990), 78-99.

22. E. H. Bowman, Production scheduling by the transportation method of linear programming *Operations Research*, **4** (1956), 100-103.

23. E. H. Bowman, Consistency and optimality in managerial decision making, *Management Science* **9** (1963), 310-321.

24. L. Breiman, J.H. Frideman, R.A. Olshen and C.J. Stone, *Classification and Regression Trees*, Wadsworth, California, 1984.

25. R.G. Brown, *Decision Rules for Inventory Management*, Dryden Press, Hinsdale, Illinois, 1967.

26. D.M. Buede, Superior design features of decision analytic software, *Computers and Operations Research* **19** (1992), 43-57.

27. E. S. Buffa, *Modern Production Management* (2nd ed.), Wiley and Sons, New York, New York, 1965.

28. W. Candler and R. Norton, Multi-level programming and development policy, *World Bank Staff Working Papers* **258**, Washington D. C., 1977.

29. W. Candler, J. Fortuny-Amat and B. McCarl, The potential role of multi-level programming in agricultural economics, *American Journal of Agricultural Economics* **63** (1981), 521-531.

30. R.G. Casey, *Allocation of copies of a file in an information network*, Proc. AFIPS Spring Joint Computer Cc AFIPS Press, 1972, 617-625.

31. S. Ceri and G. Pelagatti, *Distributed Databases: Principles and Systems*, McGraw-Hill, 1984.

32. S. Chanas, Fuzzy programming in multiobjective linear programming-A parametric approach, *Fuzzy Sets and Systems* **29** (1989), 303-313.

33. Y.L. Chan and B.E. Lynn, Performance evaluation and the analytic hierarchy process, *Journal of Management Accounting Research* **3** (1991), 57-87.

34. S. A. Charnes and W. W. Cooper, The stepping stone method of explaning linear programming calculations in transportation problems, *Management Science* **1** (1954), 49-69.

35. A. Charnes and W.W. Cooper, *Management Models and Industrial Applications of Linear Programming I and II*, New Yourk, Wiley&Sons, 1961.

36. M. C. Cheng, New criteria for the simplex algorithm, *Math Programming* **19** (1980), 230-236.

37. J.K. Cheng, Managerial flexibility in capital investment decision: insights from the real-options literature, *Journal of Accounting Literature* **12** (1993), 29-66.

38. M. C. Cheng, Generalized theorems for permanent basic and nonbasic variables, *Mathematical Programming* **31** (1985), 229-234.

39. I.S. Chien, Y. Shi and P. L. Yu, *MC^2 Program*, A Pascal Program Run on PC or VAX, School of Business, University of Kansas, Lawrence, KS 66045, USA, 1989.

40. T.S. Choi and R.R. Levary, Multi-national capital budgeting using chance-constrained goal programming, *International Journal of Systems Science* **20**(3) (1989), 395-414.

41. C.W. Churchman, *Prediction and Optimal Decision: Philosophical Issues of A Science of Values*, Prentice Hall, Englewood Cliffs, New Jersey, 1961.

42. C. W. Churchman, *The System Approach*, Delacorte Press, New York, 1968.

43. W.W. Chu, Optimal file allocation in a computer network, *IEEE Trans. Computers* **18** (1969), 885-888.

44. W.J. Conover, *Practical Nonparametric Statistics*, New York, Wiley, 1999.

45. L. Cooper and D. Steinberg, *Linear Programming*, W. B. Saunders, Pennsylvania, 1974.

46. J.L. Corner, R.F. Deckro and R.W. Spahr, Multiple-objective linear programming in capital budgeting, *Advances in Mathematical Programming and Financial Planning* **3** (1993), 241-264.

47. F.G. Cronin, E.B. Parker, E.K. Colleran and M.A. Gold, Telecommunications infrastructure and economic growth, *Telecommunications Policy* **15**(1991) 529-535.

48. H. Crowder, E. L. Johnson and M. Padberg, Solving large-scale zero-one linear programming problems, *Operations Research* **31** (1983), 803-834.

49. J. Current and H. Min, Multiobjective design of transportation networks: taxonomy and annotation, *European Journal of Operational Research* **26** (1986), 187-201.

50. R. J. Dakin, A tree search algorithm for mixed integer programming problem, *Computer Journal* **8** (1965), 250-255.

51. N. Dale and D. Orshalick, *Introduction to Pascal and Structured Design*, D.C. Heath and Company, Massachusetts, 1983.

52. G. B. Dantzig, Maximization of a linear function of variables subject to linear inequalities, in *Activity Analysis of Production and Allocation*, T. C. Koopmans (ed.), John Wiley, New York, 339-347, 1951.

53. G. B. Dantzig and P. Wolfe, Decomposition principles for linear programs, *Operations Research* (1960), 101-111.

54. G. B. Dantzig, *Linear Programming and Extensions*, Princeton Press, Princeton, New Jersey, 1963.

55. G. B. Dantzig, The diet problem, *Interfaces* **20** (1990), 43-47.

56. P. Davidson and E. Smolensky, *Aggregate Supply and Demand Analysis*, Harper and Row, New York, 1964.

57. J. A. Diaz, Finding a complete description of all efficient solutions to a multiobjective transportation problem, *Ekonomicko-Matematicky Obzor* **15** (1979), 62-73.

58. N. Dopuch and D. F. Drake, Accounting implications of a mathematical programming approach to the transfer price problem, *Journal of Accounting Research* (1964), 10-24.

59. L.W. Dowdy and D.V. Foster, Comparative models of the file assignment problem, *ACM Computing Survey* **14** (1982), 287-313.

60. R.F. Dyer and E.H. Forman, *An Analytic Approach to Marketing Decisions*, Englewood Gliffs, JF: PrenticeHall, 1990.

61. J S. Dyer, P. C. Fishburn, R. E. Steuer, J. Wallenius and S. Zionts, Multiple criteria decision making, multiattribute utility theory: the next ten years, *Management Science* **38** (1992), 645-654 S.

62. J.S. Dyer, Remarks on the analytic hierarchy process, *Management Science* **36** (1990), 249-258.

63. J.S. Dyer, A clarification of remarks on the analytic hierarchy process, *Management Science* **36** (1990), 274-275.

64. R. G. Eccles, Control with fairness in transfer pricing, *Harvard Business Review* (1983), 149-161.

65. B.L. Egan, Bringing advanced technology to rural America, *Telecommunications Policy* **16**(1992).

66. R. Elmasri and S.B. Navathe, *Fundamentals of Database Systems*, Benjamin/Cummings, 1989.

67. E. Erickson and R. House, Multiple objective policy analysis: a factor analysis and tradeoff approach applied to the agricultural sector of the dominican republic, *Journal of Policy Modeling* **6** (1984), 369-87.
68. E. Erickson, Goal tradeoffs for U. S. agricultural commodity programs: an application of multiple objective analysis, in *Decision making with Multiple Objectives,* Y. Haimes and V. Chankong (eds.), New York: Springer-Verlag, 1985.
69. E. Erickson and R. House, Multiple objective analysis for a spatial market system: a case study of U. S. agricultural policy, in *Spatial Price Equilibrium: Advances in Theory, Computation and Application,* P. Harker (ed.), New York: Springer-Verlag, 1985.
70. K.P. Eswaran, Placement of records in a file and file allocation in a computer network, *Proc. IFIP Congress* (1974) 304-307.
71. W. Fan and Y. Shi, MC^2 Integer Program: version 1.0, A C++ Program run PC, College of Information Science and Technology, University of Nebraska-Omaha, Omaha, NE 68182, USA, 2000.
72. S. C. Fang and S. Puthenpura, *Linear Optimization and Extentions: Theory and Algorithms,* Prentice Hall, Englewood Cliffs, New Jersey, 1993.
73. E.H. Fonnan, T.L. Saaty, M.A. Selly and R. Waldron. *Expert Choice,* Pittsburgh: Decision Support Software Inc., 1985.
74. E.H. Forman, AHP is intended for more than expected value calculations, *Decision Sciences* **21** (1990), 670-672.
75. E.H. Forman, T.L. Saaty, M.A. Selly, and R. Waldron, *Expert Choice,* Decision Support Software, Inc., Pittsburgh, 1985.
76. N. Freed and F.Glover, Simple but powerful goal programming models for discriminant problems, *European Journal of Operational Research* **7** (1981), 44-60.
77. M. Freimer and P. L. Yu, Some new results on compromise solutions for group decision problems, *Management Science* **22** (1976), 688-693.
78. D. Gabbani and M. J. Magazine, An interactive heuristic approach for multiobjective integer programming problems, *Journal of Operational Research Society* **37** (1986), 285-291.
79. T. Gal, A general method for determining the set of all efficient solutions to a linear vectormaximum problem, *European Journal of Operational Research* **1** (1977), 307-322.
80. T. Gal, *Postoptimal Analysis, Parametric Programming and Related Topics,* McGraw-Hill, New York, 1979.
81. D. Gale, H. W. Kuhn and A. W. Tucker, Linear programming and the theory of games, in *Activity Analysis of Production and Allocation,* T. C. Koopmans (ed.), John Wiley, New York, 317-329, 1951.
82. E.S. Gardner, Jr. and D.G. Dannenbring, Using optimal policy surface to analyze aggregate inventory tradeoffs, *Management Science* **25** (1979), 709-720.
83. R. S. Garfinkel and G. L. Nemhauser, *Integer Programming,* John Wiley & Sons, New York, 1972.
84. S. I. Gass, *Linear Programming,* McGraw-Hill, New York, 1985.
85. S. I. Gass, Model world: In the beginning there was linear programming, *Interfaces* **20** (1990), 128-132.
86. B. Gavrsh and H. Pirkul, Computer and database location in distributed computer systems, *IEEE Trans. Comput* **35** (1986), 583-590.
87. Y. S. Ge and Y. Shi, MC^2 Program: A Window NT version on PC, College of Information Science and Technology, University of Nebraska-Omaha, Omaha, NE 68182, USA, 1997.

88. A. M. Geoffrion and R. E. Marsten, Integer programming algorithms: a framework and state-of-the-art survey, *Management Science* **18** (1972), 465-491.

89. F. Glover, Improve linear programming models for discriminant analysis, *Decision Sciences* **21** (1990), 771-785.

90. A. Goicoechea, D. R. Hansen and L. Duckstein, *Multiobjective Decision Analysis with Engineering and Business Applications,* John Wiley & Sons, New York, 1982.

91. J.J. Gonzalez, G.R. Reeves and L.S. Franz, Capital budgeting decision making: an interactive multiple objective linear integer programming search procedure, *Advances in Mathematical Programming Financial Planning* **1** (1987), 21-44.

92. D. A. Goodman, A goal programming approach to aggregate planning of production and work force, *Management Science* **20** (1974), 1569- 1579.

93. H. Guo and S.B. Gelfand, Classification trees with neural network feature extraction, *IEEE Transaction on Neural Networks* **3** (1992), 923-933.

94. F.Gyetvan and Y. Shi, Weak duality theorem andcomplementary slackness theorem for linear matrix programming problems, *Operations Research Letters* **11** (1992) 249-252.

95. C. Haase, Implementation of A Transportation Problem with Multiple-Criteria and Multiple-Constraint Levels, The Thesis for Diplomarbeit, Technical University Braunschweig, D-38106 Braunschweig, Germany, 1994.

96. C. Haase and Y. Shi, MC^2 Transportation Program: Version 1.0, A C++ Program Run on PC or Unix, College of Business Administration, University of Nebraska-Omaha, Omaha, 68182, USA, 1994.

97. H. Hamacher, H. Leberling and H. J. Zimmermann, Sensitivity analysis in fuzzy linear programming, *Fuzzy Sets and Systems* **1** (1978), 269-281.

98. X. R. Hao and Y. Shi, Large-scale MC^2 Program: Version 1.0, A C++ Program run on PC or Unix, College of Information Science and Technology, University of Nebraska-Omaha, Omaha, NE 68182, USA, 1996.

99. X. R. Hao and Y. Shi, Binary Integer MC^2 Program: Version 1.0, A C++ Program run on PC or Unix, College of Information Science and Technology, University of Nebraska-Omaha, Omaha, NE 68182, USA, 1997.

100. P.T. Harker and L.G. Vargas, The theory of ratio scale estimation: Saaty's analytic hierarchy process, *Management Science* **33** (1987), 1383-1403.

101. P.T. Harker and L.G. Vargas, Reply to remarks on the Analytic Hierarchy Process by J.S. Dyer, *Management Science* **36** (1990), 269-273.

102. J. Heizer and B. Render, *Production and Operations Management* (3rd ed.), Allyn and Bacon, Boston, Massachusetts, 1993.

103. H. D. Henderson, *Supply and Demand,* Harcourt Brace Jovanovich, New York, 1922.

104. J.M. Henderson and R.E. Quandt, *Microeconomic Theory: A Mathematical Approach,* McGraw-Hall, New York, 1980.

105. F.S. Hillier, The derivation of probabilites information for the evaluation of risky investments, *Management Science* (1963), 443-457.

106. F.S. Hillier, *The Evaluation of Risky Interrelated Investments,* North-Holland Publishing Co., Amsterdam, 1969.

107. F.S. Hillier and G.J. Lieberman, *Introduction to Operations Research,* Holden-Day, 1986.

108. F. S. Hillier and G. J. Lieberman, *Introduction to Mathematical Programming,* McGraw-Hill, New York, 1990.

109. J., Hirshleifer, On the economies of transfer pricing, *The Journal of Business* (1956), 172-184.

110. F. L. Hitchcock, The distribution of a product from several sources to numerous locations, *Journal of Mathematics and Physics* **20** (1941), 224-230.

111. C. C. Holt, F. Modigliani, J.F. Muth and H.A. Simon, *Planning Production Inventories and Work Force,* Prentice Hall, Englewood Cliffs, New Jersey, 1960.

112. H. S. Houthakker, On the numerical solution of the transportation problem, *Operations Research* **3** (1955), 210-214.

113. K.M. Howe, and J.H. Patterson, Capital investment decisions under economies of scale in flotation costs, *Financial Management* **14**(3) (1985), 61-69.

114. R.A. Howell and W.A. Schwartz, Asset deployment and investments justification, *Handbook of Cost Management.* Warren Gorham Lamont, Boston, 1994.

115. T. C. Hu, *Integer Programming and Network Flows,* Addison-Wesley, California, 1969.

116. H. B. Huang and Y. Shi, MC^2 Transportation Programs: Version 2.0, A C++ Program run PC or Unix, College of Information Science and Technology, University of Nebraska-Omaha, Omaha, NE 68182, USA, 1997.

117. C. L. Hwang and M. J. Lin, *Group Decision Making under Multiple Criteria: Methods and Applications*, Spring-Verlag, Heidelberg, 1987.

118. J.P. Ignizio, An approach to the capital budgeting problem with multiple objectives, *The Engineering Economist* (1976), 259-272.

119. J.P. Ignizio, Solving large-scale problems: a venture into a new dimension, *Journal of Operational Society* **31** (1980), 217-225.

120. J.P. Ignizio, *Linear Programming in Single and Multiple Objective Systems,* Prentice-Hall, 1982.

121. H. Isermann, Duality in multiple objective linear programming, in S. Zionts (ed.), *Multiple Criteria Problem Solving,* Lecture Notes in Economics and Mathematical Systems No. 155, Springer, Berlin, 74-285, 1978.

122. H. Isermann, On some relations between a dual pair of multiple objective linear programs, Z. *Offer. Rat.* **22** (1978) 33-41.

123. H. Isermann, The numeration of all efficient solutions for a linear multiobjective transportation problems, *Naval Research Logistic Quarterly* **26** (1979), 123-139.

124. H.K. Jain and A. Dutta, Distributed computer system design: a multicriteria decision making methodology, *Decision Sciences* **4** (1986), 437-453.

125. H.K. Jain, A comprehensive model for the design of distributed computer systems, *IEEE Trans. Software Eng.* **13** (1987), 1092-1104.

126. E.A. Joachimsthaler and A. Stam, Four approaches to the classification Problems in discriminant analysis: an experimental study, *Decision Sciences* **19** (1988), 322-333.

127. C.H. Jones, Parameteric production planning, *Management Science* **13** (1967), 843-866.

128. E. L. Johnson, M. M. Kostreva and U. H. Suhl, Solving 0-1 integer programming problems arising from large scale planning models,*Operations Research* **33** (1985), 803-819.

129. P. Kall, *Stochastic Linear Programming,* Springer-Verlag, Berlin, 1976.

130. R.R. Kamath and S.Z. Khaksari, Real Estate Investment: An application of the analytic hierarchy process, *The Journal of Financial and Strategic Decisions* (1991), 73-100.

131. N. Karmarkar, A new polynomial time algorithm for linear programming, *Combinatorica* **4** (1984), 373-395.

132. L. V. Kantorovich, Mathematical methods of organizing and planning production, *Publication House of the Leningrad State University,* 1939 (in Russian), *Management Science* **6** (1959-60), 366-422 (English translation).

133. R. L. Keeney and H. Raiffa, *Decisions with Multiple Objectives: Preferences and Value Tradeoffs*, John Wiley & Sons, New York, 1976.

134. J.M. Keynes, *The General Theory of Employment, Interest, and Money*, Harcourt Brace Jovanovich, New York, 1936.

135. L. G. Khachian, A polynomial algorithm in linear programming, *Doklady Akademiia Nauk SSSR* **224**, 1093-1096 (in Russian), *Soviet Mathematics Doklady* **20** (1979), 191-194 (English translation).

136. S.H. Kim and E.J. Farragher, Current capital budgeting practices, *Management Accounting* **6** (1981), 26-30.

137. V. Klee and G. L. Minty, How good is the simplex algorithm, in *Inequalities III*, O. Shisha (ed.), Academic Press, New York, 159-179, 1972.

138. B. Kolman and R. E. Beck, *Elementary Linear Programming with Applications*, Academic Press, San Diego, California, 1995.

139. T. C. Koopmans, Optimum utilization of the transportation system, In Proceedings of the International Statistical Conference, Washington, D.C., 1947.

140. T.C. Koopmans, Analysis of production as an efficient combination of activities, in *Activity Analysis of Production and Allocation*, T. C. Koopmans (ed.), John Wiley, New York, 33-97, 1951.

141. P. Korhonen, J. Wallenius and S. Zionts, Solving the discrete multiple criteria problem using convex cones, *Management Science* **30** (1984), 1336-1345.

142. J.S.H. Kornbluth, Duality, indifference and sensitivity analysis in multiple objective linear programming, *Operational Research Quarterly*, **25** (1974), 599-614.

143. W. Kwak, Y. Shi, H. Lee and C. F. Lee, Capital budgeting with multiple criteria and multiple decision makers, *Review of Quantitative Finance and Accounting* **7** (1996), 97-112.

144. A.H. Land and A.G. Doig, An automatic method of solving discrete programming problems, *Econometrica* **28** (1960), 497-520.

145. L. Lapin, *Quantitative Methods for Business Decisions*, Dryden, New York, 1994.

146. E.L. Lawler and D.E. Wood, Branch-and-bound methods: a survey,*Operations Research* **14** (1966), 699-719.

147. D.J. Lecraw, Some evidence on transfer pricing by multinational corporations, in *Multinationals and Transfer Pricing*, New York, St. Martin's Press, A. M. Rugman & L. Eden (eds.), 1985.

148. W.N. Ledbetter, and J.F. Cox, Operations research in production management, *Production and Inventory Management* **3** (1977), 84-91.

149. C.F. Lee, *Statistics for Business and Financial Economics*, Heath, Lexington, MASS., 1993.

150. H. Lee, An evaluation method for the availability of a distributed database management system, *Information and Management* **24** (1993), 203-208.

151. H. Lee, Modeling and optimization of data assignment in a distributed information system, *Systems Science* **24** (1993), 173-181.

152. H. Lee and O.R. Sheng, A multiple criteria model for the allocation of data files in a distributed information system, *Computers and Operations Research* **21** (1992), 21-33.

153. H. Lee, S.M. Nazem and Y. Shi, Designing rural area telecommunication networks via hub cities, *Omega: The International Journal of Management Science* **22** (1994), 305-314.

154. H. Lee, S.M. Nazem and Y. Shi, Designing rural area telecommunication networks via hub cities, *Omega: International Journal of Management Science* **22**(1995), 305-314.

155. H. Lee, Y. Shi and S. Nazem, Supporting rural telecommunication networks via hub cities: a compromise solutions approach, *Annals of Operations Research* **68** (1996), 33-45.

156. H. Lee, Y. Shi and S. Nazem, Modelling the telecommunication hub decision for rural area networks, Working Paper, University of Nebraska at Omaha, 1993.

157. H. Lee, Y. Shi and J. Stolen, Allocating data flies over a wide area network: Goal setting and compromise design, *Information and Management* **26** (1994), 85-93.

158. S.M. Lee, *Goal Programming for Decision Analysis,* Auerbach, Pennsylvania, 1972.

159. S. M. Lee and L. J. Moore, Optimizing transportation problems with multiple objectives, *AIEE Transactions* **5** (1973), 333-338.

160. S.M. Lee and A.J. Lerrok, Capital budgeting for multiple objectives, *Financial Management* **3**(1) (1974), 51-53.

161. Y. R. Lee, A Linear Feasibility Transformation and Interactive Goal-Seeking Approach for Multicriteria Decision Problems, Ph.D. Dissertation, University of Kansas, 1986.

162. Y. R. Lee, Y. Shi and P. L. Yu, Linear optimal designs and optimal contingency plans, *Management Science* **36** (1990), 1106-1119.

163. W. B. Lee and B.M. Khumawala, Simulation testing of aggregate production planning models in an implementation methodology, *Management Science* **20** (1972), 903-911.

164. W. Leontief, *Structure of the American Economy,* Oxford University Press, New York, 1951.

165. R.J. Li and E.S. Lee, Fuzzy approaches to multicriteria de novo programs, *J. Math. Anal. Appl.* **153** (1990), 97-111.

166. J. Li and Y. Shi, An Integer linear programming with multi-criteria and multi-constraint levels: A branch-partition algorithm, *Working paper* at College of Information Science and Technology, University of Nebraska at Omaha, USA, 1999.

167. J. Li and Y. Shi, A dynamic transportation model with multiple criteria and multiple constraint levels, *Mathematical and Computer Modeling* **32** (2000), 1193-1208.

168. Y.N. Lien, Y.L. Chang and B.W. Wah, File allocation problems on homogeneous two-level local broadcast networks, *Proc. Fourth Int'l Conf. on Data Engineering* (1988), 92-99.

169. L. Lin, C. Lefebvre and J. Kantor, Economic determinants of international transfer pricing and the related accounting issues, with particular reference to asian pacific countries, *The International Journal of Accounting* (1993), 49-70.

170. T.W. Lin, Multiple-criteria capital budgeting under risk, *Advances in Mathematical Programming Financial Planning* **3** (1993), 231-239.

171. P. H. Lindsay and D. A. Norman, *Human Information Processing: An Introduction to Psychology*. Academic Press, New York, 1972.

172. Y. H. Liu and Y. Shi, A fuzzy programming approach for solving a multiple criteria and multiple constraint level programming problem, *Fuzzy Sets and Systems* **65** (1994), 117-124.

173. Y. Liu, Y. Shi and Y. H. Liu, Duality of fuzzy MC^2 linear programming: A constructive approach, *Journal of Mathematical Analysis and Applications* **191** (1995), 389-413.

174. M.J. Liveratore, T.F. Mohnhan and D.E. Stout, A framework for integrating capital budgeting analysis with strategy, *The Engineering Economist* **38**(1) (1992), 31-43.

175. E. A. Locke, K. N. Shaw and G. P. Lathom, Goal setting and task performance: 1969-1980, *Psychological Bulletin*, 125-152, (1981).

176. J.H. Lorie, and L.J. Savage, Three problems in rationing capital, *Journal of Business* **28** (1955), 229-239.

177. B. McCarl and T. Spreen, Price endogenous mathematical programming as a tool for sector analysis, *American Journal of Agricultural Economics* **62** (1980), 87-102.

178. E. Menipaz, *Essentials of Production and Operations Management*, 1984, Prentice Hall, Englewood Cliffs, New Jersey.

179. Y.M. Mensah and P.J. Miranti, Jr., Capital expenditure analysis and synthesis, *Journal of Accounting Literature* **8** (1989), 181-207.

180. L. J. Merville and W. J. Petty, Transfer pricing for the multinational corporation, *The Accounting Review* (1978), 935-951.

181. L. G. Mitten, Branch-and-bound methods: general formulation and Properties, *Operations Research* **18** (1970), 24-34.

182. H. Nagasawa, N. Nishiyama and K. Hitomi, "Decision Analysis for Determining the Optimum Planning Horizon in Aggregate Production Planning," *International Journal of Production Research* **20** (1982), 243-254.

183. H. Nakayama, Some remarks on trade-off analysis in multi-objective programming, The XIth International Conference on Multiple Criteria Decision Making, Coimbra, Portugal, Aug. 1-6, 1994.

184. S.M. Nazem, Telecommunications and economic development in rural America: A case for Nebraska, Research Paper No. 4, International Center for Telecommunications Management, University of Nebraska at Omaha, 1992.

185. S.M. Nazem and Y.S. Liu, The case for rural telecommunications networks and policy implications, Research Paper No. 14, International Center for Telecommunications Management, University of Nebraska at Omaha, 1992; also presented at *Bi-Annual World Conference of the international Telecommunications Society*, Nice, France, 1992.

186. S. M. Nazem, Y. S. Liu, Y. Shi and H. Lee, Supporting rural telecommunications and policy implications: Multicriteria perspective, in *Optimization*, G. Liu *et al* (eds.), Singapore, World Scientific, 1632-1641, 1995.

187. S. M. Nazem, Y.H. Liu, H. Lee and Y. Shi, Implementing telecommunication infrastructure: A rural America case, *Telematics and Informatics* **13** (1996), 23-31.

188. G. L. Nemhauser and L. A. Wolsey, *Integer and Combinatorial Optimization*, John Wiley & Sons, New York, 1988.

189. R. Nolan, Managing the advanced stages of computer technology; key research issues, in *The Information Systems Research Challenge*, E.W. McFarlan (ed.), Harvard Business School Press, Boston, Mass, 1984.

190. M.D. Oliff, and G.K. Leong, A discrete production switching rule for aggregate planning, *Decision Sciences* **18** (1987), 582-597.

191. W. Orchard-Hays, History of the development of LP solvers, *Interfaces* **20** (1990), 61-73.

192. B.N. Ouederni, and W.G.Sullivan, Semi-variance model for incorporating risk int capital investment analysis, *The Engineering Economist* **36**(2), 83-106, 1991.

193. M.T. Ozsu and P. Valdureiz, Distributed database systems: where are we now, *Computer* **24** (1991), 68-78.

194. M.T. Ozsu and P. Valdureiz, *Principles of Distributed Database Systems*, Prentice-Hall, 1991.

195. P. Y. Papalambros and D. J. Wilde, *Principle of Optimal Design*, Cambridge Univ. Press, Cambridge, 1988.

196. S. Park, Complementary slackness theorem in multiple objective linear programming, *Operations research* **30** (1982), 410-412.

197. R. G. Parker and R. L. Rardin, *Discrete Optimization,* Academic Press, California, 1988.

198. E.B. Parker, H.E. Hudson, D.A. Dillman and A.D. Roscoe, *Rural America in the Information Age: Telecommunications Policy for Rural Development,* The Aspen Institute and University Press of America, 1989.

199. E.B. Parker, H.E. Hudson, D.A. Dillman, S. Strover and F. Williams, *Electronic Byways: State Policies for Rural Development Through Telecommunications,* Westview Press, 1992.

200. D. M. Phelps, *Planning and Products,* Irwin, Illinois, 1947.

201. J. Philip, Algorithms for the vector maximization problem, *Mathematical Programming* **2** (1972), 207-229.

202. R. Pike, The capital budgeting behavior and corporate characteristics of capital-constrained firms, *Journal of Business Finance and Accounting* **10**(4), 663-671, 1983.

203. R. Pike, An empirical study of the adoption of sophisticated capital budgeting practices and decision-making effectiveness, *Accounting and Business Research* **18**(72) (1988), 341-351.

204. J.R. Quilan, Induction of decision trees, *Machine Learning* **1** (1986), 81-106.

205. S. Ram, A model for designing distributed database systems, *Information and Management* **17** (1989), 169-180.

206. S. Ram and R.E. Marsten, A model for database allocation incorporating a concurrency control mechanism, *IEEE Trans. Knowledge and Data Engineering* **3** (1991), 389-395.

207. R. Ramesh and B. Ryan, Optimal file allocation and report assignment in distributed information networks, *Naval Research Logistics Quarterly* **37** (1990), 165-181.

208. R. Ramesh, Multicriteria Integer Programming, Unpublished Ph. D. dissertation, Department of Industrial Engineering, State University of New York at Buffalo, Buffalo, NY 14260, USA, 1985.

209. R. Ramesh, M. H. Karwan and S. Zionts, Preference structure representation using convex cones in multicriteria integer programming, *Management Science* **35** (1989), 1092-1105.

210. R. Ramesh, M H. Karwan and S. Zionts, A class of practical interactive branch and bound algorithms for multicriteria integer programming, *European Journal of Operational Research* **26** (1985), 161-172.

211. R. Rymon, Search through systematic set enumeration. Proceedings of the third international conference on principle of knowledge representation and reasoning. Cambridge, MA, 539-550 (1993).

212. G.F. Reeves, K.D. Lawrence, S.M. Lawrence and J.J. Gonzalez, A multiple criteria approach to aggregate industrial capacity expansion, *Computers and Operations Research* **15**(4) (1988), 333-339.

213. G.R. Reeves and L.X. Franz, A simplified interactive multiple objective linear programming procedure, *Computers and Operations Research* **12** (1985), 589-601.

214. G.R. Reeves and S.R. Hedin, A generalized interactive goal programming procedure, *Computers and Operations Research* **20**(7) (1993), 747-753.

215. N. V. Reinfeld and W. R. Vogel, *Mathematical Programming,* Prentice-Hall, New Jersey, 1958.

216. L. Ringuest and D. B. Rinks, Interactive solutions for the linear multiobjective transportation problem, *European Journal of Operational Research* **32** (1986), 96-106.

217. W. Rödder and H. J. Zimmermann, Duality in fuzzy linear program, in *Extremal Methods and System Analysis*, A. V. Fiacco and K. O. Kortanek (eds.), Springer-Verlag, Berlin, 415-429, 1980.
218. C. Romero, *Handbook of Critical Issues in Goal Programming*, Pergamon Press, Oxford, 1991.
219. W. Rudin, *Real and Complex Analysis*, McGraw-Hill, New York, 1987.
220. M. Rueff and M. Jeger, *Sets and Boolean Algebra*, American Elsevier, New York, 1970.
221. T.W. Ruefli, A generalized goal decomposition model, *Management Science* (1971), 505 - 518.
222. R.J. Rummel, *Applied Factor Analysis*, Northwestern University Press, Evanston, IL, 1970.
223. E.J. Russell, Extension of Dantzig's algorithm to finding an initial near-optimal basis for the transportation problem, *Operations Research* **17** (1969), 187-191.
224. G. Saad, An overview of production planning models: structural classification and empirical assessment, *International Journal of Production Research* **20** (1982), 105-114.
225. T.L. Saaty, A scaling method for priorities in hierarchical structures, *Journal of Mathematical Psychology* **15** (1977), 234-281.
226. T.L. Saaty, *The Analytic Hierarchy Process*, McGraw-Hill, New York, 1980.
227. T.L. Saaty, An exposition of the AHP in reply to the paper, Remarks on the analytic hierarchy process, *Management Science* **36** (1990), 259-268.
228. H. M. Salkin, *Integer Programming*, Addison-Wesley, Massachusetts, 1975.
229. P. Samuelson, Spatial price equilibrium and linear programming, *American Economic Review* **43** (1952), 283-303.
230. R. Santhanam, K. Muralidhar and M. Schniederjans, A zero-one goal programming approach for information system project selection, *Omega: International Journal of Management Science* **17**(6) (1989), 583-598.
231. O.R. Liu Sheng and H. Lee, Data allocation design in computer networks: LAN versus MAN versus WAN, *Annals of Operations Research* **36** (1992), 125-150.
232. B. Schoner, W.C. Wedley and E.U. Choo, A rejoinder to Forman on AHP, with emphasis on the requirements of composite ratio scales, *Decision Sciences* (1992), 509-517.
233. L. Schrage, *Linear, Integer, and Quadratic Programming with LINDO: User's Manual*, The Scientific Press, 1984.
234. H. Schultz, *The Theory and Measurement of Demand*, The University of Chicago Press, Chicago, Illinois, 1938.
235. L.B. Schwarz and R.E. Johnson, An appraisal of the empirical performance of the linear decision rule for aggregate planning, *Management Science* **24** (1978), 844-849.
236. L. Seiford and P. L. Yu, Potential solutions of linear systems: the multi-criteria multiple constraint level program, *Journal of Mathematical Analysis and Applications* **69** (1979), 283-303.
237. R. Sharda, Linear programming solver software for personal computers: 1995 report, *OR/MS Today* **Oct.** (1995), 49-57.
238. Y. Shi, Optimal Linear Production Systems: Models, Algorithms, and Computer Support Systems, Ph.D. Dissertation, University of Kansas, 1991.
239. Y. Shi, A transportation problem with multiple criteria and multiple constraint levels, Working Paper #92-23, College of Business Administration, University of Nebraska-Omaha, Omaha, NE 68182, USA, 1992.
240. Y. Shi, A transportation model with multiple criteria and multiple constraint levels, *Mathematical and Computer Modelling* **21** (1995), 13-28.

241. Y. Shi, Constructing flexible dual contingency plans for optimal linear designs with multiple criteria, *Journal of Mathematical Analysis and Applications* **191** (1995), 277-304.

242. Y. Shi and C. Haase, Optimal trade-offs of aggregate production planning with multi-objective and multi-capacity-demand Levels, *International Journal of Operations and Quantitative Management*, **2** (1996), 127-143.

243. Y. Shi, A review on MC[2] linear programming: theory and applications, *International Journal of Operations and Quantitative Management* **5** (1998), 137-150.

244. Y. Shi, Optimal system design with MC[2] linear programming: a dual contingency plan approach, *European Journal of Operational Research* **107** (1998), 692-709.

245. Y. Shi, Finding dual flexible contingency plans for optimal generalized linear systems, *International Transactions in Operations Research* **5** (1998), 303-315.

246. Y. Shi, Optimal system design with multi-decision makers and possible debt: a multi-riteria de novo approach, *Operations Research* **47** (1999), 723-729.

247. Y. Shi and Z. He, Dual contingency plans in optimal linear generalized designs, *Systems Science* **25** (1994) 1267-1292.

248. Y. Shi, W. Kwak and H. Lee, Optimal trade-offs of multiple factors in transfer pricing problems, *Multi-Criteria Decision Analysis* **7** (1997), 98-108.

249. Y. Shi, and H. Lee, A binary integer linear programming with multicriteria and multiconstraint levels, Working paper #92-21, School of Business, University of Nebraska at Omaha, 1992.

250. Y. Shi and H. Lee, A binary integer linear programming with multi-criteria and multi-constraint levels, *Computers Operations Research* **24** (1997), 259-273.

251. Y. Shi and S. Lin, Optimal trade-off analysis of agricultural policy: a multiple criteria and multiple economic situation model, in *Multi-Criteria applications*, K. Lawrence, G. Reeves and R. Klimberg (eds.), Elsevier Science Inc., 165-183, 2000.

252. Y. Shi and Y. H. Liu, Fuzzy potential solutions in multi-criteria and multi constraint levels linear programming problems, *Fuzzy Sets and Systems* **60** (1993), 163-179.

253. Y. Shi, S. M. Nazem, and H. Lee, Telecommunications, in *International Encyclopedia of Business and Management*, M. Warner (editor), Thomson Business Press, England, 4824-4834, 1996.

254. Y. Shi, M. Wise, M. Luo and Y. Lin, Data mining in Credit Card Portfolio management: a multiple criteria decision making approach, Proceedings of international conference on multiple criteria decision making, Ankara, Turkey, July 10-14, 2000.

255. Y. Shi and P. L. Yu, Goal setting and compromise solutions, in *Multiple Criteria Decision Making and Risk Analysis and Applications*, B. Karpak and S. Zionts (eds.), Springer-Verlag, Berlin, 165-203, 1989.

256. Y. Shi and P.L. Yu, An introduction to selecting linear optimal systems and their contingency plans, in *Operations Research*, G. Fandel and H. Gehring (eds.), 57-77, Springer-Verlag, Berlin, 1991.

257. Y. Shi, and P.L. Yu, Selecting optimal linear production systems in multiple criteria environments, *Computer and Operations Research* **19** (1992), 585-608.

258. Y. Shi and P.L. Yu, Foundations of designing optimal systems and contingency plans, in Ravi P. Agarwal (ed.), *Recent Trends in Optimization Theory and Applications,* 371-391, World Scientific Publishing Company, Singapore, 1995.

259. Y. Shi and P.L. Yu, and D. Zhang, Generalized optimal designs and contingency plans in linear systems, *European Journal of Operational Research* **89** (1996), 618-641.

260. Y. Shi, P.L. Yu, C. Zhang and D. Zhang, A computer-aided system for linear production designs, *Decision Support Systems* **12** (1994), 127-149.

261. Y. Shi, P.L. Yu and D. Zhang, Generating new designs using union operations, *Computers and Mathematics with Applications*, **27** (1994), 105-117.

262. Y. Shi, P.L. Yu and D. Zhang, Eliminating permanently dominated opportunities in multiple-criteria and multiple-constraint level linear programming, *Journal of Mathematics Analysis and Applications* **183** (1994), 685-705.

263. Y. Shi and D. Zhang, Flexible contingency plans in optimal linear designs, *Mathematical and Computer Modelling* **17** (1993), 13-28.

264. J. Singhal, R.E. Marsten and T.L. Morin, Fixed order brand-and bound methods, for mixed-integer programming: The ZOOM System, *ORSA Journal on Computing* **1** (1989), 44-51.

265. K. Singhal and V. Adlakha, Cost and shortage trade-offs in aggregate production planning, *Decision Sciences* **20** (1989), 158-165.

266. A.J. Sofranko, Transition in rural areas of the Midwest and nation, in: *Rural Community Economic Development*, N. Waltzer (ed.), Praeger, 1991.

267. M.K. Starr, Evolving concepts in production management, in *Evolving Concepts in Management*, E. Flippo (ed.), Academy of Management Proceedings, 24th Annual Meeting (1964). Reprinted by E.S. Buffa, *Readings in Production and Operations management*, 28-35, Wiley, New York 1966.

268. M.K. Starr, *Management Production and Operations*, Prentice Hall, Englewood Cliffs, New Jersey, 1989.

269. R. Steuer, *Multiple Criteria Optimization: Theory, Computation, and Application*, John Wiley, New York, 1986.

270. G. Stigler, The cost of subsistence, *Journal of Farm Economics* **27** (1945), 303-314.

271. R. R. Stoll, *Set Theory and Logic*, W. H. Freeman and Company, San Francisco, 1963.

272. M. Stonebraker, Future trends in database system, *IEEE Trans. Knowledge and Data Engineering* (1989), 33-44.

273. C.K. Suh, E.H. Suh and I.J. Choi, A two-phased DSS for R&D portfolio selection, *Proceedings of International DSI* (1993), 351-354.

274. C.K. Sun, E.H. Suh and I.J. Choi. A two-phased DSS for R&D portfolio selection, International DSIProc. (1993), 351-354.

275. H. A. Taha, *Integer Programming: Theory, Applications, and Computations*, Academic Press, New York, 1975.

276. R.Y.W. Tang, Transfer pricing in the 1990s, *Management Accounting* (1992), 22-26.

277. W.H. Taubert, A search decision rule for the aggregate scheduling problem, *Management Science* **14** (1968), 343-359.

278. E. Thanassoulis, Selecting a suitable solution method for a multi-objective programming capital budgeting problem, *Journal of Business Finance and Accounting* **12**(3) (1985), 453-471.

279. A.L. Thomas, *A Behavioral Analysis of Joint-Cost Allocation and Transfer Pricing*, Stipes Publishing, 1980.

280. G.L. Thompson, F.M. Tonge and S. Zionts, Techniques for removing non-binding constraints and extraneous variables from linear programming, *Management Science* **12** (1996), 588-608.

281. S.T. Turney, Deterministic and stochastic dynamic adjustment of capital investment budgets, *Mathematical Computation and Modeling* **13**(5) (1990), 1-9.

282. T.J. Van Roy and L.A. Wolsey, Solving mixed 0-1 programs by automatic reformulation, *Operations Research* **35** (1987), 45-57.

283. J.L. Verdegay, A dual approach to solve the fuzzy linear programming problem, *Fuzzy Sets and Systems* **14** (1984), 131-141.

284. R.C. Vergin, Production scheduling under seasonal demand, *Journal of Industrial Engineering* **17** (1966), 260-266.

285. B. Villarreal and M.H. Karwan, An interactive dynamic programming approach to multicriteria discrete programming, *Journal of Mathematical Analysis and Applications* **81** (1981), 524-544.

286. J. von Neumann and O. Morgensterrn, *Theory of Games and Economic Behavior*, Princeton Press, Princeton, New Jersey, 1944.

287. B.W. Wah and Y.N. Lien, Design of distributed databases on local computer systems, *IEEE Trans. Software Eng.* **11** (1985), 606-619.

288. N. Waltzer, Introduction, in: *Rural Community Economic Development*, N. Waltzer (ed.), Praeger, 1991.

289. D.J.H. Watson and J.V. Baumler, Transfer pricing: a behavioral context, *The Accounting Review* (1975), 466-474.

290. H.M. Weingartner, *Mathematical Programming and the Analysis of Capital Budgeting Problems*, Prentice-Hall, Englewood Cliffs, New Jersey, 1963.

291. J.L. Whitten, L.D. Bentley and V.M. Barlow, *Systems Analysis and Design Methods*, Irwin, 1989.

292. D.J. Wilde, *Globally Optimal Design*, Wiley, New York, 1978.

293. C.E. Willis and R.D. Perlack, A comparison of generating techniques and goal programming for public investment in multiple objective decisionmaking, *American Journal of Agricultural Economics* **62** (1980), 66-74.

294. Y. Ye, Karmarkar's algorithm and the ellipsoidal method, *Operations Research Letters* **4** (1987), 177-182.

295. Y. Ye, Eliminating columns in the simplex method for linear programming, *Journal of Optimization Theory and Applications* **63** (1989), 69-77.

296. C.T. Yu, M. Siu, K. Lam and C.H. Chen, Adaptive file allocation in star computer network, *IEEE Trans. Software Eng.* **11** (1985), 959-965.

297. P.L. Yu, A class of solutions for group decision problems, *Management Science* **19** (1973), 936-946.

298. P.L. Yu, Behavior bases and habitual domains of human decisions/behavior: an integration of psychology, optimization theory and common wisdom, *International Journal of System Measurement Decisions* **1** (1981), 39-62.

299. P.L. Yu, *Multiple Criteria Decision Making: Concepts, Techniques and Extensions*, Plenum, New York, 1985.

300. P.L. Yu, and M. Zeleny, The set of all nondominated solutions in the linear cases and a multicriteria simplex method, *Journal of Mathematical Analysis and Applications* **49** (1975), 430-458.

301. P.J. Yunker, Survey study of subsidiary, autonomy, performance evaluation and transfer pricing in multinational corporations, *Columbia Journal of World Business* (1983), 51-63.

302. F. Zahei, The Analytic Hierarchy Process-A Survey of the Method and Its Applications, *Interfaces* **16** (1986), 96-108.

303. M. Zeleny, A concept of compromise solutions and the method of the displaced ideal, *Computers and Operations Research* **1** (1974), 479-496.

304. M. Zeleny, *Linear Multiobjective Programming*, Spring-Verlag, Berlin, 1974.

305. M. Zeleny, Optimal systems designs with multiple criteria: De nova programming approach, *Engineering Costs and Production Economics* **49** (1975), 430-458.

306. M. Zeleny, *Multiple Criteria Decision Making*, McGraw-Hill, New York, 1982.

307. M. Zeleny, Optimal given systems vs. designing optimal systems: the De novo programming approach, *General Systems* **17** (1990), 295-307.

308. M. Zeleny, Trade-Off-Free management via De Novo programming, *International Journal of Operations and Quantitative Management* **1** (1995), 3-13.

309. Y.H. Zhong and Y. Shi, Duality of fuzzy MC^2 linear programming: a parametric approach, *Working paper* at College of Information Science and Technology, University of Nebraska at Omaha, USA, 1999.

310. W.T. Ziemba and R.G. Vickson, *Stochastic Optimization Models in Finance*, Academic Press, New York, 1975.

311. H.J. Zimmermann, Fuzzy programming and linear programming with several objective functions, *Fuzzy Sets and Systems* **1** (1978), 45-55.

312. H.J. Zimmermann, *Fuzzy Sets, Decision Making, and Expert Systems*, Kluwer Academic, Boston, 1987.

313. S. Zionts, Size Reduction Techniques of Linear Programming and Their Applications, Ph.D. dissertation, Carnegie Institute of Technology, 1965.

314. S. Zionts, *Linear and Integer Programming*, New Jersey, Prentice-Hall, 1974.

315. S. Zionts, *Multiple Criteria Problem Solving: Proceedings*, Spring-Verlag, New York, 1978.

316. S. Zionts, A multiple criteria method for choosing among discrete alternatives, *European Journal of Operational Research* **7** (1981), 143-147.

317. S. Zionts and J. Wallenius, An interative programming method for solving the multiple criteria problem, *Management Science* **22** (1976), 625-663.

Index for References

Index for Subjects